WATER ACTIVITY:
Influences on Food Quality

ASSISTANT EDITORS

WATER ACTIVITY:
INFLUENCES ON FOOD QUALITY

A Treatise on the Influence of Bound and Free Water
on the Quality and
Stability of Foods and Other Natural Products

Edited by

Louis B. Rockland

Food Science and Nutrition
Chapman College
Orange, California

George F. Stewart

Food Science and Technology
University of California
Davis, California

ACADEMIC PRESS 1981

A Subsidiary of Harcourt Brace Jovanovich, Publishers

New York London Toronto Sydney San Francisco

Academic Press Rapid Manuscript Reproduction

ACADEMIC PRESS, INC.
111 Fifth Avenue, New York, New York 10003

United Kingdom Edition published by
ACADEMIC PRESS, INC. (LONDON) LTD.
24/28 Oval Road, London NW1 7DX

Library of Congress Cataloging in Publication Data

International Symposium on Properties of Water, 2d, Osaka,
   1978.
   Properties of water in relation to food quality and
stability.

   1.   Food—Water activity—Congresses.   2.   Food—
Microbiology—Congresses.   I.   Rockland, Louis B.
II.   Stewart, George Franklin, Date.   III.   Interna-
tional Union of Food Science and Technology.   IV.   Title.
TX553.W3157   1978   664'.07   79-26632
ISBN 0-12-591350-8

PRINTED IN THE UNITED STATES OF AMERICA

81 82 83 84   9 8 7 6 5 4 3 2 1

Contents

Section 1.   Characterization of Moisture Sorption Isotherms

Section 2.   Bound Water

Section 3.   Solute Mobility in Aqueous Systems and Water – Membrane Interactions

Section 4.   Influence of Water on Chemical Structure and Reactivity

Section 5.   Influence of Water Activity on Chemical Reactivity and Stability of Foods

Contributors

Numbers in parentheses indicate the pages on which authors' contributions begin.

D. C. Abbott *(179)*, Department of Biochemistry, Oklahoma State University, Stillwater, Oklahoma 74074

S. Arai *(489)*, Department of Agricultural Chemistry, University of Tokyo, Bunkyo-ku, Tokyo 113, Japan

E. Berlin *(467)*, ARS, Beltsville Agricultural Center, USDA, Beltsville, Maryland 20705

M. Bizot *(179)*, Laboratoire de Biophysique des Aliments, I. N. R. A. Chemin de la Géraudière, 44072 Nantes Cedex, France

B. Blanc *(791)*, Federal Dairy Research Center, ARS, CH-3097 Liebefeld, Switzerland

S. Bruin *(1)*, Department of Food Science, Agricultural University, Biotechnion de Dreijen 12, Wageningen, Netherlands

M. Caurie *(63)*, Food Research Institute, P.O. Box M.20, Accra, Ghana

J. H. B. Christian *(825)*, CSIRO, P.O. Box 52, North Ryde, N.S.W. 2113, Australia

M. Ciner-Doruk *(567)*, Institut für Lebensmitteltechnologie und Verpackung E. V., an der technischen Universität, München 50, West Germany

R. L. D'Arcy *(111)*, Division of Textile Physics, CSIRO, 338 Blaxland Road, Ryde, Sidney N.S.W. 2112, Australia

J. L. Doublier *(179)*, Laboratoire de Biophysique des Aliments, I. N. R. A. Chemin de la Géraudière, 44072 Nantes Cedex, France

R. B. Duckworth *(295)*, Department of Nutrition and Food Science, University of Starthclyde, Glasgow, Scotland

K. Eichner *(567)*, Institut für Lebensmitteltechnologie und Verpackung E.V., an der technischen Universität, München 50, West Germany

O. Fennema *(713)*, Department of Food Science, University of Wisconsin, Madison, Wisconsin 53706

M. Fujimaki* *(489)*, Department of Agricultural Chemistry, University of Tokyo, Bunkyo-ku, Tokyo 113, Japan

S. Gal *(89)*, Beethovenstrasse 11, CH-3073 Gumligen, Switzerland

A. S. Ginzburg *(679)*, Moscow Technological Institute for Food Industry, Volokolamska Shosse, II, Moscow A-80, USSR

B. Halphen *(319)*, Laboratoire de Biologie Physico-Chemique, Ecole Nationale Supérieure de Biologie Appliqée à la Nutrition et à l'Alimentation, Université de Dijon, France

J. G. Kapsalis *(143)*, Food Sciences Laboratory, U.S. Army Natick Research and Development Command, Natick, Massachusetts 01760

*Present address: Department of Food and Nutrition, Ochanomizu University, Bunkyo-ku, Tokyo 112, Japan.

M. *Karel (511),* Department of Nutrition and Food Science, Massachusetts Institute of Technology, Cambridge, Massachusetts 02139

J. R. *Kirk* (531),* Department of Food Science and Human Nutrition, Michigan State University, East Lansing, Michigan 48824

S. *Koga (813),* The Institute of Applied Microbiology, University of Tokyo, Bunkyo-ku, Tokyo 113, Japan

I. *Kojima (765),* Department of Agricultural Chemistry, The University of Tokyo, 1-1-1 Yayoi, Bunkyo-ku, Tokyo, Japan

M. *Krispien (855),* Bundesanstalt für Fleischforschung, Oskar von Miller Strasse 20, 8650 Kulmbach, German Federal Republic

T. P. *Labuza (605),* Department of Food Science and Nutrition, University of Minnesota, St. Paul, Minnesota 55108

M. *Le Meste (319),* Laboratoire de Biologie Physico-Chemique, Ecole Nationale Supérieure de Biologie Appliqué à Nutrition et à l'Alimentation, Université de Dijon, France

H. T. *Lechert (223),* Universität Hamburg, Institut für Physicalische Chemie, Laufgraben 24, 2 Hamburg 13, German Federal Republic

J. *Lefebvre (179),* Laboratoire de Biophysique des Aliments, I. N. R. A. Chemin de la Géraudière, 44072 Nantes Cedex, France

L. *Leistner (855),* Bundesanstalt für Fleischforschung, Oskar von Miller Strasse 20, 8650 Kulmbach, German Federal Republic

M. *LeMaguer (347),* Department of Food Science, Alberta University, Edmonton 7, Canada

F. *Lindeløv* (651),* Food Technology Laboratory, Technical University of Denmark, Building 221, DK-2800 Lyngby, Denmark

W. A. P. *Luck (407),* Fach Physikalische Chemie, Philips Universität auf de Lahnbergen Marberg, 3550 Marburg (Lahn), German Federal Republic

Y. *Maeda (813),* The Institute of Applied Microbiology, University of Tokyo, Bunkyo-ku, Tokyo 113, Japan

J. L. *Multon (179),* Institut National de la Recherche Agronomique, Laboratoire de Biophysique des Aliments, I. N. R. A. Chemin de la Géraudière, 44072 Nantes Cedex, France

N. *Nagashima (247),* Central Research Laboratory, Ajinomoto Co., Suzuki-cho, Kawasaki-ku, Kawasaki, Japan

E. E. *Neuber (199),* Universität Hohenheim, Institut für Agrartechnik, Garbenstrasse 25, 700 Stuttgart 70 (Hohenheim), German Federal Republic

H. *Noguchi (281),* Department of Food Science and Technology, Nagoya University, Chikusa-ku, Nagoya 464, Japan

---

*Present address: Department of Food Science and Human Nutrition, University of Florida, Gainesville, Florida 32661.

*Present address: Co-op Denmark, Central Laboratory, 65 Roskildevej, DK-2620 Albertslund, Denmark.

D. Petroff (319), Laboratoire de Biologie Physico-Chemique, Ecole Nationale Supérieure de Biologie Appliquée à la Nutrition et à l'Alimentation, Université de Dijon, France

K. P. Poulsen (651), Food Technology Laboratory, Technical University of Denmark, Building 221, DK-2800 Lyngby, Denmark

W. Rodel (855), Bundesanstalt für Fleischforschung, Oskar von Miller Strasse 20, 8650 Kulmbach, German Federal Republic

R. Rüegg (791), Federal Dairy Research Center, ARS, CH-3097 Liebefeld, Switzerland

M. Saltmarch* (605), Department of Nutrition and Food Science, University of Minnesota, St. Paul, Minnesota 55108

A. S. Schneider (377), Memorial Sloan Kettering Cancer Center and Cornell University, Graduate School of Medical Sciences, New York, New York 10021

D. Simatos (319), Ecole Nationale Supérieure de Biologie Appliquée à la Nutrition et à l'Alimentation, Campus Universitaire, Université de Dijon, France

S. T. Soekarto (265), Bogor Agricultural University, Jalan Gunung Gede, Bogor, Indonesia

M. P. Steinberg (265), Department of Food Sciences, University of Illinois, Urbana, Illinois

E. Suzuki (247), Central Research Laboratory, Ajinomoto Co., Suzuki-cho, Kawasaki-ku, Kawasaki, Japan

T. Suzuki (748), Tokyo Regional Fisheries Research Laboratory, Ministry of Agriculture and Forestry, Kachidoki, 5-5-1, Tokyo 104, Japan

Y. Torikata (765), Department of Agricultural Chemistry, The University of Tokyo, 1-1-1 Yayoi, Bunkyo-ku, Tokyo 113, Japan

C. van den Berg (1), Department of Food Science, Agricultural University, Biotechnion, De Driejen 12, Wageningen, Netherlands

D. T. Warner (435), Research Laboratories, The Upjohn Company, Kalamazoo, Michigan 49001

T. Watanabe (738), Kyoritsu Women's College, Chyoda-ku, Tokyo 101, Japan

I. C. Watt (111), Division of Textile Physics, CSIRO, 338 Blaxland Road, Ryde, Sidney N.S.W. 2112, Australia

M. Yamashita (489), Department of Agricultural Chemistry, University of Tokyo, Bunkyo-ku, Tokyo 113, Japan

T. Yano (765), Department of Agricultural Chemistry, University of Tokyo, Bunkyo-ku, Tokyo 113, Japan

S. Yong (511), Department of Nutrition and Food Science, Massachusetts Institute of Technology, Cambridge, Massachusetts 02139

*Present Address: Kelco Division of Merck, 8225 Aero Drive, San Diego, California 92123.

Preface

The importance of moisture content in controlling certain properties of food has been recognized since pre-Biblical times. Both natural drying methods and mechanical dehydration have been used to preserve food for centuries, probably long before it was recognized that the preservation effect is due to loss of water through evaporation or sublimation. Adjustment and control of moisture content have now become important quality assurance measures in the food processing industry.

Intuitive understanding of the effect of water vapor in the environment on personal comfort is reflected in the old, well-known cliche "It isn't the heat, it's the humidity." Data on the nonlinear, direct relationship between water content and equilibrium relative humidity, for a number of systems expressed as a moisture sorption isotherm, have been published in the International Critical Tables. However, only within recent decades has it been recognized that the chemical, physical, and biological properties and, hence, the quality and stability of food products are related directly to the systems' equilibrium relative humidity (ERH) or their water activity (a_w).

A small, but growing group of "water activists," many of whom contributed to this volume, have had profound influences on the development of new and improved processing technologies, packaging practices, and storage conditions which optimize or maximize retention of aroma, color, texture, nutrient properties, and biological stability of foods.

Anomalous changes in chemical, physical, and biological properties of food and other natural products, which do not appear to be related directly to either moisture content or water activity, often can be explained in terms of the interrelationships between the two parameters expressed as a moisture sorption isotherm (MSI). Changes in isotherm characteristics, related to temperature, pressure, hysteresis, etc., provide insight concerning the physical and chemical changes that influence food quality and stability. For example, in earlier times it was believed that food product stability increases directly with progressive decreases in moisture content. It is now recognized (and is consistent with the newer concepts of moisture–sorption theory) that for many foods there is an optimum moisture content and a corresponding a_w at which maximum stability can be realized. Modern moisture–sorption theory explains satisfactorily the basis for the induction or development of rancidity at high or low moisture levels and, contrary to the popular belief, also during freezing storage. These unfavorable reactions occur with or without the loss of moisture. In contrast, at constant moisture content raising a_w simply by increasing temperature accelerates enzyme activity and the growth of microorganisms; on the other hand it can either increase or decrease browning reactions. Conversely, lowering the temperature can induce the growth of microorganisms by allowing condensation of water on the product (jams and jellies) with a resultant localized high a_w at which the microorganisms can proliferate. Because of the hysteresis phenomenon, the stability of a system can be influenced by the method (i.e., adsorption or desorption) employed to adjust moisture content and water activity.

The papers presented in this treatise provide insight into the effects of water on the changes that occur in foods during storage. On the basis of this insight, more rational bases are established for improving processing, packaging, and storage conditions for both raw and processed foods.

The treatise has been divided into seven sections. Each section is designed to complement and/or supplement information and data contained in "Water Relations in Foods" (R. B. Duckworth, ed., Academic Press, London, 1975).

Section 1 contains chapters primarily concerned with characterization of moisture sorption isotherms on the basis of both theoretical and applied considerations. Critical presentations are also made on the methodology employed to estimate both moisture content and water activity (a_w). The nature and importance of hysteresis in establishing rheological and other properties of food products are covered by two research groups, each with a different perspective.

Section 2 is concerned with bound water and its relationship to the physical and chemical properties of natural products, including foods. The measurement of bound water and the differentiation of free and the several types of bound water are presented, using nuclear magnetic resonance (NMR), differential thermal analysis (DTA), and thermogravimetric analysis (TGA).

The structure of water and the influence of solutes and solute mobility on water activity are considered in Section 3, together with a discussion of a new electron spin resonance (ESR) technique for characterizing the properties of water in complex systems.

Section 4 contains a critical evaluation of the influence of water and water activity on the structural and functional characteristics of certain natural polymers i.e., the proteins and carbohydrates, as well as on the reactivity of proteins.

Consistent with the principal theme of the treatise, Section 5 and 6 cover the influence of water activity and temperature on the rates of several important chemical reactions (i.e., lipid oxidation, vitamin decomposition, browning, and other reactions), which affect the chemical, physical, and nutritional properties of food. Also the influence of water activity on food processing and storage practices are discussed from both theoretical and applied viewpoints. Specifically the application of water activity principals to the processing and preservation of leafy vegetables, cheese, dried fish, and other products are covered.

The final section, Section 7, is composed of contributions concerned with the influences of water activity on the behavior of food-related microorganisms. Special attention is given to the role of solvents in controlling water activity and the related survival of certain microorganisms.

Each chapter contains an extensive bibliography and citations that support the newer concepts but would not normally be accommodated on the basis of classical concepts of moisture–stability relationships.

L. B. Rockland

G. F. Stewart

Acknowledgments

The concept of a small, international meeting concerned with the influence of water activity on the properties of food was conceived by Dr. R. B. Duckworth who organized the first symposium entitled "Water Relations in Foods" held at the University of Strathclyde, in Scotland, September 9–14, 1974 under the sponsorship of the International Union of Food Science and Technology (IUFOST). The proceedings, edited by Dr. Duckworth, were published by Academic Press.

The second International conference (ISOPOW-II), also sponsored by IUFOST, was organized to complement and supplement information reported in Scotland in 1974. Held in Osaka, Japan, September 1978, it was broadened by contributions from Japanese scientists who have had a traditional interest in the preservation of foods through control of their water content and water activity. ISOPOW-II attracted the participation of the most distinguished scholars and industrial workers in this field. Since attendance at ISOPOW-II was limited to less than one hundred, it appeared appropriate to prepare a treatise that would permit documentation of new information on the subject, including much of that presented at ISOPOW-II.

This volume, therefore, is not a proceeding of the conference. Individuals presenting papers were asked to submit updated and revised manuscripts for review by the editorial committee which regrets that all could not be accepted. The editors wish to express their gratitude to all who have contributed to this effort.

The editors wish to express their appreciation and gratitude to the following list of coordinating committee members for their unique individual efforts: S. Arai, R. B. Duckworth, O. Fennema, M. Fujimaki, S. Gal, A. S. Ginsburg, J. G. Kapsalis, M. Karel, J. L. Multon, W. E. L. Spiess, I. Takahashi, and K. Yasumatsu. The interest, support, and cooperation of D. Bone, J. Hawthorn, N. Getschell, W. Martinez, H. Mitsuda, M. J. Pallansch, and B. S. Schweigert are also gratefully acknowledged.

Financial support for the symposium was provided by contributions from Amatil, Arnott's Biscuits Pty Ltd., CSR Ltd., Cadbury Schweppes Pty Ltd., H. J. Heinz Co. Ltd., Mauri Bros & Thomson Ltd., The Nestle Co. Ltd., Quaker Products Ltd., Roche Products Ltd., White Wings Pty Ltd. (Australia). Canada Packers Ltd. (Canada); Dibona Markenvertrieb Kg, C. H. Knorr, Meggle Milchindustries & Co. Kg, Melitta-Werke, Pfanni-Werk Otta Eckart Kg (German Federal Republic); D. E. J. B. V. (Holland); Alpha Food Co., Ltd., Kikkoman Shoyu Co. Ltd. (Japan); Battelle Memorial Institute, Beratungsgesellschaft für Nestle Produkte Ag, Nova-Sina Ag (Switzerland); Beecham Group Ltd., Cadbury Schweppes Ltd., Corn Products Co., The Distillers Co. Ltd., H. J. Heinz Co. Ltd., Londreco Limited, J. Lyons & Co., Ltd. Metal Box Ltd., Rank Hover McDougall Ltd., Unilever

Research (United Kingdom); Beatrice Foods Co., Campbell Co., The Coca Cola Co., Florasynth, Inc., Foremost-McKesson Foundation, Inc., Fritzsche Dodge & Olcott, Inc., General Mills, Inc., Gerber Co., Griffith Laboratories, Hershey Foods Corp., Hoffman–La Roche, Inc., Hunt-Wesson Foods, Inc., ITT Continental Baking Co., Inc., Kellog Co., Kraft, Inc., Thomas J. Lipton, Inc., McCormick & Co., Inc., Mead Johnson Research Center, Oregon Freeze-dry Foods, Inc., The Proctor & Gamble Co., Ralston Purina Co., Swift and Co. (United States).

WATER ACTIVITY AND ITS ESTIMATION IN FOOD SYSTEMS:
THEORETICAL ASPECTS

C. van den Berg
S. Bruin

I. INTRODUCTION

One of man's earliest discoveries for food preservation must
have been the fact that fresh foods become much less perishable
when their water content is drastically reduced. It is general
knowledge that not all biological products show the same behavior
in this respect. The various biological products as well as
their components show great differences in water-binding properties. For example, potato starch containing 24% moisture on dry
basis is perfectly stable, whereas crystalline white sugar
(sucrose) may spoil, albeit slowly, even at only 4% moisture.

These and comparable phenomena in the area of food deterioration have been poorly understood for many decades until it was
realized that not the actual water content, but some other factor
related to the "nature" or "state" of the constituent water, determined eventual spoilage. It was a problem to find a convenient
expression for this state of the water or better the "availability"
of water for deterioration reactions in foods.

A. Historical

This general lack of understanding remained until about 25
years ago, even though physical chemists, physicists, and chemical
engineers studying drying processes and dryer design had long used

1

the relation between water content and equilibrium vapor pressure
of their products, because they knew that vapor pressure differences are the essential driving forces during the drying process,
at least in the gas phase. Measurements of this relation between
water content and equilibrium relative humidity at a certain temperature, i.e., the water vapor (or moisture) sorption isotherm,
for (some) food and biological products date back to the last
century. Schloesing (1893) reported on complete water vapor sorption isotherms for textile fibers in 1893. Other complete sorption isotherms for starches were obtained by Rakowski (1911), who
observed sorption hysteris of organic colloids. His measurements
may still be regarded among the best in this area. But these
types of finding were not considered in depth and applied then,
and in practice, food products were still dried to a water content
that had empirically been found to be safe.

Research into the relation between storage stability and water
content for foods was also begun early (e.g. Spieckermann and
Bremer, 1901). The first systematic studies on growth of microorganisms in relation to relative humidity or a directly related
physical property, usually osmotic pressure or concentration of
the solution (NaCl or H_2SO_4) maintaining atmospheric humidity,
were reported independently by Walderdorff (1924) and Walter
(1924). In particular, the German botanist Walter, who studied
swelling of protoplasma and plant growth, carried out experiments
with many different microorganisms on substrates equilibrated at
different relative humidities, over periods that were short compared with food storage times. In that year he made the following conclusion quoted below (translated):

> To preserve moulding materials by drying, it is sufficient to
> decrease the water content to such an extent that the relative
> vapor pressure is not over 85%. This is a general rule, the
> general validity of which is very probably true for different
> materials, although this needs thorough research and proof.
> The generalization allowed by these data is a great advantage
> towards the data for water contents, which are much different
> for different materials at a relative vapor pressure of 85%....

Whether there are some organisms that may develop at lower
relative vapor pressures remains open. For practice those
organisms will not be of too great an importance....

Walter's last statement, we know today, was certainly not correct:
85% for relative humidity is not safe. Xerophilic molds and osmophilic yeasts growing at relative humidities down to about 65%
can cause great damage through spoilage of stored food commodi-
ties. However, in principle, his insight and the correctness of
his results deserve being mentioned.

Later in Australia, Scott (1936, 1937, 1938), who knew of
Walter's work, studied the growth of microorganisms on chilled
beef. He recorded relative humidity as an important growth
parameter. In 1953, he called the equilibrium relative humidity,
i.e., the availability of water in a food medium, water activity
and described the concept in the food literature (1953, 1957).
He showed clearly that there is a direct correlation between microbial growth on a substrate and its thermodynamic water activity.
Perhaps a paper on the physiology of microbial spoilage in foods
by Mossel and Westerdijk (1949, p. 193) gave Scott this idea.
These authors used in passing the term "water activity" of a food
as being an independent characteristic of a substrate and one of
the intrinsic factors determining spoilage.

This brief history would certainly not be complete without
dating the origin of the basic concept of activity of a substance.
Lewis (1907a,b) introduced this concept while applying thermodynamic concepts (as defined earlier by J. W. Gibbs) in order to
make them workable in practice. He defined the activity of a component in a mixture and also the activity of water in a food system
as the ratio of fugacities (which will be dealt with below). Thus
it took more than 40 years before the great practical utility of
this basic concept was realized in food science.

Concerning usage and definitions, water activity was used to
indicate an intrinsic parameter of a food system and equilibrium
relative humidity a property of the surrounding atmosphere in equilibrium with the food system under consideration.

[The fact that Scott's publication appeared some 25 years ago was celebrated in April 1978 at a three day symposium, "Water Activity and Food Microbiology," in Parma, Italy, organized by the Institute for the Canning Industry (Christian *et al.*, 1978).]

B. Food Deterioration and Water Activity

Scott's work was widely acclaimed and led to a rapid expansion in the use of water activity, denoted a_w, which has now become one of the major control variables in food preservation technology. Research in this area in the last two decades finally led to general and more fundamental knowledge of its influence on all types of food deterioration. Examples of its influence on various food systems are given by Karel (1974). With some reservation, a generalized figure (Fig. 1) can be drawn up from these and comparable literature data. This figure, first designed by Labuza

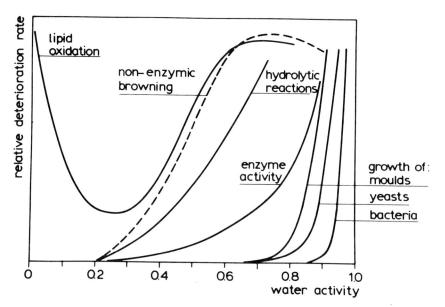

Fig. 1. *Generalized deterioration reaction rates in food systems as a function of water activity (room temperature).* [*Adapted from Fennema (1976) and Heiss and Eichner (1971).*]

(1970), was termed (Labuza *et al.*, 1972) the "food stability map
as a function of a_w." Since all food systems possess their own
special features, chemically and physically, with respect to
spoilage reactions, the information in this diagram can give only
relative data. However, in general it supplies a part of the required insight into the different deterioration phenomena as a
function of a_w. Starting from $a_w = 1$, a decrease in water activity
slows down all types of chemical deterioration reactions and microbial growth until, at a certain level, all reactions are almost
completely inhibited except for chemical oxidation of lipids,
which is strongly favored by a further decrease. [As we are concerned with the properties of water in relation to food quality
and stability, the principles illustrated by this figure play a
vital role.]

It should be realized, however, that in this figure a_w is a
"phenomenological" concept. For the theoretical estimation of
a_w, we reduce the idea of water activity (i.e., how we encounter
it when it acts in a food system) to a basic thermodynamic concept.
Because a_w cannot act solely by means of the gas phase as a vapor
pressure, the practical effect of a_w cannot be separated from the
nature of the substances adjusting the a_w or those reacting to
the imposed a_w. Thus the practical a_w is always a "mixed" property, as illustrated, for example, by the fact that a microorganism
usually reacts differently to the same a_w in different media.
Glycerol, sugar, salt, or other compounds at the same a_w do not
act in the same way. The activity of water in a food in the sense
of its effect on microorganisms (Mossel, 1975), as well as on
chemical and enzymatic spoilage reactions, is a more complex function. This is well illustrated by Fig. 2, adapted from the results
of Corry (1975). In fact, the reaction pattern of a microorganism
is determined by properties of its cytoplasmic membrane, itself a
result of complex functions of the entire physiology of the microorganism. This mixed character of a_w has often been a source of
confusion. Realizing this, Mossel (1975, p. 348) and Tracey (1975)

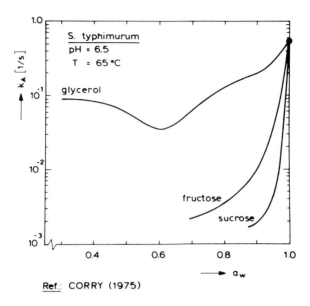

Ref.: CORRY (1975)

Fig. 2. Influence of water activity, adjusted by different solutes: sucrose, fructose, and glycerol on the killing of Salmonelle typhimurium. (Adapted from Corry, 1975).

in retrospect, even doubted whether the choice of the term "water activity" was indeed as good as initially thought.

C. Use of Water Activity in Food Systems

A brief analysis may clarify this confusion a little. Apart from physical changes, all types of deteriorations, including microbial growth, are based on chemical and biochemical reactions. Formally it can be stated that except for diffusion limitations the speed of any reaction is determined by its substrate concentration, the reaction order, and the reaction rate constant. In its simplest form (conversion of A):

$$r_A = -K_A c_A^n \tag{1}$$

where r_A is the rate of conversion of A per unit volume, c_A the concentration of A, n the reaction order, K_A the (first-order) reaction rate constant, itself a function of temperature, pressure, and composition of the medium, and $K_A = f(P; T; C_A, C_B, C_C, \ldots, C_W)$. If water is a reactant, it acts more directly via its concentration together with its influence on K_A.

A secondary (but at lower water contents the major) influence is the effect that water content has on the mobility of the reactants. This mobility is greatly retarded by a decreasing amount of water. At low water concentrations, diffusion coefficients of most food components decrease by several orders of magnitude, further decreasing water content (e.g., Bomben *et al.*, 1973).

In conclusion, a deterioration reaction in a food system turns out to be always a multifactorial process of which a_w is only one factor. Only when the water component is the rate-limiting factor, which in fact is often the case, can a_w be expected to have a direct influence. Therefore, it would be more sound to keep the concept of thermodynamic a_w strictly separated from kinetics of detailed deterioration reactions and if necessary take only the broader view as indicated in Fig. 1.

The situation with a_w is even more complicated due to the complex nature of natural products. We have already seen that activity of a component in a one-phase multicomponent system is a basic quantity. Foods are multicomponent systems—sometimes one phase (homogeneous), often two or more phases (heterogeneous). Of course, the activity of each component can be defined for each phase if this phase internally is in equilibrium. However, only when also thermodynamic equilibrium exists between the phases is the activity of a particular component the same throughout this heterogeneous system. With foods this condition is not always fulfilled. Many agricultural products are multicomponent multiphase systems, partially dissolved and in gel condition. The fibrous components constitute the solid phase, holding a liquid phase, dispersed or continuous. When oil and fat droplets are

present even more liquid phases exist. The internal structure of
these systems is generally very difficult to characterize. This
picture is further complicated by dissolution and delayed crystallization effects. Hysteresis between drying and humidification is
commonly observed. These facts suggest that in these systems
there exists no true thermodynamic equilibrium but some "pseudo-
equilibrium," which is not clearly definable. As a consequence
the thermodynamic activity of the component water as such, which
by its strict definition is an equilibrium value, does not exist
throughout the system. Instead we often deal with an empirical
pseudo-water activity.

With this delicate problem we should be very careful. A possible adaptation of the definition for practice should be made
only after very careful consideration, because this would certainly result in further confusion. Moreover, the numerical value of
this practical pseudo-water activity will in many cases not deviate
much from the strict thermodynamic equilibrium water activity.

Exceptions to this where the numerical value for a_w deviates
much from the equilibrium value are those cases where a much-
delayed crystallization phenomenon causes a great change in the
ingredient distribution of a system, over a longer period of time.
This has been observed with amorphous sugar (Karel, 1975a; Roth,
1976). For example, at a_w = 0.15 it takes a pure amorphous
sucrose 3 years to attain equilibrium. Throughout this chapter,
we continue the use of the thermodynamic concept of a_w, whose
definition is dealt with in Section III.

II. WATER VAPOR SORPTION AND THE NATURE OF INTERACTION BETWEEN
    WATER AND FOOD SOLIDS

Before proceeding to the actual estimating methods we recall
some of the general knowledge about moisture sorption isotherms of
foods and perhaps provide some new information. Regarding the

nomenclature on sorption, use the general term <u>sorption</u> as coined
by McBain (1909) to indicate all processes wherein food solids
<u>reversibly</u> combine with water molecules. These processes embrace
real physical <u>ad</u>sorption and capillary condensation to surfaces
as well as the formation of liquid and solid solutions.
To indicate the direction of the process, we prefer the terms
<u>desorption</u> (drying) and <u>resorption</u> (humidification). The latter
indicates that all systems in which we are interested originally
have been born under wet conditions. For this purpose the term
was previously used by MacKenzie and Rasmussen (1972).

Figure 3 shows moisture sorption isotherms for a few biological
materials under resorption conditions. We observe two essentially
different types of isotherms, namely, the sigmoid type usual for
most food systems and also the gradually increasing one that is
seen much less frequently. They are isotherm types two and three
of the classification of Brunauer *et al.* (1940). The last type is
obtained with foods, like coffee extract, fruits, and sweets, that
are high in sugars and other soluble low-molecular-weight compounds
and low in polymers.

Figure 4 gives a general sigmoid sorption isotherm for a hypothetical food system. For interpretation purposes, this curve is
usually divided into three different parts. Since the sorption
equilibrium is a dynamic one (De Boer, 1953), the distinction between these parts or ranges necessarily cannot be strict, but
rather gives an indication about the nature of most of the water
present in the particular range.

A. Range 1.

This represents strongly bound water, with an enthalpy of
vaporization H_{vap} considerably larger than that of pure water.
In many respects it behaves as essentially part of the solid. It
is generally understood that these first water molecules are
sorbed at the active polar groups in the food solids (Pauling,

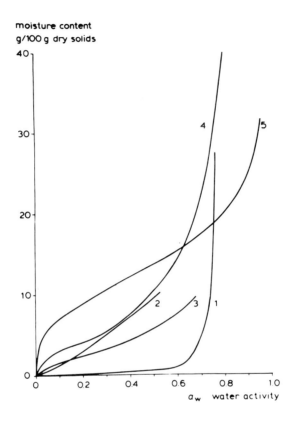

Fig. 3. Different types of resorption isotherms obtained
with some biological materials (unpublished results). Temperature,
20°C, except for no. 1, which was measured at 40°C. (1) Composed
sweet (main component: powdered sucrose), (2) spray-dried chicory
extract, (3) roasted Columbian coffee, (4) pig pancrease extract
powder, (5) native rice starch.

1945; McLaren and Rowen, 1952). A high water uptake at these low
a_w normally reflects a high content of hydrophilic macromolecular
materials such as proteins and polysaccharides. At these low activities water cannot support solution processes or organic solids
and is not able to act as a plasticizer on the biopolymer. Thus
in this range food solids act as rigid physical adsorbents that
sorb water on active sorption sites (localized adsorption!). That

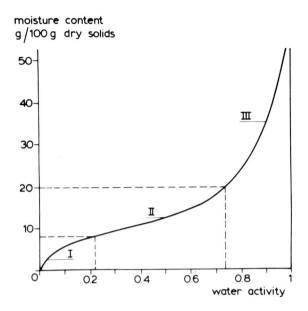

Fig. 4. Idealized water vapor sorption isotherm of a food system, showing different ranges of sorption.

this plasticizing effect of water on the biopolymer structure, e.g., with starch, does not appear at these low water activities is effectively illustrated by some unpublished results of Muetgeert (1978). Ordinary dried starches, as well as other biopolymers, as a rule show low (up to a few m^2/gram) specific surface areas for nitrogen adsorption, indicating that the substance is not porous for nitrogen molecules. Muetgeert physically transformed a starch material into a very porous compound with a specific surface of more than 100 m^2/g (measured with nitrogen). While water vapor was sorbed into this porous compound, a drastic decrease in specific surface area to about 4 m^2/g could be noticed (Fig. 5) starting at approximately 10-11% for water content, indicating the weakening and new arrangement of the organic structure. The latter value for water content is about the BET monolayer value for starches (Van den Berg et al., 1975). Comparable cases with very porous biomaterials are mentioned by Berlin elsewhere in this volume.

residual specific surface area
% of original area.

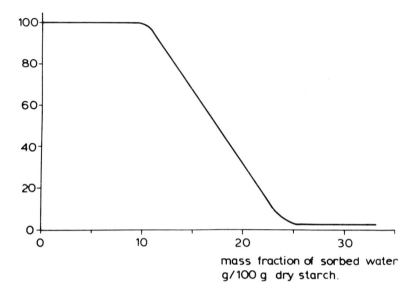

*Fig. 5. Specific surface area of modified starch as a func-
tion of water content. (From Muetgeert, 1978.)*

B. Range 2

This represents a water fraction less firmly bound than the
first. The enthalpy of vaporization in this range is little
greater than the enthalpy of vaporization for pure water. These
water molecules sorb near or on the top of the first molecules or
penetrate into newly created holes of the already swollen structure. Solution processes and chemical and biochemical reactions
are enhanced by part of this water, depending on the nature of the
solids present. This class of constituent water can be looked
upon as the continuous transition of the first to the third type
of water.

C. *Range 3.*

This is more or less free water, mechanically entrapped in
the void spaces of the system, having nearly all properties similar to those of bulk water. Here virtually no excess heat of
binding can be detected. Especially high water uptake at these
high water activities indicates dissolution of major components
of the system.

When brought back to a submicroscopic level of observation,
roughly seven types of physical interactions between water and
food solids may be distinguished: (1) coulomb forces (dissociated
groups, ions), (2) hydrogen bonds (polar groups), (3) London-Van
der Waals forces, (4) steric effects, (5) solution effects, (6)
mobility change of polymer segments (plasticizing of macromolecular
structure), and (7) capillary condensation (hydrodynamic curvature
effect of liquid sorbate surface). The last three interaction
types are of a different nature than the others. These forces do
not act between single molecules but between greater assemblies of
molecules.

As a consequence of the ability of water molecules to come up
with hydrogen bonds and build structures, water is structuring itself around hydrophobic molecules and parts of molecules. This effect is called somewhat erroneously "hydrophobic bond."

D. *Theoretical Approach*

The resulting aggregate effect of these interactions on a
macroscale is conveniently expressed by the Gibbs free energy
function, which for each component is directly related to its activity. It follows that the theoretical estimation of a_w can be
based on three approaches depending on the existence of a specific
sorption surface: (1) theoretical models of Gibbs free energy in
complete mixtures (solution, one phase systems); (2) theories
about physical adsorption if a surface can be distinguished; (3)

combinations of (1) and (2). Applying this approach, we shall see
with multiphase food systems that roughly physical sorption theories apply better at low a_w and solution theories are more successful at higher a_w.

III. FREE ENERGY AND DEFINITION OF ACTIVITY

   The main purpose of this chapter is to present a survey of the
current theoretical tools that are available to estimate the water
activity a_w of a food system of well-known composition without actually measuring it. Together with a presentation of equations
that are available to compute water activity, the main fundamentals
of the concepts and underlying theory leading to them are also
briefly explained.

   The thermodynamic concept of activity of any component i, including water, basically is defined by Lewis (1907a,b) as the
ratio of its actual fugacity and the fugacity of the component in
a standard state. To explain what is meant by fugacity, we note
that any system in equilibrium is characterized by a minimum value
of "free energy." This energy when not at its minimum, will drive
any reaction (or work) in order to attain equilibrium. Most practical systems including foods, which are subject to change, undergo this change at a constant or nearly constant pressure. Thus
they are ruled by the Gibbs free energy, also called Gibbs function
or free enthalpy, generally denoted by G. For a closed[*] system
this function is defined as

   $$G = U + pV - TS \tag{2}$$

where U is total energy, p pressure, V volume, T absolute temperature, and S entropy. The term $U + pV$ is called enthalpy, denoted

[*]A closed system does not exchange matter with its environment.

by H. In liquid and solid phases the term pV is usually negligible
at ambient pressures. For more details about those functions and
their use, the interested reader is referred to textbooks on chemical and engineering thermodynamics (e.g., Prausnitz, 1969; Sandler,
1977).

Expressed on a molar basis the Gibbs free energy is called the
chemical or thermodynamic potential, denoted by μ. For n moles of
pure molecular species i,

$$\mu_i = (\frac{G}{n_i})P, \; T \quad \text{(pure species)} \qquad (3)$$

In mixtures each component has its own partial molal free energy
or thermodynamic potential. This thermodynamic potential is the
change in Gibbs free energy of the system resulting from the addition of one mole of pure component i to the system at constant
temperature, pressure and amounts of other components n_j $(j \neq i)$:

$$\mu_i = (\frac{\partial G}{\partial n_i}) \; P, \; T, \; n_j \quad \text{(mixture)} \qquad (3a)$$

When a closed system, consisting of more than one single phase
and one or more components, is in thermodynamic equilibrium, molecules interchange continuously between phases. Since there is no
net change in the amounts of the various phases, the rates of interchange of molecules of all species between the phases are
equal. Thus there is no net difference in free energy between
the phases. The chemical potentials of all components in the
phases in mutual thermodynamic equilibrium are equal. This equali-
ty of chemical potentials at thermodynamic equilibrium is a basic
fact of systems under constant temperature and pressure.

Unfortunately, thermodynamic potential does not have an imme-
diate equivalent in the physical world and therefore Lewis (1901)
expressed it in terms of some other more tangible function, which
might be more easily identified with physical reality. Such an
auxiliary function is supplied by the concept of fugacity.

The well-known Gibbs-Duhem equation for an open homogeneous system (for example, one phase in a heterogeneous system) is

$$S \, dT - V \, dp + \sum_{i=1}^{m} n_i \, d\mu_i = 0 \tag{4}$$

where n_i is the number of moles of components i. For a system of only one single component i, at constant temperature ($dT = 0$) it follows directly that

$$d\mu_i = \frac{V}{n_i} \, dp_i \tag{5}$$

Here V/n_i stands for the molar volume of i. For a perfect gas ($pV = nRT$) the following relation is obtained:

$$d\mu_i = \frac{RT}{p} \, dp_i = RT \, d \ln p_i \quad (T \text{ constant}) \tag{6}$$

where R is the gas constant and p_i the pressure of the pure component under consideration. This derivation can be generalized to mixtures of perfect gases. Then p_i is the partial pressure of component i in the mixture. For a liquid in equilibrium, p_i would be the vapor pressure at the given temperature. Thus Eq. (6) relates the abstract concept of thermodynamical potential of a component i in an ideal gas mixture or a pure ideal gas to a simple function of a physically real quantity of that component or gas, pressure.

Lewis (1901) generalized this relation and made it applicable to real systems by defining a corrected pressure function called fugacity, denoted f. Fugacity can be considered as a measure of the tendency of a component to excape (from Latin *fugare*, to escape). It has the unit and dimension of vapor pressure, which has been corrected for any nonideality of that vapor. An infinitesimal isothermal change in thermodynamic potential for any component in any mixture whether solid, liquid, or gas, pure or mixed, ideal or not, is according to this definition given by

$$d\mu_i = RT \, d \ln f_i \quad (T \text{ constant}) \tag{7}$$

This equation is of conceptual aid in performing the translation from thermodynamic to real physical variables.

Integration of Eq. (7) at constant T yields

$$\mu_i = RT \ln f_i + C \tag{8}$$

Since the absolute value of chemical potential is not known, the integration constant C cannot be measured.   Thus we have to define a standard or reference state at the temperature under considera-tion.   The relative value of chemical potential in relation to this reference can now be measured.   For the arbitrarily chosen reference state we can write

$$\mu_i^\theta = RT \ln f_i^\theta + C \tag{9}$$

Subtracting Eq. (9) from Eq. (8) yields

$$\mu_i - \mu_i^\theta = RT \ln (f_i / f_i^\theta) \tag{10}$$

For a pure perfect gas, fugacity equals pressure; for a component i in a mixture of perfect gases, it equals partial pressure.   For real gases at very low pressures, fugacity equals vapor pressure also, since all systems then approach ideal behavior:

$$f_i / p \rightarrow 1 \quad \text{when } p \rightarrow 0 \tag{11}$$

For liquids a physical representation of the fugacity may be seen in the tendency of molecules to escape from the liquid into the gas phase.

The ratio of fugacities (f_i / f_i^θ) for a component is called the activity (or better, relative activity, to distinguish between other defined activities; Lewis et al., 1961) a_i of component i:

$$a_i = f_i / f_i^\theta \tag{12}$$

The reference state for fugacity is partly arbitrary.   The limita-tions are that the temperature of the reference state equals the temperature of the mixture under consideration and the total pres-

sure is defined. The last factor influences the fugacity only to
a minimum extend and is usually neglected.

It is common practice to take the vapor pressure of the pure
component i in vacuum or at 1 atm as the fugacity reference state.
If so, the reference pressure varies with temperature and slightly
with total pressure. The numerical value of the activity normally
ranges from 0 (absolute dry) to 1 (pure component). At equilibrium
between different phases the fugacity of each component is the same
throughout the heterogeneous system. In this case the activity is
the same throughout the system when the reference fugacities are
defined equally for each phase.

Water Activity

Hence the activity of water in a mixture, whether in gas
phase, in solution, "bound" to a solid or otherwise, is conveniently expressed as the fugacity of water in the mixture divided
by the fugacity of pure water at the same temperature and its own
pressure.

Comparison of this a_w with the equilibrium relative humidity
(p_w/p_w^θ) of the same system (Gàl, 1972) showed that there is only
a relative difference of at most about 0.2% at ambient temperatures and pressures. Remember that fugacity is a "corrected pressure," and water vapor pressures at ambient temperatures are still
low, i.e., water vapor behaves about perfectly. This justifies
for normal cases the commonly used definition

$$a_w = p_w/p_w^\theta \tag{13}$$

Here p_w is equilibrium water vapor pressure over the system and
p_w^θ the vapor pressure of pure water at the same temperature and
(usually) atmospheric pressure.

Three important consequences of the given definition of water
activity must be realized: (a) the water activity refers only to
the true equilibrium state, dealing with foods, it should be

realized that real food systems do not fulfill this requirement always; (b) the water activity is defined at a specific tempera- ture and total pressure; (c) the reference state must be well specified, since this may be a matter of choice.

IV. MIXTURES

A. Perfect Mixtures

When studying the vapor pressure of ether solutions, Raoult (1888) observed that for dilute solutions the depression of vapor pressure is a direct function of the mole fraction of the solved component and the vapor pressure of the pure solvent and indepen- dent of temperature. He formulated Raoult's law, which is valid for ideal mixtures. For water as a component of a mixture under ambient conditions, Raoult's law reads

$$p_w = x_w p_w^\theta$$

or

$$p_w/p_w^\theta = a_w = x_w = n_w/(n_w + n_s) \tag{14}$$

where x_w is mole fraction of water, n_w total moles of water, and n_s total moles of solute. Rearranged and expressed in terms of fugacity for a mixture of two components it gives

$$f_1 = x_1 f_1^\theta, \quad f_2 = x_2 f_2^\theta \tag{15}$$

In words, in a perfect behaving mixture of two real components 1 and 2, the tendency of 1 to escape from the mixture will be a fraction x_1 (mole fraction of 1) of the escaping tendency of pure 1 (f_1^θ) at that temperature.

Dilute aqueous solutions approach this perfect behavior. For example, Eq. (14) predicts accurately (difference <1%) a_w for glu-

cose and glycerol solutions up to 4 M, for saccharose solutions
up to 2 M, and for many electrolytes up to 1 M, provided that the
ions are considered as separate molecular species (Karel, 1975b;
Reid, 1976).

B. *Nonideality*

Nonideality of mixtures is generally described by excess functions and is incorporated into the equations for activity through
activity coefficients. The activity coefficient of component i,
γ_i, is unity for perfect behavior, i.e., at mixing of components
i in the system there are no volume changes, heat effects, or excess entropy changes. The real solution, when mixable in all
ratios, thus behaves like a perfect solution of mole fraction
$\gamma_i x_i$:

$$a_i = \gamma_i x_i \tag{16}$$

It follows that the possible estimation of activity coefficients
of the water component appears to be crucial in the theoretical
prediction of water activity in food systems.

The observed enthalpy in a nonideal case is divided into two
additive parts: one part resulting from perfect behavior, H_{id},
and the second part H^E the excess enthalpy (which may be negative
or positive) resulting from nonideal behavior:

$$H = H_{id} + H^E \tag{17}$$

These excess functions obey common thermodynamic rules, e.g.,

$$G^E = H^E - TS^E \tag{18}$$

For the excess partial molal Gibbs free energy of component i it
follows that

$$\left[\frac{\partial (G^E)}{\partial n_i}\right]_{T,P,n_{k(k \neq i)}} = g_i^E = RT \ln \frac{f_i}{f_i(id)} \tag{19}$$

where $f_i(\text{id})$ is the component fugacity at ideal behavior.

By combining Eqs. (19), (16), (14), and (12), we obtain

$$g_i^E = RT \ln \gamma_i \qquad (20)$$

This partial molal excess Gibbs function, like other Gibbs functions, cannot be measured simply or predicted from thermodynamics. Assumptions about the relation of this partial excess Gibbs free energy to the molecular behavior of component i in mixtures have to be made. If one roughly makes the assumption that the excess entropy is very much smaller than the excess enthalpy divided by $T(S^E \ll H^E/T)$ the concept of regular solution theory is obtained (Hildebrand et al., 1970). If on the contrary H^E is approximately zero, we have the concept of athermal mixtures.

The interaction of a liquid organic component with water depends mainly on its ability to influence the formation or interruption of hydrogen bonds. Rowlinson (1969) made a useful attempt to classify binary aqueous-organic mixtures by the strength of the hydrogen bonds from the hydroxy group in water. In the series nitriles, ketones, ethers, alcohols, polyethers, amines, polyalcohols, and polyamines, the first two show with water an upper critical solution temperature (i.e., where phase separation occurs), the last two are completely miscible with water, and others show behavior in between.

In this field several solution theories and a considerable number of equations for the computation of activity coefficients of liquid components have been developed by many workers. Some of these theories have been reviewed by Prausnitz (1969) and Bruin (1969).

Table I summarizes four useful equations for the estimation of activity coefficients in binary mixtures. Recently progress in this field has been reported by Prausnitz's group (Fredenslund et al., 1975, 1977) in using group contribution methods for the estimation of activity coefficients. Gmehling and Onken (1977) published a large data collection of vapor-liquid equilibria with fit-

C. VAN DEN BERG AND S. BRUIN

TABLE I. Some Equations for Activity Coefficients of Binary
Systems Containing Components 1 and 2

Type of equation	Adjustable parameters	$\ln \gamma_1$, $\ln \gamma_2$
Margules	A_{12}	$[A_{12} + 2(A_{21} - A_{12})x_1]x_2^2 = \ln \gamma_1$
	A_{21}	$[A_{21} + 2(A_{12} - A_{21})x_2]x_1^2 = \ln \gamma_2$
van Laar	A_{12}	$A_{12}\left(\dfrac{A_{21}x_2}{A_{12}x_1 + A_{21}x_2}\right)^2 = \ln \gamma_1$
	A_{21}	$A_{21}\left(\dfrac{A_{12}x_1}{A_{12}x_1 + A_{21}x_2}\right)^2 = \ln \gamma_2$
Wilson	$\Lambda_{12} - \Lambda_{11}$	$\ln(x_1 + \Lambda_{12}x_2) + x_2\left(\dfrac{\Lambda_{12}}{x_1 + \Lambda_{12}x_2} - \dfrac{\Lambda_{21}}{\Lambda_{21}x_1 + x_2}\right) = \ln \gamma_1$
	$\Lambda_{21} - \Lambda_{22}$	$-\ln(x_2 + \Lambda_{21}x_1) - x_1\left(\dfrac{\Lambda_{12}}{x_1 + \Lambda_{12}x_2} - \dfrac{\Lambda_{21}}{\Lambda_{21}x_1 + x_2}\right) = \ln \gamma_2$
NRTL	$g_{12} - g_{22}$	$x_2^2\left[\tau_{21}\left(\dfrac{G_{21}}{x_1 + x_2 G_{21}}\right)^2 + \left(\dfrac{\tau_{12}G_{12}}{(x_2 + x_1 G_{12})^2}\right)\right] = \ln \gamma_1$
	$g_{21} - g_{11}$ α_{12}	$x_1^2\left[\tau_{12}\left(\dfrac{G_{12}}{x_2 + x_1 G_{12}}\right)^2 + \left(\dfrac{\tau_{21}G_{21}}{(x_1 + x_2 G_{21})^2}\right)\right] = \ln \gamma_2$

ting parameters, including many binary and ternary aqueous-organic systems. Wichterle *et al.* (1973) compiled all the literature data on vapor-liquid equilibria until 1972. Le Maguer (elsewhere in this volume) uses the uniquac equations for description of the activity coefficients of diluted flavor compounds.

Except for the one constant Margules equations, the use of these theories in food science for the estimation of a_w of liquid food components is still a largely nondeveloped area. In principle, Margules type equations, NRTL equations, and Wilson equations (see Table I) are very suitable for describing the water activity of water in liquid organic mixtures (Chandrasekharan and King, 1972; Bruin and Prausnitz, 1971; Gmehling and Onken, 1977). Especially the description of humecant-water mixtures as being used now for the formulation of intermediate moisture foods can benefit importantly from this approach.

For a binary mixture one of the simplest expressions for a description of the excess Gibbs free energy as a function of composition is

$$G^E = kx_1x_2 \tag{21}$$

From the partial specific excess Gibbs free energy for component 1, and rearrangement (e.g., Sandler, 1977) the one-constant Margules equation ($C_{12} = C_{21}$) for this component is derived:

$$RT \ln \gamma_1 = C_{12}x_2^2 \tag{22}$$

where C_{12} is the interaction term of both components 1 and 2. Rearrangement leads to

$$\log_{10}\gamma_1 = Ax_2^2 \tag{22a}$$

which Taylor and Rowlinson (1955) used to describe the activity coefficients of water in glucose and sucrose solutions. Later the same equation was proposed by Norrish (1966) to fit sugar solutions generally:

$$\log_{10}\gamma_w = \log(a_w/x_w) = k_2 x_s^2 \qquad (22b)$$

where k_2 (negative) is the interaction term multiplied by constants and x_s is the mole fraction of sugar. He generalized this equation also for multicomponent solutions. For simplification, ternary interactions were neglected and a physically unspecified assumption concerning the interaction terms was adopted, yielding

$$\log a_w = \log x_w - (k_2^{1/2} x_2 + k_3^{1/2} x_3 + k_4^{1/2} x_4 + \dots)^2 \qquad (23)$$

where k_2, k_3, ... are the binary interaction terms of water with components 2, 3, ...

Fitting Eqs. (22) and (23) to solutions of various sugars and sugar mixtures, Norrish observed good correlation between the predicted values and experimental data of himself and other investigators. He also reported that the k_2 values of simple sugars correlated with their respective number of OH groups. Absolute values for k_2 are 0.38 for glycerol, 0.7 for glucose and fructose, 0.85 for sorbitol, and 2.6 for sucrose.

Two empirical relations to calculate practically a_w in sugar confectionary solutions were proposed earlier by Grover (1947) and Money and Born (1951). These equations, which have been used in practice, were reviewed and annotated by Karel (1975b).

C. Equation of Ross

Another possibility of deriving the activity of components in a mixture is given by the Gibbs-Duhem equation [Eq. (4)]. The isothermal, isobaric version of this equation for a binary system may be used as an approach, only for ambient conditions because the phase rule is violated (Ibl and Dodge, 1953; Prausnitz, 1969). It gives with Eqs. (10) and (12) for one component in a multicomponent mixture:

$$\sum_i n_i\, d(\ln a_i) = 0 \quad (P, T \text{ constant}) \qquad (24)$$

For a binary solution containing only one solute in water it is useful now to change to molality m_i,[*] instead of mole fraction. Equation (24) becomes

$$55.5 \, d(\ln a_w) = -m_i \, d(\ln a_i) \tag{25}$$

Since by definition, 1 kg of water (55.5 moles) contains m_i moles of solute. The solute activity (a_i) in this binary mixture is expressed as (Robinson and Stokes, 1970)

$$a_i = m_i \gamma_i^0 \tag{26}$$

where γ_i^0 is the solute activity coefficient in the binary system, adapted for molalities. Substitution gives

$$d \, \ln(a_w) = -\frac{1}{55.5} \, m_i \, d \, \ln(m_i \gamma_i^0) \tag{27}$$

In a multicomponent solution of water + n components, Eq. (27) generalizes into

$$d \, \ln(a_w) = -\frac{1}{55.5} \sum_{i=1}^{n} d \, \ln(m_i \gamma_i') \tag{28}$$

Here γ_i' is the activity coefficient of the solute component i in the system. As a result of the interaction among the different solute species, in the complex mixture the activity coefficients of the single solutes take another value than that of the one-component solution. These very many interactions, of which detailed knowledge is usually not available, make Eq. (28) very difficult to solve without simplification.

If solutions are not too concentrated, the interaction between different solute components may be assumed to be negligible. Different solute components do not "see" each other, i.e., with respect to each other they behave perfectly or, as Ross (1975) assumed, the interaction effects between those components "cancel on

[*] m_i = moles i/solvent (kg).

the average," which means mathematically

$$\gamma'_i = \gamma^0_i$$

Equations (29) and (28) yield

$$d \ln(a_w) = \frac{1}{55.5} \left[m_1 d \ln(m_1 \gamma^0_1) + m_2 d \ln(m_2 \gamma^0_2) + m_3 d \ln(m_3 \gamma^0_3) + \ldots \right]$$

Substitution of Eq. (25) and integration yields

$$\ln(a_w) = \ln(a_w)_1 + \ln(a_w)_2 + \ln(a_w)_3 + \ldots$$

which is the equation as derived by Ross (1975):

$$a_w = (a_w)_1 (a_w)_2 (a_w)_3 (a_w)_4 \ldots \tag{30}$$

Thus the water activity of a complex solution becomes simply the product of the water activity values of the aqueous solutions of each component, when measured at the same molality as in the complex solution. Fitting Eq. (30) with real mixtures, Bone *et al.* (1974) and Ross (1975) showed with different mixtures of sugar, salts, and nonsoluble food ingredients the error in calculations at $a_w > 0.8$ being smaller than 2% relative.

At high a_w, as long as the mixture remains moist Eq. (30) of Ross gives a reasonable estimate for its a_w. At lower a_w for solid mixtures the equation gives large deviations. An example of this has been described by Chuang and Toledo (1976). The Ross equation must be considered essentially a mixture equation. For the prediction of a_w of mixtures in the intermediate moixture range (0.60-0.95), it has proved to be a useful tool.

D. Influence of Temperature

Ideal solutions obey Raoult's law: the activity of the solvent is equal to its mole fraction ($a_i = x_i$) independent of pressure and temperature. In case of nonideality, however, the solvent activity coefficient γ_i changes with temperature (and constant pressure), according to (e.g., Prausnitz, 1969)

$$\left[\frac{d \ln \gamma_i}{dT}\right]_p = - \frac{h_i^E}{RT^2} \tag{31}$$

where h_i^E denotes the excess partial molal enthalpy, or partial molal excess heat of mixing. Integration at a constant pressure and h_i^E assumed to be independent of T yields

$$\ln \gamma_i = \frac{h_i}{RT} + C \tag{32}$$

where C is an integration constant.

Writing Eq. (32) for two different temperatures and subtracting those equations from each other after rearrangement yields

$$\ln \frac{\gamma_{i,1}}{\gamma_{i,2}} = \frac{h_i^E}{R}\left(\frac{1}{T_1} - \frac{1}{T_2}\right) \tag{33}$$

and since $\gamma_w = a_w/x$ and $(x)_{T_1} = (x)_{T_2}$, we obtain for water as component

$$\ln \frac{(a_w)_{T_1}}{(a_w)_{T_2}} = \frac{h_w^E}{R}\left(\frac{1}{T_1} - \frac{1}{T_2}\right) \tag{34}$$

For food systems, Eq. (34) is the most practical form of the well-known equation of Clausius-Clapeyron. This equation is usually derived from the temperature dependence of the pressure of condensing vapors, while applying the ideal gas law and neglecting the volume of condensed vapor. It may be applied only when the heat term h_w^E is constant over the observed temperature range. For foods this equation relates a_w and h_w^E in the right order of magnitude, but no precise calculation may be expected.

Perfect solutions have no heat of mixing and also $h_i^E = 0$. The solvent activity is independent of temperature. In nearly all moist food systems, however, h_w^E has a negative value. The water is sorbed exothermally, which gives a water activity increasing with temperature. In the ambient temperature range the observed change in water activity per °C is for common sugar solutions of the order of 0.0001 (Ross, 1975) and for some cereals a far less "ideal" system, about 0.0023 (Van den Berg and Leniger, 1976).

At subfreezing temperatures all systems have the same water activity,

$$a_w = p_{ice}/p_w^\theta \tag{35}$$

where p_w^θ is the vapor pressure of supercooled water at the system temperature. At ordinary frozen storage temperature (-18°C) an aqueous system has a water activity of 0.839. Freezing and drying are comparable processes in this respect. The practical validity of this rule was confirmed in accurate measurements by Fennema and Berny (1974). More detailed information on the a_w of food systems at subfreezing temperatures is given by Fennema elsewhere in this volume.

E. Electrolytes

So far we considered only mixtures of electrically neutral molecules. However, aqueous solutions containing ionic species occur frequently in food systems. Charged particles interact with coulombic forces, which act as larger separation distances than those forces interacting between neutral molecules. Consequently, ions in solution show a nonideal behavior already at very low concentrations, i.e., the electrolyte activity coefficient significantly deviates from unity. The formation of ion clouds around each ion favors this nonideality importantly. Nevertheless, separate ions follow Raoult's law up to about 0.4 molality.

The thermodynamics of more concentrated electrolyte solutions is rather complicated. Here we restrict ourselves to a brief outline of the a_w prediction methods for aqueous electrolyte solutions. For more background information the interested reader is referred to Robinson and Stokes (1970) and related literature.

An arbitrary electrolyte molecule MX dissociates in water like

$$MX = \nu_+ M^{z_+} + \nu_- X^{z_-}$$

where ν_+, ν_- are the respective numbers of positive and negative ions, and z gives the charge number of the respective ions. As a consequence the condition of electrical neutrality $\nu_+ z_+ = \nu_- z_-$ must be fulfilled.

For an electrolyte the mean ionic activity coefficient is defined as (Lewis et al., 1961)

$$\gamma_\pm = \left(\gamma_+^{\nu_+} \gamma_-^{\nu_-} \right)^{1/\nu}$$

where γ_+, γ_- are the activity coefficients of the respective ions, and $\nu = \nu_+ + \nu_-$.

Debye and Hückel (1923) used a statistical mechanical method to describe the average ion distribution in solution. They derived for the dependence of the mean ionic activity coefficient on electrolyte concentration

$$\ln \gamma_\pm = -a \left| z_+ z_- \right| I^{1/2} \tag{36}$$

where   is a constant involving only fundamental parameters, the dielectric constant of the solvent, and the absolute temperature. I is the ionic strength of the solution, defined by

$$I = \frac{1}{2} \sum_i z_i^2 m_i$$

where m_i is the molality of ion i in solution.

Equation (36), usually referred to as the Debye-Hückel limiting law, is exact only at very low ionic concentrations ($I < 0.01$). At higher concentrations, the deviations from ideality depend upon the precise nature of the ion, especially its size.

Another proposed approximate equation valid for a wide range of ionic strengths and with a semitheoretical basis is

$$\ln \gamma_\pm = -\frac{a \left| z_+ z_- \right| I^{1/2}}{1 + \beta a I^{1/2}} + \delta I \tag{37}$$

where α is a constant related to the average hydrated radius of ions, usually about 4 Å. This equation exists in some variations. Sometimes it is used in a simplified form when δ is taken zero. The product $\beta\alpha$ is sometimes set equal to 1 or treated as an adjustable parameter. Sandler (1977) gives a table with values of α and β for aqueous solutions at different temperatures between 0 and 100°C. At 25°C $\alpha = 1.178$ (moles/liter)$^{1/2}$ and $\beta = 0.3291$ (moles/liter)$^{1/2}$ Å. The parameter   is sometimes set equal to $0.1|z_+z_-|$ and sometimes taken to be an adjustable parameter. Equation (37) and its variations to improve Eq. (36), but its fitting ability is still limited to dilute solutions ($I < 0.03$). In general the accuracy of these equations is best for a 1:1 electrolyte ($z_+ = 1$, $z_- = 1$), and becomes progressively less satisfactory for 1:2, 2:2, etc., electrolytes, which are increasingly more nonideal.

For aqueous solutions at still higher concentrations Bromley (1973) derived the following equation:

$$\ln\left[\gamma_\pm^{1/(|z_+z_-|)}\right] = \frac{-1.176\ I^{1/2}}{1 + I^{1/2}} + \frac{(0.138+1.38B)\,I}{\left[1 + (1.5/|z_+z_-|)I\right]^2} + \frac{2.303BI}{|z_+z_-|}$$

(38)

where B (kg mole^{-1}) is a constant related to the type of electrolyte and depending on temperature (e.g., at 25°C, NaCl:$B = 0.0574$; CaCl$_2$:$B = 0.0948$; MgCl$_2$:$B = 0.1129$; HCl:$B = 0.1433$). Bromley (1973) gives B values for many electrolytes and also indicates how to calculate B in case this value is unknown. Even very uncommon electrolytes can be calculated this way.

When $\gamma_\pm$ is known as a function of the mean ionic molality ($m = m_+^{\nu_+}m_-^{\nu_-}$), then the activity coefficient of the solvent γ_w (water) can be derived by application of the Gibbs-Duhem equation. This Eq. (4) at a constant pressure and temperature [Eq. (24)] yields for molalities instead of mole fractions

$$\frac{1000}{M_w}\, d\ln a_w - \nu m_\pm\, d\ln(\gamma_\pm m_\pm) = 0$$

(39)

where M_w is the molal weight of the solvent (water). To calculate
the a_w of 1:1 electrolyte solutions Kusik and Miessner (1973) developed this procedure into a useful graph (Fig. 6). This graph
relates for one temperature, a family of curves for constant
$\log_{10} a_w$ with ionic strength I and $\log_{10} \Gamma$, where Γ is the <u>reduced
mean ionic activity coefficient</u>:

$$\Gamma = \gamma_{\pm}^{1/(|z_+ z_-|)}$$

Also electrolytes with higher valences than 1:1 can be handled by
this method. The following equation (Kusik and Meissner, 1973)

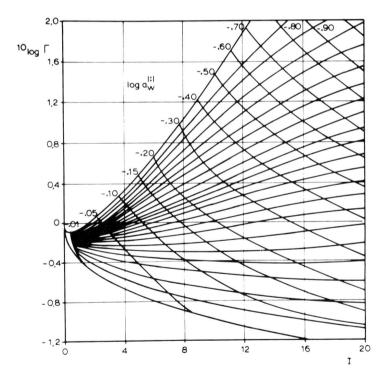

Fig. 6. *Generalized graph showing the logarithm of reduced
mean ionic activity coefficient as a function of ionic strength
with $\log_{10} a_w$ as a parameter at 25°C for 1:1 electrolytes ($M^+ X^-$).
[Reprinted with permission from Kusik and Meissner (1973). Copy-
right by American Chemical Society.]*

relates the derived water activity for the 1:1 electrolyte to the a_w in demand:

$$\ln \left[\frac{(a_w)_{z_+z_-}}{(a_w)_{1:1}} \right] = 0.0359 \left(1 - \frac{1}{|z_+z_-|} \right) I \tag{40}$$

Also Ross (1975) briefly outlined this method.

While using rigorous thermodynamics, Bromley (1973) derived from Eq. (38) and the Gibbs-Duhem equation an expression for the direct calculation of a_w. Upon rearrangement this equation yields

$$\ln a_w = \frac{M_w \Sigma m_i}{1000} \left[1.177 |z_+z_-| I^{1/2} \psi_1 - (0.138 + 1.38B) \right.$$
$$\left. |z_+z_-| \psi_2 - 1.1515BI - 1 \right] \tag{41}$$

where

$$\psi_1 = \frac{1}{I^{3/2}} \left[1 + I^{1/2} - \frac{1}{1+I^{1/2}} - 2 \ln(1+I^{1/2}) \right]$$

$$\psi_2 = \frac{1}{Z} \left[\frac{1+2ZI}{(1+ZI)^2} - \frac{\ln(1+ZI)}{ZI} \right]$$

$$Z = 3/2 |z_+z_=|$$

Using Eq. (41) we composed Fig. 7. These graphs give for 1:1, 1:2, and 2:1 electrolyte at $25°C$ a_w as a function of ionic strength with B as a parameter. When B of the electrolyte considered is known we can use this graph to estimate a_w directly.

Water activities of solutions of volatile weak electrolytes (NH_3, H_2S, CO_2, and SO_2) can be accounted for by methods presented by Beutier and Renon (1978).

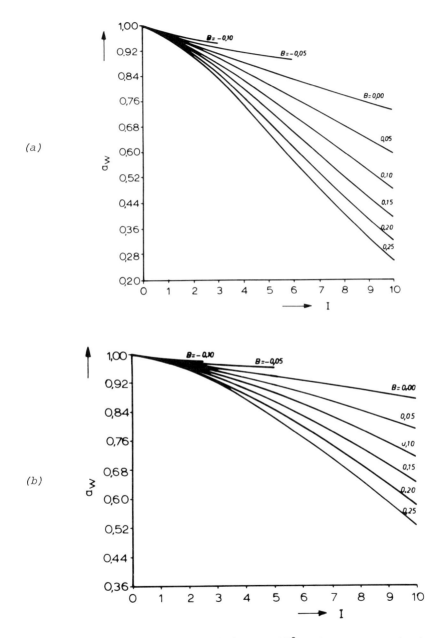

Fig. 7. (a) Water activities at 25°C of aqueous solutions of strong 1:1 electrolytes (M^+X^-) as a function of ionic strength. (b) Water activities at 25°C of aqueous solutions of strong 1:2 and 2:1 electrolytes ($M^+_2 2^-$, $M^{2+}X^-_2$) as a function of ionic strength.

V. SORPTION MODELS

A. *Undissolved Components and Sorption Models*

    The undissolved components of a food usually constitute the
major part of the total dry solids in the food system. Their influence on a_w cannot be estimated from the foregoing theory.
These components are mostly carbohydrates and proteins or related
compounds. They are mainly polymers, often with high molecular
weights. Their concentration when expressed on a mole fraction
basis is low. However, water is bound to each monomer unit, so
that these compounds help to lower the a_w significantly. The
numerical effect on a_w starts at mass fractions of water (on a
dry basis) of less than ∿50%. Per polar group, high-molecular-
weight compounds are sometimes more effective in depressing a_w
than low-molecular-weight compounds with a similar molecular
structure, as was reported for polyethylene glycols (Bone, 1969;
Karel, 1975b).

    It should also be remarked that fats and oils as food compo-
nents do not depress a_w significantly. Because the water solubility in fats and oils is very low and equal to the reciprocal activity coefficient at infinite dilution, the water activity is about
equal to the ratio of the actual water content of the fat or oil
and the solubility of water in the particular fat or oil under consideration. This solubility, being a function of temperature, is
small: from 0.2% at room temperature it increases to about 1% at
100°C. This result has been confirmed by Loncin (1955).

B. *Mathematical Description*[*]

    For the mathematical description of the moisture sorption iso-
therm of undissolved food components and virtually whole food systems, many different models can be found in the literature. In

    [*]*For equations, see Appendix.*

this survey of theoretical methods of a_w estimation, a compilation of these models seems in place. It is not our intention to go in depth into details of theories about sorption models. We present merely the available water sorption isotherm equations showing their origin, together with a few general comments about their mutual relations and use. In no way do these comments cover the field adequately, or do they give full justice to all the equations mentioned.

In the field of food engineering it is hardly necessary to stress the importance of a relatively simple equation describing the moisture sorption isotherm of a food reliably with a limited number of parameters (calculation of drying and storage processes; fewer measurements are needed to obtain the complete isotherm). These parameters preferably should have a reasonably clear physical significance and a known temperature dependence, together with possible corrections for the influence of hysteresis. This equation does not exist at present and is not expected to be found in the near future. The already mentioned difficulties around the physical characterization of food systems or even mere food components (e.g., the sorbing surface), their chemical complexity, the plasticizing effect of water on the food structure, and the apparent lack of equilibrium in some cases, together with general problems in mathematical description of physical sorption models (e.g., binding energy distribution of the active sorption sites) makes the problem of predicting a_w of a food a very delicate one to handle from a theoretical point of view.

On this problem Young and Crowell (1962, p. 157) stated earlier: "the perfect sorption theory which takes all factors into account would lead to an isotherm of such complexity that it could not fail to describe any isotherm shape. Such an equation would also be quite useless because none of its constants could be evaluated unequivocally."

From a mathematical point of view, it is easy to fit any number of experimental data points to a sigmoid curve. A suitable exponential function with a few logarithms and sufficient parameters will always be able to fit with very good accuracy, provided the experimental data points do not scatter too wildly. Some of the model sorption equations to be presented owe their success at least partially, to this fact. Therefore the consequent better agreement between calculated and observed isotherms must be accepted with the reservation that it may be due solely to the added flexibility conferred on the equation by the improved mathematical function and the new parameters. Bakker-Arkema *et al.* (1977) were able to fit sorption isotherm data with a set of smoothed cubic polynomials with seven parameters more accurately than with any of the available equations that have some physicochemical basis. However, as "there can be no true physical science which looks first to mathematics for the provision of a physical model" (A. N. Whitehead), we should always try to find some compromise to meet our practical demands as well as our need for a theoretical justification. Nonetheless, we can never prove the validity of a sorption model just by its ability to fit the observed isotherm.

C. Compilation of Sorption Models

Going through the literature we found altogether more than 75 isotherm equations that have been proposed or merely used to fit moisture sorption isotherms of biological materials over a smaller or larger range of a_w or even over the whole isotherm. The equations are shown in Appendix I as applied to moisture sorption relating W (mass fraction of water on dry basis, or in some cases W', mass fraction of water on total basis) and a_w. It was not practicable to write all equations explicit in a_w or W. They have been categorized into four groups, the headings of which were chosen according to their origin:

(1) localized monolayer sorption models, only able to describe the first convex part of the sigmoid isotherm at low a_w;

(2) localized multilayer sorption and condensed film models;

(3) sorption models from polymer science;

(4) sorption models and ad hoc fitting equations from food and agricultural engineering.

The first two groups originate from the field of physical adsorption. Subdivisions were made with respect to type and origin of the models. Part of these equations have been reviewed previously by Labuza (1968, 1975). Spiess and Wolf (1969) and recently Iglesias and Chirife (1975) summarized 23 of those equations together with surveying their use in food science. Barrie (1968) reviewed several water sorption models for polymers.

D. Fitting Parameters

Tables with fitting parameters for various food and biological materials were produced by Lykow (1955) for the Lykow equation (A71),[*] Rounsley (1961) for the Pickett equation (A9), Keey (1972) for the Henderson equation (A52), Spiess *et al.* (1969) for the BET equation (A7), Pfost *et al.* (1976) (see Bakker-Arhema *et al.*, 1977) for the Henderson-Thompson equation (A55) and the Chung-Pfost equation (A56), and very extensively by Iglesias and Chirife (1975, 1976a) for the Frenkel-Halsey-Hill equation (A21) as well as the Henderson equation (A52). Due to natural variation of food products one should not rely on these given parameters for obtaining precise sorption data, but they are useful for supplying a rapid estimate for the product under consideration.

Upon rearrangement not all equations were found to be different. For example, the equations of Chen (A57) and De Boer and Zwikker (A16) are identical, and also the equations of Strohman and Yoerger (A56), simplified Chen and Clayton (A59) and the equation of Bradley (A17).

[*]*All equation numbers preceded by the letter A refer to equations in Appendix I to this chapter.*

E. Monolayer Models (A)

To fit the sigmoid isotherm at very low a_w (0.15) a heterogeneous model is to be preferred above a homogeneous model such as
the Langmuir equation. As there are usually many different polar
groups in a food system, the moisture sorption in this low a_w range
may be expected to be of a heterogeneous type. Even pure starch,
which is composed of one monomer only, may not be expected to sorb
water only homogeneously. Due to steric hindrance and preferential
hydrogen bonding, different water molecules will be bound by different energies. This in general is confirmed by experiences with
the Freundlich equation (A3). This initially empirical equation
(Freundlich, 1926) was shown by Appel (1973) to be based on a
heterogeneous sorption model. The Freundlich equation is especially well able to describe the lowest part of the isotherm of many
food materials accurately (Katz, 1917; Gàl *et al.*, 1962; Spiess and
Wolf, 1969; Tomka, 1973).

The sorptive capacities of different polar groups are additive.
This was shown by Watt and Leeder (1968), who were successful in
predicting the moisture sorption isotherm for several proteins at
low a_w from stoichiometric analysis of the available polar groups
and their sorptive capacity. See also Watt and D;Arcy (1976).

For the practice of food engineering these monolayer models
are of no great significance as hardly any food products are being
dried to these low water contents.

F. Multilayer Models (B)

Among the multilayer models for sorption it should be remarked
that the BET equation (A7), since its conception in 1938, acted as
the useful compromise between theory and practice. Its parameters
have physical significance and the model itself gives an oversimplified (homogeneous), but valuable picture of the sorption process,
together with the ability to fit all sigmoid sorption isotherms up

to about a_w = 0.40. Especially for surface area determination the BET method found wide acceptance.

Many corrections were proposed to correct the obvious short-comings of the BET model, but nearly all destroyed its simplicity. Only the equations of Pickett (A9) and Anderson (A10) add favorably to the model, in physical significance as well as in fitting ability (Rounsley, 1961; Van den Berg et al., 1975; Van den Berg and Leniger, 1976). Anderson's equation was put on a firm theoretical basis by Guggenheim (1966). It is interesting to note that also Brunauer (Brunauer et al., 1969) considered the Anderson-Guggenheim equation (A10) to be an important improvement of the BET equation (A7).

Polarized layer model equations, especially Bradley's version, gained importance in cases where sorbed water molecules are assumed to structure themselves in several subsequent layers on the sorbent surface. Examples are proteins (e.g., Kuntz and Kauzmann, 1974) and polyvinylpyrrolidone (McKenzie and Rasmussen, 1972).

The theoretically well-based sorption isotherm equation of Frenkel-Halsey-Hill (A20) has been applied successfully to many food materials by Chirife's group (Iglesias and Chirife, 1975, 1976a; Boquet et al., 1978).

The importance of capillary condensation (A23) as a mechanism to depress a_w in food systems has been overemphasized for many years. Perhaps apart from systems with very strong thick-walled capillaries, such as in wood, capillary condensation in its strict sense cannot play any role of significance at a_w < 0.95. This is due to the high pressures, involved in the physical process, acting on the capillary wall. Theoretically this is hard to establish with certainty, because water is a plasticizer for the food structure, and appropriate methods to prove this, such as the t plot method cannot be applied without important assumptions for simplification (Van den Berg et al., 1975).

G. Polymer Sorption Models (C)

For the development of most of the isotherm equations in this
group, both surface (ad)sorption and solution theories have been
combined. In these models, e.g., those leading to isotherm equations (A36)-(A46), distinction has been made between two [in the
case of (A43) and (A44) in three] types of water binding: one
type firmly bound to polar sites of the polymer and one type loosely bound or trapped in the voids of the polymer. The last type of
water causes the polymer to weaken or to dissolve partially.

The most complicated theoretical sorption model for swelling
biopolymers was developed by Tomka (1973). This sorption isotherm
equation (A46) contains as many as seven parameters and may be applied only when two other equations as boundary conditions are fulfilled. The model was summarized by Gàl *et al*. (1976). This equation is able to fit the entire sorption isotherm of casein accurately. This model might be considered as an approach to the "perfect sorption theory," mentioned before.

The model behind the isotherm equation (A43) of D'Arcy and Watt
(1970) as well as its use is explained in detail elsewhere in this
volume by D'Arcy.

H. Food Sorption Models (D)

In this group theoretically interesting sorption models, such
as the isotherm equations (A63) (Ngoddy, 1969; Ngoddy and Bakker-
Arkema, 1970) and (A62) (Young and Nelson, 1967) can be found.
Equation (A52) of Henderson (1952) especially has found much practical application, mainly in the United States. On the basis of
this equation, Rockland (1969) developed his practical local isotherm concept to estimate storage stability for some products.
Henderson's equation is one of the only equations with a direct
temperature term. In order to improve its ability to fit sorption
data at different temperatures, Day and Nelson (1965) and Thompson
(1967) modified the equation empirically in this respect.

Recently Boquet *et al.* (1978) evaluated the usefulness of eight two-parameter isotherm equations [in Appendix I equations (A17), (A21), (A22), (A52), (A69), (A73), (A74)] in describing moisture sorption isotherms of 39 foods of different nature. It was found that among these eight equations the most versatile are those of Frenkel-Halsey-Hill (A21) and Oswin (A69). More systematic work in this area remains to be done.

V. SPECIAL METHODS

For the calculation of the final a_w of a mixture of components each with a different a_w and a known sorption isotherm, Salwin and Slawson (1959) developed a simple approximation method. In this procedure, portions of the isotherms of the components were approximated by straight lines, (Fig. 8), and the final water activity $(a_w)_f$ was calculated by the following equation:

$$(a_w)_f = \frac{a_1 s_1 m_1 + a_2 s_2 m_2 + a_3 s_3 m_3 + \cdots}{s_1 m_1 + s_2 m_2 + s_3 m_3 + \cdots} \tag{42}$$

where a_1, a_2, ... are water activities of components 1, 2, ..., a_1, a_2, ... are isotherm slopes of components 1, 2, ..., m_1, m_2, ... are masses of solids of components 1, 2, ... A variation of this method is the practical way to calculate the isotherm of the mixture linearly out of the known isotherms of the composing substances and their proportions in the mixture.

VI. CONCLUDING REMARKS

In the foregoing a survey is given of the most important theoretical aspects of water activity in food systems together with the available equations for the calculation of the numerical value of

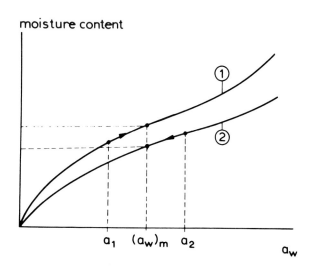

*Fig. 8. Graphical representation of moisture sorption rela-
tions used in calculation of water activity in dry mixtures. (Procedure of Salwin and Slawson, 1959.)*

a_w from the composition of the food system. It is not easy to derive practical recommendations for use in our food laboratories.

Besides actual measurement of the required a_w and the few data
for solutions we have in reference tables (e.g., glucose and sucrose; Norrish, 1967), in the laboratory we are limited to the
equations and methods referred to here:

1. Raoult's law for very dilute solutions of low-molecular-
weight compounds.

2. Different liquid theories for more concentrated solutions
of water and organic compounds, e.g., the equation of Norrish
(1966) for sugar solutions.

3. Methods of Bromley (1973) and Kusik and Meissner (1973) for
electrolyte solutions.

4. The equation of Ross (1975) for moist mixtures.

5. The approximation method of Salwin and Slawson (1959) for

dry mixtures.

6. Fit of the isotherm with one of the many indicated sorption isotherm equations.

Acknowledgments

The authors wish to thank Prof. Dr. D. A. A. Mossel (Veterinary Faculty, University of Utrecht, Netherlands), Dr. S. Gàl, (Haco AG, Gümligen, and University of Bern, Switzerland), Prof. Dr. W. A. P. Lück (Dept. of Physical Chemistry, University of Marburg, West Germany), and Prof. Dr. J. Lyklema (Dept. of Physical Chemistry, Agricultural University, Wageningen, Netherlands) for their very useful comments. Mr. H. Bizot (Institut Nationale de la Recherche Agronomique, Nantes, France) was so kind as to draw the first author's attention to the isotherm equation of Troesch (A15).

44   C. VAN DEN BERG AND S. BRUIN

APPENDIX

TABLE II. Equations for Fitting Water Vapor Sorption Isotherms of Biological Materials

---

a_w *water activity*

W *mass fraction of water on dry basis (g/100 g)*

W' *mass fraction of water on total basis*

W_m *mass fraction of water on dry basis at fully occupied active sorption sites with one molecule of water (monolayer) (g/100 g)*

T *temperature (unit depending on equation)*

Constants

  C_L *Langmuir constant*

  C_B *BET constant*

  C_B $C_L + 1$

  C_G *Guggenheim Anderson constant*

  $A, B, C, K, K_1, K_2, K_3,$ *etc.*

A. *Localized Monolayer Sorption Models*

  A1. Langmuir (1918), independent, identical sorption sites, homogeneous:

$$\frac{W}{W_m} = \frac{C_L a_w}{1 + C_L a_w} \quad or \quad W = \frac{W_m a_w}{K + a_w}$$

  A2. McGavack and Patrick (1920), adapted Freundlich equation:

$$W = \delta^{1/n} K(a_w)^{1/n}, \quad \delta = \text{surface tension of sorbed phase}$$

  A3. Freundlich (1926), Appel (1973), independent, different sorption sites, heterogeneous sorption $n > 1$:

$$W = C\, a_w^{1/n}, \quad n > 1$$

  A4. Virial development:

$$W/W_m = K_1 a_w + K_2 (K_1 a_w)^2 + K_3 (K_1 a_w)^3 + \ldots$$

A5.   Hill and De Boer (De Boer, 1953), two-dimensional Van der Waals condensation:

$$a_w = K_2 \frac{W/W_m}{1 - W/W_m} \exp\left(\frac{W/W_m}{1 - W/W_m}\right) \exp(-K_1\, W/W_m)$$

A6.   Toth (1971), heterogeneous, partially empirical:

$$W/W_m = \frac{a_w}{\left(A + a_w^m\right)^{1/m}} \,, \quad 0 < m \leq 1$$

B1. *Homogeneous Sorption (BET and Close Modifications)*

A7.   Brunauer *et al.* (1938), Cassie (1945), independent identical sorption sites, ∞ sorbed layers

$$\frac{W}{W_m} = \frac{C_B a_w}{(1-a_w)\left[1 + (C_B-1)a_w\right]} \,, \quad \text{or} \quad \frac{W}{W_m} = \frac{C_L a_w}{1 + C_L a_w} + \frac{a_w}{1 - a_w}$$

A8.   Brunauer *et al.* (1940), independent with restricted number of layers (n):

$$\frac{W}{W_m} = \frac{C_B a_w\left[1 - (n+1)a_w^n + na_w^{n+1}\right]}{(1-a_w)\left[1 + (C_B-1)a_w - C_B a_w^{n+1}\right]}$$

A9.   Pickett (1945), Anderson (1946), Rounsley (1961), layers built up in orderly manner:

$$\frac{W}{W_m} = \frac{Ca_w\left(1 - a_w^n\right)}{(1-a_w)\ 1 + (C-1)a_w}$$

A10.   Anderson (1946), Guggenheim (1966), Dent (1977), independent with correction of multilayer properties:

$$\frac{W}{W_m} = \frac{C_G Ka_w}{(1-Ka_w)\left[1 + (C_G-1)Ka_w\right]}$$

A11. Anderson and Hall (1948):

$$\frac{W}{W_m} = \frac{Ca_w}{(1-Aa_w)\left[1 + (C-a)a_w\right]}$$

A12. Anderson and Hall (1948), variant:

$$\frac{W}{W_m} = \frac{CKa_w}{(1-Aka_w)\left[1 + (C-A)Ka_w\right]}$$

A13. Anderson (1946), variant:

$$\frac{W}{W_m} = \frac{Ca_w}{(1-Aa_w)\left[1 + (C-1)a_w\right]}$$

A14. Gascoyne and Pethig (1977) (rearranged), correction of multilayer properties:

$$\frac{W}{W_m} = \frac{CKa_w}{(1-Ka_w)\left[1 + (C-K)a_w\right]}$$

A15. Troesch (1951):

$$\frac{W}{W_m} = \frac{Ca_w\left[1 + (1-K)a_w\right]}{(1-Ka_w)\left[1 + (C-K)a_w\right]}$$

B2. *Polarized Layers*

A16. De Boer and Zwikker (1929), De Boer (1931):

$$\ln\left(\frac{a_w}{K_3}\right) = K_2K_1^{W/W_m}$$

A17. Bradley (1936):

$$\ln a_w = K_2K_1^{W/W_m}$$

A18. Hoover and Mellon (1950):

$$K \ln a_w = \exp(-W/W_m)$$

B3. Liquid Film and Derivatives

A19. Polanyi (1928), adsorption potential:

$$W = \frac{K_1}{(\ln a_w)^{1/3}} + K_2$$

A20. Harkins and Jura (1944) (rearranged):

$$W^2 = \frac{K_1}{K_2 - \ln a_w}$$

A21. Frenkel-Halsey-Hill (ca. 1948) (e.g., Young and Crowell, 1962):

$$\left[\frac{W}{W_m}\right]^m = -K \ln a_w, \quad K = RT/K_1$$

A22. Kühn (1964) (rearranged), multilayer film sorption + capillary condensation:

$$W = \frac{K_1}{(\ln a_w)} + K_2$$

B4. Condensation Models

A23. Kelvin equation, Zsigmondy (1911):

$$\ln a_w = -2 \, \delta \, \cos \theta \, v_w/rRT$$

δ, liquid surface tension, θ, contact angle of liquid in pores; v_w, molar volume of liquid; r, radius of capillary.

A24. Corrected Kelvin equation (e.g., Broekhoff and Linsen, 1970) corrections for thickness and curvature of adsorbed layer.

A25. Othmer plot, Othmer *et al.* (1940, 1942; Othmer and Sawyer, 1943): graphical method based on modified Clasius-Clapeyron equation.

B5. *Other Multilayer Models*

A26. Huttig (1948), heterogeneous sorption:

$$\frac{W}{W_m} = \frac{Ca_w}{1 + Ca_w} (1 + a_w)$$

A27. Dubinin-Radushkevich (1947) (rearranged), condensation
in micropores:

$$\frac{W}{W_m} = A \exp[B(RT \ln a_w)^2]$$

A28. Jaroniec (1975), heterogeneous sorption, generalized
Dubinin-Radushkevich:

$$\frac{W}{W_m} = \exp\left[\sum_{n=0}^{m} B_n (\ln a_w)^n\right]$$

C. *Sorption Models from Polymer Science*

C1. *Solution-Type Models*

A29. Tester (1946), osmotic mechanism:

$$W/(W+K) = a_w, \quad \text{for } a_w > 0.5$$

A30. White and Eyring (1947):

$$(W-W_m)/W = Ka_w, \quad \text{for } W > W_m$$

A31. Dole (1948, 1949):

$$W = \frac{Ka_w}{1 - a_w}$$

A32. Barrer (1947), Langmuir sorption + solution:

$$a_w = \frac{V_1}{(1 - 2/zV_2)^{z/2}}$$

z, coordination number; V_1, V_2, volume fractions of
water, polymer.

A33. Flory and Huggins (Flory, 1953), lattice model not for
      low a_w:

$$\ln a_w = \ln V_1 + V_2 + \chi V_2^2$$

V_1, V_2, volume fractions of water, polymer; χ, interaction parameter polymer-solvent

A34. Rogers *et al.* (1959):

$$\ln a_w = \ln V_1 + V_2 + \chi V_2^2 + C V_2^{1/3}$$

for crystalline polymers: V_1, V_2 expressed relative to amorphous phase.

A35. Heil and Prausnitz (1966), complicated function of segment-solvent local volume fractions with two parameters.

C2. Localized Sorption and Solution Models

A36. Peirce (1929) (rearranged), binding + solution:

$$a_w = 1 - \exp\left[-K(W-W_m)\right]$$

A37. id., with correction term for evaporation:

$$1 - a_w = (1-K_1 W_m) \exp -K_2(W-W_m)$$

A38. Smith (1947), sorption + solution:

$$W = K_1 - K_2 \ln(1-a_w)$$

A39. id., corrected for swelling:

$$\frac{W}{1 + W} = K_1 - K_2 \ln(1-a_w)$$

A40. Hailwood and Horrobin (1946), localized sorption + solid solution:

$$\frac{W}{W_m} = \frac{K_1 a_w}{1 + K_1 a_w} + \frac{K_2 a_w}{1 - K_2 a_w} \quad \text{or} \quad \frac{a_w}{W} = A + B a_w - C a_w^2$$

A41. McLaren and Rowen (1952), BET model with two types of
sorption sites:

$$W \frac{1-a_w}{a_w} = \frac{W_m C}{1 + (C-1)a_w} + \frac{W_m' C'}{1 + (C'-1)a_w}$$

A42. Enderby (1955), King (1960), binding + solid solution:

$$W = \frac{W_m K_1 a_w}{1 + K_1 a_w} + \frac{W_m' K_2 a_w}{1 - K_2 a_w}$$

A43. D'Arcy and Watt (1970), localized sorption of monolayer
+ weak binding + solid solution:

$$W = \frac{K_2 K_1 a_w}{1 + K_1 a_w} + K_3 a_w + \frac{K_5 K_4 a_w}{1 - K_4 a_w}$$

A44. Kollmann (1962), Freundlich model + capillary condensation in statistically distributed pore sizes:

$$W = K_1 a_w^m + K_2 \exp\left[-K_3 \left[a_w - \left(1 + \frac{1}{\sqrt{2} \, K_3}\right)\right]^2\right]$$

A45. Rowen (1958):

$$\log a_w = \frac{1}{B}\left[\frac{\beta}{\frac{W}{W_m} + \beta}\left(1 - \frac{1}{S}\right) + \frac{\beta}{\frac{W_{1m}}{W_m} + \beta} \log\left(\frac{\frac{W}{W_m} - \frac{W_{1m}}{W_m}}{\frac{W}{W_m} + \beta}\right)\right.$$

$$\left. + \frac{1}{2} C \left(\frac{\frac{W_{1m}}{W_m} + \beta}{\frac{W}{W_m} + \beta}\right)^2 \right]$$

B, ratio monomer volume to sorbate molecular volume; C
heat of mixing coefficient.

A46. Tomka (1973), Freundlich sorption + polymer model:

$$\ln a_w = \frac{V_{1m} + 1}{V_1 + 1} \left(1 + X \frac{1}{V_1 + 1} \right) + \ln \left(\frac{V_1 - V_{1m}}{V_1 + 1} \right)$$

$$+ \left[\left(\frac{1}{V_1 + 1} \right)^{1/3} - \frac{1}{2} \frac{1}{V_1 + 1} \right] \frac{\nu}{s\nu\beta}$$

with two additional equations as boundary equations (seven parameters in total): $V_1 = V_{1m} + V_{11}$; ν, number of polymer chains; s, number of segments in polymer chain; β, volume ratio of polymer segment to sorbate molecule.

C3. *Miscellaneous*

A47. Zimm (1953):

$$\frac{G}{v_w} = -V_2 \left[\frac{\partial \left(\frac{a_w}{V_1} \right)}{\partial a_w} \right]_{P,T}$$

v_w, partial molar volume of sorbate; B/v_w, "clustering" function of sorbate molecules.

A48. Schwarz (1970), "cooperative binding" function relating water concentration, occupied sorption sites, and activity.

D1. *Partially Theoretically Based*

A49. Miniowitsch (1937) (rearranged) (see Lykow, 1958):

$$K_1 a_w = \log W' - \log(K_2 + K_2 T), \quad T \text{ in } °K$$

A50. Riedel (1965), Langmuir sorption + solution:

$$W = \frac{K_1 a_w}{K_2 + a_w} + \frac{K_3 a_w^m}{1 - a_w}$$

A51. Fugassi and Ostapchenco (1959):

$$W = \frac{K_1 a_w}{K_2 a_w (1-a_w) + K_3 a_w}$$

A52.　Henderson (1952):

$$\ln(1-a_w) = KTW^m, \quad T \text{ in } °R$$

A53.　Day and Nelson (1965), simplified Henderson equation:

$$\ln(1-a_w) = -K'W^m$$

A54.　id., modified Henderson equation:

$$\ln(1-a_w) = -K_1 T^{h_1} W^{k_2^{h_2}}$$

A55.　Henderson and Thompson (Thompson, 1967):

$$\ln(1-a_w) = -K_1 (T+K_2) W^m$$

A56.　Chung and Pfost (1967) (=Bradley eq.):

$$\ln a_w = -\frac{A}{RT} \exp(-BW)$$

A57.　Strohman and Yoerger (1967) (is Bradley eq.) derived from Othmer plot:

$$\ln a_w = K_1 \exp(K_2 W) + K_3 \exp(K_4 W)$$

A58.　Chen (1971) (is de Boer and Zwikker eq.):

$$\ln a_w = K_1 + K_2 \exp(K_3 W)$$

A59.　Chen and Clayton (1971):

$$\ln a_w = -K_1 T^{m_1} \exp(-K_2 T^{m_2} W)$$

A60.　Simplified (is Bradley eq.):

$$\ln a_w = -K_1 \exp(-K_2 W)$$

A61. Iglesias and Chirife (1976b), Frenkel-Halsey-Hill equation empirically modified:

$$\ln a_w = -\exp\left[(K_1 T + K_2) W^{-m}\right]$$

A62. Young and Nelson (1967), combination of BET + Smith models:

$$\text{resorption:} \quad W = \frac{\rho V_m}{D} (\theta + a) + \frac{\rho V}{D} \phi$$

$$\text{desorption:} \quad = \frac{\rho V_m}{D} (\theta + a) + \frac{\rho V}{D} a_{w\ max}$$

ρ, spec. weight of water; D, dry weight of sorbens; V_m, volume of water sorbed in a full monolayer; θ, $f(a_w$, heat of sorption); ϕ, $f(a_w, \theta)$; a, $f(a_w$, heat of sorption).

A63. Ngoddy and Bakker-Arkema (1969, 1970), combination of BET + capillary condensation + adsorption potential:

$$W = \frac{\rho K}{N} (Z^N - \lambda^N)$$

K and N, characteristic parameters of the physical structure of the sorbent; Z and λ, functions of a_w.

D2. Fully Empirical Isotherm Equations

A64. Linear:

$$W = K_1 a_w + K_2$$

A65. Polynome:

$$W = K_1 + K_2 a_w + K_3 a_w^2 + K_4 a_w^3 + \cdots$$

A66. Franchuk (1941) [see Luikov (1966)]:

$$C_w \text{ is water concentration } (kg/m^3)$$

A67. Posnow (1953) [see Lykow (1958)]:

$$\frac{1}{W'} = \frac{1}{W'_{max}} + K \ln a_w$$

A68. Rode (1952) [see Lykow (1958)]:

$$0 \quad a_w < 0.35, \quad W' = Ka_w^{0.5}$$

$$0.35 < a_w < 0.94, \quad W' = K_1 + K_2 a_w^2$$

A69. Oswin (1946), Pearson's series expansion:

$$W = K \left(\frac{a_w}{1-a_w}\right)^m$$

A70. De Boer (1953):

$$\text{not for low } a_w \begin{cases} \dfrac{W}{W_m} = \dfrac{1}{1 - a_w} \\[2ex] \dfrac{W}{W_m} = \dfrac{a_w}{1 - a_w} \end{cases}$$

A71. Lykow (1955):

$$W' = \frac{K_1 a_w}{K_2 - a_w} , \quad 0.1 < a_w < 1$$

A72. Haynes (1961) (rearranged):

$$\ln a_w = K_1 + K_2 W + K_3 W^2$$

A73. Mizrahi *et al.* (1970):

$$W = \frac{K_1 + K_2 a_w}{1 - a_w} \quad \text{or} \quad a_w = \frac{K_1 + W}{K_2 + W}$$

A74. Caurie (1970) (rearranged):

$$\ln \left(\frac{100 - W''}{W''}\right) = K_1 - K_2 a_w , \quad \text{in which } W'' = \frac{W}{100 + W}$$

A75.   Iglesias and Chirife (1976c):

$$\ln\left[W + (W^2 + W_{0.5})^{1/2}\right] = K_1 a_w + K_2, \quad W_{0.5} = W \text{ at } a_w = 0.5$$

A76.   Bakker-Arkema *et al.* (1977); set of cubic polynomials
       with seven parameters $(F_1, \ldots, F_7)$ as $f(T)$ and four
       parameters as $f(F_1, \ldots, F_7)$.

A77.   Filonenko and Chuprin (1967), extension of Lykow's equa-
       tion (A71)

$$W_1' = \frac{K_1 + a_w}{K_2}, \quad W_2' = \frac{K_3 \Delta a_w}{K_4 - \Delta a} + K_5, \quad W' = W_1' + W_2'$$

References

Anderson, R. B. (1946). *J. Am. Chem. Soc. 68*, 686.
Anderson, R. B., and Hall, K. W. (1948). *J. Am. Chem. Soc. 70*, 1727.
Appel, J. (1973). *Surface Sci. 39*, 237.
Bakker-Arkema, F. W., Brook, R. C., and Lerew, L. E. (1977). *Adv. Cereal Sci. Technol. 2*, 1.
Barrer, R. M. (1947). *Trans. Faraday Soc. 43*, 3.
Barrie, J. A. (1968). *In* "Diffusion in Polymers" (Crank, J., and Park, G. S., eds.), p. 259. Academic Press, London.
Beutier, D., and Renon, H. (1978). *Ind. Eng. Chem. Process Des. Dev. 17*, 220.
Bomben, J. L., Bruin, S., and Thijssen, H. A. C. (1973). *Adv. Food Res. 20*, 1.
Bone, D. P. (1969). *Food Prod. Dev. 3*, 81.
Bone, D. P., Shannon, E. L., and Ross, K. D. (1975). *In* "Water Relations of Foods" (Duckworth, R. B., ed.), p. 613. Academic Press, London.
Boquet, R., Chirife, J., and Iglesias, H. A. (1978). *J. Food Technol. 13*, 319.
Bradley, R. S. (1936). *J. Chem. Soc.*, 1467, 1799.
Broekhoff, J. C. P., and Linsen, B. G. (1970). *In* "Physical and Chemical Aspects of Adsorbents and Catalysts" (Linsen, B. G., ed.), p. 1. Academic Press, London.
Bromley, L. A. (1973). *Am. Inst. Chem. Eng. J. 19*, 313.
Bruin, S. (1969). Ph.D. Thesis. Agricultural Univ., Wageningen.
Bruin, S., and Prausnitz, J. M. (1971). *Ind. Eng. Chem. Process Des. Dev. 10*, 562.
Brunauer, S., Emmett, P. H., and Teller, E. (1938). *J. Am. Chem. Soc. 60*, 309.
Brunauer, S., Deming, L. S., Deming, W. E., and Teller, E. (1940). *J. Am. Chem. Soc. 62*, 1723.
Brunauer, S., Skalny, J., and Bodor, E. E. (1969). *J. Colloid Interface Sci. 30*, 546.
Cassie, A. B. D. (1945). *Trans. Faraday Soc. 41*, 450.
Caurie, M. (1970). *J. Food Technol. 5*, 301.
Chandrasekaran, S., and King, C. J. (1972). *Am. Inst. Chem. Eng. J. 18*, 513.
Chen, C. S. (1971). *Trans. Am. Soc. Agri. Eng. 14*, 924.
Chen, C. S., and Clayton, J. T. (1971). *Trans. Am. Soc. Agri. Eng. 14*, 927.
Chirife, J., and Iglesias, H. A. (1978). *J. Food Technol. 13*, 159.
Christian, J. H. B., Michener, H. D., and Jarvis, B. (eds.) (1978). "Safety of Food as Influenced by Water Activity." Parma, Italy.
Chuang, L., and Toledo, R. T. (1976). *J. Food Sci. 41*, 922.
Chung, D. S., and Pfost, H. B. (1967). *Trans. Am. Soc. Agri. Eng. 10*, 549, 552, 556.

Corry, J. E. L. (1975). *In* "Water Relations of Foods" (Duckworth, R. B., ed.), p. 325. Academic Press, London.

D'Arcy, R. L., and Watt, I. C. (1970). *Trans. Faraday Soc. 66,* 1236.

Day, D. L., and Nelson, G. L. (1965). *Trans. Am. Soc. Agri. Eng. 8,* 293.

De Boer, J. H. (1931). *Z. Phys. Chem. B14,* 457.

De Boer, J. H. (1953). "The Dynamical Character of Adsorption." Clarendon Press, Oxford.

De Boer, J. H., and Zwikker, C. (1929). *Z. Phys. Chem. B3,* 407.

Debye, P., and Hückel, E. (1923). *Phys. Z. 24,* 185.

Dent, R. W. (1977). *Text. Res. J. 47,* 145.

Dole, M. (1948). *J. Chem. Phys. 16,* 25.

Dole, M. (1949). *Ann. N.Y. Acad. Sci. 51,* 705.

Dubinin, M. M. and Radushkevich, L. V. (1947). *Dokl. Akad. Nauk SSSR 55,* 327.

Enderby, D. H. (1955). *Trans. Faraday Soc. 51,* 106.

Fennema, O. (1976). *In* "Principles of Food Science," Part 1 (Fennema, O., ed.), p. 13. Marcel Dekker, New York.

Fennema, O., and Berny, L. A. (1974). *Proc. 4th Int. Congr. Food Sci. Technol., Madrid 1974, II,* 27.

Filonenko, G. K., and Chuprin, A. E. (1967). *J. Eng. Phys.* (transl. from *Inzh.-Fiz. Zh. 13,* 98).

Flory, P. J. (1953). "Principles of Polymer Chemistry." Cornell Univ. Press, Ithaca, New York.

Fredenslund, A., Jones, R. L., and Prausnitz, J. M. (1975). *Am. Inst. Chem. Eng. J. 21,* 1086.

Fredenslund, A., Gmehling, J., and Rasmussen, P. (1977). Vapor-Liquid Equilibria Using Unifac, a Group Contribution Method." Elsevier, Amsterdam.

Freundlich, H. (1926). "Colloid and Capillary Chemistry." Methuen, London.

Fugassi, P., and Ostapchenco, G. (1959). *Fuel 38,* 271.

Gàl, S. (1972). *Helv. Chim. Acta 55,* 1752.

Gàl, S., Arm, H., and Signer, R. (1962). *Helv. Chim. Acta 45,* 748.

Gàl, S., Tomka, I., and Signer, R. (1976). *Chimia 30,* 65.

Gascoyne, P. R. C., and Pethig, R. (1977). *J. Phys. Chem. Faraday Trans. 1,* 171.

Gmehling, J., and Onken, U. (1977). Vapor-liquid equilibrium data collection. *Dechema Chem. Data Ser. 1,* Part 1.

Grover, D. W. (1947). *J. Soc. Chem. Ind. 66,* 201.

Guggenheim, E. A. (1966). "Applications of Statistical Mechanics." Clarendon Press, Oxford.

Hailwood, A. J., and Horrobin, S. (1946). *Trans. Faraday Soc. 42B,* 84.

Harkins, W. D., and Jura, G. (1944). *J. Am. Chem. Soc. 66,* 1366.

Haynes, B. C. (1961). *U.S. Dept. Agri. Tech. Bull.,* 1229.

Heil, J. F., and Prausnitz, J. M. (1966). *Am. Inst. Chem. Eng. J. 12,* 678.

Heiss, R., and Eichner, K. (1971). *Chem. Mikrobiol. Technol. Lebensm. 1,* 33.

Henderson, S. M. (1952). *Agri. Eng. 33*, 29.
Hildebrand, J. H., Prausnitz, J. M., and Scott, R. L. (1970). "Regular and Related Solutions." Van Nostrand Reinhold, New York.
Hoover, S. R., and Mellon, E. F. (1950). *J. Am. Chem. Soc. 72*, 2562.
Huttig, G. F. (1948). *Monatsh. Chem. 78*, 177.
Ibl, N. V., and Dodge, B. F. (1953). *Chem. Eng. Sci. 2*, 120.
Iglesias, H. A., and Chirife, J. (1975). *J. Food Technol. 10*, 289.
Iglesias, H. A., and Chirife, J. (1976a). *J. Food Sci. 41*, 984.
Iglesias, H. A., and Chirife, J. (1976b). *J. Food Technol. 11*, 109.
Iglesias, H. A., and Chirife, J. (1976c). *Can. Inst. Food Sci. Technol. J. xx.*
Jaroniec, M. (1975). *Surface Sci. 50*, 553.
Karel, M. (1974). "Advances in Preconcentration and Dehydration of Foods" (Spicer, A., ed.), p. 45. Applied Science Publ., New York.
Karel, M. (1975a). *In* "Freeze Drying and Advanced Food Technology" (Goldblith, S. A., Rey, L., and Rothmayr, W. W., eds.), p. 643. Academic Press, London.
Karel, M. (1975b). *In* "Water Relations of Foods" (Duckworth, R. B., ed.), p. 639. Academic Press, London.
Keey, R. B. (1972). "Drying, Principles and Practice." Pergamon Press, Oxford.
King, G. (1960). *In* "Moisture in Textiles" (Hearle, J. W. S., and Peters, R. H., eds.), p. 59. Butterworths, London.
Kollmann, F. (1962). *Die Naturwissenschaften 49*, 206.
Kühn, I. (1964). *J. Colloid Sci. 19*, 685.
Kuntz, I. D. Jr., and Kauzman, W. (1974). *Adv. Protein Chem. 28*, 239.
Kusik, G. L., and Meissner, H. P. (1973). *Ind. Eng. Chem. Process Des. Dev. 12*, 112.
Labuza, T. P. (1968). *Food Technol. 22*, 263.
Labuza, T. P. (1970). *Proc. 3rd Int. Conf. Food Sci. Technol., SOS 70*, p. 618. I.F.T., Washington, D.C.
Labuza, T. P. (1975). *In* "Theory, Determination and Control of Physical Properties of Food Materials" (Cho Kyun Rha, ed.). D. Reidel, Dordrecht/Boston.
Labuza, T. P., McNally, L., Gallagher, D., Hawkes, J., and Hustado, F. (1972). *J. Food Sci. 37*, 154.
Langmuir, I. (1918). *J. Am. Chem. Soc. 40*, 1361.
Lewis, G. N. (1901). *Z. Phys. Chem. 38*, 205.
Lewis, G. N. (1907a). *Proc. Am. Acad. Sci. 43*, 259.
Lewis, G. N. (1907b). *Z. Phys. Chem. 61*, 129.
Lewis, G. N., Randall, M. E., Pitzer, K. S., and Brewer, L. (1961). "Thermodynamics," 2nd rev. ed. McGraw-Hill, New York.
Loncin, M. (1955). *Fette u. Seifen 57*, 41.

Luikov, A. V. (also transliterated Lykow, A. W.). (1966). "Heat and Mass Transfer in Capillary-Porous Bodies." Pergamon Press, Oxford.

Lykow, A. W. (also transliterated Luikov, A. V.). (1955). "Experimentelle und Theoretische Grundlagen der Trocknung." V.E.B. Verlag, Berlin.

Lykow, A. W. (also transliterated Luikov, A. V.). (1958). "Transporterscheinungen in Kapillarporösen Körpern." Academie Verlag, Berlin.

McBain, J. W. (1909). *Phil. Mag. 18,* 916.

McGavack, J. Jr., and Patrick, W. A. (1920). *J. Am. Chem. Soc. 42,* 946.

McKenzie, A. P., and Rasmussen, D. H. (1972). *In* "Water Structure at the Water-Polymer Interface" (Jellinek, H. H. G., ed.), p. 146. Plenum Press, New York/London.

McLaren, A. D., and Rowen, J. W. (1952). *J. Polymer Sci. 7,* 289.

Mizrahi, S., Labuza, T. P., and Karel, M. (1970). *J. Food Sci. 35,* 799.

Money, R. W., and Born, R. (1951). *J. Sci. Food Agri. 2,* 180.

Mossel, D. A. A. (1975). *In* "Water Relations in Foods" (Duckworth, R. B., ed.), p. 347. Academic Press, London.

Mossel, D. A. A., and Westerdijk, J. (1949). *Anthonie van Leeuwenhoek 15,* 190.

Muetgeert, J. (1978). Project no. UR-E19-(10)-18, unpublished results.

Ngoddy, P. O. (1969). Ph.D. Thesis, Michigan State Univ. Univ. Microfilms, Ann Arbor, Michigan.

Ngoddy, P. O., and Bakker-Arkema, F. W. (1970). *Trans. Am. Soc. Agri. Eng. 13,* 612.

Norrish, R. S. (1966). *J. Food Technol. 1,* 25.

Norrish, R. S. (1967). Selected tables of physical properties of sugar solutions. *Scient. Tech. Surv. 51,* B.F.M.I.R.A., Leatherhead (Surrey), U.K.

Oswin, C. R. (1946). *J. Chem. Ind. (London) 64,* 419.

Othmer, D. F. (1940). *Ind. Eng. Chem. 32,* 841.

Othmer, D. F. (1942). *Ind. Eng. Chem. 34,* 1072.

Othmer, D. F., and Sawyer, F. G. (1943). *Ind. Eng. Chem. 35,* 1269.

Pauling, L. (1945). *J. Am. Chem. Soc. 67,* 555.

Peirce, F. T. (1929). *J. Textile Inst. 20,* T 133.

Pickett, G. (1945). *J. Am. Chem. Soc. 67,* 1958.

Polanyi, M. (1928). *Z. Phys. Chem. A138,* 459.

Prausnitz, J. M. (1969). "Molecular Thermodynamics of Fluid Phase Equilibria." Prentice Hall, Englewood Cliffs, New Jersey.

Rakowski, A. (1911). *Kolloid Z. 9,* 225.

Raoult, F. M. (1888). *Z. Phys. Chem. 2,* 353.

Reid, D. S. (1976). "Intermediate Moisture Foods" (Davies, R., Birch, G. G., and Parker, K. J., eds.), p. 54. Applied Science Publishers, London.

Riedel, L. (1965). *In* "Hanbuch der Lebensmittelchemie" (Schormuller, J., ed.), Volume 1, p. 114. Springer-Verlag, Berlin.

Robinson, R. A., and Stokes, R. H. (1970). "Electrolyte Solutions,"
    2nd rev. ed. Butterworths, London.
Rockland, L. B. (1969). *Food Technol.* *23*, 1241.
Rogers, C. E., Stannett, V., and Szwarc, M. (1959). *J. Phys. Chem.*
    63, 1406.
Ross, K. D. (1975). *Food Technol.* *29*, 26.
Roth, D. (1976). Ph.D. Thesis, Universität Karlsruhe, West Ger-
    many.
Rounsley, R. R. (1961). *Am. Inst. Chem. Eng. J.* *7*, 308.
Rowen, J. W. (1958).• *J. Polymer Sci.* *31*, 199.
Rowlinson, J. S. (1969). "Liquids and Liquid Mixtures," 2nd ed.
    Butterworths, London.
Salwin, H., and Slawson, V. (1959). *Food Technol.* *13*, 715.
Sandler, S. I. (1977). "Chemical and Engineering Thermodynamics."
    Wiley, New York.
Schloesing, M. T. Jr. (1893). *Bull. Soc. Encour. Ind. Nat.* *92*,
    717.
Schwarz, G. (1970). *Eur. J. Biochem.* *12*, 442.
Scott, W. J. (1936). *J. Council Sci. Ind. Res. Aust.* *9*, 177.
Scott, W. J. (1937). *J. Council Sci. Ind. Res. Aust.* *10*, 338.
Scott, W. J. (1938). *J. Council Sci. Ind. Res. Aust.* *11*, 266.
Scott, W. J. (1953). *Aust. J. Biol. Sci.* *6*, 549.
Scott, W. J. (1957). *Adv. Food Res.* *7*, 83.
Smith, S. E. (1947). *J. Am. Chem. Soc.* *69*, 646.
Spieckermann, X., and Bremer, X. (1901). *Landw. Jahrbuch 31*, 81.
Spiess, W. E. L., and Wolf, W. (1969). "Forschungsberichte Luft-
    technik," Vorträge, p. 74. Forsch. Luft- und Trocknungstech-
    nik, Frankfurt.
Spiess, W. E. L., Solé, C. P., and Pritzwald-Stegmann, B. F.
    (1969). *Deutsche Lebensmitt.-Rundschau 4*, 115.
Strohman, R. D., and Yoerger, R. R. (1967). *Trans. Am. Soc. Agri.
    Eng. 10*, 675.
Taylor, J. B., and Rowlinson, J. S. (1955). *Trans. Faraday Soc.*
    51, 1186.
Tester, D. A. (1946). *J. Polymer Sci.* *19*, 535.
Thompson, T. L. (1967). Ph.D. Thesis, Purdue University. Univer-
    sity Microfilms, Ann Arbor, Michigan.
Tomka, I. (1973). Ph.D. Thesis, Universität Bern, Switzerland.
Tóth, J. (1971). *Acta Chim. Acad. Sci. Hung.* *69*, 311.
Tracey, M. V. (1975). *In* "Water Relations of Foods" (Duckworth,
    R. B., ed.), p. 661. Academic Press, London.
Troesch, A. (1951). *J. Chim. Phys.* *48*, 454.
Van den Berg, C., and Leniger, H. A. (1976). *Proc. Intern. Congr.
    Engng. Food, Boston*, 9-13 August.
Van den Berg, C., Kaper, F. S., Weldring, J. A. G., and Wolters,
    I. (1975). *J. Food Technol.* *10*, 589.
Walderdorff, M. G. (1924). *Bot. Arch.* *6*, 84.
Walter, H. (1924). *Z. Bot.* *16*, 353.
Watt, I. C., and D'Arcy, R. L. (1976). *J. Polymer Sci.: Symp.* *55*,
    155.

Watt, I. C., and Leeder, J. D. (1968). *J. Polymer Sci.: Symp.* *55*, 155.
White, H. J., and Eyring, H. (1947). *Text. Res. J. 17*, 523.
Wichterle, J., Linek, J., and Hala, E. (1973). "Vapour Liquid Equilibrium Data Bibliography." Elsevier, Amsterdam.
Young, D. M., and Crowell, A. D. (1962). "Physical Adsorption of Gases." Butterworths, London.
Young, J. H., and Nelson, G. L. (1967). *Trans. Am. Soc. Agri. Eng. 10*, 260.
Zimm, B. H. (1953). *J. Chem. Phys. 21*, 934.
Zsigmondy, R. (1911). *Z. Anorg. Chem. 71*, 356.

DERIVATION OF FULL RANGE MOISTURE
SORPTION ISOTHERMS

M. Caurie

I. INTRODUCTION

The moisture content of dehydrated foods has formed the basis
of many studies (Acker, 1969; Rockland, 1969; Labuza *et al.*,
1970; Duckworth, 1975). These studies indicate the basic practical importance that water holds in food processing and storage
operations. For many of these operations a precise description
of moisture in the final product is required to enable certain
predictions to be made on its chemical, physical, mechanical, and
microbiological characteristics.

Many mathematical models have been used to describe the states
of moisture believed to exist in foods. Some important isothermal models relevant to foods have been reviewed by Labuza (1968).
In this chapter we consider some other isotherm models that have
been applied to food materials, and some comments on other well-
known models. The greater part of this chapter is, however,
devoted to a consideration of Caurie's (1979) recently derived
adsorption equation based on the equation of Brunauer *et al.*
(1938). In the following discussion, unless otherwise stated,
identical symbols have the same significance.

II. SOME ADSORPTION EQUATIONS APPLIED TO FOODS

The idea that the amount of moisture that may be adsorbed by food and other materials is more than will cover a monomolecular layer (monolayer) has been widely held for a long time. de Boer and Zwikker (1929) developed a multimolecular (multilayer) adsorption equation based on the assumption of electrostatic polarization of the polar solid surface. Bradley (1936) developed the equation

$$\ln 1/a - K_3 = K_2 K_1^{m}$$

which is identical with de Boer and Zwikker's (1929) equation except that $\ln a$ is replaced here by $\ln 1/a$. In the above equation a is the relative vapor pressure or water activity (a_w), m the moisture content, K_1, K_2 constants that are functions of the field of the sorptive polar groups, the dipole moment of the sorbed vapor, the temperature, and the polarizability of the vapor, and K_3 a constant that equals the difference between the heat of evaporation from the polarized surface and from the bulk liquid.

Since this equation was developed, it has been realized that electrostatic polarization forces are not nearly strong enough to account for multimolecular adsorption. Recent work by Mackenzie (1975) shows that the linearity of the Bradley plot may depend on a number of factors including temperature, direction of sorption, and the purity of the material with respect to salt content.

An adsorption equation developed for high polymers to describe the final curved portion of sorption isotherms between 0.50 a_w is that of Smith (1947):

$$m = b + b_1 \ln(1-a)$$

where b is bound water, and b_1 the weight of normally condensed water required to saturate the first layer. Becker and Sallans (1956) have reported a successful application of this equation to

wheat. Tested against other food materials, it has given systematic deviations considerably greater than the estimated experimental error (Ayerst, 1965).

More recently Caurie (1970) has suggested the empirical equation

$$\ln 100/m = (1/0.045m_0) - ba$$

where b is a constant, and m_0 the weight of moisture adsorbed on
the monolayer/100 g of solids. The author used this equation for
a wide range of food materials with success up to 0.85 a_w.

Becht and Steinberg (1977) have offered an empirical equation
to cover the upper section of sorption isotherms:

$$1/m = K - ba$$

where K is a constant, and b the reciprocal of the degree of
water binding. This equation applies between 0.50 and 0.95 a_w
and the suggestion that it can provide an estimate of the amount
of moisture adsorbed at saturation pressure or unit water activity is not valid.

A common factor to note with all adsorption equations currently available for application to food materials is that none
covers the entire sorption range. Due to the complexity of chemical, mechanical, and physical characteristics of foods, most isotherm models account for one or the other of the various states
of moisture existing in foods.

In view of these difficulties, Rockland (1957, 1969) suggested
that sorption isotherms may not, in fact, be smooth curves at all
but composites of a number of isotherms. He supported this opinion with a modified Henderson (1952) equation

$$\log \log[1/(1-a)] = Y + n \log m$$

where n and Y are constants. A plot of this equation gives two
or three straight lines or "localized isotherms," which are believed to define the limits of the different states of moisture
existing in food and other materials.

By far the most popular theory to account for multilayered
adsorption is that of Brunauer *et al.* (1938). The supporting
equation popularly referred to as the BET equation states that

$$\frac{a}{m(1-a)} = \frac{1}{cm_0} + \frac{c-1}{cm_0}\, a$$

where c is a constant related to the heat of adsorption in the
first layer. This equation was derived based on a number of simplifying assumptions, including the idea that the number of adsorbed layers at saturation pressure is infinite as well as the
assumption that the properties of the second and higher adsorbed
layers of vapor are the same as bulk liquid. These assumptions
are, of course, not entirely correct and therefore result in an
equation that is applicable between 0.05 and 0.35 a_w in the monolayer range, which provides approximate estimates of constants of
interest in sorption phenomena.

Because of its limitations, the BET theory has been the subject of modification, extension, or critical analyses (Cassie,
1945; Cook, 1948; Halsey, 1948; Hill, 1946; Weller, 1947), but
the general picture it offers of multilayered adsorption has not
been invalidated and the simple form in which the BET equation is
presented also makes it highly attractive. After considerable
work on the theory, Hill (1946) formed the opinion that any future
improvement on it must be in the form of a refinement rather than
a modification on the basic theory.

The area of the BET theory least criticized has been the arrangement of the supporting BET equation. Caurie (1979) has shown
that this equation is identical with

$$\frac{a}{m(1-a)} - \frac{1}{m} = \frac{1}{m_0} + \frac{1}{cm_0}\, \frac{1-a}{a}$$

or

$$\frac{a}{m(1-a)} = \frac{1}{cm_0} \frac{1-a}{a} - \frac{m_0-m}{mm_0} \qquad (1)$$

From this equation it may be observed that when c is small and
the difference (m_0-m) is not large, the second term on the right-
hand side of the equation may be neglected as small, and the
equation simplifies to

$$\frac{a}{m(1-a)} = \frac{1}{cm_0} \frac{1-a}{a}$$

or

$$\frac{1}{m} - \frac{1}{cm_0} \left(\frac{1-a}{a}\right)^2 \qquad (2)$$

It may be observed that this equation will restrict adsorption to
a narrow range of a_w within the monolayer coverage, and indeed
the form of the equation also represents a special case of a more
general equation of the form

$$\frac{1}{m} = \frac{1}{cm_0} \frac{1-a}{a}^{2B}$$

where B is an unknown constant. This general equation may be
used to describe multimolecular adsorption.

Starting from this equation, Caurie (1979) has derived the
following equations to describe multimolecular adsorption:

$$\frac{1}{m} = \frac{1}{cm_0} \left(\frac{1-a}{a}\right)^{2c/m_0} \qquad (3a)$$

$$\ell n\ 1/m = -\ell n\ cm_0 + 2c/m_0 \ell n\ \frac{1-a}{a} \qquad (3b)$$

Equation (3b) has been tested here using Bull's (1944) sorption
data for a number of selected protein materials and on data for
wheat from Becker and Sallans (1956). Some of the plots are shown

in Fig. 1. These plots show straight lines between 0.05 and
0.95 a_w, except for nylon, in which the equation predicts lower
sorption values above 0.70 a_w.

Calculated monolayer moisture (m_0) values from these plots
may be observed to be higher than the BET estimates (Table I) but
these higher values may be more accurate because reports by various workers in the field show that the BET monolayer underesti-
mates the moisture considered safe for the storage stability of
dehydrated foods (Caurie, 1971).

Values of c (Table I), in agreement with our assumptions, are
indeed small. This is in sharp contrast to the rather large
values of c estimated from the classical BET formulation.

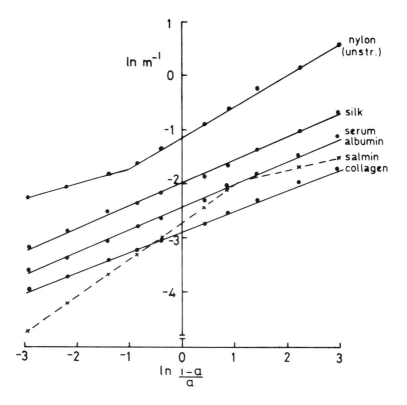

Fig. 1. *Typical water sorption isotherm plots of new
equation.*

TABLE I.   Comparison of Estimated c, Q, and m_0 Values Using Eq. (3) and the Classical BET Equation on the Sorption Data of a Number of Adsorbents at 25°C.[a]

Adsorbent	Eq. (3)			BET equation		
	m_0 (%)	c	-Q (cal)	m_0 (%)	c	-Q (cal)
Nylon (Stretched)	3.41	0.900	2254	1.76	5.97	1058
Wool	7.53	1.494	2553	6.58	11.13	1427
B.-Zein	5.63	1.163	2406	4.10	13.20	1528
C.Zein	5.55	1.193	2421	3.78	12.30	1486
Elastin	8.06	1.410	2520	6.22	11.90	1467
Gelatin	9.07	1.814	2669	8.73	17.40	1692
Egg albumin (lyophilized)	6.65	1.456	2538	5.65	11.58	1450
Egg albumin (unlyophilized)	6.93	1.543	2574	6.15	11.60	1451
Egg albumin (coagulated)	7.25	1.208	2429	4.97	13.62	1547
β-Lactoglobulin (lyophilized)	6.57	1.566	2582	5.93	9.44	1329
β-Lactoglobulin (wet crystals)	6.85	1.643	2611	6.67	8.57	1017
α-β-Pseudoglobulin	7.62	1.649	2613	7.15	12.20	1481
γ-Pseudoglobulin	7.47	1.632	2607	7.16	7.16	1463
Wheat[b]	8.56	1.497	2556	7.80	22.90	1854

[a]Data from Bull (1944).
[b]Data from Becker and Sallans (1956).

As previously reported (Bull, 1944; Pauling, 1945), salmine again showed an indication that adsorption occurred on two mono-layers.   Results in Table II show that moisture begins to con-dense on the second monolayer before the first is complete.   The large difference in the a_w corresponding to the bound water capa-cities, which also mark the beginning of condensation on the two surfaces, shows that at the exposure of the second monolayer the first had not been fully covered.   Indeed, from the higher heat of sorption calculated for the second monolayer (Caurie, 1979) we expect it to be preferentially covered after its exposure.

TABLE II. Total Number of Adsorbed Monolayers and Corresponding Levels of Bound or Unfreezable Moisture in Various Protein Materials, Wheat, and Cellulose at 25°C.[a]

Adsorbent	Slope S	Adsorbed mono-layers N	Bound or nonfreezable water			Other values for bound or nonfreezable water from the literature	
			%H_2O (m_0)	a_w	a_w $(N/2)$	H_2O %	Method
Nylon (unstr.)	0.573	3.49	11.31	0.911	0.776		
Nylon (str.)	0.528	3.79	12.92	0.937	0.802		
Silk	0.358	5.58	34.76	0.982	0.897	37.0	Gravimetric[f]
Wool	0.397	5.04	37.95	0.958	0.789	35.0	Gravimetric[c]
B-Zein	0.413	4.84	27.25	0.970	0.853		
C-Zein	0.430[d]	4.64	25.81	0.962	0.840		
Salmine	0.242[d]	8.26	72.77	0.999	0.779		
	0.375[e]	5.33	48.66	0.843	0.659		
Elastin	0.350	5.71	46.02	0.983	0.885		
Collagen	0.387	5.16	49.48	0.935	0.695	36-46	DTA[g]
						45.0	NMR[h]
Gelatin	0.400	5.00	45.35	0.921	0.719		
Egg albumin (lyoph.)	0.438	4.57	30.39	0.932	0.731		
Egg albumin (unlyoph.)	0.445	4.49	31.33	0.920	0.698	36.0	Direct[i]
Egg albumin (coag.)	0.335	6.00	43.50	0.970	0.902	30.0	Direct[i]

β-Lactoglobulin (lyoph.)	0.477	4.19	27.53	0.895	0.650	
β-Lactoglobulin (wet crystals)	0.480	4.17	28.56	0.874	0.622	
Serum albumin (horse)	0.430	4.65	33.81	0.930	0.715	37.0 (bovine) NMR[h]
α-β-Pseudoglobulin	0.433	4.62	35.20	0.918	0.686	
γ-Pseudoglobulin	0.437	4.58	34.21	0.915	0.681	
Wheat	0.350	5.71	48.88	0.981	0.868	50.0 NMR[j]
Cellulose (cotton)[c]	0.442	4.91	22.51	0.960	–	26.0 Gravimetric[f]

(at 0.995 a_w)

[a] Bull (1944).
[b] Becker and Sallans (1956).
[c] Morrison and Dzieciuch (1959).
[d] First monolayer.
[e] Second monolayer.
[f] Morrison (1963).
[g] Duckworth (1971).
[h] Kuntz et al. (1969).
[i] Moran (1935).
[j] Steinberg and Leung (1975).

III. ISOTHERM SHAPES

Five types of isotherms are recognized in physical adsorption (Brunauer, 1945). By varying the constant c in Eq. 3a BET type I, II, III, and IV isotherm shapes have been reproduced (Fig. 2) from a single set of sorption data on gelatin. Large values of c resulted in a type I isotherm shape characteristic of the Langmuir (1918) monolayer isotherm.

With small values of c, a type III isotherm shape resulted, while at moderate values the sigmoid type II and IV isotherm

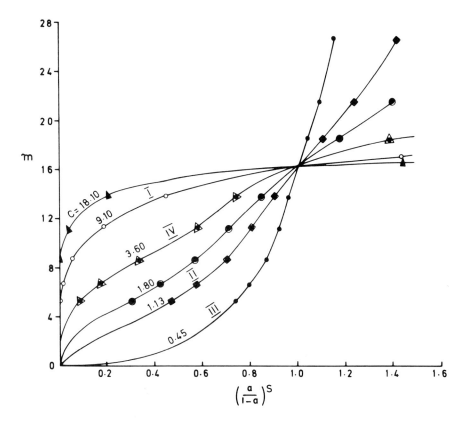

Fig. 2. BET type I-IV isotherm shapes derived from a single set of isotherm data on gelatin (Bull, 1944) by altering the size of C in Eq. (3a).

shapes were observed, with the type IV representing a higher energy level. Figure 2 clearly shows that the various isotherm shapes
are determined by the magnitude of c related to their first monolayers. It is of interest here to note (Table I) that all the
calculated c values of the adsorbents used in this work are of the
order that give sigmoid isotherms. It may also be observed from
Fig. 2 that Brunauer's (1945) classification of adsorption isotherms does not follow a natural order; Brunauer's type IV, II,
and III isotherm shapes may be reclassified, respectively, as
type II, III, and IV. Figure 2 also does not offer any evidence
for the existence of a fifth isotherm shape as classified by
Brunauer.

IV. FINITE ADSORPTION AND LIMITS OF MULTILAYERED ADSORPTION

At saturation pressure the classical BET equation predicts
that solids would adsorb an infinitely large quantity of moisture.
Experience shows, however, that such an unlimited adsorption of
moisture does not occur and that solids generally adsorb strictly
limited amounts of moisture, in saturated atmospheres. A modified
BET equation based on limited adsorption (Brunauer et al., 1938)
fitted experimental data much better than with the two constant
classical BET equation, but the introduction of an additional
constant rendered the equation unwieldy and therefore difficult to
apply.

It may be noted that close to saturation pressure Eq. (3b) expands approximately into the infinite series

$$\ln m = \ln cm_0 + (2c/m_0)(a+a^2/2 + a^3/3 + ...) \qquad (4)$$

If it is recognized that the successive terms in this equation
relate to the contents of the first, second, and higher layers it
will be clear that there is an increasing limitation placed on

the development of these layers and therefore on the total amount
of moisture adsorbed in multilayers.

A major difficulty with the BET theory is that, aside from
the monolayer, the BET equation does not offer any indication of
how much of the finite moisture adsorbed is taken up in multilayers and in capillary condensation. Since condensation follows
multilayered adsorption, a knowledge of the limits of the latter
determines the beginning of the former.

It may be observed that Eqs. (2) and (3a) describing adsorption in the monolayer region and multilayered adsorption, respectively, differ from each other only by their exponents (S).
From these two equations it may be noted from the magnitudes of
S, estimable from the slopes of plots using Eq. (3b) (Table II),
that in moving from monolayered to multilayered adsorption the
exponent of Eq. (2) decreases from a value of two to fractional
values. This trend suggests an inverse relationship between the
number of monolayers of adsorbed moisture (N) and S. Symbolically
this may be expressed as

$$S = k_1/N$$

where k_1 is a constant of proportionality. Since from Eq. (2),
$N = 1$ when $S = 2, k_1$ must equal two, and therefore

$$S = 2/N$$

Substituting into Eq. (3b),

$$\ln\ 1/m = -\ln\ cm_0 + 2/N\ \ln(1-a/a) \qquad (5)$$

Thus the gradient of Eq. (3b) provides an estimate of the number
of monolayers actively adsorbed by a given material and also provides a rule for differentiating between monolayered and multilayered adsorption; when the gradient of a plot using Eq. (3) is
less than two, the adsorption is multilayered, while when it is
equal to or more than two it is monolayered.

From the calculated values of N for the wide range of protein materials and wheat examined (Table II) except for salmine, which adsorbed eight layers of moisture on its first monolayer surface, most of the materials examined here adsorbed only four to six monolayers of moisture. This limited adsorption estimable with N supports the idea inherent in Eq. (4) that the development of second and higher layers during moisture sorption in foods and other materials is severely limited and does not extend to saturation pressure, as opposed to the unlimited sorption stretching up to saturation pressure predicted with the classical BET equation. Following the procedure described for applying the three-constant BET equation, Bushuk and Winkler (1957) and Lee (1970) found that an N value of 5 or 6 provided a good representation of wheat flour-water isotherm in agreement with our estimated value of 5.71 (Table II).

The point on the isotherm at which active adsorption, as indicated by N, ceases should mark the beginning of moisture condensation on an adsorption surface, and this generally occurred well above 0.90 a_w (Table II). Beyond this point, surface forces do not apply. In this area Raoult's law applies, and we show in a later section that though the variables in the Raoult's law equation are the same as in Eq. (3b) the constants differ, which must account for a discontinuity in the straight-line isotherm using Eq. (3b). It may be concluded from this that due to differences in forces operating at high and low ranges of sorption isotherms no single straight-line isotherm can legitimately cover the whole a_w range. Strictly since the slopes of sorption isotherms relate to the heat of adsorption there must be only one discontinuity in the adsorption isotherm and this must occur in the very high a_w range in the areas of influence of Raoult's law for dilute solutions.

V. DENSITY OF ADSORBED MOISTURE AND THE CONSTANT c

Moisture adsorbed in soils and in wheat is different from normal free water. In particular it is found, for example, that the density of adsorbed moisture in soil-water and in wheat flour-water systems is higher than that of normal liquid water. Gur-Arieh et al. (1967) observed that water in wheat flour up to the BET monolayer moisture content of 7% had a density of 1.482 g/cm^3; the second layer moisture up to 14% had a density of 1.111 g/cm^3, while the third and higher layers had a density less than unity (0.967 g/cm^3). In soil-water systems, moisture levels up to 30% (dry basis) have been found to have densities equal to or higher than 1 g/cm^3, while at higher moisture levels the density of adsorbed water is reported to be less than unity (De Witt and Arens, 1950; Mooney et al., 1952; Anderson and Low, 1958; Mackenzie, 1958).

The density of normal free water is determined by distances between water molecules, which are determined by forces acting between these molecules. The closer the intermolecular distances, the greater the forces acting between the molecules and therefore the higher the density of water. When a monolayer of moisture is adsorbed on the surface of an adsorbent the intermolecular distances between the water molecules are reduced and at the same time they are affected by surface forces of high magnitude, which together alter the density of the adsorbed moisture. A second-layer molecule overlying another in the monolayer is separated from the adsorbent surface by one molecular diameter and consequently surface forces will have a correspondingly diminishing effect on molecules in this and subsequent layers until at the beginning of condensation a density equal to that of normal liquid water is reached.

In the work of Gur-Arieh et al. (1967), the density of water molecules associated with the monolayer in a wheat flour-water system was 1.482 g/cm^3. Significantly the calculated value of c

for Becker and Sallans' (1956) wheat flour-water system, reported
in this work, of 1.497 is close to this value. It is also interesting to note from Table I that calculated values of c from Eq.
(3) for the other materials examined are of the same order of magnitude as the density of water associated with the monolayer of
wheat flour-water systems. From this close numerical agreement
between the magnitude of c and the density of adsorbed moisture
associated with the monolayer it is suggested that the magnitude
of c calculated from Eq. (3b) estimates the density of adsorbed
moisture on the first layer (D_0).

One of the criticisms of the BET theory (Cassie, 1945) is
that it does not consider lateral interaction between adsorbate
molecules. This criticism has been refuted by Brunauer (1953,
1961). The results here agree with Brunauer since the magnitude
of c, which is identified with the density of adsorbed moisture,
depends on adsorbate-adsorbate interaction. This interaction explains the low value for the density of moisture adsorbed on nylon
and reflected in its c value (Table I). Even though surface
forces are quite high on nylon (Table I) because of the relative
paucity of sorption sites on this material, adsorbate-adsorbate
distances are relatively large and this has the observed diminishing effect on c and therefore on the density of adsorbed moisture on the monolayer of this material.

VI. NONFREEZABLE OR BOUND WATER

Water that is so closely united with other food constituents
that its properties are different from those of free water is considered bound (Kuprianoff, 1958). A basic assumption in the BET
theory is that only the first-layer moisture has properties different from bulk water and therefore by this definition is bound.
With the development of instrumental methods of analysis it is
becoming increasingly clear that the BET definition is inadequate.

It is now widely known that moisture corresponding to a considerable proportion of the a_w range of sorption isotherms cannot
be frozen at subfreezing temperatures and therefore by definition
should be considered bound along with the first layer.

Current adsorption equations are silent on how much of the
moisture content of food and other materials is actively adsorbed
and therefore bound. It is in this context that the usefulness
of Eq. (5) may be viewed. The gradient of this equation estimates
the number of adsorbed layers and from it the total amount of
moisture actively bound. Table II shows the amount of bound moisture and corresponding a_w levels calculated from plots of Eq. (5)
for a number of protein materials, cellulose, and wheat. These
estimated values are of the same order of magnitude and agree
quite well with corresponding literature values of nonfreezable
or bound water. It must be borne in mind, however, that these
values are not absolute and are dependent upon a number of factors
including pretreatment of the adsorbent. There can therefore be
considerable variation in bound water capacity between samples.
For example, it is observed here (Table II) that coagulated or
denatured egg albumin is able to bind more moisture than the un-
coagulaated egg albumin, while according to Kuntz *et al*. (1969) a
change of hydrogen ion concentration from pH 4.5 to 2.0 alters
the bound water capacity from 37 to 28% moisture.

Points on the isotherm corresponding to half the total number
of layers (*N*) adsorbed corresponded to the upper inflection point
of sigmoid isotherms and these generally occurred above 0.60 a_w
(Table II). This a_w is, microbiologically, the critical point
below which moisture is bound sufficiently tightly that it is unavailable to all microorganisms.

Three species of adsorbed or bound water may thus be recognized. The first is the monolayer moisture, which is generally
not available for either chemical or microbiological activities.
The second is that species of moisture beyond the monolayer up to
moisture, corresponding to approximately 0.60 a_w and bound suffi-

ciently tightly that it is unavailable to microorganisms but available for chemical activity. The third species of adsorbed or bound water is what may be described as the loosely bound fraction occurring beyond 0.60 a_w and available for chemical and microbiological activity.

Bound water capacity has recently been proposed as a term (Steinberg and Leung, 1975) to estimate moisture corresponding to saturation pressure. It is clear from the a_w corresponding to bound water capacities reported in Table II that this may not be correct; bound water capacities were always attained before saturation pressure.

VII. NATURE OF THE ADSORBED MOISTURE

The second species of adsorbed or bound water immediately following the first layer is reported by Duckworth and Smith (1963) to have some solvent properties. In this connection it is pertinent to note from the exponent of multilayered Eq. (3a) that the monolayer Eq. (2) results when the constant c is highest and numerically equal to m_0. That is when $c = m_0 = D_0V_0$ = weight of water vapor adsorbed in the first monolayer/100 g solids. Here V_0 and D_0 are, respectively, the volume of density of moisture adsorbed in the first monolayer. This is an interesting result because we have already noted that numerically, in multilayered adsorption, the constant c related to the heat of adsorption in the first monolayer is equal only to D_0 and not to D_0V_0. This leads to the suggestion that in multilayered adsorption the volume of moisture adsorbed in the first layer is contracted, indicating a possible change in the state of adsorbed water molecules from a gas or vapor to some form of bulk liquid state, which for the same weight adsorbed should have a higher density than normal liquid water, as has indeed been observed by Gur-Arieh *et al.* (1967).

It is reasonable to assume from this that the second-layer
molecules forming over the first in multilayered adsorption systems also exist in some bulk liquid state, which we learn from
the work of Duckworth and Smith (1963) possess very limited solvent
properties. It is of interest to note that this assumption is in
close agreement with the original simplifying assumption proposed
in the BET theory.

The foregoing discussion clearly suggests that water molecules
in adsorption systems giving type I isotherm shapes are gaseous
and are held by the highest energy of adsorption while all the
molecules in multilayered adsorption systems are liquidlike but
with less mobility than the bulk liquid state as has been observed
from infrared (Yates and Sheppard, 1957) and nuclear magnetic
resonance (Graham and Phillips, 1957) studies.

VIII. SURFACE AREA OF ADSORBENTS

A parameter closely related to the phenomenon of adsorption
is the surface area on which the adsorbate is held. The cross-
sectional area of a water molecule is given as 10.6×10^{-20} m^2.
Assuming a spherical molecule, this is equal to $\pi d^2/4$, where d
equals the diameter of a water molecule. Hence

$$d^2 = \frac{10.6 \times 10^{-20} \times 4 \text{ m}}{\pi}$$

and

$$d = 3.673 \times 10^{-10} \text{ m}$$

The weight of water adsorbed in the monolayer is known from Eq.
(3) to equal m_0 g/100 g solids. Now if the magnitude of c in
Eq. (3) equals the density of adsorbed moisture in the monolayer
as suggested, then the volume (V_0) of adsorbed water molecules on
the first monolayer becomes

$$V_0 = \frac{m_0}{100c} \ cm^3/g = \frac{m_0}{100c} \frac{1}{10^6} \ m^3/g$$

It may be recalled from Eq. (3b) that m_0/c equals twice the reciprocal value of the slope (S). Hence

$$V_0 = \frac{2}{10^8 S} \ m^3/g$$

For a horizontal layer of spherical water molecules spread over
the monolayer surface, the area (A) occupied by these molecules
will equal V_0 divided by d, and therefore

$$A = V_0$$

$$d = \frac{2}{10^8 S} \frac{10^{10}}{3.673} = 54.45$$

$$S \quad m^2 \tag{6}$$

If c indeed equals D_0 as suggested here, Eq. (6) must give surface
area values in good agreement with values determined by other
methods. This indeed is found to be the case. The surface areas
of materials examined in this paper calculated with the above
equation are given in Table III. These results provide interesting comparisons with corresponding values reported by Bull (1944).
They show, for example, that stretched nylon had a slightly larger
surface area than unstretched nylon. They also show that there is
practically no difference between the surface area of lyophilized
and unlyophilized egg albumin. On the other hand, coagulation,
which is a principal cause of denaturation in proteins, considerably increased the surface area of egg albumin available for adsorption with a corresponding increase in the amount of moisture
adsorbed. The ability of denatured proteins to adsorb larger
amounts of water than undenatured proteins has also been recorded
by Kuntz et al. (1969) with respect to denatured tRNA. These
results differ from Bull's (1944) surface area values of the same
data (Table III) calculated on BET estimates of m_0 and which in-

TABLE III. *A Comparison of Surface Area Values of Various Protein Materials and of Wheat Calculated with Eq. (6) and with the BET Equation.*

Adsorbent	Surface area (m^2/g)		
	Eq. (6)	BET equation[a]	
Nylon (unstr.)	95	68	(73)
Nylon (str.)	103	62	(69)
Silk	152	144	(129)
Wool	137	233	(156)
B-Zein	132	145	(125)
C-Zein	127	138	(116)
Salmine	225[b]	187	(175)
	145[c]	–	–
Elastin	156	220	(156)
Collagen	141	337	(181)
Gelatin	136	309	(170)
Egg albumin (lyoph.)	124	200	(137)
Egg albumin (unlyoph.)	122	218	(141)
Egg albumin (coag.)	163	176	(146)
β-Lactoglobulin (lyoph.)	114	210	(134)
β-Lactoglobulin (wet crystals)	113	236	(144)
Serum albumin (horse)	127	238	(152)
α-β-Pseudoglobulin	126	254	(154)
γ-Pseudoglobulin	125	254	(156)
Wheat	156	–	–

[a] Bull (1944). (Corrected for c.)
[b] First monolayer.
[c] Second monolayer.

dicated stretched nylon to have a smaller surface than unstretched nylon and coagulated egg albumin to have a reduced surface.

It may be observed from Eq. (6) that the surface area of solids depends on both the density of molecules c adsorbed in the monolayer and on the magnitude of m_0. It may also be observed (Table III) that when the BET surface areas are divided by their corresponding values of c estimated with Eq. (3b) there is closer agreement between BET surface areas and those calculated using Eq. (6). It would appear from this that the BET surface area values are calculated based on the density of bulk water instead of the density of the adsorbed molecules in the monolayer that interact directly with the absorbent surface.

IX. SPECIAL ISOTHERM EQUATIONS

We cannot leave this subject without mentioning Raoult's and
Henry's laws, which provide two special equations relating to adsorption at the extreme ends of sorption isotherms.

A. *Raoult's Law*

Raoult's law equation, which normally applies to the upper
region of sorption isotherms, is reported (Arnold, 1949) to apply
better than the BET equation in certain cases of adsorption on
solids at lower a_w. Indeed, most nonelectrolytes are adequately
represented by Raoult's law up to several molal concentrations
(Ross, 1975).

Raoult's law for dilute solutions as may occur in the capillary condensation region of sorption isotherms states that

$$\frac{1-a}{a} = \frac{M_1}{M_2} \frac{W_2}{W_1}$$

or

$$\frac{1}{W_1/W_2} = \frac{1}{M_1/M_2} \frac{1-a}{a}$$

where M_1 is the molecular weight of solvent, M_2 the molecular
weight of solute, W_1 the weight of solvent, W_2 the weight of
solute, and

$$\frac{1}{m} = \frac{1}{B} \frac{1-a}{a} \tag{7}$$

where $m = 1000$ g H_2O/weight of dissolved solute, and $B = M_1/M_2$
= const. If the similarity between Eqs. (7) and (2) is recognized,
then Raoult's law equation, with its unit exponent, may be seen to
be a special case of Eq. (3) that applies in nonelectrolyte solutions to describe multilayered adsorption as has been observed by
Hansen *et al.* (1949) and by Arnold (1949). The unit exponent in

this equation imposes, according to Eq. (5), a limitation on the
number of layers of moisture that can build up around solute particles in solution to a maximum of two.

B. *Henry's Law*

It is believed that at very low a_w or pressures and correspondingly low adsorptions, all models of adsorption should sim-
plify to Henry's law, according to which the solubility of a gas
in a liquid or the amount of water vapor adsorbed on a solid is
directly proportional to its pressure or a_w:

$$m = k_2 a$$

where k_2 is a constant. It is interesting to note here that this
equation, which applies to the low a_w or pressure end of sorption
isotherms, is similar to the simplified form of the reciprocal of
Raoult's law Eq. (7), which applies at high a_w or pressure end of
sorption isotherms.

It is also to be noted that at these low a_w or pressure
ranges, the reciprocal of the general Eq. (3) also simplifies to

$$m = k_3 a^S \tag{8}$$

where $k_3 = cm_0$. This equation shows that at very low a_w m is not
directly proportional to a as should be expected from Henry's law,
but to a power of a. Thus like Raoult's law equation, Henry's
law equation may also be looked upon as a special case of Eq. (3),
which is also restricted to a maximum of two monolayers of adsorbed moisture. Equation (8) removes this restriction and predicts more layers of moisture at low a_w. Indeed, there are many
reports in the published literature of adsorption isotherms at low
coverages (Lopez-Gonzalez *et al.*, 1961; Hobson and Armstrong,
1963; Fedorova, 1963) that are curvilinear and do not therefore
obey Henry's law. So far such curvilinear deviations from Henry's
law have been interpreted to mean that a sufficiently low coverage

has not been attained (Deitz, 1965). Clearly such curves may indicate the existence of greater multilayering at very low coverages, which may be more adequately described with Eq. (6) than
with Henry's law equation. Indeed, the existence of greater multilayering of adsorbed moisture at low coverages is in accord with
the BET theory.

Equation (3) may be put in the form

$$\frac{1-\frac{1}{m}}{1} = \frac{(cm_0)m_0/2c}{m_0/2c}a = K^1am^{-B} \tag{9}$$

where $K^1 = (cm_0)^B$ $B = m_0/2c$. In this form Eq. (9) suggests a
common law underlying physical adsorption of water on solid surfaces that states "The relative lowering of the water activity of
pure water on solid surfaces is directly proportional to the
product of the water activity and the $(m_0/2c)$th power of the reciprocal moisture content of the solid" of which both Raoult's
law Eq. (7) and Henry's law Eq. (8) are special cases.

In summary, we have noted that most existing sorption equations only describe fractions of the bound or unfreezable water
capacity. Caurie's (1979) adsorption equation reveals that beyond
the bound water capacity, though the same basic law applies, because of changed constants there is a discontinuity in the
straight line isotherm. The equation also reveals that monolayered adsorption is gaseous as opposed to the adsorption of
various states of bulk water in multilayered adsorption. The
new equation provides, in addition to other parameters, a method
for estimating the number of layers actively adsorbed and therefore the bound or unfreezable water capacity, density of adsorbed
moisture in the monolayer, as well as the area of the surface
available for adsorption. Raoult's and Henry's law equations are
shown to be special cases of this new adsorption equation.

Acknowledgment

I am grateful for the useful discussions with Professor R. B. Duckworth, Head, Department of Nutrition and Food Science, University of Ghana, during the preparation of this manuscript.

References

Acker, L. W. (1969). *Food Technol. 23,* 1257.
Anderson, D. M., and Low, P. F. (1958). *Soil Sci. Soc. Am. Prod. 22,* 29.
Arnold, J. R. (1949). *J. Am. Chem. Soc. 71,* 104.
Ayerst, G. (1965). *J. Sci. Food Agric. 16,* 71.
Becker, H. A., and Sallans, H. R. (1956). *Cereal Chem. 33,* 79.
Bradley, R. S. (1936). *J. Chem. Soc.,* 1467.
Brunauer, S. (1945). "The Adsorption of Gases and Vapors," Vol. 1. Princeton Univ. Press, Princeton, New Jersey.
Brunauer, S. (1953). *In* "Structure and Properties of Solid Surfaces" (R. Gomer and C. S. Smith, eds.), p. 395. Univ. Chicago Press, Chicago.
Brunauer, S., Emmett, P. H., and Teller, E. (1938). *J. Am. Chem. Soc. 60,* 309.
Bull, H. B. (1944). *J. Am. Chem. Soc. 66,* 1499.
Bushuk, W., and Winkler, C. A. (1957). *Cereal Chem. 24,* 73.
Cassie, A. B. D. (1945). *Trans. Faraday Soc. 41,* 450.
Caurie, M. (1970). *J. Food Technol. 5,* 301.
Caurie, M. (1971). *J. Food Technol. 6,* 193.
Caurie, M. (1979). *J. Food Sci. (in press).*
Cook, M. A. (1948). J. Am. Chem. Soc. 70, *2925.*
de Boer, J. H., and Zwikker, C. (1929). Z. Phys. Chem. B3, 407.
Deitz, V. R. (1965). *Ind. Eng. Chem. 57,* 49.
De Witt, C. T., and Arens, P. L. (1950). *Trans. 4th Int. Congr. Soil Sci. 2,* 59.
Duckworth, R. B. (1971). *J. Food Technol. 6,* 317.
Duckworth, R. B., ed. (1975). "Water Relations of Foods." Academic Press, London.
Duckworth, R. B., and Smith, G. (1963). *Proc. Nut. Soc. 22,* 182.
Fedorova, M. F. (1963). *Sov. Phys. Tech. Phys. 8,* 434.
Graham, D., and Phillips, W. D. (1957). *Proc. 2nd Int. Congr. Surface Activity 2,* 22.
Gur-Arieh, C., Nelson, A. I., and Steinberg, M. P. (1967). *J. Food Sci. 32,* 442.
Halsey, G. (1948). *J. Chem. Phys. 16,* 931.
Hansen, R. S., Fu, Y., and Bartell, F, E. (1948). *J. Phys. Colloid Chem. 53,* 769.
Henderson, S. M. (1952). *Agric. Eng. 33,* 29.
Hill, T. L. (1946). *J. Chem. Phys. 14,* 263.
Hobson, J. P., Jr., and Armstrong, R. A. (1963). *J. Phys. Chem. 67,* 2000.

Kuntz, I. D., Brassfield, T. S., Law, G. D., and Purcell, G. V. (1969). *Science 163*, 1329.

Kuprianoff, J. (1958). *In* "Fundamental Aspects of Dehydration of Foodstuffs," p. 14. Soc. Chem. Industry, London.

Labuza, T. P. (1968). *Food Technol. 22*, 263.

Labuza, T. P. (1975). *In* "Water Relations of Food" (R. B. Duckworth, ed.), p. 155. Academic Press, London.

Labuza, T. P., Tannenbaum, S. R., and Karel, M. (1970). *Food Technol. 24*, 543.

Langmuir, I. (1918). *J. Am. Chem. Soc. 40*, 1361.

Lee, F. A. (1970). *Food Technol. Austr. 22*, 516.

Lopez-Gonzalez, J. de D., Carpenter, F. G., and Dietz, V. R. (1961). *J. Phys. Chem. 65*, 112.

Mackenzie, A. P. (1975). *In* "Water Relations of Foods" (R. B. Duckworth, ed.), p. 477. Academic Press, London.

Mackenzie, R. C. (1958). *Nature 181*, 334.

Mooney, R. W., Keenan, A. G., and Wood, L. A. (1952). *J. Am. Chem. Soc. 74*, 1371.

Moran, T. (1935). *Proc. Roy. Soc. B118*, 548.

Morrison, J. L. (1963). *Nature 198*, 84.

Morrison, J. L., and Dzieciuch, M. A. (1959). *Can. J. Chem. 37*, 1379.

Pauling, L. (1945). *J. Am. Chem. Soc. 67*, 555.

Rockland, L. B. (1957). *Food Res. 22*, 604.

Rockland, L. B. (1969). *Food Technol. 23*, 1241.

Ross, K. D. (1975). *Food Technol. 29*(3), 26.

Smith, S. E. (1947). *J. Am. Chem. Soc. 69*, 646.

Steinberg, M. P., and Leung, H. (1975). *In* "Water Relations of Foods" (R. B. Duckworth, ed.), p. 238. Academic Press, London.

Weller, S. (1947). *J. Chem. Phys. 15*, 336.

Yates, D. J. C., and Sheppard, W. (1957). *Proc. 2nd Int. Congr. Surface Activity 2*, 27.

RECENT DEVELOPMENTS IN TECHNIQUES

FOR OBTAINING COMPLETE SORPTION ISOTHERMS

S. Gal

I. INTRODUCTION

Water vapor sorption isotherms represent the sorption equilibrium at constant temperature between the vapor pressure of
water in the gas phase and the moisture content of the material
in question. Due to the great individuality of materials and
objectives of sorption, methodology has always been characterized
by an exceptional diversity of apparatus and methods. Since the
first comprehensive treatise on this subject (Gal, 1967), which
itself had been preceded by an excellent thesis work (Hofer,
1962), several more or less detailed review articles and progress
reports have been published showing the great variety of possible
approaches to this end (Gal, 1975; Loncin and Weisser, 1977;
Troller and Christian, 1978). As a valuable and abundant source
of information the four-volume work of Wexler (1965) must also
be mentioned in this context.

This review has been prepared to summarize the present state
of the art of sorption methodology in the food field, to show the
progress made in the past few years, and to emphasize the
necessity for elaborating (1) a precise reference method for
comparative and research purposes and (2) a simple and efficient
method for routine work.

As opposed to many previous reviews, however, methods for
obtaining complete sorption isotherms and for measuring water
activity are treated separately in this volume with regard to
the rapid developments in both fields. In fact, water activity
determination has become a well-established instrumental technique in the last decade. Therefore, short references to such
methods should suffice in the present context, as made, for
example, in Section II, B.

This chapter is based on information from the IFST Abstracts
and other abstract periodicals, but also, and to a considerable
part, on private communications from colleagues. It is a
progress report, i.e., references to publications prior to 1973
will only be made if necessary for understanding or if they
are chronologically or causatively related to developments
made since then.

Sorption methodology has been subjected to two collaborative
tests in the period under consideration. The first one was
organized by J. L. Multon (private communication, 1973) with the
participation of ten laboratories from six countries. The
sorption isotherm of a well-defined sample of tapioca starch had
to be determined by the different methods practiced by the
participants. Multon and Bizot (1976) gave the reproducibility
figures of the water content within ±3% absolute and those of
the water activity within ±5% absolute, though extreme values
were spread over an even larger area. This was a very dis-
couraging result, which nevertheless provided valuable information on the sources of error in measuring water vapor sorption
equilibria.

The second collaborative test is still in progress, W. Wolf and W. E. L. Spiess (private communications, 1978) being the organizers of this trial. Five laboratories from three countries are determining the sorption isotherm of three thoroughly standardized substances of very different sorption capacity. Results will be published later by the organizers and they are expected to deliver starting data for a further, more extensive interlaboratory study. The early (and so far the only) attempt of the Packaging Institute (1952) to standardize working conditions for obtaining sorption isotherms and water activities is in many respects out of date. However this study delineates all important elements for executing such measurements along with equilibrium and precision considerations.

Recent years considerable attention has been paid to the proper characterization of materials and methods in reporting on sorption measurements. As generally accepted now, a detailed description of all relevant data is absolutely essential for comparative purposes. Such data have been indicated in the comprehensive collection of sorption isotherms of Wolf *et al.* (1973) and Gough and Bateman (1977).

Recently a thesis work has been devoted to the study of the optimum conditions for the determination of reproducible sorption isotherms of grains (E. Neuber, private communications, 1978). This very thorough investigation revealed a great many factors influencing the results of such measurements and led to the compilation of a catalog of virtually all relevant data. Such a list of characteristics of materials and methods should accompany all published results of sorption investigation in the future.

In recognition of the importance of sorption data, in May 1978, the EC finalized a cooperative research program for the collection and production of sorption isotherms and water activity values of foods and related materials. This project, sponsored during three years by the EC within the scope of COST (Cooperation on Science and Technology in Europe) will certainly contribute a great deal to our knowledge of the sorption behavior of foodstuffs and their raw materials, as well as the techniques for obtaining precise and comparable sorption data.

II. APPARATUS AND METHODS

Sorption isotherms can be determined according to two basic principles, the gravimetric and manometric or hygrometric. Nevertheless, the addition of a third supplementary group is necessary to accommodate special methods applicable under extreme conditions, like very high water activities, which usually do not fit with the other two main groups of the classification (see Table I).

TABLE I.   Methods for Determination of Sorption Isotherms

A. Gravimetric methods
   1. Methods with continuous registration of weight changes
     (a) Evacuated systems
     (b) Dynamic systems
   2. Methods with discontinuous registration of weight changes
     (a) Static systems
     (b) Dynamic systems
     (c) Evacuated systems

B. Manometric and hygrometric methods
   1. Manometric methods
   2. Hygrometric methods

C. Special methods

A. *Gravimetric Methods*

Gravimetric methods have been preferred for obtaining complete
sorption isotherms in the period reviewed. Therefore, some
general remarks are first made on this type of sorption measurement. Irrespective of construction and mode of operation of the
apparatus, such methods involve the execution of the following
five basic functions:

(a) keeping temperature constant and, as an even more
important task, minimizing temperature fluctuations between
samples and their surroundings or the source of water vapor, respectively;

(b) determining the exact dry weight of the samples;

(c) maintaining predetermined relative water vapor pressures
in the space around the samples;

(d) achieving hygroscopic and thermal equilibrium between
samples and water vapor source;

(e) register the weight change of the samples in equilibrium
with the respective water vapor pressures;

These functions are now discussed briefly in the light of
the recent literature.

(a) The temperature of the samples must be kept constant
during the whole experiment, which is a basic requirement for
obtaining an isotherm. With respect to the accuracy of the
isotherms, i.e., to the variance of the equilibrium moisture
contents at a given relative humidity, however, it is even more
important to minimize temperature fluctuations between samples
and their surroundings and/or the source of water vapor. In
(c) two precision classes are defined for routine and reference
purposes, respectively. From the temperature dependence of the
saturated water vapor pressure one can readily compute that a
temperature constancy of better than ±0.2°C for routine work and

±0.02°C for reference purposes throughout the whole system should
be aimed at in constructing sorption apparatus.

(b) The dry weight of the samples must be exactly known because this is the reference basis of the isotherm. The equilibrium moisture contents also will preferably be expressed as
percentages on a water-free basis. The dry weight of the samples
should be determined, whenever possible, at the temperature of
the isotherm. If this is not feasible because of the long time
necessary for complete dehydration (often also in vacuum), the
drying temperature may be elevated slightly but only to levels
to allow a reasonable dehydration time.

The dry weight of samples can be determined at the beginning
of the sorption experiment with separate samples from the same
lot, or at the end of it in the equilibrated samples themselves.
Automatic apparatus generally provides for the drying of the
sample on the balance pan, thus recording the dry weight and, at
the same time, making possible true adsorption isotherm determinations. For most practical purposes, however, samples with
moisture content as received from the processing are most suitable
to start with, whereby points of the isotherm below this represent desorption values and those above it adsorption values, both
branches lying somewhere inside the hysteresis loop.

(c) The review literature cited at the outset contains practically all aspects of maintaining preselected water vapor pressures
in the space around the samples in sorption measurements. For
the sake of better understanding, however, the most important
methods are outlined briefly as follows.

(1) Mixing two air streams, one of which is permanently kept
dry, the other being saturated with water vapor. This is the
preferred method in dynamic systems allowing a convenient and
continuous humidity control at moderate precision (Best and
Springler, 1972).

(2) Two-temperature system. Water or ice is kept in a thermostated container at a lower temperature than the sample. Thus the temperature of the vapor source defines its partial pressure within the apparatus. This is the most commonly used method for maintaining constant vapor pressures in research work because of the ease and precision of controlling temperature instead of humidity, and the continuous relative humidity scale available (e.g., J. L. Multon, private conversation, 1978; Berlin and Anderson, 1975; Hermansson, 1977). Near saturation this method becomes less precise due to the very large dependence of the saturated vapor pressure on temperature. Consequently, the highest precision can be achieved only in the low relative vapor pressure range. The two-temperature method can be combined with the two-pressure system and also with saturated solutions (see below) to extend their range of applicability beyond that under isothermal conditions.

(3) Two-pressure system. This is a comparatively new way to maintain constant vapor pressures in dynamic sorption measurements. Air or any other inert gas is saturated with water vapor at the same temperature but at a higher pressure than that existing in the experimental space. The gas-vapor mixture expands consecutively to atmospheric pressure and becomes unsaturated, as expressed by Dalton's law. The control of relative humidity is thus replaced by pressure control, which can be performed more precisely especially in the high relative vapor pressure range (Multon et al., 1971). An interesting modification of this principle has been developed by Lowe et al. (1974) for humidity control in testing chambers. According to this, a measured amount of liquid water is continuously evaporated into an air stream saturated previously at $0^{\circ}C$ ad 5 psig.

(4) Aqueous solutions. Solutions of electrolytes or non-electrolytes, saturated or unsaturated, have been used for decades for maintaining constant vapor pressures in sorption

measurements (see also the literature cited in Section I). An
early attempt to lay down standardized procedures in humidity control by means of solutions has been made by the ASTM (1971).
Sulfuric acid, glycerol, and 16 saturated salt solutions are
recommended for this purpose and detailed instructions are given
with respect to suitable containers and working at, above, and
below room temperature as well as to the most important point
of avoiding temperature fluctuations. With the same objective
a German standard is currently in work (FNM, 1977). Eleven
saturated solutions and glycerol should be proposed for keeping
constant the relative vapor pressures of water in closed
containers with a precision $\pm 1\%$ absolute for salt solutions and
up to $\pm 1.5\%$ absolute for glycerol solutions, respectively.

Recently, considerable doubt has arisen about the accuracy
of vapor pressures of saturated salt solutions. The most recent
review compiled by Greenspan (1976) reflects the considerable
uncertainties of published values by various authors at different
temperatures. Even if very thorough precautions are met, an
average deviation of ± 0.005 or more from the expected values
must be taken into account.

Unsaturated solutions change their concentration when used
over longer times but they allow the choice of any desirable
vapor pressure. The most precise values are still those of
analytical-grade sulfuric acid (Gröninger, 1972). Based on the
approximate formula of Lewis and Randall, M. Rüegg (private
conversation, 1976) has developed a computer program for the
calculation of the water activity of sulfuric acid solutions in
a limited temperature range from the best published values at
25ºC, as compiled by Gal (1976).

As a novelty, the Chemcotec AG (Baden, Switzerland) has
recently introduced noncorrosive and nonirritating solutions for
humidity control under the trade name Wet-Dry Humidity Standards.

The solutions are reportedly standardized to ±0.001 a_w and cover a field of a_w = 0.040 to 0.990 and T = 10 to 80°C.

(5) As regards accuracy, the methods for humidity control according to the preceding paragraphs can be roughly classified into two groups as follows. The reproducibility of relative humidities controlled by the two-pressure system by mixing two air streams as well as with saturated salt solutions amounts to ±0.01 relative humidity (RH). This is, at the same time, the overall reproducibility of monitoring and controlling relative humidities by means of electrical hygrometers, as found in different collaborative tests. Under very strict working conditions, on the other hand, the two-temperature system and some electrolyte solutions offer a reproducibility to ±0.001 RH. These two accuracy figures, therefore, can be assigned to routine and reference precision levels, respectively.

(d) Achieving hygroscopic and thermal equilibrium between samples and environment is generally known as a very slow process. Several measures can be taken to accelerate the attainment of the equilibrium state, as follows:

(1) The apparatus can be evacuated, which is often done in both types of gravimetric techniques. High vacuum hastens the mass transfer but acts as thermal insulation around the samples, so that the rate-controlling process in such cases is usually the heat transfer. The Cahn Co. offers for this reason a "sample positioner" to allow thermal contact between the sample pan and a metallic conductor, as used for example by J. L. Multon (private conversation, 1978).

(2) A high rate of heat and mass transfer can be maintained, of course, by applying the dynamic principle, i.e., circulating air of constant temperature and humidity over, or even better, through the sample. This method is usually combined with mixing two air streams but also with electrolyte solutions (e.g., Audu

et al., 1978) and the two-temperature humidity control system
(e.g., Bolliger *et al.,* 1972).

Referring to continuous methods, vacuum systems offer higher
precision but demand more expensive construction and instrumentation.

One of the most important aspects of performing sorption
measurements is, without doubt, the definition and registration
of the equilibrium state between samples and the water vapor
source. Theoretically, this state can only be expected after
an infinitely long period of time. Practically, the sorption
process could be stopped when the weight difference left to the
final equilibrium becomes less than the sensitivity of the
balance used. But even this stage often requires an impractically
long waiting time. Therefore an apparent equilibrium state is
commonly defined in terms of a maximum tolerable weight change
during an arbitrary time interval. The automatic registration
of this apparent equilibrium is the main feature of some modern
sorption apparatus, of which the latest model has been constructed by the Instrumatic Export SA, as described in the next
section.

(e) With respect to the construction and operation of sorption
apparatus the most important characteristic is whether the weight
change of the sample can be observed continuously or discontinuous-
ly, as discussed in detail previously (Gal, 1967, 1975).

For continuous recording and for automation purposes by far
the most frequently used instruments are electrobalances from
Cahn, Sartorius, and Mettler, whereas other types may also be
suitable, e.g., the Beckmann Model LM 600 (Drexler, 1972). Very
good experience has also been gained with an Ainsworth recording
vacuum balance, Model RVA (J. F. Kapsalis, private conversation,
1978), used previously in meat research by Strasser (1969). The
moderate sensitivity (0.1 mg) and the inductance transducer-type

sensing of beam reflection (in contrast to the usual light beam-photomultiplier tube of other balances) as well as several other features render this balance especially suitable for heavy-duty, long term sorption studies.

Still very popular and widely used for gravimetric sorption studies is the spring balance. A modern construction has been developed recently by Weldring et al. (1975) and used in sorption measurements on potato starch (van den Berg et al., 1975). Similar instruments described earlier have been used in the period under consideration by Tomka (1973), Walker et al. (1973), and Das (1974).

The great advantage of this system is the possibility of multiplying its capacity at a comparatively low price by using several springs with a single cathetometer. With standard spring models, however, no automation is possible, and therefore such assemblies require steady attendance.

A very important aspect of the choice of the right balance is the sensitivity required for the problem in hand. Provided the temperature control works within the limits indicated in (a), the overall precision is next defined by the error in relative humidity values. Corresponding figures are given in (c). Now, by definition, the sorption isotherm (or more precisely its slope) reflects the relation between water content and water activity. The slopes of isotherm vary widely as shown, for example, by the recent collection of sorption isotherms of foods (Wolf et al., 1973), covering a range of 10^{-2} to 10^{0} (weight fraction of water per unit water activity). It varies also along the water activity scale, being larger at the low end of the isotherm and particularly in the high water activity range. Thus, as a first approximation, an average slope of 10^{-1} may be assumed for further calculations. Combining the precision figures of water activities with this average slope of isotherms, one can readily estimate the resulting reproducibilities of weight changes in sorption measurements. On this basis the following maximum deviations can be defined as general guidelines along with the temperature limits given above: For routine work, $\pm 0.2^\circ C$, $\pm 0.01 a_w$, and $\pm 0.1\%$ moisture content (absolute); for research purposes, $\pm 0.02^\circ C$, $\pm 0.001 a_w$, and $\pm 0.01\%$ moisture content (absolute). These figures may vary depending on the particular sorption behavior of the substance(s) to be investigated.

1. Methods with Continuous Registration of Weight Changes. Sorption assemblies with built-in electrobalances have established themselves in sorption research in the last decade as outlined in the foregoing review (Gal, 1977). Recently, J. L. Multon (private communication, 1978) has constructed a new sorption apparatus with a Cahn R 100 electrobalance based on the two-temperature vacuum principle. In the Swedish Food Institute a similar apparatus has been installed and used in investigations on sorption properties of proteins (Hermansson, 1977). Further work with continuous gravimetric sorption apparatus described earlier has been published by Berlin and Anderson (1975), Rüegg et al. (1974), Lüscher et al. (1973), Roth (1976, 1977), and Kopelman et al. (1977).

In the last few years a highly important development has taken place in sorption methodology, i.e., the step into the microprocessor age. Figure 1 shows the block diagram of a novel construction called Autosorb (Instrumatic Export SA, Corsier/GE, Switzerland). The sample is suspended from a Cahn model 1000 electrobalance. Air of controlled relative humidity ($\pm 2\%$) and temperature ($\pm 0.5^\circ C$) passes through the sample tube at a rate (50-400 ml/min) to allow rapid equilibration but without disturbing the performance of the balance. The computer executes a number of different functions. The program contains a start routine and separate routines for time, temperature, relative humidity, and balance readout. Three temperatures can be chosen

SYSTEM FLOW DIAGRAM

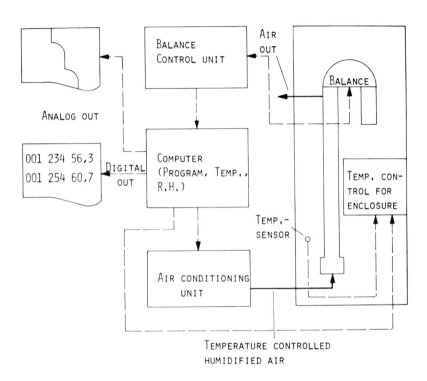

Fig. 1.   Block diagram of the Autosorb.

manually for the isotherm procedure: 20°, 25°, and 30°C.   The
relative humidity can be programmed from 30 to 70% in both directions with fixed steps of 5%.

The move to the next relative humidity level is performed
automatically after reaching constant weight of the sample.   This
state is defined in terms of three parameters:

(a) Maximum tolerated weight change during several successive printouts expressed as a percentage of the dry sample weight. Three values may be selected: 0.5, 1.0, and 2.0%.

(b) The time interval between two printouts of weight can be chosen from 10 min, 1 hr, and 6 hr. The printed weights are sums of several individual weighings in order to spare printer strip paper.

(c) The number of consecutive printouts that must lie within the maximum tolerable weight change can also be set manually for each experiment. Three values are programmed: 5, 10, and 20.

It can easily be computed that the extreme time values of reaching the apparent equilibrium state, after which the microprocessor gives the order to switch to the next relative humidity, are 5 printouts in 10 min intervals, totaling 50 min, or 20 printouts in 6 hr each, totaling 120 hr during which one of the three weight changes given under (a) must not be exceeded.

Weighing and drying the sample manually before starting the program are not necessary as these operations are carried out automatically by the start routine. The normal sorption cycle consists of an adsorption isotherm followed by desorption and the automatic stop.

2. *Methods with Discontinuous Registration of Weight Changes*. The explanations given at the outset of this section about the five basic functions apply also for the case when the equilibration and weighing are carried out in a discontinuous way, i.e., in separate operations. In a critical evaluation of literature data regarding such methods, Gal (1976) has shown that evacuation and dynamic systems offer similar efficiencies in achieving sorption equilibrium.

The most efficient method, but also the most complicated respective construction, is forcing air through the samples, as realized by Gur-Arieh *et al.* (1965). In the desiccator and iso-

piestic technique, stirring the head space and the electrolyte
solution is a sufficient measure for most purposes to accelerate
equilibration.

A large number of laboratories have been using such methods
for obtaining sorption isotherms. Vacuum desiccators with
saturated salt solution have been used, for example, in studies
on beet root components by Iglesias *et al.* (197a, b) and in the
investigation of air-dried beef by Iglesias and Chirife (1977).

Despite its time-consuming mode of action, the simple static
desiccator principle has also been given preference in numerous
sorption studies, for example, on soy sauce (Hamano *et al.*, 1972),
soy beans (Alam and Shove, 1973), tea (Jayaratnam and Kirtisinghe,
1974), and dried vegetables with and without added salt (Speck,
1976).

The respective heat and mass transfer improved desiccator
method with stirred headspace and salt solution has also found
increasing application (Gustafson and Hall, 1974; Mauch and
Asseley, 1975; Gal, 1977; W. Wolf and W. E. L. Spiess, private
communication, 1978). Quite recently Chemcotec AG (Baden,
Switzerland) started to market a simple and handy desiccator set
called Wet-Dry Sorbostat for routine sorption measurements. As
depicted in Fig. 2 the sorbostat consists of a small desiccator
in which a perforated aluminum disk carrying up to eight weighing
bottles is inserted together with a small magnetically operated
ventilator. The electrolyte solution is accommodated in the
bottom of the desiccator and can be stirred by means of a magnetic
stirrer driven by a low-voltage synchronous type motor running at
a constant speed of about 150 rpm. A comprehensive operation
manual contains details of carrying out sorption measurements
with this device.

The precise and efficient isopiestic techniques revived a
few years ago by Gal (1974) and described in detail by Gal and
Hunziker (1977), have been extensively used in studies on starch

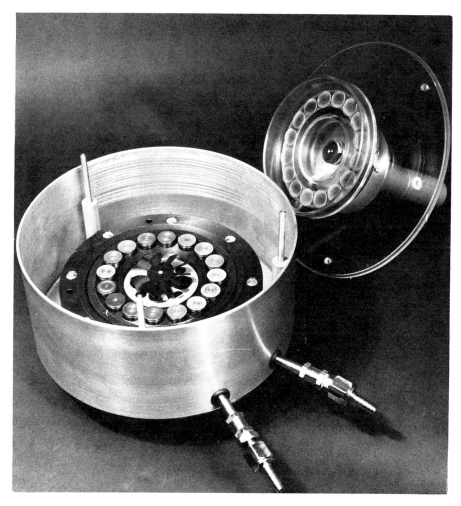

Fig. 2. Wet-dry Sorbostat of the Chemcotec AG.

(Nirkko, 1973), tropocollagen (Lüscher *et al.*, 1974), casein
(Rüegg and Häni, 1975; Rüegg and Blanc, 1976), and cheese
(Rüegg and Blanc, 1977). Chemcotec AG is reportedly starting
to produce a similar apparatus for 16 samples in one thermostated
container (K.G. Gröninger, private communication, 1978). Besides
meeting all important requirements for thermal stability and

high equilibration rate, the main feature of this device is the magnetically operated opening and closing mechanism for the weighing bottles inside the closed desiccator. Figure 3 is a picture of this novel type of isopiestic apparatus, called Sorbostat de Luxe.

All these discontinuous methods offer the additional advantage of working simultaneously with many samples concurrently observing important "marking points" of the isotherm, like caking, shrinking, and the onset of rapid chemical and microbiological deteriorations.

B. Manometric and Hygrometric Methods

For obtaining complete sorption isotherms with the aid of direct or indirect water vapor pressure measurements, samples with different water contents must be available. This happens often when nature or an industrial process produces samples of varying moisture content. Otherwise, subsamples of the same lot

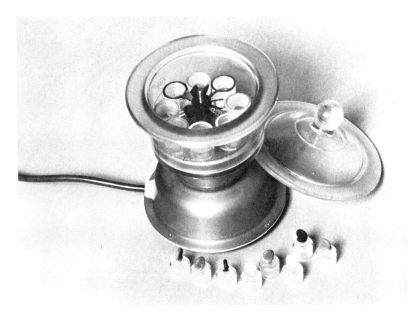

Fig. 3.   Sorbostat de Luxe of the Chemcotec AG.

must be conditioned to different water contents. This work is
usually very time consuming due to the comparatively large sample
size and difficulties in achieving homogeneous distribution of
moisture. Therefore, such an operation is hardly ever carried
out with isotherm determinations alone in mind. Rather, if
such samples have to be prepared for storage stability or packaging studies and their moisture content and a_w values are
determined the sorption isotherm is obtained, in a sense,
as a by-product.

As for the preparation of samples, Pixton and Warburton (1973)
and Multon and Bizot (1976) have given some practical guidelines.
Occasionally small samples are subjected to equilibration following a procedure similar to the gravimetric desiccator sorption
method. Another convenient technique is mixing dry and moist
lots of the same substance. Isotherm points in this case lie
inside the hysteresis loop. Adding water to increase the
moisture content is generally not advisable because of phases
changes, partial dissolution and uneven distribution of
soluble components due to migration during wetting and
drying. Despite this, in a series of sorption studies this
method has been applied with apparent success, as, for example,
with rapeseed (Pixton and Warburton, 1977) and dried figs
(Pixton and Warburton, 1976).

For such studies water content determination following
national or international standard procedures can be best recommended with due reference to the working conditions. Different
and well-known instruments have been used for measuring a_w values
in pertinent studies in the last years, e.g., manometers and
hygrometers.

C. Special Methods

The author is unaware of any development during the period reviewed of special techniques for obtaining complete sorption isotherms other than those discussed by Caurie (this volume).

III. CONCLUSIONS

The actual state of water vapor sorption methodology presents itself in a great variety of apparatus and methods along with a poor reproducibility between laboratories. International programs are currently underway to identify and eliminate the sources of excessive interlaboratory variance in sorption measurements and to gather reproducible and comparable sorption data on foods and related materials.

The gravimetric technique has been preferred during the period reviewed for obtaining complete sorption isotherms. Sorption units with built-in recording electrobalance have established themselves in sorption research. The most modern assembly of this type applies microprocessor control for every operation including the automatic registration of successive equilibria. The respective heat and mass transfer improved desiccator and isopiestic methods have also found increasing application.

In the future, greater attention should be paid to the thorough characterization of materials and methods, i.e., to the listing of all relevant data in reporting on sorption measurements. It seems very desirable to envisage an international standardization of methods for reference purposes, and simple, precise, and efficient methods for routine work.

ACKNOWLEDGMENTS

The author is deeply indebted to colleagues and companies
for providing information, personal views, experiences, and
material for demonstration purposes.

REFERENCES

Alam, A., and Shove, G. C. (1973). *Trans. Am. Soc. Agri. Eng.*
    16, 707.
American Society for Testing and Materials (1971). Standard
    Recommended Practice for Maintaining Constant Relative
    Humidity by Means of Aqueous Solutions. Philadelphia.
Audu, T. O. K., Loncin, M., and Weisser, H. (1978). *Lebensmittel-*
    Wiss. Technol. 11, 31.
Berlin, E., and Anderson, B. A. (1976). *J. Dairy Sci. 58,* 25.
Best, R., and Spingler, E. (1972). *Chem. Ing. Techol. 44,* 1222.
Bolliger, W., Gal, S., and Signer, R. (1972). *Helv. Chim. Acta*
    55, 2659.
Das, B. (1974). *Deutsche Lebens. Rdsch. 70,* 139.
Drexler, H. (1972). Untersuchunger zur Technologie und Qualitäts-
    beurteilung zerstäubungsgetrockneter Fruchtpulver. Ph.D.
    Thesis, Univ. of Hohenheim, FRG.
Fachnomenausschuss Materialprüfung (FNM) im Deutsches Insti-
    tut für Normung e.V. (1977). Konstantklimate über wässriger
    Lösungen, Entwurf. DIB 40 008.
Gal, S. (1967). "Die Methodik der Wasserdampf-Sorptionsmessunge."
    Springer-Verlag, Berlin.
Gal, S. (1975). Recent advances in techniques for the determina-
    tion of sorption isotherms. *In* "Water Relations of Foods"
    (R. B. Duckworth, ed.), pp. 139-154. Academic Press, London.
Gal, S. (1976). "Einfache und Schnelle Methoden zur Bestimmung
    von Wasserdampf-Sorptionsisothermen." Lecture, Institute
    of Food Technology and Packaging, Munich.
Gal, S. (1977). "Die Bestimmung der Wasseraktivität und der Sorp-
    tionsisothermen." Lecture, ETH, Zürich.
Gal, S., and Hunziker, M. (1977). *Makromol. Chem. 178,* 1535.
Gough, M. C., and Bateman, G. A. (1977). *Trop. Stored Prod. Inf.*
    33, 25.
Greenspan, L. (1976). *J. Res. NBS 81A,* 89.
Gröninger, K. G. (1972). Hygrometrie und Untersuchung über die
    Elektrische Messung der Feuchte. Ph.D. Thesis, Univ. of
    Hannover, FRG.
Gustafson, R. J., and Hall, G. E. (1974). *Trans. Am. Soc. Agri.*
    Eng. 17, 130.

Hamano, M., Aoyama, Y., and Yokotsuka, T. (1972). *J. Food Sci. Technol. (Tokyo) 19,* 503.

Hermansson, A. M. (1977). *J. Food Technol. 12,* 177.

Hofer, A. A. (1962). Zur Aufnahmetechnik von Sorptionsisothermen und ihre Anwendung in der Lebensmittel-Industrie. Ph.D. Thesis, Univ. of Basel.

Iglesias, H. A., and Chirife, J. (1977). *Lebensmitt. Wiss. Technol. 10,* 151.

Iglesias, H. A., Chirife, J., and Lombardi, J. L. (197a). *Food Technol. 10,* 299.

Iglesias, H. A., Chirife, J., and Lombardi, J. L. (1975b). *J. Food Technol. 10,* 385.

Jayaratnam, S., and Kirtisinghe, D. (1974). *Tea Quart. 44,* 164, 170.

Kopelman, I. J., Meydav, S., and Weinberg, S. (1977). *J. Food Technol. 12,* 403.

Loncin, M., and Weisser, H. (1977). *Chem. Ing. Technol. 49,* 312.

Lowe, E., Durkee, E. L., and Farkas, D. F. (1974). *J. Food Sci. 39,* 1072.

Lüscher, M., Giovanoli, R., and Hirter, P. (1973). *Chimia 27,* 112.

Lüscher, M., Rüegg, M., and Schindler, P. (1974). *Biopolymers 13,* 2489.

Mauch, W., and Asseley, S. (1975). Sorptionsverhalten von Fructose-, Glucose- und Saccharoseschmelzen unterscheidlicher Erhitzungsgrade. Forschungsbericht 1, Institut für Zuckerindustrie, Berlin.

Multon, J. L., and Bizot, H. (1976). *In* "Biodeterioration Investigation Techniques" (Walters, A. H., ed.), pp. 149-183. Applied Science Publ. Barking, Essex.

Multon, J. L., Trentesaux, E., and Guilbot, A. (1971). *Lebensm. Wiss. Technol. 4,* 184.

Nirrko, P. (1973). Wasserdampfsorption von Kartoffelstärke, ihren Bestandteilen und Abbauprodukten. Ph.D. Thesis, Univ. of Berne.

Packaging Institute (1952). Procedure for Determination of Humidity-Moisture Equilibria of Food Products. PI Food lp-52. New York.

Pixton, S. W., and Warburton, S. (1973). *J. Stored Prod. Res. 9,* 189.

Pixton, S. W., and Warburton, S. (1976). *J. Stored Prod. Res. 12,* 87.

Pixton, S. W., and Warburton, S. (1977). *J. Stored Prod. Res. 13,* 35.

Rangaswamy, J. R. (1973). *J. Food Sci. Technol. (India) 10,* 59.

Roth, D. (1976). Amorphisierung bei der Zerkleinerung und Rekristallisation als Ursachen der Agglomeration von Puderzucker und Verfahren zu deren Vermeidung. Ph.D. Thesis, Univ. of Karlsruhe, FRG.

Roth, D. (1977). *Zucker 30,* 274.

Rüegg, M., and Blanc, B. (1976). *J. Dairy Sci. 59,* 1019.

Rüegg. ,M. and Blanc, B. (1977). *Milchwissenschaft 32,* 193.

Rüegg, M., and Häni, H. (1975). *Biochim. Biophys. Acta 400,* 17.

Rüegg, M., Lüscher, M., and Blanc, B. (1974). *J. Dairy Sci. 57,* 387.

Speck, P. (1976). Herstellung und Lagerung von Trockengemüse mit Kochsalzzusatz. Ph.D. Thesis, ETH, Zürich.

Strasser, J. (1969). *J. Food Sci. 34,* 18.

Tomka, I. (1973). Untersuchungen über die Wasserdampfsorption quellbarer Körper. Ph.D. Thesis, Univ. of Berne.

Troller, J. A., and Christian, J. H. B. (1978). "Water Activity and Food." Academic Press, New York.

van den Berg, C., Kaper, F. S., Weldring, J. A. G., and Wolters, I. (1975). *J. Food Technol. 10,* 589.

Varshney, N. N., and Ojha, T. P. (1977a). *J. Food Sci. Technol. 14,* 69.

Varshney, N. N., and Ojha, T. P. (1977b). *J. Dairy Res. 44,* 93.

Walker, J. E., Wolf, M., and Kapsalis, J. G. (1973). *J. Agri. Food Chem. 21,* 878.

Weldring, J. A. G., Wolters, I., and van den Berg, C. (1975). *Med. Landbouw. Wageningen, 75,* 18.

Wexler, A. (ed.)(1965). "Humidity and Moisture." Reinhold, New York.

Wolf, W., Spiess, W. E. L., and Jung, G. (1973). Wasserdampf-Sorptionsisothermen von Lebensmitteln. No 18, Fachgemeinschaft Allgemeine Lufttechnik VDMA, Frankfurt/Main, FRG.

WATER VAPOR SORPTION ISOTHERMS ON MACROMOLECULAR SUBSTRATES

R. L. D'Arcy
I. C. Watt

I. INTRODUCTION

The interaction of water, and other vapors, with biopolymers
is typified by the wool-water vapor system, which exhibits the
general phenomena observed in polymer and protein sorption systems. Consequently; data for this system have contributed greatly
to an increased understanding of the mechanisms operating in
macromolecular sorption systems. The sorption process is of prime
importance since many physical properties of macromolecular materials are greatly modified by the presence of sorbed moisture.
For instance, in the case of wool, a form of keratin, the Young's
modulus can decrease by a factor of 2.7:1 (Woods, 1940), the torsional rigidity by 10:1 (Speakman, 1929), while the length may increase by 1-2% and the diameter by 16% (Bendit and Feughelman,
1968) as the dry fiber is transformed to the wet state.

Although the sigmoidal shape of the wool-water vapor isotherm,
in common with those of macromolecular substrates in general, has
been well established, no simple expression to satisfactorily relate the amount of water incorporated into the substrate at any
specified relative humidity between zero and saturation is available. A number of equations have achieved success in describing
the adsorption isotherm over a limited humidity range, but none
has accommodated the phenomenon of hysteresis.

In the absence of a satisfactory isotherm equation, a great deal of attention has been directed toward the mechanisms by which water is adsorbed by keratin. Speakman (1944) postulated that the sorbate attaches to specific sites within the substrate at low humidities, while a second mechanism of adsorption, considered by Barrer (1947) to be a mixing of water with the polypeptide chains, becomes involved at higher humidities. From variations in the mechanical properties, Windle (1956) proposed that adsorbed water exists in three states:

(1) Localized water attached by hydrogen bonds to specific hydrophilic polar groups is bound with the highest energy.

(2) Water attached by hydrogen bonds with a lower binding energy, to this primarily adsorbed and localized water, constitutes a second type of bound water.

(3) Finally, some of the adsorbed water is loosely incorporated into the system in a form that is somewhat analogous to, but distinct from liquid water.

Of the many sorption models postulated on the basis that the sorbed water is present in different forms, the most widely used are probably the BET (Brunauer, Emmett, and Teller, 1938) localized sorption model and the solid-solution model of Hailwood and Horrobin (1946).

A stoichiometric analysis (Watt and Leeder, 1968) of the wool-water system established that specific hydrophilic sites such as carboxylic, amino, and hydroxyl (both aromatic and aliphatic) residues, in addition to the backbone peptide groups, are hydrated at low relative humidities (RH). With increasing humidities the amount of water attached, with a relatively low binding energy, to peptide bonds becomes proportionately greater. At 80% RH it accounts for almost half of the total amount of adsorbed water. Although the average mobility of the hydrogen-bonded network of sorbed water increases with the humidity, there is no evidence that any of the sorbed water in keratin can be considered as free or liquid water even at humidities approaching saturation. Although much of the sorbed water in the network at higher humidities has no direct association with specific sorption sites, it is associated with other water molecules that are directly attached to these sites.

In considering the interaction of adsorbed water with wool the tacit assumption is made that each morphological component and all the hydrophilic sites within these components are accessible to water vapor. There are, however, differences in chemical composition and structure between morphological components, and the distribution of sorbed water between these components may vary with humidity.

A further characteristic of water vapor sorption, by substrates of biological origin, is the phenomenon of hysteresis over the entire humidity range. Numerous qualitative theories have been proposed to account for hysteresis but a satisfactory thermodynamic formulation of the effect is lacking. Since thermodynamic treatments are normally applied to reversible equilibrium states, the problem remains as to how to treat hysteresis changes so that they are amenable to thermodynamic reasoning.

II. DETERMINATION OF ISOTHERMS

The experimental determination of accurate sorption isotherms is a necessary prerequisite to elucidating the sorption mechanism. Most sorption systems are not athermic; evolution of heat, as water vapor is adsorbed, results from an interaction of the substrate with sorbed water molecules. Therefore, isotherms determined at different temperatures may be utilized to estimate the strength of such interactions.

The contributions made by early workers in the field have been summarized by Speakman (1930), who published complete adsorption-desorption isotherms demonstrating hysteresis effects. Further attempts to obtain reliable isothermal data for higher temperatures

have highlighted the fact that the reproducibility of data for
the sorption of water vapor by keratin, at increasing temperatures
and humidities, is very poor. Complete isotherms could not be obtained at temperatures above about 55°C (Speakman and Cooper,
1936); this was attributed to thermal degradation of the wool.
It has since been suggested (Jeffries, 1960) that textile fibers
should be stabilized by the application of a sorption cycle at
90°C before making an isotherm determination. Such a treatment,
however, constitutes a set of annealing conditions (Delmenico
and Wemyss, 1969) that are sufficient to significantly modify any
subsequently determined isotherms. Since Speakman (1930) established the sigmoidal nature of the sorption isotherm, subsequent
determinations by numerous workers have shown reasonable agreement with the values he obtained for adsorption isotherms at 25°C.
However, isotherms determined from a desorption process, or at
temperatures significantly higher than 25°C, have given rise to
significant differences between results of various studies. The
present authors (Watt and D'Arcy, 1978) have determined wool-water
vapor isotherms in the temperature range 20°-100°C taking cognizance
of a number of factors that may modify the isotherm measurements.

When isotherm data are used for the calculation of thermodynamic parameters, or for the elucidation of the mechanism of the
sorption process, the following factors that may modify the isotherm must be taken into consideration:

1. The source of the substrate may have a significant effect
on the absolute values of water content and the overall shape of
the isotherm. Averaging the values from a number of studies will
merely provide a result within the range of observed values and,
particularly at the higher humidities, it may even be less accurate than a single result.

2. Since long-term changes can occur to the substrate-water
complex, any experimental isotherm depends upon the experimental
conditions and the definition of attainment of equilibrium. Although this may not unduly affect the values of the absolute water

content, any interpretation of the mechanisms involved may be significantly influenced.

3. Variations of water content resulting from experimental
factors become more pronounced as the temperature is increased.
Hence, it is more difficult to obtain reproducible isotherms at
elevated temperatures and it is of critical importance to accurately define and reproduce the conditions of measurement before
comparing isotherms.

III. THEORETICAL ANALYSIS OF ISOTHERMS

Classical sorption theory formally distinguishes between adsorption and ab-sorption. In the former case the phenomena that
constitute the sorption process are considered to be restricted
to the physical interface between a solid and its environment.
The absorption process, on the other hand, is concerned with
phenomena occurring within the solid phase of the substrate.
Apart from making this distinction, and pointing out that the
rates for the two processes usually differ by some order of magnitude, classical theory is concerned almost exclusively with
adsorption.

Hayward and Trapnell (1964) have suggested that the absorption
process can be regarded essentially as a process in which a sur-
face-like adsorption occurs in the interior of the substrate, with
sorbate molecules being bound by physical forces at specific sites
within the interior of the substrate. Diffusion of the sorbate
from the surface of the substrate to its interior occurs via fine
capillaries, crystal grain boundaries, interfaces between different
morphological components, and penetration of sorbate molecules
between atoms in the substrate. The driving force for such diffusion originates from the concentration gradient established when
sorbate molecules are initially sorbed at the surface of the substrate.

Sorption systems in practice may differ from classical theoretical systems in three important aspects: (a) the sorption
process is not necessarily restricted to the formation of a monolayer of sorbate; (b) sorption processes frequently involve both
surface adsorption and internal absorption; (c) the sorbent is
often heterogeneous in the sense that considerable variability
exists among the available sorption sites.

For a theoretical isotherm to be fully satisfactory it must
not only accurately describe the experimentally observable relationship between the substrate and sorbate, but it must also be
consistent with the various modes of attachment of sorbate molecules to the substrate. Any general theoretical isotherm must be
applicable to a wide range of systems. Therefore, it should be
capable of accommodating the experimental facts that some systems
are capable of forming true solution, thus implying an infinite
amount of sorbed water, at humidities at or even below 100% RH,
while for other systems the uptake of sorbate is limited to a
finite amount at all humidities.

A. Development of a Theoretical Isotherm

In order to develop a general isotherm, the following model of
the sorption process is postulated:

1. The overall sorption process is comprised of both the clas-
sically distinguishable adsorption and absorption processes.

2. The absorption process can be regarded as a special case of
adsorption as proposed by Hayward and Trapnell (1964).

3. Initially a monolayer of adsorbed molecules is incorporated
into the substrate at specific sorption sites.

4. Additional molecules of sorbate may be incorporated into
the substrate by being sorbed onto the primarily adsorbed monolayer
to form a multilayer.

Since multilayer formation is fundamentally concerned with
sorbate-sorbate interactions between a single species of sorbate,

it would be unrealistic to regard such interactions as anything other than a single mechanism despite the existence of a number of possible mechanisms for forming the primary monolayer. The formation of a monolayer involves sorbate-substrate interactions occurring at specific sites that may be quite dissimilar. On the assumption that a number of sites for the sorbate-substrate inter-action can be regarded as being identical, or sufficiently similar, they can be considered to constitute a type, or class of homoge-neous sites for the adsorption of a monolayer. The binding energy at each site can be expressed as the average of the energy of binding at all the sites within the appropriate class.

The free energy for a sorption system to which the assumptions in Table I apply can be defined by

$$G_n = \sum_{i=1}^{s} G_{r_i} + G_{(n-\Sigma r_i)} + G_{(\text{interactions})}$$

Assuming that the interactions between molecules of sorbate at different sites make a negligible contribution to the free energy of the system, the last term in the above equation can be ignored. Application of a statistical thermodynamic treatment leads to

TABLE I. *Assumptions Relating to the Sorption Model*

---

1. *There are s classes of distinguishable primary sorption sites within the substrate.*
2. *Each class of primary sorption sites is characterized by a heat of sorption w_i, which can be defined as the limiting value of the differential heat of adsorption at zero relative vapor pressure.*
3. *There are r_i molecules of sorbate primarily adsorbed at sites of the ith class.*
4. *There are α_i sites of the ith class.*
5. *A total n molecules of sorbate are adsorbed at equilibrium.*
6. *A total D of the $\Sigma_{i=1}^{s} r_i$ primary sorption sites are utilized for multilayer formation.*
7. *The average heat of adsorption of a molecule into the multi-layer is w_m.*
8. *The vapor phase of the sorbate can be regarded as an ideal gas.*

$$n = \sum_{i=1}^{s} \frac{S_i k_i p}{1+k_i p} + \frac{Dk_d p}{1-k_d p} \tag{1}$$

where the symbols are defined in Table II.

The first term in Eq. (1) is the sum of an as yet indefinite number of Langmuir-like terms describing the formation of a monolayer of primarily adsorbed molecules at specific sorption sites. It can be shown analytically that if there are two or more classes of site onto which the sorption of vapor to form a monolayer can be described by a Langmuir-type expression, then the sum of the individual Langmuir expressions can only be combined into a single expression of the same form if the parameters of the independent variables in the denominators of the individual expressions are identical, i.e., the values for all the k_i must be identical. If they are not, and their sum is represented by a single Langmuir expression, an error term must be included. In carrying out the summation indicated for the first term of Eq. (1) the value of k_i for a particular class may be so slow that the product $k_i p$, at the maximum vapor pressure of concern, is negligible compared to unity. In such cases the particular term can be approximated by a linear expression,

$$\frac{s_i k_i p}{1+k_i p} \sim s_i k_i p$$

Conditions that enable this linear approximation to be made only occur when the sorption sites are very weakly adsorbing and the characteristic heat of adsorption for the site is sufficiently low. When such an approximation is valid for one or more classes of site, the series of linear approximations concerned can be accurately represented by a single linear term of the form Kp, where $K = \sum_{i=1}^{i} s_i k_i$. Hence Eq. (1) can now be rewritten as

$$n = \sum_{i-1}^{s-j} \frac{s_i k_i p}{1+k_i p} + Kp + \frac{Dk_d p}{1-K_d p} \tag{2}$$

TABLE II. Key to Symbols Used in Equations

Symbol	Meaning
n	number of molecules of sorbate adsorbed at equilibrium
s	number of separable classes of sorption sites
S_i	number of sorption sites of the ith class
k_i	temperature-dependent function of the standard chemical potential u_0 of the sorbate in the vapor phase, the molecular partition function J_i of the sorbate primarily adsorbed at specific sites of the ith class, and the characteristic differential heat of adsorption ω_i of the sorbate at specific sites of the ith class, $$k_i = J_i \, e^{\left(\dfrac{u_0 + \omega_i}{RT}\right)}$$
k_d	similar to k_i except that it refers to the secondarily adsorbed sorbate constituting the multilayered sorbate; subscript d indicates the multilayer material so that $$k_d = J_d \, e^{\left(\dfrac{u_0 + \omega_d}{RT}\right)}$$
D	total number of primary sorption sites that may be involved in multilayer formation
p	equilibrium vapor pressure of the sorbate vapor
M	mass of sorbate adsorbed at equilibrium by 100 g of initially dry substrate
m	molecular weight of the sorbate
ℓ	number of sensibly different classes of specific sorption sites
A_i	molar number of primary sorption sites of the ith class per 100 g of initially dry substrate.
B_i	modified form of k_i defined by $B_i = p_0 k_i / 100$, where p_0 is the saturation vapor pressure of the sorbate

Table II (Continued)

Symbol	Meaning

| C | slope of the linear approximation of the Langmuir adsorption onto weakly binding sites, defined by |

$$C = p_{0_{100}} S_j J_j e\left(\frac{u_0 + \omega_j}{RT}\right)$$

where the subscript j refers to weakly binding sites whose Langmuir adsorption can be approximated by a linear expression.

| E | modified form of k_d defined by $E = (p_{0_{100}})k_d$ |
| χ | relative equilibrium vapor pressure of the sorbate expressed as a percentage |

The first term in this expression is the sum of a series of Langmuir expressions and describes monolayers being formed at the different classes of specific sorption sites. Each of these classes is characterized by a heat of sorption for each molecule of sorbate incorporated into the monolayer.

The second term, which has a linear form, approximates the sum of Langmuir terms formally describing the adsorption of a monolayer of sorbate at weakly binding sites. Since the vapor pressure of the sorbate does not exceed its saturation value in practice, and the sites concerned only weakly attract the sorbate, the advantages of using a simpler linear expression instead of the more formally appropriate Langmuir expressions seem justified.

The third term defines the contribution made to the overall sorption process by further adsorption, either at primarily adsorbed molecules of sorbate to initiate a multilayer or into the multilayer already formed by such secondarily adsorbed molecules. The driving force for this multilayer formation originates from the failure of a single layer of sorbed material to fully neutralize the available sorption energy of the specific sites. The residual sorption energy is more or less uniformly distributed throughout the primarily adsorbed monolayer rather than being quantized at specific sites. No formal restriction is imposed on the number of secondarily adsorbed molecules that may be associated, in a number of tiers as the multilayer, with any particular sorption site. The presence of a monolayer, which does not necessarily have to be complete, is a necessary precursor to any multilayer formation. The amount of multilayer formation is determined by constraints to swelling imposed by the substrate as more sorbate is incorporated, and the amount of residual attraction for water exerted by the substrate after primary adsorption of a monolayer has occurred. The isotherm represented by Eq. (2) can be normalized by expressing the number of molecules adsorbed as the equilibrium weight of sorbate per 100 g of substrate and the vapor pressure as the percentage relative humidity to give

$$M = m\left(\sum_{i=1}^{\ell} \frac{A_i B_i X}{1+B_i X} + CX + \frac{DEX}{1-EX} \right) \tag{3}$$

where the symbols are given in Table II. This expression is a general one and for a particular system one or more of the terms may be inoperative. In such a case, the constituent process represented by the inoperative term makes no contribution to the overall sorption process and the appropriate coefficients in Eq. (3) assume zero values.

    The development of this isotherm equation is similar to the treatment by Hill (1946) but differs in that it specifically makes provision for the presence within the substrate of a number of classes of specific sorption sites characterized by different energies of adsorption. Identical expressions can also be derived for the same postulated model from a kinetic treatment, similar to that used in the derivation of the BET equation, or by a statistical mechanical approach.

B. Determination of the Isotherm Coefficients

Evaluation of the coefficients in Eq. (3) is made from data
obtained experimentally for the sorption system of interest. The
deviation between the experimentally observed equilibrium water
content and that predicted from Eq. (3) with arbitrarily assigned
values for the coefficients (A, B, C, D, E) is determined for a
number of points on the isotherm. The values of these coefficients
are then systematically modified so as to minimize the sum of
squares of the deviations between the observed and predicted values
of the water content. However, the curve of best fit is not necessarily obtained by minimizing the sum of squares of deviations in
the water content at nominated humidities since the errors tend to
become appreciably greater as the slope of the isotherm curve increases. This tends to place an undue emphasis on the results at
humidities approaching saturation. If the deviation between the
observed and calculated results is divided by the slope of the
curve at the humidity concerned and the deviation modified in this
way is used as the basis for the sum of squares the undue emphasis
can be avoided.

The sum of squares of the modified deviations is minimized by
iteratively applying a simplex method (Nelder and Mead, 1965) of
function minimization), which is most conveniently implemented on
a computer. The number of Langmuir expressions constituting the
first term in Eq. (3) and that must be summed so as to adequately
describe the monolayer formation on high-energy-specific sites is
initially assumed to be unity. On this assumption, values of the
coefficients are calculated to give a modified least-squares best
fit to the available data. The deviations between the observed
and calculated equilibrium water contents are then examined.
Since the Langmuir terms preferentially modify the low-humidity
region of the isotherm, any systematic variation between the two
sets of values at the lower end of the vapor pressure spectrum is
indicative of the need to provide an additional Langmuir expression.

The values for all the coefficients, including those for the additional Langmuir expression, are then redetermined by repeating the function minimization procedure described above. In practice, it has not been necessary to utilize more than a single Langmuir term in Eq. (3).

C. Application to Experimental Isotherms

Values of the coefficients in Eq. (3), for a number of isotherms from various sources are presented in Table III. The data from which the coefficients have been evaluated represent a number of diverse systems and illustrate that the isotherm equation can be applied to a large number of situations.

Figure 1 shows a series of water vapor isotherms, at 35°C, for four of the substrates from Table III. These materials, keratin, bacterial spores, skin collagen, and nylon, have been selected to demonstrate variations in the typically sigmoidal-shaped curves characteristic of many macromolecular systems that may be encountered in practice. The solid curves represent Eq. (3), with values of the coefficients evaluated from the expermentally determined points.

It is possible to isolate the contributions made by the three constituent processes implied by Eq. (3). In Fig. 2, the isotherm, obtained for wool keratin is analyzed into three constituent components, shown as the broken curves A, B, and C. Curve A, corresponding to the first term in Eq. (3), represents the contribution made to the overall sorption process by the Langmuir-like monolayer sorption onto highly reactive sorption sites. Curve B represents the formation of a monolayer of sorbate on weakly reactive sorption sites. Curve B represents the formation of a monolayer of sorbate on weakly reactive sites within the substrate. Formally, the middle linear term of Eq. (3), which defines this curve, should be of a Langmuir form similar to that for curve A, but as previously discussed, saturation of this monolayer cannot

TABLE III. *Values of Isotherm Parameters for Various Sub-strates*

Substrate	ℓ	A	B $(\times 10^{-3})$	C $(\times 10^{-3})$	D	E $(\times 10^{-3})$
Bacillus subtilis[a] spores	1	6.442	112.740	32.261	2.831	9.860
Wool keratin[a]	1	3.582	246.950	128.540	1.895	9.134
Skin collagen[a]	1	11.426	68.266	8.788	8.796	8.404
Nylon[a]	0	–	–	79.530	0.646	8.366
Jute[a]	1	4.730	71.484	30.861	4.732	8.706
Serum albumin[b]	1	10.245	46.002	32.633	2.697	9.393
α- and β-pseudo-globulin[b]	1	6.347	109.491	76.006	3.156	9.283
γ-Pseudoglobulin[b]	1	4.585	160.310	127.713	1.686	9.797
Collagen[b]	1	5.219	300.917	197.241	2.726	9.323
β-Lactoglobulin (crystalline)[b]	1	2.979	175.450	128.300	1.749	9.888
β-Lactoglobulin (lyphilized)[b]	1	1.833	423.338	138.415	1.655	9.661
Egg albumin (unlyphilized)[b]	1	4.810	115.741	86.824	2.098	9.503
Egg albumin (lyphilized)[b]	1	3.218	186.965	105.111	1.691	9.550
Egg albumin (coagulated)[b]	1	2.552	266.490	110.978	0.843	9.799
Gelatin[b]	1	10.311	75.679	95.419	2.5223	9.741
Elastin[c]	1	2.123	296×10^{5}	167.439	0.626	9.994
Salmin[b,c]	1	3.483	378×10^{7}		12.547	9.317
C Zein[b]	1	4.206	88.104	20.555	2.380	8.978
β Zein[b]	1	3.473	129.378	29.614	2.209	8.899
Silk[b]	1	2.734	145.469	73.092	0.932	9.735
Cotton[d]	1	1.725	337.199	57.806	0.958	9.411

[a]*Results obtained at 35°C (D'Arcy and Watt, 1970).*
[b]*Results obtained at 40°C (Bull, 1944).*
[c]*Published results for these isotherms (Bull, 1944) would suggest that some of the reported points are seriously in error, since the isotherm does not appear to pass through the origin.*
[d]*Results obtained at 25°C (Hailwood and Horrobin, 1946).*

be approached at the relative vapor pressures under investigation. Curve C represents the formation of a multilayer of sorbed vapor on the primarily adsorbed monolayer. The extent to which this multilayer formation can progress is limited by mechanical con-

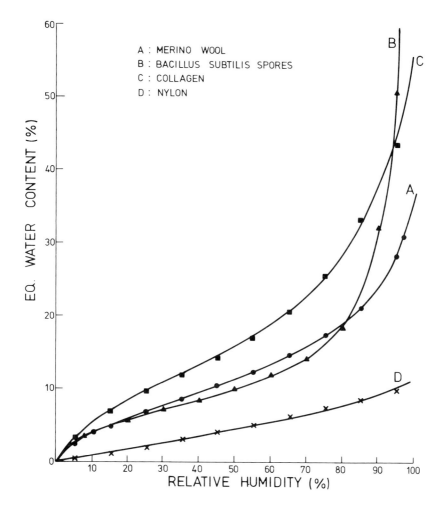

Fig. 1. *Water vapor adsorption isotherms at 35°C for (A) merino wool, (B) Bacillus subtilis spores, (C) collagen, and (D) nylon.*

straints to swelling of the substrate and by interfacial tension effects.

The prime requirement for any theory of adsorption is that the experimentally determined isotherm is completely described by the theory. It is also desirable that the theoretical model should have physical significance with respect to the actual sorption

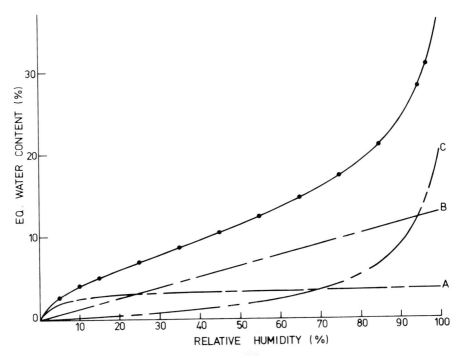

Fig. 2. Water vapor isotherm at 35 C for merino wool. Full curve represents overall adsorption process; broken curves represent component processes. (A) Monolayer adsorption at strongly hydrophilic sites; (B) monolayer adsorption at weakly hydrophilic sites; (C) multilayer formation.

system, so that an adequate interpretation of the sorption processes can be made. In the sorption model presented here, there is an initial surface adsorption at the physical interface between the sorbent and its environment, which establishes a concentration gradient between the interior of the sorbent and its physical surface. Consequently, diffusion of sorbate from the solid-gas interface to the interior then occurs. Further sorption takes place within the interior, at what can be regarded, following Hayward and Trapnell (1964), as an internal surface.

Experimental observations (Watt and Leeder, 1968) verify that in keratin the sorption consists, at least in part, of the summation of a number of processes occurring independently on different

types of sorption sites within the substrate. Further, they have
shown that analysis of the experimental isotherm in terms of a
single Langmuir isotherm fails to provide an adequate fit to the
experimental data, but that each type of hydrophilic side chain is
responsible for a component of the total adsorption, which can be
adequately described by a Langmuir isotherm at low humidities. A
similar conclusion for the cellulose-water system was reached by
Enderby (1955), who proposed that the isotherm for this system
could best be described by an equation comprised of a Langmuir
term plus a second term that at low humidities approximates a
straight line.

The model developed here differs fundamentally from previous
postulates in that it accommodates the existence of a number of
primary sorption processes and a secondary sorption process all
of which are each simultaneously and continuously contributing to
the overall process at all vapor pressures. This approach is in
agreement with the fundamental requirements of the adsorption
process as enunciated by Gilbert (1946). The isotherm represented
by Eq. (3) is therefore of a complex form and invokes, in general,
up to five disposable parameters. Consequently, it may appear to
be less attractive than other expressions with fewer disposable
parameters. However, it must be stressed that the sorption process
on nonhomogeneous substrates is intrinsically a complicated process, and the emergence of a complicated expression is not necessarily a disadvantage. In fact, the use of a simpler empirical
expression could well be a disadvantage since an approximate expression, although it may adequately describe the total amount of
sorbed material present, frequently fails to emphasize and describe
the sorption process actually occurring. In this context it should
be stressed that the expression being proposed is equally applicable
to simpler systems, since the absence of one or more of the three
constituent processes is readily accommodated by setting the appropriate constants in Eq. (3) to zero, when there is direct experimental evidence to show that the appropriate processes are inoperative.

Results obtained by the application of this new isotherm analysis to the values published by Bull (1944) for the wool-water isotherm at 25 C are compared with those obtained by the application of the Hailwood and Horrobin (1946) isotherm equation to the same data, since this isotherm is frequently used to describe sorption isotherms. Results of this comparison are given in Table IV, and there is an obvious overall improvement in the agreement between observed values and those calculated from Eq. (3). Results of an analysis of variance carried out to test the significance of this improved agreement are presented in Table V. Comparison of the variance ratio of 9.86 with tabulated F values of 7.98 and 4.10 at the 0.01 and 0.05 probability levels, respectively, for 9/6 degrees of freedom shows that the improvement over the isotherm of Hailwood and Horrobin has a 99% confidence level that it is a real and not a random chance improvement.

The excellent agreement between the experimental and calculated isotherm is insufficient to fully verify the correctness of an isotherm equation. Two additional factors that need to be considered are whether the assumptions made in the derivation of the isotherm are reasonable and whether the isotherm can be used

TABLE IV. *Comparison of Equilibrium Water Contents Calculated by the Isotherm Eq. (3) and the Hailwood-Horrobin Isotherm on Data Reported by Bull (1944)*

	Equilibrium water content (%)		
	Calculated from isotherm equation		
Relative humidity	Eq. (3)	Hailwood-Horrobin	Observed
0	0	0	0
5	2.96	2.64	2.97
10	4.28	4.29	4.25
20	6.28	6.46	6.25
30	8.07	8.14	8.22
40	9.86	9.72	9.90
50	11.62	11.43	11.43
60	13.49	13.42	13.47
70	15.54	15.91	15.61
80	18.08	19.21	18.05
90	22.55	23.91	22.54
95	29.13	27.14	29.15

TABLE V. Analysis of Variance for Derived Isotherms[a]

Source of variation	Degrees of freedom	Single layer site	Multilayer site
Due to Model 1	5	2470.7141	494.142
Residual 1	6	0.1447	0.024116
Total 1	11	2470.8588	224.623
Due to Model 2	2	2468.7182	1234.359
Residual 2	9	2.1406	0.237844
Total 2	11	2470.8588	224.623

[a]*Model 1 [Eq. (3)]*

$$M_0 = \frac{ABP/P_0}{1+BP/P_0} + CP/P_0 + \frac{DEP/P_0}{1-EP/P_0}$$

Model 2 (Hailwood-Horrobin):

$$M_0 = \frac{acP/P_0}{1-aP/P_0} + \frac{abcP/P_0}{1+abP/P_0}$$

Variance ratio:

$$\left(\frac{residual\ 2}{residual\ 1}\right) = \frac{0.2378}{0.0241} = 9.86$$

to make meaningful interpretations of the sorbent-sorbate relationships. The concept that some water is adsorbed with high affinity
onto hydrophilic groups in the substrate while the remainder of the
water is held less firmly has been widely accepted. At low humidities a Langmuir isotherm is generally acceptable as a valid approximation in which small discrepancies between theoretical and experimental data can be attributed to mutual interactions between
sorbed water molecules. Experimental data for the keratin-water
system (Watt and Leeder, 1968) and the cellulose-water system (Enderby, 1955) both suggest the presence of a water uptake that can
be expressed as a linear function of the vapor pressure superimposed on a Langmuir-type isotherm.

Considerations of the amount of water capable of being adsorbed by wool, and the number of potential sites for direct attachment of water to the substrate, show quite conclusively that

it is impossible for all the water to be accommodated in the form
of a monolayer. Consequently, some multilayering of the adsorbed
water must occur. When a Langmuir isotherm is applied to the association of water molecules with strongly binding sites, a limiting value for the water uptake is asymptotically approached at
vapor pressures below saturation. Such a Langmuir analysis of the
water associated with basic side chains in keratin is experimentally valid up to approximately 25% RH. Since additional water
molecules are associated with these sites at higher vapor pressures (Leeder and Watt, 1965), it is clear that these water molecules must be associated with water molecules previously adsorbed
at lower humidities. The conditions necessary for multilayer formation are simply that there should be residual attractive forces
from the primary sorption site that are transferres to the primarily adsorbed monolayer to provide for sorbate-sorbate attraction as distinct from the sorbate-sorbent attraction operative in
the monolayer formation. In the absence of this residual attraction the entire isotherm could be adequately described by the
Langmuir and linear terms.

The residual attraction of the substrate for water molecules
in the multilayer, coupled with the constraints to swelling imposed by the substrate, determine the extent of the multilayer
formation. Water molecules in the multilayer are not to be
equated with liquid water since there is a diffuse residual attraction from the substrate that is not present in liquid water.
Although an infinite amount of water, indicating complete solubility of the substrate, is possible, it is not necessarily inferred, since the mechanical constraints to swelling by the substrate can serve to limit the extent of multilayer formation at
saturation.

An examination of Eq. (3) shows that the curve represented
by the isotherm equation has two points of discontinuity, as illustrated in Fig. 3, which shows schematically the overall shape
of the entire function. The first discontinuity can be ignored

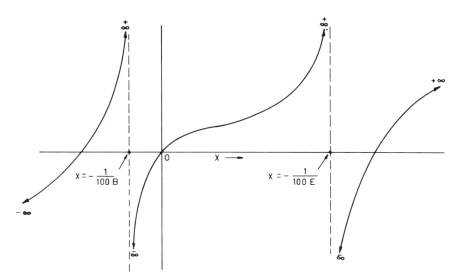

Fig. 3. Schematic curve of complete function for sorption process given by Eq. (3).

since it occurs at negative values of the relative vapor pressure. In the case of the second discontinuity, which occurs at a positive vapor pressure, the curve indicates that the amount of sorbed vapor asymptotically approaches infinity as the vapor pressure approaches $1/100E$. Since an infinite amount of sorbed water is synonymous with solubility of the substrate, the value of E should be indicative of the vapor pressure at which the substrate-sorbate complex forms a solution. If the asymptote occurs at or below saturation vapor pressure it would be expected that the material should be water soluble. In general, proteins whose values of E are approximately 10^{-2} are water soluble, while those with lower values are not.

In the case of wool, whose E value is 9.134×10^{-3}, the asymptote corresponding to infinite water adsorption would occur somewhere in the vicinity of 110% RH. Since such a humidity is unattainable in practice the material is regarded as being insoluble. Compression of liquid water by the application of a large hydrostatic pressure will increase the number of molecules striking a unit area, and

conditions equivalent to relative vapor pressures in excess of
100% RH could be realized so that partial solubility of insoluble
substrates may be achieved.

D. Thermal Effects

Use of isotherm data to calculate isosteric heats of adsorption normally yields results of poor accuracy. Calculation of the
isosteric heat of adsorption, a quantity whose value depends on
the water content of the substrate, can be carried out by applying
the Clausius-Clapeyron equation to isotherm data obtained at two
different temperatures. Although the Clausius-Clapeyron equation
is in a differential form, it can be integrated over a finite temperature range to give

$$q_{st} = \frac{RT_1T_2}{T_2-T_1} \ln \frac{X_2}{X_1}$$

where q_{st} is the isosteric heat for adsorption of vapor by a substrate with a specified constant water content and X_2 is the relative humidity at temperature T_2 that is in equilibrium with the
same amount of adsorbed material at humidity X_1 and temperature T_1.
Since the isosteric heat is, in general, temperature dependent,
the temperature interval between T_1 and T_2 must be kept small for
the heat to be sensibly constant. Furthermore, the heat calculated will pertain to a system at a temperature and humidity that
are the limiting values of the temperature and humidity as the
temperature difference $T_2 - T_1$ approaches zero. It is extremely
difficult in practice to measure the change in relative humidity
necessary to maintain a constant water content in the substrate
as the temperature is changed through the small interval required
to apply the above expression. An additional difficulty arises
from the fact that since the isotherm curve is lowered by increased
temperatures, there will be some water contents in the vicinity of
saturation that will not be able to be maintained as the temperature

is increased.

The availability of an analytical expression such as Eq. (3), that accurately defines the isotherm throughout the entire humidity spectrum enables the differential of the water content with respect to the equilibrium relative humidity to be readily calculated. It also enables the change in relative humidity necessary to maintain a constant amount of sorbate in the substrate as the temperature is changed to be accurately calculated. The ability to analytically define the isotherm curve as the differential of the water content is of vital importance in carrying out calculations of a number of thermodynamic parameters, such as the Gibbs free energy, for the sorption process.

Examination of Eq. (3) indicates that the effect of any change in temperature will be reflected in the parameters B, C, and E. Since each of these parameters involves a molecular partition function of sorbed material and an exponential term raised to the inverse power of the absolute temperature, it can be readily shown that the isotherm curve will indicate a lowering of the equilibrium water content at all humidities as the temperature is increased. Consequently, Fig. 4, illustrating such behavior, contrasts with previously reported data (Jeffries, 1960) suggesting that with increased temperatures the isotherms intersect in the vicinity of 85% RH.

IV. VARIATIONS OF SORPTION BEHAVIOR

A. Transitions in the Substrate

In some cases, particularly where the substrate has been modified by chemical treatments, isothermal data may not appear to follow a normally shaped isotherm curve. In Fig. 5 the kinetic behavior of a single wool fiber, determined by a vibroscope technique (D'Arcy, 1976), in which the change in the resonant frequency of a

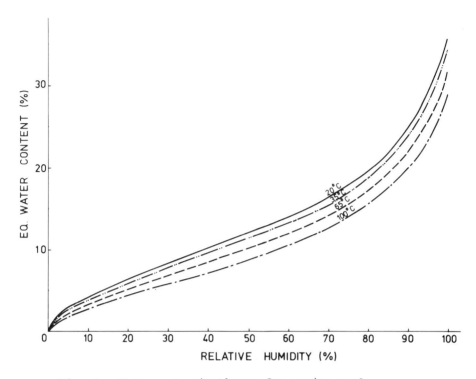

Fig. 4. Water vapor isotherms for merino wool.

vibrating fiber is used to monitor the mass change as sorption occurs, is shown for both adsorption and desorption steps in the range between 96 and 98% RH. The adsorption kinetics for the fiber, modified in this case by the deposition of approximately 14.5% by weight of the additive, dodecyl diphenyl ether 2,4'-disulfonic acid, are initially normal. After approximately 15 min the increase in weight, which had originally been a linear function of the square root of time, reached a maximum of 0.7% uptake before decreasing to an equilibrium value 0.5% less than the initial weight.

Decreasing the relative humidity from 98 to 96% reverses the process just described. Initially, a weight loss is recorded, but a minimum weight is reached and an eventual net weight gain of 0.5% results. Thus, the weight of the sample becomes identical with its weight at the start of the initial adsorption step from

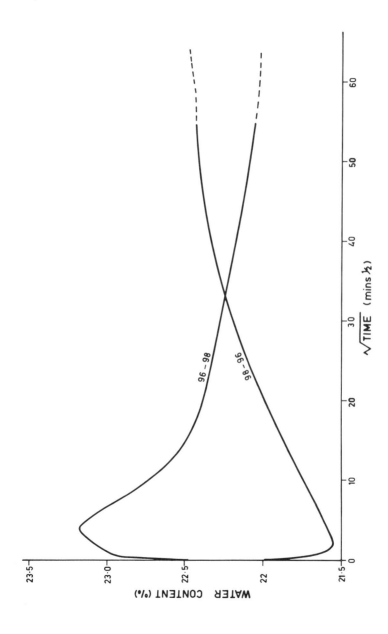

Fig. 5. Kinetic curves for sorption of water vapor by chemically modified single wool fibers. (A) Adsorption; (B) desorption.

96 to 98% RH. Similar effects were observed on other fibers from
the same sample, but the anomalous behavior was found to occur at
various humidities within the range of 90 to 98% RH. The isotherm
for a large number of fibers is an average and the anomalous behavior occurring at different humidities in individual fibers appears as a reduction in slope of the isotherm in this region.

B. *Hysteresis*

   A prominent feature of many isotherms published in the literature is for hysteresis to be exhibited over the entire humidity
range, i.e., the equilibrium water content at a particular humidity is higher when a sample is equilibrated by desorption from a
higher humidity than if the same sample had been subjected to an
adsorption step change in humidity from a lower value. In the
case of wool at ambient temperatures, equilibrium water contents
for desorption have been reported as being some 2 to 3% higher
than for adsorption to the same humidity. At higher temperatures
the width of the hysteresis loop lessens. Detailed examination of
published data shows that, in general, there is good agreement between the adsorption isotherms obtained by different workers, but
that a great deal of variation in the results for desorption does
exist. Consequently, much of the variability in the reported hysteresis can be attributed to differences in the equilibrium water
contents of substrates subjected to desorption steps.
   The water contents obtained from desorption processes, especially from one intermediate humidity to another, depend upon
the precise experimental technique adopted. This is demonstrated
by a number of desorption isotherms obtained for wool in the form
of (a) a single fiber, (b) a teased lock of fiber, and (c) a piece
of fabric woven from similar wool. Use of these three forms of
wool from the same source ensures that any variations observed are
attributable to effects arising from the packing density, which
modifies the observed kinetic behavior. Isotherms were constructed

from data obtained by means of a vibroscope in the case of the
single fiber, and by a sorption balance when a lock of fibers or
piece of fabric was used.

Results of these experiments are presented as Fig. 6, where
solid curve A is the common adsorption isotherm for the three
samples. Broken curves B, C, and D are the desorption isotherms
obtained from a single fiber, a lock of teased fibers, and a piece
of fabric, respectively. These results clearly demonstrate the increased hysteresis with increased packing density of the fibers.
Superimposed on this effect is the influence of the desorption
step size as illustrated by the data in Table VI, in which the
tabulated values represent the additional amount of water (expressed as a percentage of the weight of dry wool) in the desorp-

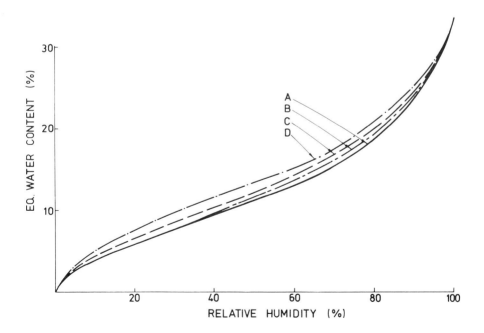

Fig. 6. *Water sorption isotherms for merino wool exhibiting
varying amounts of hysteresis. (A) Adsorption; (B-D) desorption
under different conditions.*

TABLE VI. Sorption Hysteresis in Wool Fibers

equilibrium (%)	No. of steps	Observed hysteresis (% water content)		
		Single fiber	Teased lock	Fabric
70	1	0.6	1.2	2.0
50	1	0.4	0.9	2.2
50	2	0.7	1.1	2.3
25	1	0.0	0.9	2.1
25	2	0.2	1.1	2.0
25	3	0.4	1.3	2.1

tion isotherm at equilibrium. At any nominated humidity the observed hysteresis is least when desorption from saturation to the
designated humidity is effected in a single step. As the number
of desorption steps (of equal magnitude) in reaching the nominated
humidity from saturation is increased, the observed equilibrium
water content also increases.

The most notable feature in these results is that under some
conditions (e.g., single-step desorption of a single fiber from
saturation to humidities less than 25% RH) no hysteresis could be
detected. It is significant that the complete elimination of hysteresis at 25% RH occurs in the humidity region where Jeffries
(1960) reported a maximum amount of hysteresis and that this region is generally accepted as being the region in which the proportional hysteresis is greatest. Such results clearly demonstrate that the occurrence and magnitude of any observed hysteresis
is a function of experimental technique and that the rapidity with
which the water content of the substrate changes is important.
For a single fiber that responds rapidly to changes in the humidity
of its surrounding atmosphere, the hysteresis effect can be eliminated below about 25 to 30% RH. In the case of a teased lock of
fibers, in which an increase in the packing density will decrease
the speed with which the sample can respond to humidity changes,
the complete elimination of hysteresis is not evident. At higher
packing densities, as in the case of woven fabric, the hysteresis
effect becomes more pronounced and is, in fact, of the same order
as the hysteresis values commonly reported in the literature.

It can be shown by kinetic studies that for small humidity in-
crements there is a rapid initial adsorption followed by a slower
second stage during which additional water is taken up by the substrate to maintain thermodynamic equilibrium with the external
vapor pressure, while relaxation of stresses, introduced by swelling
of the substrate during the initial sorption stage, is occurring.
Hysteresis only becomes apparent when the substrate is allowed sufficient time for a significant amount of stress relaxation to occur at the higher humidity before desorption to a lower humidity is
initiated. Thus, the phenomenon of sorption hysteresis in wool can
be associated with relaxation processes taking place within the
fibers. This interpretation is consistent with a study of the
mechanical properties of wool by Feughelman and Robinson (1967).
They point out that for a wool-water system in equilibrium with an
external vapor pressure there is a dynamic balance between the number of water molecules entering and leaving the fibers. This number is directly proportional to the vapor pressure, so that properties involving movement of chain segments should be related to the
relative humidity rather than water content. Hysteresis therefore
arises from the stress relaxation taking place in the substrate at
higher humidities to form a system with lower entropy.

V. CONCLUSION

The amount of water incorporated into a substrate at equilibrium under specified conditions is not an absolute quantity, but
it can be modified by factors such as temperature, relative humidity of the atmosphere at equilibrium, previous sorption history of
the sample, the manner in which the humidity is attained, the time
taken for the experiment and the method adopted to determine the
dry weight of the sample. Consequently, quantitative data must be
considered in the context of the experimental conditions under
which they were obtained.

The theoretical isotherm, Eq. (3), appears to be soundly based on thermodynamic principles and will accurately describe the amount of water present in the substrate at equilibrium. Although it was developed in conjunction with studies on keratin substrates, it should be applicable to sorption systems in general. Although the isotherm equation does not readily yield a quantitative description of the temperature dependence of the isotherm, examination of the temperature dependence of the parameters B, C, and E in Eq. (3) shows that the equilibrium water content at all humidities is lessened as the temperature is increased.

The presence of five disposable parameters in the isotherm equation emphasizes that the sorption process is a complex one. Equations with fewer parameters cannot be expected to describe the sorption process with the same precision as this form since they do not take into account the three processes constituting the over-all sorption process; monolayer adsorption on strongly hydrophilic sites, monolayer adsorption on weakly hydrophilic sites, and multi-layer adsorption onto the primarily adsorbed monolayers, which are concurrently occurring at all humidities. The use of simpler ex-pressions, such as the widely accepted BET equation, may give a reasonable approximation to the isotherm over a restricted range, but fail to cover the full humidity range from zero to 100% RH with-out exhibiting significant deviations in some regions at least.

The excellent agreement between observed and calculated values, obtained from Eq. (3), for the wool-water system can be reproduced with a wide range of sorption systems. In some cases one or more of the three-component systems may be inoperative and the corres-ponding terms in Eq. (3) may be omitted.

Some substrates may undergo a transition at the temperature and humidity being investigated, and having passed through that transition, the capacity of the substrate for water is reduced. Such a transition would account for the reversibility observed in Fig. 5. The "equilibrium" water content observed after desorption is a variable quantity depending upon the way the desorption is performed. Thus desorption studies have demonstrated that the amount of observed hysteresis is not an intrinsic feature of the system. The magnitude of the desorption step and the rate at which water can be removed from the substrate both decrease the amount of water retained. Moreover, provided that the desorption process can be effected sufficiently quickly, the hysteresis effect can be completely eliminated.

It appears that the desorption isotherms normally presented do not represent a true equilibrium condition, or if so, the equilibrium refers to a substrate that has been modified by the sorption of water. A criticism frequently made of isotherm analyses is that they fail to account for sorption hysteresis. Since isotherm analyses refer to equilibrium conditions, it is necessary to restrict such analyses to adsorption isotherms that fulfill such reproducible equilibrium conditions.

References

Barrer, R. M. (1947). *Trans. Faraday Soc. 43,* 3.
Bendit, E. G., and Feughelman, M. (1968). *Encyclopedia Polymer Sci. Technol. 8,* 30.
Brunauer, S., Emmett, P. H., and Teller, E. (1938). *J. Am. Chem. Soc. 60,* 309.
Bull, H. B. (1944). *J. Am. Chem. Soc. 66,* 1499.
D'Arcy, R. L. (1976). M.Sc. Thesis, Univ. of New South Wales, pp. 72, 102.
D'Arcy, R. L., and Stearn, A. E. (1968). *Rev. Sci. Instrum. 39,* 1875.
D'Arcy, R. L., and Watt, I. C. (1970). *Trans. Faraday Soc. 66,* 1236.
Delmenico, J., and Wemyss, A. M. (1969). *J. Text. Instrum. 60,* 78.
Enderby, J. A. (1955). *Trans. Faraday Soc. 51,* 106.
Feughelman, M., and Robinson, M. S. (1967). *Text. Res. J. 37,* 441.
Gilbert, G. A. (1946). *In* "Symposium on Fibrous Proteins," p. 96. Soc. of Dyers and Colourists, London.
Hailwood, A. J., and Horrobin, S. (1946). *Trans. Faraday Soc. 42B,* 84.
Hayward, D. O., and Trapnell, B. M. W. (1964). "Chemisorption," p. 5. Butterworths, London.
Hill, T. J. (1946). *J. Chem. Phys. 14,* 263.

Jeffries, R. (1960). *J. Text. Instrum.* *51,* T399.

Leeder, J. D., and Watt, I. C. (1965). *J. Phys. Chem.* *69,* 3280.

Nelder, J. A., and Mead, R. (1965). *Computer J.* *7,* 308.

Speakman, J. B. (1929). *Trans. Faraday Soc.* *25,* 92.

Speakman, J. B. (1930). *J. Soc. Chem. Ind.* *49,* T209.

Speakman, J. B. (1944). *Trans. Faraday Soc.* *40,* 6.

Speakman, J. B., and Cooper, C. A. (1936). *J. Text. Instrum.* *27,* T191.

Speakman, J. B., and Stott, E. (1936). *J. Text. Instrum.* *27,* T186.

Walker, I. K. (1963). *N.Z. J. Sci.* *3,* 127.

Walker, I. K., and Jackson, F. H. (1973). *N.Z. J. Sci.* *16,* 281.

Watt, I. C., and D'Arcy, R. L. (1978). *J. Text. Instrum.* In press.

Watt, I. C., and Leeder, J. D. (1968). *J. Text. Instrum.* *59,* 353.

Watt, I. C., Kennett, R. H., and James, J. F. P. (1959). *Text. Res. J.* *29,* 975.

Weigerink, J. G. (1940). *U.S. Bur. Stand. J. Res.* *24,* 645.

Windle, J. J. (1956). *J. Polymer Sci.* *21,* 103.

Woods, H. J. (1940). *Proc. Leeds Phil. Lit. Soc. Sci. School 3,* 577.

MOISTURE SORPTION HYSTERESIS

John G. Kapsalis

I. INTRODUCTION

In the field of water vapor sorption by a solid sorbent,
moisture sorption hysteresis is the phenomenon according to which
two different paths exist between the adsorption and desorption
isotherms.

If one plots the amount of water per unit mass of solid in
the ordinate and the corresponding relative vapor pressure in the
abscissa, usually the desorption isotherm lies above the adsorption isotherm and a close hysteresis loop is formed.

Moisture sorption hysteresis has important theoretical and
practical implications in foods. The theoretical implications
range from general considerations of the irreversibility of the
sorption process to the question of validity of thermodynamic
functions derived therefrom. The practical implications deal
with the effects of hysteresis on chemical and microbiological
deterioration, and with its importance in low and intermediate
moisture·foods.

Due to hysteresis, a much lower vapor pressure is required
to reach a certain amount of water by desorption than by adsorption. In nature, hysteresis may be considered as a built-in
protective mechanism against extremities, such as loss of water
due to a dry atmosphere, frost damage, and freezer burn.

In comparison to theoretical work, information on the prac-

tical aspects of hysteresis in foods is limited, although some
important findings have been reported in the last 10 years. The
purpose of this chapter is to examine both the theoretical and
practical aspects of moisture sorption hysteresis, especially as
they apply to foods. For comprehensive treatments of hysteresis
the reader is referred to the reviews of Arnell and McDermot
(1957), Everett (1967), McLaren and Rowen (1951), and to the
earlier work by Rao (1941).

II. THE THEORETICAL BASIS OF SORPTION HYSTERESIS

A. Types of Hysteresis

Figure 1 shows four types of hysteresis according to the
classification of Everett (1967). In type A, the loop occurs over
a limited range of relative pressures; an example is the adsorption of benzene on silica gel. In type B, the loop extends from
the saturation vapor pressure down to a well-defined closure
point, which is characteristic mainly of the kind of the vapor
adsorbed; an example is the adsorption of polar and nonpolar
gases on certain zeolitic materials. In type C, the loop extends
over the entire range of vapor pressure; this is the case of the
adsorption of water on protein or cellulosic fibers. In type D,
which is the mixture of types B and C, the desorption curve follows the pattern of type B loop, but before meeting the adsorption curve it bends away downward to the zero vapor pressure as
in type C; an example is the adsorption of vapors on certain carbons. In type D, if at point X the desorption is reversed a
smooth, reproducible (without hysteresis) curve is obtained.

In foods, a variety of hysteresis loop shapes can be observed. Wolf *et al.* (1972) found wide differences in magnitude,
shape, and extent of hysteresis of dehydrated foods, depending on
the type of food and the temperature. Variations could be grouped

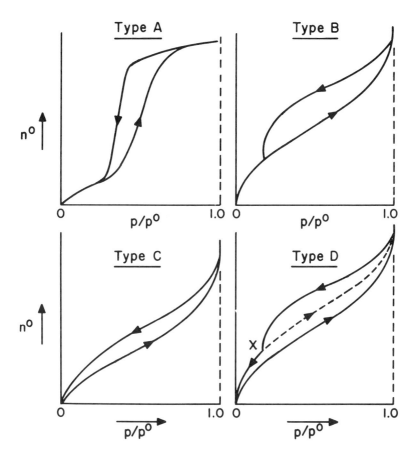

Fig. 1. Main types of sorption hysteresis (Everett, 1967).

into three general types, shown in Fig. 2 for hysteresis loops
obtained at 4.4°C. In high-sugar-high pectin foods, exemplified
by the air-dried apple (top of figure), hysteresis occurs mainly
in the monomolecular layer of water region. Although the total
hysteresis is large, there is no hysteresis above a_w = 0.65.

In high-protein foods, exemplified by the freeze-dried pork
(middle), a moderate hysteresis begins at about a_w = 0.85, i.e.,
in the capillary condensation region, and extends over the rest
of the isotherm to the zero water activity. In both adsorption
and desorption, the isotherms retain the characteristic sigmoid

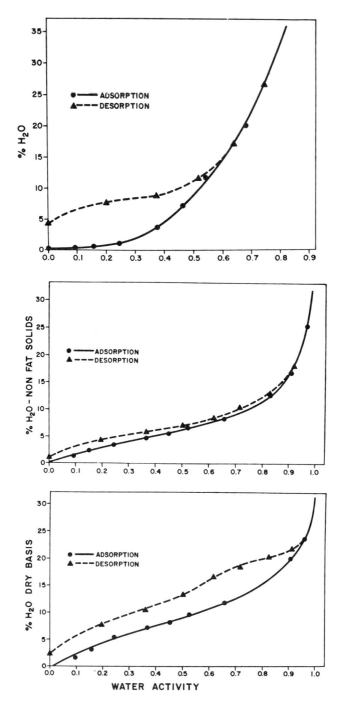

Fig. 2. Examples of sorption hysteresis in foods. Top, air-dried apple slices; middle, freeze-dried cooked pork; bottom, freeze-dried rice. All measurements at 4.4°C (Wolf et al., 1972).

shape for proteins. In starchy foods, as in freeze-dried rice
(bottom of figure), a large hysteresis loop occurs with a maximum at about a_w = 0.70, which is within the capillary condensation
region.

In the same work, increasing temperature decreased the total
hysteresis and limited the span of the loop along the isotherm
(Fig. 3). The latter change was dramatic for the apple, where
the beginning of hysteresis was shifted from a_w = 0.65 to a_w =
0.20, and for the pork, where the beginning of hysteresis shifted
from a_w = 0.95 to a_w = 0.60. The temperature dependence of hys-

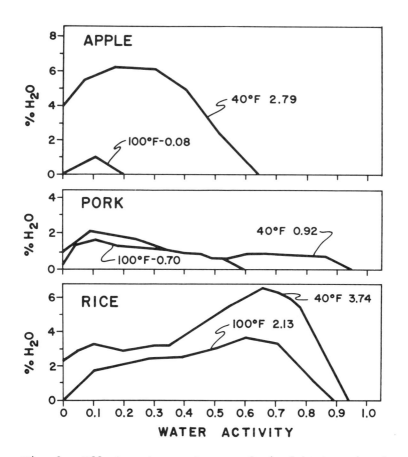

Fig. 3. Effect on temperature on derived hysteresis, i.e.,
differences between desorption and adsorption (Wolf et al., 1972).

teresis observed here was at variance with the results on bovine
serum albumin reported by Seehof *et al.* (1953). These workers
found that the amount of hysteresis was practically independent
of temperature and constant over the entire range of relative
vapor pressures. This was probably due to the small temperature
differential covered by the Seehof group, and to the presence in
the pork (in the work of Wolf *et al.*, 1972) of other, non-protein-
absorbing species.

The change in the apple isotherms from a type III for the adsorption branch to a type II for the desorption branch (classification by Brunauer, 1945) indicates a definite change in surface
structure. It has been attributed by Labuza (1975, 1976) to the
transition of the sugar from the crystalline into the amorphous
phase, which on desorption holds more water due to supersaturation.

A different type of sorption behavior has been reported by
Berlin *et al.* (1968) for a number of dried milk products exemplified by the foam-spray-dried whole milk in Fig. 4. The drop in
the adsorption curve at about $a_w = 0.60$, where the powder undergoes desorption at constant vapor pressure, has been attributed
to the conversion of the original amorphous hygroscopic lactose
glass into the relatively nonhygroscopic α-monohydrate, which beyond this a_w adsorbs in the form of a type II isotherm. Upon desorption, a smooth sigmoid curve is obtained that at $a_w = 0$ continues to hold some water irreversibly. Subsequent adsorption
data yield a reversible smooth isotherm. In a later study on
lactose crystallization, Berlin *et al.* (1971) found that when
anhydrous α-lactose adsorbs water, a Langmuir-type isotherm associated with the formation of the monohydrate is obtained; this
form at very high relative pressures ($P/P_0 = 0.98$ to 1.00) is converted into glass. Upon redrying, the amorphous form is obtained.
(For further details on this and related work, see the chapter by
Berlin in this volume.)

Desorption isotherms usually give a higher water content than
adsorption isotherms, with exceptions only in specific cases (Ber-

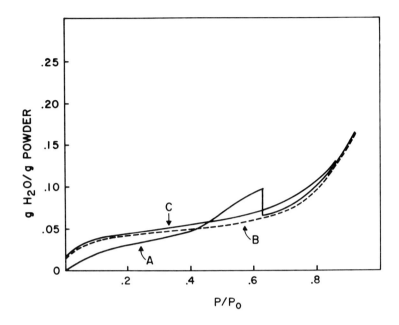

Fig. 4. Water sorption isotherm for foam-spray-dried whole milk at 24.5 C. Curve A, adsorption; curve B, desorption; curve C, readsorption (Berlin et al., 1968).

lin and Anderson, 1975; Rutman, 1967). In general, the type of changes encountered upon adsorption and desorption will depend on (a) the initial state of the sorbent (amorphous versus crystalline), (b) the transitions taking place during adsorption (c) the final a_w adsorption point, and (d) the speed of desorption. With regard to (c) and (d), if the saturation point has been reached and the material has gone into solution, rapid desorption may preserve the amorphous state due to supersaturation. (For phase transition and hysteresis, see Mellon and Hoover, 1951.)

Hysteresis seems to be reproducible and quite persistent over many adsorption-desorption scans, especially at low temperatures and over relatively short periods of time (Benson and Richardson, 1955; Strasser, 1969). However, at higher temperatures this may not be the case (Chung, 1966), due probably to denaturation.

The hysteresis loop may decrease with increasing temperature, pass through a minimum, and then increase again (Amberg *et al.*, 1957). Elimination of hysteresis upon the second or subsequent cycles may take place for a variety of reasons, such as "mixed" hysteresis (type D in Fig. 1), "time-dependent hysteresis" when the experiment is carried out very slowly, change in the crystalline structure when a new crystalline form persists upon subsequent cycles, swelling, and increased elasticity of capillary walls resulting in a loss of power of trapping water (Rao, 1939a,b), and surface active agents (Rutman, 1967). In an experiment reported by Block and Bettelheim (1970), an increase in the sorptive capacity of hyaluronic acid II with successive isotherms was observed. This was attributed to rupturing of hydrogen bonds, which provided a greater number of sorption sites.

The importance of the detailed structure of the adsorbent in determining the size and shape of the hysteresis loop cannot be overemphasized. For example, certain porous materials may show changes in hysteresis due to activation, compression, or other treatments. However, the relative pressure at which the loop closes depends only on the gas adsorbed and on the temperature. Mechanical treatment may also eliminate hysteresis (Everett, 1967).

B. *Scanning of the Hysteresis Loop*

The sorption path between the boundaries and within the hysteresis loop depends on the point of the adsorption or desorption branch where the ~~detection~~ *direction* of sorption was reversed. The crossing of the loop by scanning curves on desorption has been attributed to the entrapment of water in small capillaries until the main desorption boundary is reached. At this point water is trapped mainly in the larger diameter pores (Rao, 1939b). Figure 5 is a representative example of spiral paths within the hysteresis loop. The characteristics of the scanning paths are important in testing the predictive power of any theory of hysteresis.

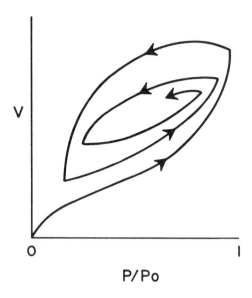

V

O   I

P/Po

Fig. 5.   Scanning paths in sorption hysteresis.

C. Sorption Isobar Hysteresis

The moisture sorption isobar provides a new way of obtaining
and examining sorption data.   In contrast to a sorption isotherm
where the temperature remains constant and the vapor partial pressure changes, the sorption isobar is obtained by holding the vapor
pressure constant and varying the temperature.   Figure 6 shows the
adsorption and desorption isobars of samples of freeze-dried cooked
beef obtained by Strasser (1969), using the Natick Ainsworth
vacuum balance.   At the lower or higher vapor pressures no hysteresis was observed, the loop being largest between 0.5 and 2.0
torr.   At the lower relative humidities and higher temperatures,
the moisture sorptions isobars tend to come closer together and
the sensitivity of the method is reduced.

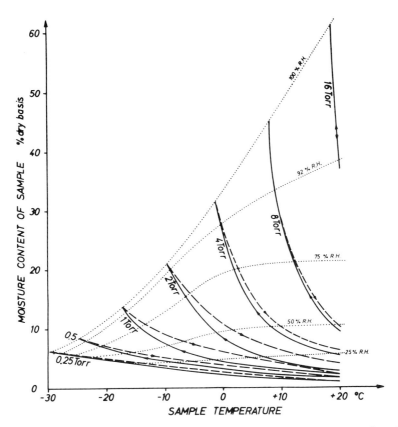

Fig. 6. Adsorption isobars (solid lines) and desorption isobars (dashed lines) of cooked, freeze-dried beef (Strasser, 1969).

D. Theories of Sorption Hysteresis

The interpretations proposed for sorption hysteresis can be
classified under one or more of the following categories, based
on the structure of the sorbent (Arnell and McDermot, 1957):

(a) Hysteresis on porous solids. In this category belong the
theories based on capillary condensation.

(b) Hysteresis on nonporous solids. It includes interpretations based on partial chemisorption, surface impurities, or phase
changes.

(c) Hysteresis on nonrigid solids. It deals with interpreta-

tions based on changes in structure, as these changes hinder penetration and egress of the adsorbate.

To the above we could add the case of polar sorption in biological materials, which may involve one or more processes, especially in categories (a) and (c).

1. Incomplete Wetting Theory (Fig. 7)

Suggested by Zsigmondy (1911), the theory represents the earliest attempt at explaining hysteresis. As all theories of capillary condensation, it is based on the Kelvin equation

$$RT \ln P/P_0 = -2\sigma V \cos \theta/r_m$$

where P is the vapor pressure of liquid over the curved meniscus,
P_0 the saturation vapor pressure at temperature T, σ the surface
tension, θ the angle of contact (in complete wetting $\theta = 0$ and
$\cos \theta = 1$, V the molar volume of the liquid, r_m the mean radius
curvature of the meniscus defined as $2/r_m = 1/r_1 + 1/r_2$, where
r_1 and r_2 are the principal radii of curvature of the liquid-vapor
interface, and R the gas constant. Due to the presence of impurities (dissolved gases, etc.) the contact angle θ of the receding
film upon desorption is smaller than that of the advancing film
upon adsorption. Therefore, capillary condensation along the
adsorption branch of the moisture sorption isotherm will be at a
higher relative vapor pressure.

The Zsigmondy theory fails to explain adsorption results at
low relative pressure, where in some cases it requires r to be
smaller than the diameter of the adsorbed molecule. It may have
some application in the limited case of a hysteresis open at the
lower end, in contrast to most cases in foods where the most common type of hysteresis is the closed-end, retraceable loop.

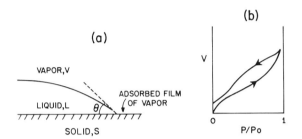

Fig. 7. Incomplete wetting theory of hysteresis: (a) contact angle; (b) open hysteresis.

2. Ink Bottle Neck Theory (Kraemer, 1931; McBain, 1935; Rao, 1941, Chapter IV)

This theory explains hysteresis on the basis of the difference in radii of the porous structure of the sorbent. The latter consists of large-diameter pores simulated by the main body of an ink bottle, equipped with narrow passages simulated by the neck of the ink bottle (Fig. 8).

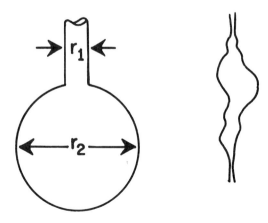

Fig. 8. Ink-bottle-neck theory of hysteresis. Left, schematic representation; right, actual pore.

In adsorption, condensation first takes place in the large diameter cavity at

$$P/P_0 = \exp(-2\sigma V/r_2 RT)$$

which results from substituting r_2 for r_m in the Kelvin equation. In desorption, the neck of the pore is blocked by a meniscus, which can evaporate only when the pressure has fallen to

$$P_d = \exp(-2\sigma V/r_1 RT)$$

at which point the whole pore empties at once. Therefore, for a given amount of water adsorbed, the pressure is greater during adsorption than during desorption.

3. Open-Pore Theory (Fig. 9)

The theory, elaborated by Cohan (1938, 1944) extends the ink bottle theory by including considerations of multilayer adsorption. It is based on the difference in vapor pressure between adsorption P_a and desorption P_d, as affected by the shape of the meniscus. The latter is cylindrical on adsorption, where the Cohan equation applies, and hemispherical on desorption, where the Kelvin equation applies. Thus

$$P_a = P_0 \exp(-\sigma V/rRT) = P_0 \exp\left[-\sigma V/(r_c-D)RT\right]$$

where r_c is the radius of the pore and D the thickness of the adsorbed film. Once condensation has taken place, a meniscus is formed and the desorption pressure P_d is given by the Kelvin equation. Assuming that wetting is complete, $\cos \theta = 1$ and

$$P_d = P_0 \exp(-2\sigma V/r'RT)$$

When $r_c > 2D$, $P_a > P_d$, and hysteresis will occur. When $r_c = 2D$,

$$P_a = P_d = P_0 \exp\left[-\sigma V/(r_c-D)RT\right]$$

and no hysteresis will take place (Gregg and Sing, 1967).

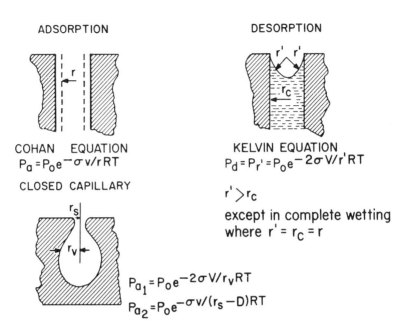

Fig. 9. Open-pore theory of hysteresis (Cohan, 1938, 1944).

The pressure at which hysteresis begins should correspond to equilibrium pressure for $r_c = 2D$, i.e., for these pores that are of two molecules diameter.

4. *The Domain Theory* (Fig. 10).

This theory has been discussed in its latest form by Everett (1967) as a reformulation of his earlier concept, which was more generalized to include the phenomena of magnetic hysteresis [see e.g., Everett and Whitton (1952) and Everett (1955)]. It is an effort not only to explain and predict the irreversibility of the sorption process, but also to interpret sorption curves resulting by crossing the hysteresis loop group through repeated scanning cycles.

The term "domain" as first introduced with regard to magnetic hysteresis applies to a group of atoms that can exist in one of two thermodynamically metastable states. These states are separated from each other by a small but finite gap. The group of

atoms should be both sufficiently small to show discrete steps in
the thermodynamic curve on a small scale and sufficiently large
to discount the possibility that thermal fluctuation will overcome
the potential barrier separating the two states (Enderby, 1955).
In Everett's theory a pore domain is a region of pore space accessible from neighboring regions through pore restrictions. An
isolated pore domain has well-defined condensation-evaporation
properties involving one or more spontaneous irreversible steps.
The pore domain is divided into elements of volume dV that contain
either liquid or vapor (Fig. 10). In adsorption, a liquid-vapor
interface will sweep through this element of volume at a certain
relative pressure x_{12} and in desorption the same interface will
pass back through it leaving it empty at x_{21}. In an irreversible
pore domain situation $x_{12} > x_{21}$. All elements of volume are
classified in terms of x_{12} and x_{21}, which form the coordinates of
the lower part of Fig. 10. Due to the inequality defined above,
the points representing all elements of volume will lie in the
triangle OAB. Each area $dx_{12} \, dx_{21}$ of this diagram is associated
with a quantity $v(x_{12}, x_{21})$, such that $v(x_{12}, x_{21}) \, dx_{12} \, dx_{21}$ is
the volume of the pore domain associated with values of x_{12} and
x_{21} in the ranges x_{12} to $x_{12} + dx_{12}$ and x_{21} to $x_{21} + dx_{21}$. The
distribution function $v(x_{12}, x_{21})$ characterizes the properties of
the pore domain. Thus, an adsorption process may be represented
by the movement of a vertical line across the triangle from left
to right, and at any point the state of the system may be represented by a diagram similar to the lower row of Fig. 10 ("domain
complexion"). The diagram functions as "memory" in systems exhibiting hysteresis.

The desorption process can be described in terms of the movement of a horizontal line from top to bottom of the triangle.
The properties of the system can always be expressed as the sum
of integrals over triangles of the same hypotenuse OB, in a dia-
gram of x_{21} vs. x_{12} between the limits $x_{21} = x_{\ell}$ and $x_{12} = x_{u}$, as
follows:

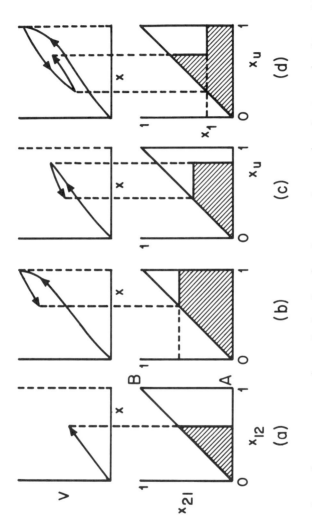

Fig. 10. Schematic representation of the domain theory of sorption hysteresis (Everett, 1967).

$$V(x_\ell, \ x_u) = \int_{x\ell}^{x_u} \int_{x\ell}^{x_u} v(x_{12}, \ x_{21}) dx_{12} \ dx_{21}$$

In the independent-domain theory each pore domain interacts with the vapor as an isolated pore. The theory fails to account quantitatively for the behavior of many systems, especially those related to capillary condensation processes. It was subsequently reexamined and broadened by Everett by considering pore-blocking effects. Whereas an isolated pore will fill and empty at given values of relative vapor pressures (according to the independent-domain theory), when the pore is interconnected with others, this same element may or may not undergo these changes at the same relative vapor pressures. This will depend on the way the pore is interconnected and on whether the adjacent pores are full or empty. Stated briefly, it will depend on the history of the system. For more details, the reader is referred to the original discussion by Everett (1967).

E. Polar and Other Interpretations of Sorption Hysteresis in Biological Materials

These interpretations do not constitute a "theory" in the literal sense of the word. Instead they are explanations that could be categorized under the main groups listed above, especially under the group relating to deformability and elastic stresses of the sorbent. Hysteresis is attributed to a deformation of the polypeptide chains within the protein molecule as the polar adsorbates (in our case water) occupy suitable positions for hydrogen bonding or ion dipole interaction. In view of our interest in biological materials, we briefly discuss them.

Seehof et al. (1953), on the basis of literature and their own data, supported a polar group interpretation of hysteresis where binding involves mainly the free basic groups of the protein.

TABLE I. *Correlation between Maximum Hysteresis and Polar Groupings in Proteins, Compiled from Literature Data*[a]

	Arginine 1	Histidine 2	Lysine 3	Cysteine 4	Cystine 5	RNA 6	Σ 1-6	Maximum hysteresis
				mmol/g protein				
Casein	0.2	0.2	0.6				1.0	1.1
TMV	0.5		0.1	0.1		2.7	3.4	3.5
Insulin	0.2	0.3	0.2	0.1	0.5		1.3	1.2
Collagen	0.5	0.1	0.3				0.9	0.8
BovSAlb	0.3	0.2	0.8		0.3		1.6	1.5
BovPAlb								1.4

[a]From Seehof et al., 1953.

Table I shows a correlation of the maximum amount of hysteresis
(last column) with the sum of arginine, histidine, lysine, and
cystine groupings (next to last column). Besides the free basic
groups of the protein molecule, sulfur linkages are also of prime
importance in hysteresis (Speakman and Stott, 1936). In contrast
to this work, hysteresis in casein was found independent of the
content of free amino groups by Mellon *et al.* (1948). A twofold
nature of hysteresis is proposed: constant hysteresis, independent of the RH desorption point, and hysteresis proportional
to the amount adsorbed above the upper absorption break of the
isotherm.

Van Olphen (1965) attributes sorption hysteresis on vermiculite clay to a retardation of adsorption due to the development
of elastic stresses in crystallites during the initial peripheral
penetration of water between the unit layers. The shift toward
higher relative vapor pressure on adsorption is caused by the activation energy required to open the unit layer stacks.

Bettelheim and Ehrlich (1963) suggested that in a swelling
polymer, the hysteresis cannot be interpreted by capillary condensation. Rather hysteresis seems to depend on the mechanical
constraints contributed by the elastic properties of the material
and on the ease with which the polymer swells. In one case, typified by calcium chondroitin sulfate C, the large sorptive and
swelling capacity, associated with polymer chains weakly bound
between each other and weakly bound water, led to a small hysteresis loop. In the opposite case, typified by the calcium chondroitin sulfate A, a small sorptive and swelling capacity, associated
with a tightly bound matrix and strongly bound water, led to a
large hysteresis. In the latter case, the sharp maxima in the
entropy and enthalpy curves indicate strong interchain attractions.

In general, hysteresis seems to be the net result of reinforc-
ing or competing variables between adsorbend and adsorbate.

From the standpoint of the adsorbate, important factors in
hysteresis are the hydrogen-bonding ability, the amount adsorbed,

and the molar volume (Benson and Richardson, 1955). Adsorbates
of greater hydrogen-bonding ability (H_2O, EtOH) develop large hysteresis loops, but those of little or no hydrogen-bonding ability
(Et_2O and EtCl) give very small loops. For illustration, EtOH
causes a large hysteresis loop and EtCl a very small one, in spite
of the about equal size of the molecule and the very close dipole
moments of these adsorbates. Adsorbates that are adsorbed in
greater amounts cause greater hysteresis loops (greater total deformation of the protein molecule) than adsorbates adsorbed in
small amounts. A typical example is water, which is adsorbed in
relatively large amounts and which is strongly hydrogen bonding.
Important in hysteresis is the molar volume of the adsorbate.
Thus EtOH causes large hysteresis as does water, in spite of the
fact that comparatively fewer EtOH molecules are sorbed per gram
of protein then water. In this case, the network deformation per
molecule is larger for EtOH than H_2O. On the other hand, the
increasing bulk of the adsorbate molecule may result in decreased
sorption due to too large local deformation, where sorption becomes energetically unfavorable, due to the increased work required by the larger deformation. For example, the amount of
sorption decreases with increasing size in the R group of the MeOH,
EtOH, and i-C_4H_9OH.

F. *Thermodynamic Considerations*

The presence of a persistent hysteresis indicates that the
system, though reproducible, is not in true equilibrium. It is
doubtful whether either the adsorption or the desorption branch
of the isotherm has greater thermodynamic significance (Amberg,
1957; Amberg *et al.*, 1957). The question of validity of thermodynamic quantities calculated from such a system was brought into
sharp focus by the classical debate of La Mer (1967) with Bettelheim.

The isoteric heats of sorption calculated from the Clausius-

Clapeyron equation

$$\frac{d\ \ln\ P}{d\,(1/P)} = \frac{\overline{\Delta H}^0\ \text{isosteric}}{R}$$

and the subsequently calculated differential entropy of sorption

$$\overline{\Delta S}^0 = \frac{\overline{\Delta H}^0 - \overline{\Delta G}^0}{T}$$

refer to a reversible process between two defined initial and
final states, which the water vapor sorption process is not.
Hysteresis shows that the process is irreversible. Hence, there
is an entropy production in the system that indicates the degree
of irreversibility, including structural changes, occurring in
the sorbing solid matrix. Therefore, the isosteric heat and
hence the calculated "entropy of sorption" has a term $T\ \Delta S$ irreversible included and it does not strictly refer to heat transfer of the sorption process.

In order to evaluate the true heat effects, calorimetric experiments have to be performed. This measures the integral heats
of sorption at different water uptakes. In order to compare with
the isosteric heats of sorption, the differential heats of sorptions are calculated from the integral heats:

$$\left(\frac{\partial\ \Delta H}{\partial n}\right) = q_{\text{diff}}$$

The isosteric heats of sorption now will be equal:

$$\overline{\Delta H}^0_{\text{iso}} = q_{\text{diff}} + RT + T\ \Delta S_{\text{irrev}}$$

where $T\ \Delta S = q_{\text{irrev}}$ (Bettelheim, 1970). The RT term under room
temperature conditions amounts to 0.6 kcal/mole, a small correc-
tion. Thus, from a comparison of calorimetric and isosteric
heats, the entropy production of the system during the sorption
process can be calculated.

According to LaMer the Clausius-Clapeyron equation could be
used only to compare the heat of sorption q_{st}, calculated from

the equation, with the corresponding ΔH values obtained by direct
calorimetry. This makes it possible to calculate the magnitude
of the irreversibly created entropy and evolution of heat. The
symbols ΔH and ΔS should be reserved for thermodynamic quantities
calculated for a reversible process between defined states. The
application of the Clausius-Clapeyron equation to sorption data
should not be called a thermodynamic analysis, unless corrections
are made for loss of work and the resultant creation of entropy
or heat due to hysteresis.

Bettelheim agrees with the relative nature of the magnitude
of heat and entropy changes calculated on the basis of the
Clausius-Clapeyron equation using sorption data, and he restricts
its use to the definition of the monolayer by means of the maximum of the ΔS function. A maximum in this function implies the
completion of a "monolayer" (Block and Bettelheim, 1970). He
believes that for a swelling polymer network that eventually goes
into solution only the adsorption branch of the isotherm is
. meaningful, since the single reference state is the dry state.

III. PRACTICAL ASPECTS OF MOISTURE SORPTION HYSTERESIS IN FOODS

A. *Hysteresis and Chemical Deterioration*

Relatively very little work has been done on the relationships
between moisture sorption hysteresis and shelf life of foods.
This is probably due to the fact that hysteresis has been considered a phenomenon of theoretical interest, for the reasons
discussed above.

Strasser (1969) using his sorption isobar approach found the
isobar loop of freeze-dried beef generally decreases upon storage.
Figure 11 shows this change by plotting the temperature difference ΔT in the hysteresis vs. the moisture content of the sample.
The stored beef had a maximum of $\Delta T = 6.5°C$, whereas the unstored

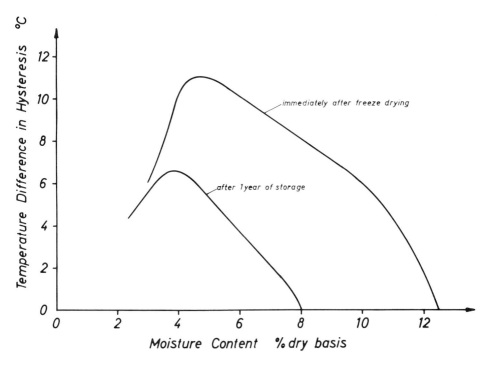

Fig. 11. Temperature difference in the sorption isobar hysteresis of cooked, freeze-dried beef as a function of the moisture content, before and after storage (Strasser, 1969).

product reached a maximal value of $\Delta T = 11°C$. Important was the observation that besides the change in the isobar hysteresis loop, a decrease of about 20% in the total water sorption capacity occurred during storage. This is reflected in the free energy change ΔF, shown in Fig. 12. Generally, the negative ΔF values decreased upon storage, with the difference between desorption and adsorption being greater in the sample immediately upon freeze-drying than after storage. A considerable decrease in the sorptive capacity of proteins and other sorbing moieties of the food has taken place as a result of storage.

Using the sorption isotherm approach, Wolf *et al.* (1972) investigated the usefulness of hysteresis as an "index of quality" of dehydrated foods. They found that storage resulted in a general increase of the hysteresis loop (Fig. 13). This was more pro-

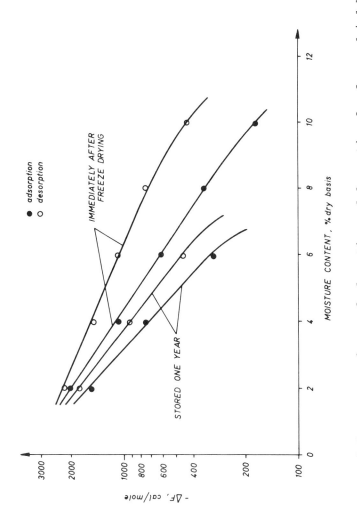

Fig. 12. Integral free-energy change of adsorption and desorption for freeze-dried beef (Strasser, 1969).

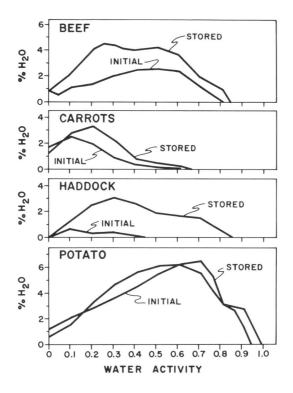

*Fig. 13.   The 7.2°C hysteresis values of beef, carrots, had-
dock, and potato plotted against water activity, initially and
after storage of samples at 37.7 C for 6 months (Wolf et al.,
1972).*

nounced in beef and haddock (high protein foods) in contrast to

the smaller increase in carrots and to the practically no change

in the potato. In the haddock a considerable change in the dis-

tribution of the hysteresis loop along the isotherm also occurred.

Table II indicates that foods that exhibited a large increase of

hysteresis upon storage showed a drastic decrease in the capacity

of the food to sorb water, resulting in an apparent increase of

TABLE II. *Comparison of Integrals before and after storage of Hysteresis, Adsorption, and Desorption Branches of the Isotherm[a]*

	Hysteresis		Initial minus storage	
	Initial	Storage	Adsorption	Desorption
Beef	1.32	2.11	2.38	1.50
Carrot	0.615	1.006	0.649	0.049
Haddock	0.115	1.374	1.715	0.455
Potato	3.50	3.60	0.15	0.17

[a]*From Wolf et al., 1972.*

the difference between desorption and adsorption. Sensory examination of the stored foods showed decreases in quality attributes of color, taste, and rehydratability.

The increase of the hysteresis loop upon storage when the data are obtained by the isotherm method is opposite to the decrease observed by Strasser (1969) using the isobar method. No clear explanation for this difference exists. Probably the two methods involve basically different physical processes and derive different types of information. In view of the importance of temperature in hysteresis, the isotherm method, depending on constant temperature, and the isobar method, depending on a change in temperature, are apt to give different results.

We know of no further sorption work based on the isobar method; this may be partly due to the sophisticated instrumentation required. More research along these lines, especially experiments performed on the same type of sample using the two methods, is clearly desirable.

The relationship between chemical deterioration and sorption hysteresis in intermediate moisture foods (IMF), was examined by Labuza and his co-workers. In a first paper on lipid oxidation and hysteresis (Labuza et al., 1972b) they showed that food systems are highly oxidizable at IMF conditions. In most cases, the desorption system is oxidized three to six times faster than the adsorption system. This was attributed to the better mobility of

catalysts as the available liquid volume increased, and to swelling, which exposes new catalyst sites. It was concluded that IMF
prepared by adsorption, although they may be more expensive, will
have a longer shelf life.

In subsequent papers (Chou *et al.*, 1973; Chou and Labuza,
1974; and Labuza and Chou, 1974) new findings and certain reversals in the expected role of water activity and water upon adsorption and desorption in model systems were explained in terms of
changes in viscosity and associated changes in effective metal ion
concentration and catalyst mobility and binding. This is illustrated in Fig. 14 (Labuza and Chou, 1974). At low metal content,
increasing a_w increases the rate of oxidation, and the system that
was prepared by desorption (DM), oxidizes faster than the system

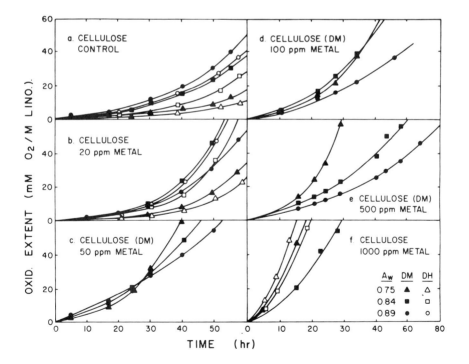

*Fig. 14. Oxidation extent at 35°C as a function of sorption
hysteresis for various trace metal contents. DM refers to desorption systems; DH to adsorption systems (Labuza and Chou, 1974).*

prepared by adsorption. At a metal concentration of 50 ppm all systems oxidize at about the same rate. At 500 ppm the oxidation pattern is completely reversed, with the rate of oxidation being faster at the lower water content and lower a_w. The results were explained in terms of a "dilution effect." At high metal concentration, increasing a_w and water content decreases the free metal concentration in contact with the lipid phase. This dilution factor more than compensates for the increased mobility due to the lowering of viscosity. Figure 15 shows that at similar water content and metal concentration there is a very little effect of hysteresis on the oxidation rate. Therefore, at high a_w it is the moisture content that controls oxidation rate.

With regard to the effect of hysteresis on other, nonlipid type of chemical deterioration in foods, published information is very limited. Warmbier (1974) studied the rate of nonenzymatic browning as a function of a_w and moisture content in systems prepared by adsorption and desorption. Although there was a substantial effect of a_w and water, with the rate increasing from the BET monolayer value to a maximum at $a_w = 0.43$ and then decreasing steeply, there was practically no effect of the hysteresis. Lee and Labuza (1975) reported a significant effect of hysteresis on ascorbic acid destruction, the half-life (inverse of the rate of destruction) being consistently lower for the system prepared by desorption in the range of $a_w = 0.32 - 0.93$ and temperature $I = 23° - 45°C$ studied.

B. Hysteresis and Microbiological Deterioration

The effect of a_w on microorganisms was reviewed by Scott (1957) and more recently by Troller (1973). Generally, there is a limiting a_w below which microorganisms will not grow. This depends on the species of the microorganism, type of food, presence of humectants, temperature, pH, and other factors. In this chapter we discuss only the effect of moisture sorption hysteresis on microorganisms.

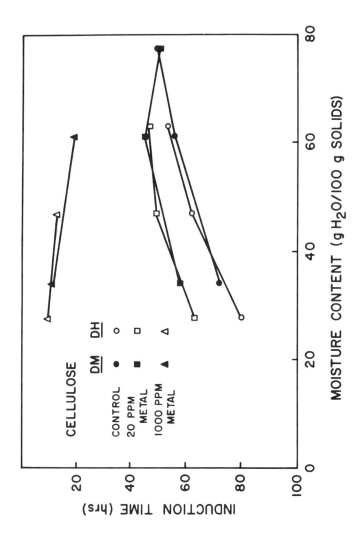

Fig. 15. Oxidation induction time as a function of moisture content for cellulose model systems (Labuza and Chou, 1974).

In general, microorganisms grow more rapidly at a water activity which has been reached by desorption than by adsorption.
As in the case of lipid oxidation, the greater amount of water
that is held by the food on desorption, promotes greater chemical
reactivity, including enzymatic activity involved in cellular
metabolism and division. The effect of water, in addition to the
a_w, on microorganisms was first reported by Labuza *et al.* (1972)
in studies of intermediate moisture foods. Table III from this
paper shows that the minimum water activity for growth of different organisms is lower in a system prepared by desorption (higher moisture content) than by adsorption (lower moisture content).
This can be taken advantage of in processing, by using the effect
of hysteresis (especially the higher a_w limits of growth in adsorption) to prepare stable intermediate moisture foods.

Figure 16 shows the effect of water activity and sorption hysteresis on the growth pattern of four microorganisms: (a) *Pseudomonas fragi,* (b) *Candida cypolytica,* (c) *Staphylococcus aureus,*
and (d) *Aspergillus niger* (Acott and Labuza, 1975).

Desorption (open symbols) is associated with a higher viable
fraction than adsorption (solid symbols). This is most pronounced
at a_w = 0.93 in *Pseudomonas fragi,* which is most sensitive to
stress, and at a_w = 0.79 for *Aspergillus niger.* It is less pronounced in the other two species. The difference between desorp-

TABLE II. *Summary of Minimum Growth a_w in IMF System[a]*

	Desorption	Adsorption
Banana IMF		
Mold	~0.68	>0.9
Pork IMF		
Mold	*0.75-0.68*	>0.9
Yeast	*0.84-0.75*	>0.9
Pseudomonads	*0.84-0.75*	>0.9
Staphylococci	*0.84-0.75*	>0.9

[a]*From Labuza et al., 1972a.*

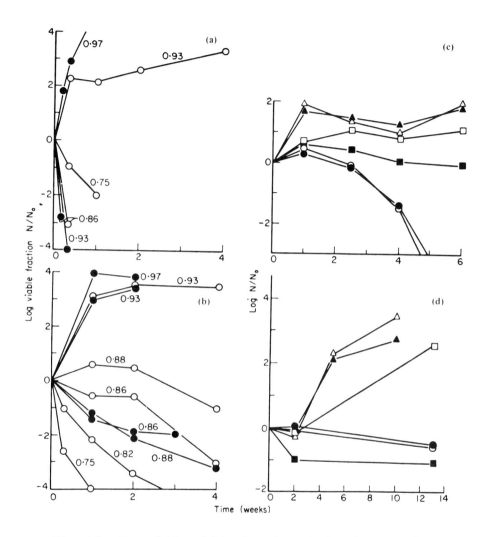

Fig. 16. Log of the viable fraction of the microorganism at
21°C in a chicken cube IMF system as a function of time at different $a_w^{\cdot}$. DM refers to a desorption prepared system; FDR refers
to an adsorption system. (a) Pseudomonas fragi; (b) Candida cypolytica; (c) Staphylococcus aureus; (d) Aspergillus niger. (a)
and (b) O , DM; ● , FDR. (c) A_W: 0.86: DM, O , FDR, ● ; 0.90:
DM, □ , FDR, ■ : 0.93: DM, Δ , FDR, ▲ . (d) A_W: 0.75: DM, O ,
FDR, ● ; 0.79, DM, □ , FDR, ■ ; 0.83, DM, Δ , FDR, ▲ . (c) and
(d), chicken cubes at 23°C (Acott and Labuza, 1975).

tion and adsorption seems to be greatest at the limit of growth
(0.88-0.90).

A similar effect of sorption hysteresis was reported by Plitman
et al. (1973) with regard to the viability of *Staphylococcus aureus*
in intermediate moisture strained chicken and pork dices. The
maximum a_w that inhibited growth in adsorption systems was higher
than 0.88, whereas a_w values lower than 0.88 were needed to inhibit growth in the desorption system of intermediate moisture pork
dices with glycerol as the water-binding agent. At the same a_w,
death rates were higher in adsorption than in desorption samples.
An important bactericidal effect of the humectants 1,2-propanediol
or 1,3-butanediol, which could not be explained solely by their
water binding properties, was observed.

IV. FUTURE RESEARCH

The foregoing discussion of the theoretical and practical aspects of moisture sorption hysteresis points to certain areas
where research is particularly desirable.

In the theoretical area a predictive, inclusive theory of
hysteresis to account for the multiplicity of phenomena of size
and shape of the loop, scanning paths, changes, and reversals, remains to be developed. This can lay the groundwork for much
practical work in the area of low and intermediate moisture foods,
where hysteresis may have an important bearing on chemical reactivity, microbial growth, and sensory acceptance.

In the practical area, further work is necessary on the role
of hysteresis in enzymatic activity, lipid oxidation, amino-car-
bonyl browning, and other reactions. There is an urgent need for
more information on the effect of hysteresis on the viability and
growth of individual microbial species.

The same is true with regard to the effect of hysteresis on
physical properties (see paper by Multon in this volume).

References

Acott, K. M., and Labuza, T. P. (1975). Microbial growth response
   to water sorption preparation. *J. Food Technol. 10*, 603.
Amberg, C. H. (1957). Heats of adsorption of water vapor on bovine
   serum albumin. *J. Am. Chem. Soc. 79*, 3980.
Amberg, C. H., Everett, D. H., Ruiter, L., and Smith, F. W. (1957).
   The thermodynamics of adsorption and adsorption hysteresis.
   In "Surface Activity," Vol. 2 (J. H. Schulman, ed.), p. 3.
   Butterworths, London.
Arnell, J. C., and McDermot, H. L. (1957). Sorption hysteresis.
   In "Surface Activity," Vol. 2 (J. H. Schulman, ed.), p. 113.
   Butterworths, London.
Benson, S. W., and Richardson, R. L. (1955). A study of hysteresis
   in the sorption of polar gases by native and denatured pro-
   teins. *J. Am. Chem. Soc. 77*, 2585.
Berlin, E., and Anderson, B. A. (1975). Reversibility of water
   vapor sorption by cottage cheese whey solids. *J. Dairy Sci.
   58*, 25.
Berlin, E., Anderson, B. A., and Pallansch, M. J. (1968). Water
   vapor sorption properties of various dried milks and wheys.
   J. Dairy Sci. 51, 1339.
Berlin, E., Kliman, P. G., Anderson, B. A., and Pallansch, M. J.
   (1971). Calorimetric measurement of the heat of desorption
   of water vapor from amorphous and crystalline lactose.
   Thermochim. Acta 2, 143.
Bettelheim, F. A., and Ehrlich, S. H. (1963). Water vapor sorption of mucopolysaccharides. *J. Phys. Chem. 67*, 1948.
Block, A., and Bettelheim, F. A. (1970). Water vapor sorption of
   hyaluromic acid. *Biochim. Biophys. Acta 201*, 69.
Brunauer, S. (1945). "The Adsorption of Gases and Vapors," Vol.
   1. Princeton Univ. Press, Princeton, New Jersey.
Chou, H. E., and Labuza, T. P. (1974). Antioxidant effectiveness
   in intermediate moisture model systems. *J. Food Sci. 39*, 479.
Chou, H. E., Acott, K., and Labuza, T. P. (1973). Sorption hys-
   teresis and chemical reactivity: Lipid oxidation. *J. Food
   Sci. 38*, 316.
Chung, D. S. (1966). Thermodynamic factors influencing moisture
   equilibrium of cereal grains and their products. Ph.D.
   Thesis, Kansas City Univ.
Cohan, L. H. (1938). Sorption hysteresis and the vapor pressure
   of concave surfaces. *J. Am. Chem. Soc. 60*, 433.
Cohan, L. H. (1944). Hysteresis and the capillary theory of ad-
   sorption of vapors. *J. Am. Chem. Soc. 66*, 98.
Enderby, J. A. (1955). The domain model of hysteresis. Part I.
   Independent domains. *Trans. Faraday Soc. 51*, 835.
Everett, D. H. (1955). A general approach to hysteresis. Part
   IV. An Alternative formulation at the domain model. *Trans.
   Faraday Soc. 51*, 1551.
Everett, D. H. (1967). Adsorption hysteresis. *In* "The Solid-Gas
   Interface" (E. A. Flood, ed., p. 1055. Marcel Dekker, New York.

Everett, D. H., and Whitton, W. I. (1952). A general approach to hysteresis. *Trans. Faraday Soc.* 48, 749.

Gregg, S. J., and Sing, K. S. W. (1967). "Adsorption, Surface Area, and Porosity." Academic Press, New York.

Kraemer, E. O. (1931). *In* "A Treatise on Physical Chemistry" (H. S. Taylor, ed.), p. 1661. Van Nostrand, New York.

Labuza, T. P. (1975). Storage stability and improvement of intermediate moisture foods. Final Report (August 1974-August 1978), Contract NAS 9-12560. NASA, Food and Nutrition Office, Houston, Texas.

Labuza, T. P. (1976). Storage stability and improvement of intermediate moisture foods. Final Report (September 1975-September 1976), Contract NAS 9-12560, Phase IV. NASA, Food and Nutrition Office, Houston, Texas.

Labuza, T. P., and Chou, H. E. (1974). Decrease of linoleate oxidation rate due to water at intermediate water activity. *J. Food Sci.* 39, 112.

Labuza, T. P., Cassil, S., and Sinskey, A. J. (1972a). Stability of intermediate moisture foods. 2. Microbiology. *J. Food Sci.* 37, 160.

Labuza, T. P., McNally, L., Gallagher, D., Hawkes, J., and Hurtado, F. (1972b). Stability of intermediate moisture foods. 1. Lipid oxidation. *J. Food Sci.* 37, 154.

La Mer, V. K. (1967). The calculation of thermodynamic quantities from hysteresis data. *J. Colloid Interface Sci.* 23, 297.

Lee, S., and Labuza, T. P. (1975). Destruction of ascorbic acid as a function of water activity. *J. Food Sci.* 40, 370.

McBain, J. W. (1935). An explanation of hysteresis in the hydration and dehydration of gels. *J. Am. Chem. Soc.* 57, 699.

McLaren, A. D., and Rowen, J. W. (1951). Sorption of water vapor by proteins and polymers: A Review. *J. Polymer Sci.* 7, 289.

Mellon, E. F., and Hoover, S. R. (1951). Hygroscopicity of amino acids and its relationship to the vapor phase water absorption of proteins. *J. Am. Chem. Soc.* 73, 3879.

Mellon, E. F., Korn, A. H., and Hoover, S. R. (1948). Water absorption of proteins. II. Lack of dependence of hysteresis in casein on free amino groups. *J. Am. Chem. Soc.* 70, 1144.

Plitman, M., Park, Y., Gomez, R., and Sinskey, A. J. (1973). Viability of *Staphylococcus aureus* in intermediate moisture meats. *J. Food Sci.* 38, 1004.

Rao, K. S. (1939a). Hysteresis in the sorption of water on rice. *Curr. Sci.* 8, 256.

Rao, K. S. (1939b). Hysteresis loop in sorption. *Curr. Sci.* 8, 468.

Rao, K. S. (1941). Hysteresis in sorption. I-IV. *J. Phys. Chem.* 45, 500.

Rutman, M. (1967). The effect of surface active agents on sorption isotherms of model systems. M.S. Thesis, MIT, Cambridge, Massachusetts.

Scott, W. J. (1957). Water relations of food spoilage microorganisms. *Adv. Food Res.* 7, 83.

Seehof, J. M., Keilin, B., and Benson, S. W. (1953). The surface areas of proteins. V. The mechanism of water sorption. *J. Am. Chem. Soc. 75,* 2427.

Speakman, J. B., and Stott, C. J. (1936). The influence of drying conditions on the affinity of wool for water. *J. Textile Inst. 27,* T.186.

Strasser, J. (1969). Detection of quality changes in freeze-dried beef by measurement of the sorption isobar hysteresis. *J. Food Sci. 34,* 18.

Troller, J. A. (1973). The water relations of food-borne bacterial pathogens. A Review. *J. Milk Food Technol. 36,* 276.

Van Olphen, H. (1965). Thermodynamics of interlayer adsorption of water in clays. I. Sodium vermiculite. *J. Colloid Sci. 20,* 822.

Warmbier, H. C. (1974). Non-enzymatic browning of an intermediate moisture model food system. Ph.D. Thesis, Univ. of Minnesota.

Wolf, M., Walker, J. E., and Kapsalis, J. G. (1972). Water vapor sorption hysteresis in dehydrated foods. *J. Agr. Food Chem. 20,* 1073.

Zsigmondy, R. (1911). Structure of gelatious silicic acid. Theory of dehydration. *J. Anorg. Chem. 71,* 356.

EFFECT OF WATER ACTIVITY AND SORPTION HYSTERESIS ON RHEOLOGICAL BEHAVIOR OF WHEAT KERNELS

J. L. Multon
H. Bizot
J. L. Doublier
J. Lefebvre
D. C. Abbott

It is well known that moisture content of grains influences their rheological behavior. Because of our interest in grain fitness for milling (Shelef and Moshenin, 1969; Multon *et al.*, unpublished, 1974; Beullier, 1975), we conducted a study, with emphasis on water sorption hysteresis, the importance of which has apparently been neglected.

Water vapor sorption phenomena are expressed graphically by sorption isotherms; these are obtained by plotting moisture content vs. water activity (a_w), which is equivalent to the air relative humidity in equilibrium with the sample (Multon, 1977).

The adsorption curve (corresponding to a moistening process) always lies under the desorption curve (corresponding to a dehydrating process). This demonstrates that the adsorption-desorption process is not thermodynamically reversible and that hydration properties of a product depend on its history.

Our purpose in this work was to investigate differences in rheological properties of wheat grains, evaluated by uniaxial compression, in connection with water sorption hysteresis.

I. MATERIALS AND METHODS

A. *Biological Materials*

Wheat used for this experiment is the French variety Talent harvestedin 1977 in southern France.

B. *Methods*

1. *Moisture Content Determination*

The method used for ordinary moisture content measurement [ICC (1960-1975) standard 109-1 or ISO (1968) standard R-712] consists of drying ground wheat grains at 130°C, for 2 hr under atmospheric pressure. All moisture contents are expressed on a dry basis.

2. *Water Activity (a$_w$) Determination*

The A_w determination was made by measuring the water vapor pressure in equilibrium with a grain sample, after evacuation of most noncondensable gases, at -80°C, according to Bizot and Multon (1977), but without freezing in order to minimize sorption state disturbance.

3. *Measurement of Apparent Modulus of Elasticity of Grains*

According to Shelef and Moshenin (1967) and Arnold and Roberts (1969), we used the Hertz equation to compute an apparent modulus of elasticity, with respect to the geometrical charac-teristics of the kernels (length, width, and thickness). The general form of this equation, proposed for two convex-faced ma-terials, has been simplified in the following expression:

$$E = \frac{0.5P(1-\mu^2)}{D^{3/2}} \left(\frac{1}{R_1} + \frac{1}{R'_1} \right)$$

where E is the apparent modulus of elasticity, P the applied compressive load, μ the Poisson ratio, D the total grain deformation, and R_1, R_1' principal radii of the grain at the contact surface.

For each sample, each of 50 individual kernels was submitted to five successive identical compressions on a standard universal testing Instron machine. The loading rate was 0.05 cm/min and the maximum load reached 3 kg (for a sample of moisture content up to 20%), 1 kg or less for more humid and plastic grains.

Preliminary loading-unloading tests confirmed the earlier observation (Shelef and Mohsenin, 1967) that the deformation of wheat kernels within the linear deformation range consists of two parts: an elastic deformation and a residual deformation. The residual deformation reaches an essentially constant value in the third loading-unloading cycle. Thus, subsequent deformations represent elastic deformations. Therefore, data from only the last three deformations were used in the calculation of $\bar{E}$ for each grain tested. $\bar{E}$ (mean value with its standard deviation) was directly computed from the load deformation curves.

4. Milling Trials

Brabender (LST 7170) laboratory mill (designed for Zeleny test) was used for milling grains samples. After measuring flour yields (grams of flour per gram of total wheat, dry basis) and flour protein content, particle size distribution was studied with a Coultronic Coulter counter, in the range 4-160 μm.

II. EXPERIMENTAL PROCEDURE

After cleaning and sifting, grains from the middle-size fraction were tempered. Instead of an equilibration through the vapor phase in air-conditioned cabinets, we preferred a method that consisted of mixing wet and dry grains in various proportions. Such a procedure described by Hart (1964) and shown in Fig. 1 respects fairly well the hysteresis conditions of adsorp-

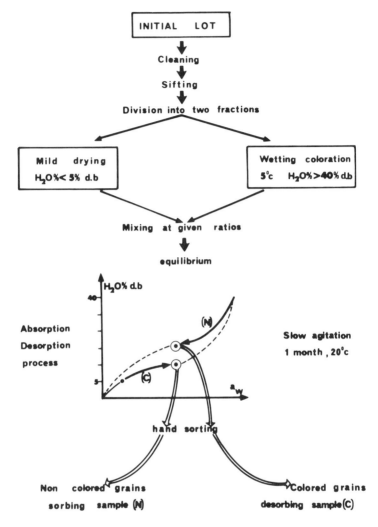

Fig. 1. Scheme of grain samples conditioning for rheological study.

tion and desorption, and avoids difficulties inherent in conditioning cabinets.

In practice, one-half of the initial lot of grain (16% d.b.) was dried with great care under vacuum without heating in order to avoid any physical, chemical, or thermal damage; the residual moisture content was brought to about 5%.

The other half was moistened by direct mixing with water and then mixing on an intermittent rotating blender for two weeks at 23°C. An antiseptic (HgCl$_2$, 20 mg/liter) to minimize bacterial and mold development and gentian violet (67.5 mg/liter) as a grain colorant were included in the moistening water. The final moisture content of the moistened grain was brought to about 33% d.b.

After equilibration was achieved in the moistened portion, five different blends were prepared (650 g total) by mixing dried and wetted grain in a tight container in definite ratios; after mixing, each sample was agitated slowly for a period of three weeks. By this procedure dry grains absorbed water (sorption process) while wet grains were redried (desorption process). The final expected moisture content of grains and the equilibrium relative humidity of air depended on the relative ratio of dry to wet grain in each mixture: these ratios were chosen to provide equilibrium relative humidity levels regularly distributed between 20 and 90%. The whole process was followed by periodic water content determinations on small aliquots of grain of both types from each mixture.

After three weeks, equilibrium was reached and for each mix small samples of noncolored (N) grains, in adsorption state, and colored (C) grains, in desorption state, were quickly hand-separated to achieve the following determination: (1) for (N) and (C) fractions of each mixture; moisture content measurement; a_w measurements and milling trials; and (2) on 50 individual kernels of each (N) and (C) fractions from each mixture, modulus of elasticity E (rheological behavior), length, width, and thickness measurements.

III. RESULTS AND DISCUSSION

A. *Adsorption-Desorption State of Wheat*

Figure 2 shows the adsorption-desorption isotherms (23°C) resulting from plotting sample moisture content vs. a_w for the absorbing (N) and desorbing (C) grains after hand sorting of the
noncolored absorbing (N) and colored desorbing (C) grains from
each mixture. If points 6N and 6C are disregarded, this diagram
shows the usual and well-known hysteresis loop comparable to results obtained by other methods, such as the isopiestic procedure

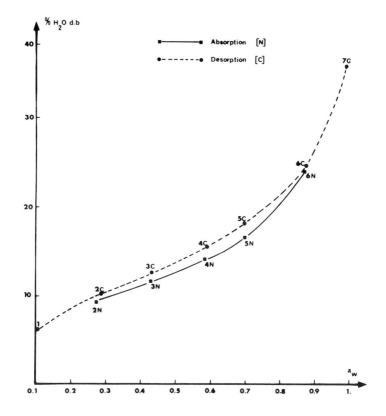

*Fig. 2.   Adsorption and desorption isotherms obtained by a_w
and moisture content measurement on conditioned grains.*

with sulfuric acid solutions or two pressure air-conditioned cabinets (Multon and Bizot, 1976; Multon, 1977). This indicates that
the desorbed colored grains (C) and the adsorbed noncolored grains
(N) have achieved a true adsorption and desorption "equilibrium"
essentially identical to those obtained by air tempering. The
proximity of the 6N and 6C points might indicate that our tempering
method does not respect desorption and adsorption states when high
equilibrium relative humidities are reached.

B. Rheological Behavior of (C) and (N) Grains

The plot of the apparent modulus of elasticity $(\bar{E})$ as a function of a_w for both desorbing (C) and adsorbing (N) grains is
shown in Fig. 3, while the plot of $\bar{E}$ as a function of water content is shown in Fig. 4. In both plots two different curves are
obtained: one concerning grains in the desorption state, the
other concerning grain in the adsorption state. In both cases
(adsorption and desorption), $\bar{E}$ decreases with increasing a_w and
moisture content.

In the 0.2-0.6 a_w range the apparent modulus of elasticity
for desorbing grains was about 13% less than for the adsorbing
grains. The student t test of differences showed significant differences over this range. This means that grains following a desorption process have a smaller apparent modulus of elasticity
than grains in a adsorption process for the same a_w or for the
same moisture content.

Quantitatively the decrease of apparent modulus of elasticity
between sorption and desorption (about 13%) is equivalent of a
water activity increase of 0.25 in either state or a water content
increase of about 3.5%.

Beyond $a_w = 0.7$ hysteresis effect on rheological behavior is
apparently reversed, but this phenomenon requires further investigations to be confirmed. Let us notice however that this in-
version occurs simultaneously with the appearance of the "solvent
water" in the sense defined by Guilbot and Lindenberg (1960).

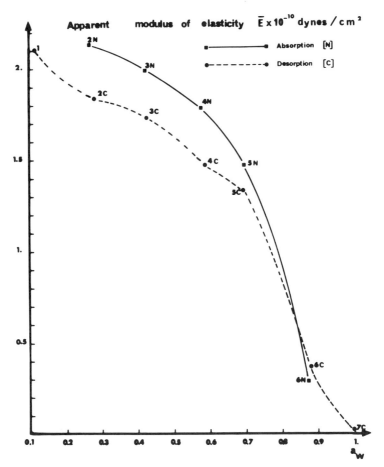

Fig. 3.   Changes of mean apparent modulus of elasticity $\bar{E}$ of wheat grains vs. a_w.

Most recent interpretations of rheological properties of wheat kernels are based on endosperm structure. More precisely, the state of the protein matrix and its continuity as a starch granules trap was recently related to grain hardness (Stenvert and Kingswood, 1977).

Referring to Fig. 2, at any given moisture content the a_w for absorbing grains is greater than the a_w for desorbing grains. This means that, since the total water content is the same, more

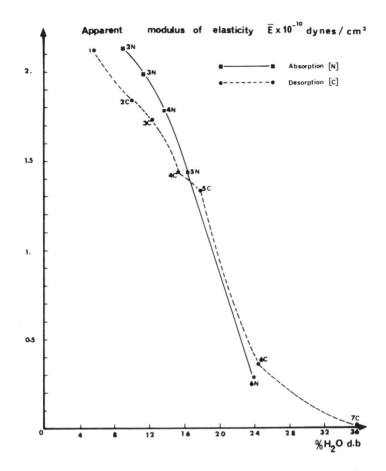

Fig. 4. Changes of mean apparent modulus of elasticity $\bar{E}$ of wheat grains vs. moisture content.

of this water is "bound" to the grain constituents (protein, starch, pentosans, etc.) in the desorbing grains. At the same time, from Fig. 4 we find that at the same given moisture content, the value of $\bar{E}$ is lower for the desorbing grains than for the adsorbing grains. It seems reasonable to say that the lower $\bar{E}$ in the desorbing grains is associated with the greater amount of "bound" water in these as compared to the adsorbing grains at this moisture content. The same conclusion is reached if one considers Figs. 2 and 3. According to Fig. 2, for the same a_w

the moisture content of the desorbing grains is higher than that for adsorbing grains, while for the same a_w the value of $\bar{E}$ for the desorbing grains is less than that for the adsorbing grains. From this one could conclude that the lower value of $\bar{E}$ for desorbing grains as compared to adsorbing grains at the same a_w is associated with a greater amount of "bound" water in the desorbing grains as reflected by their higher moisture (total water) content.

A more precise analysis of the apparent hysteresis effect on rheological properties has been attempted.

From an initial lot mildly dried (at 20°C, under vacuum, with P_2O_5 as desiccant) three grain samples were prepared by conditioning them in a cabinet the relative humidity of which was regulated by a saturated salt solution (Fig. 5):   (1) Sample S was submitted to direct adsorption, (2) sample SD was soaken to more than 40% moisture content, and subsequently desorbed to the same a_w as sample S, and (3) sample SDS resulted from redrying (as above mentioned) sample SD below 5% moisture content, and subsequent readsorption to the same a_w as S and SD.

Sample S (without soaking) gave the highest $\bar{E}$ value (even higher than the corresponding noncolored sample), while both samples SD and SDS (with soaking) gave approximately the same $\bar{E}$ value (much lower than the corresponding colored sample 2C) as shown in Table I.

TABLE I

Sample	a_w	$\bar{E} \times 10^{-10}$ dynes/cm^2
S	0.255	2.64
SD	0.288	1.01
SDS	0.289	0.97

Then the influence of the hydric pretreatment (i.e., soaking) seems to overlap the hysteresis effect.   This phenomenon might be related to the state of internal grain vitreousness, which was

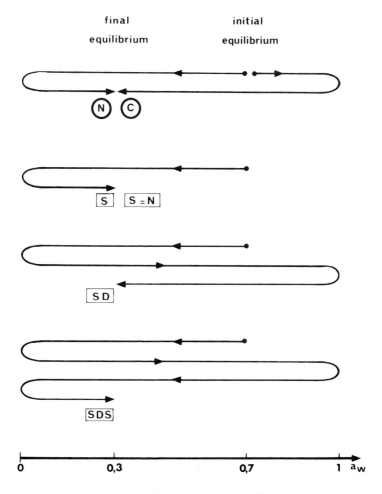

Fig. 5. Scheme of grain sample conditioning, with or without soaking.

judged visually on 50 transversal sections of grains: Sample S appeared clearly vitreous, while samples SD and SDS had a mealy aspect.

The same test led on samples N and C of the first trials is not as demonstrative, but colored samples (which also went through a soaking treatment) present a slightly but significantly more mealy cut than N samples, as shown in Fig. 6a.

Scanning electron microscopy observations (JEOL, JSM-50A) on
3N and 3C samples (Fig. 6b,c) have shown that more vitreous adsorbing kernels present a coherent protein matrix closely en-
traping starch grains (Fig. 6b), while desorbing mealy grains
present a looser structure with free starch grains and more frag-
mented proteins (Fig. 6c).

The influence of a liquid humidification has already been reported (Stenvert and Kingswood, 1977). It is considered that humidification up to high moisture levels followed by drying at low

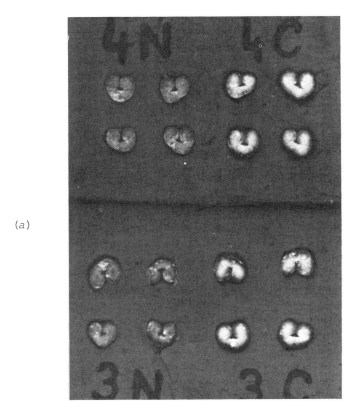

(a)

*Fig. 6. (a) Grain sections, showing a more vitreous appear-
ance for adsorbing grains (N), and a more mealy structure for desorbing grains (C). (b) Scanning electron microscope observations
of 3N sample. (c) Scanning electron microscope observations of 3C
sample.*

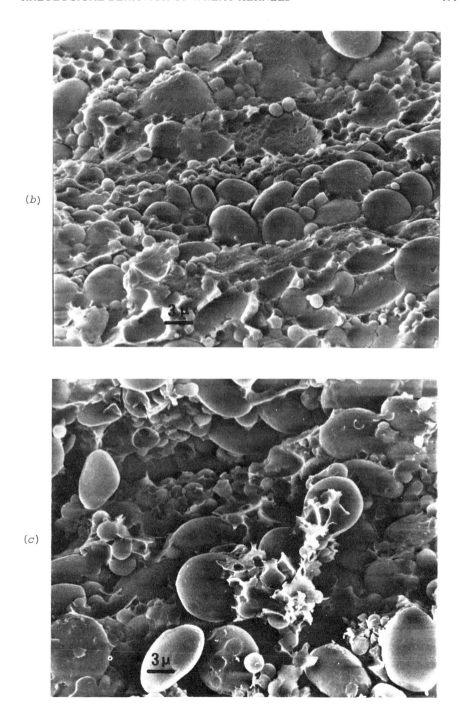

(b)

(c)

Fig. 6b,c

temperatures modifies the structure of internal protein matrix.
Dense, vitreous, hard grains become softer, more mealy, and less
dense when returned to the same water content after soaking; these
results are independent of proper sorption hysteresis cycle consideration; they outline the great importance of the soaking phase
in the conditioning, but it should be also interesting to investigate similar processes, without any soaking.

Moreover it has been reported that changes that occur in the
protein matrix during drying depend on the kinetic and the temperature of the process (Stenvert and Kingswood, 1977).

Consequently for further experiments repeated adsorption-
desorption cycles should be studied from rheological and hydration
points of view. Starch-protein interactions should also be taken
in consideration.

C. Milling Trials

The consequences of $\bar{E}$ changes in connection with water
sorption state have been studied on milling behavior. After
hand sorting all N and C samples in the range $a_w = 0.40\text{-}0.70$,
about 100 g were ground in a laboratory mill, and flour yield,
flour protein content, and particle size distribution were
determined (Fig. 7). No clear correlation could be drawn from
two groups of only three points each, but general trends were apparent. For (C) desorbing grains, we observe that, for the same
a_w as (N) grains: (1) Flour yields are higher (about 14% more in
relative value), (2) average particle size is lower (about 35%
less in relative value), (3) flour protein content is lower (about
4% less in relative value), (4) apparent modulus of elasticity is
lower, as shown above (cf. Fig. 3).

Differences observed between N and C grains cannot be solely
attributed to modification of the apparent modulus of elasticity,
since at equal $\bar{E}$ values desorbing grains give higher flour yields
than adsorbind grains (Fig. 8).

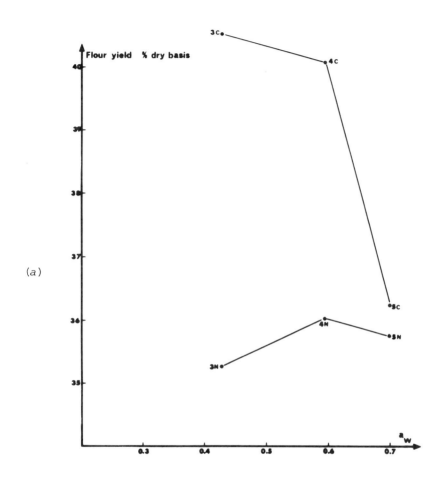

Fig. 7. Milling trials: (a) flour yield, (b) flour average particle size, (c) flour protein content, vs. a_w for adsorbing and desorbing samples.

To sum up, we can say that, in the related experimental procedure, N adsorbing grains (more vitreous) give less flour with
larger particles and more proteins than C desorbing grains, these
having a lower modulus of elasticity.

Since protein content is higher in the outer layers of grain
(Pyler, 1974) we may consider that the first stage of industrial
milling (comparable to milling for the Zeleny test) could be

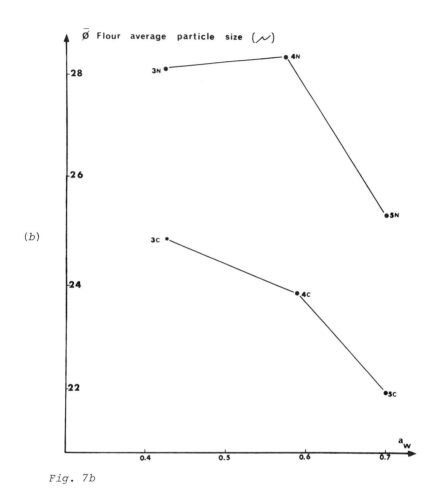

Fig. 7b

easier for C desorbing grains; the core of the kernels is readily ground to particles smaller and poorer in protein.

Concerning sorbing grains, the more homogeneous and vitreous structure, with higher $\bar{E}$ value, may render grinding more difficult, thereby affecting proportionately more the peripheral layers of grain; the flour resulting has a higher protein content and larger average particle size.

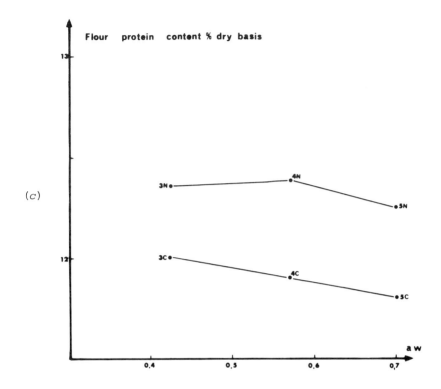

Fig. 7c

IV. CONCLUSION

This short study must be considered just as a first approach
to the effect of sorption hysteresis on rheological behavior of
foods. We have clearly shown that the sorption history (adsorption or desorption) affects the apparent modulus of elasticity
of wheat grains, and consequently their milling behavior.

This result is also important in solids rheology, since it
appears necessary to know the hydric history and the sorption
characteristics of the product when measuring a rheological property, or when comparing different rheological behaviors. In other
words, the sorption state measurements must be part and parcel of
the experimental procedure of solid food rheological determinations.

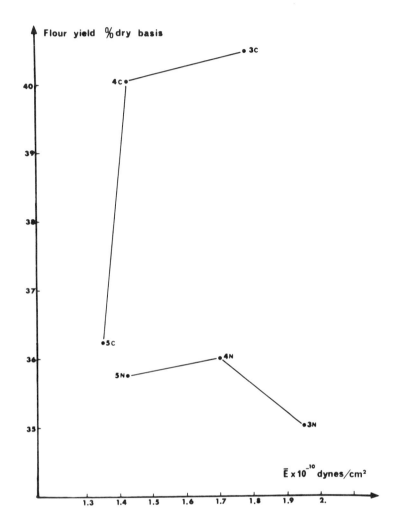

Fig. 8. Flour yield vs. apparent modulus of elasticity for adsorbing and desorbing samples.

A lot of questions remain, particularly the effects of maturation and aging of grains, the role of different parts of the grains (bran, aleurone layer), and the functional role of different components (e.g., protein, starch, outer layer components).

Taking into account the results we have obtained more information on mechanical behavior related to sorption state will certainly be helpful in food texture problems, particularly in milling technology and preparation of intermediate moisture foods.

Acknowledgments

The authors express their thanks to Dr. Gallant for ESM observations, to Mrs. Riou and Miss Gadet for their technical collaboration, and to Mr. Varoquaux, who loaned us an Instron apparatus.

References

Arnold, P. C., and Roberts, A. W. (1969). Fundamental aspects of load-deformation behavior of wheat grains. *Trans. Am. Soc. Agri. Eng.* 104-108.

Beullier, G. (1975). Contribution aux méthodes de dosage de l'eau dans les grains et graines; application au dosage de l'eau dans les populations de grains de maïs. Ph.D. Thesis, Univ. of Paris-Sud (Orsay).

Beullier, G., and Multon, J. L. (1976). Distribution statistique des teneurs en eau individuelles entre les grains de maïs: méthode d'étude et exemples d'application pratique. *Ann. Technol. Agri. 25*, 1-27.

Bizot, H., and Multon, J. L. (1977). Méthode de référence pour la mesure de l'activité de l'eau dans les produits alimentaires. *Ann. Technol. Agri.*, to be published.

Guilbot, A., and Lindenberg, A. (1960). Eau non solvante et eau de sorption de la cellule de levure. *Biochem. Biophys. Acta 39*, 389-397.

Hart, J. R. (1964). Hysteresis effects in mixtures of wheat taken from the same sample but having different moisture contents. *Cereal Chem. 41*, 340-350.

International Association of Cereal Chemists (1960-1975). Moisture Content Determination. Standard 110-1.

I.S.O. (1968). Moisture content determination in wheat. Standard R-712.

Multon, J. L. (1977). The state of water in less hydrated biological and natural products. *7th Cours Int. Lyophil.*, INSAM, Lyon.

Multon, J. L., and Bizot, H. (1976). Method for studying water vapour sorption and water activity of materials. *In* "Biodeterioration Investigation Techniques" (Walter, A., ed.). Applied Interscience Pub. Ltd., London.

Multon, J. L., Beullier, G., and Martin, G. (1974). Determination of moisture content distribution between maïze grains: reference method and example of application. *AACC 59th Annu. Meeting, Montreal.*

Pyler, E. J. (1974). "Baking Science and Technology," Vol. I, pp. 286-366. Siebel Publ. Co., Chicago.

Shelef, L., and Mohsenin, N. N. (1967). Evaluation of the modulus
    of elasticity of wheat grain. *Cereal Chem.* *44,* 392.
Shelef, L., and Mohsenin, N. N. (1968). Effect of moisture con-
    tent on mechanical properties of corn horny endosperm.
    Cereal Chem. *40,* 242-252.
Stenvert, M. L., and Kingswood, K. (1977). The influence of the
    physical structure of the protein matrix on wheat hardness.
    J. Sci. Food Agri. *28,* 11-19.

EVALUATION OF CRITICAL PARAMETERS FOR DEVELOPING
MOISTURE SORPTION ISOTHERMS OF CEREAL GRAINS

Egbert E. Neuber

A limited amount of critical work has been conducted to standardize techniques and procedures for obtaining reliable and reproducible moisture sorption isotherms (Neuber, 1976; 1977). Further
major efforts will be required to develop standardized methods
that will permit comparable results to be obtained in different
laboratories. Using corn and other cereal grains as examples, the
following subjects are considered in the present text:

(1) characteristics of cereal grains;

(2) published sorption isotherms for corn;

(3) critique of previous work;

(4) a definition for sorption isotherms;

(5) factors influencing equilibrium moisture content; and

(6) future outlook.

I. CHARACTERISTICS OF CEREAL GRAINS

Plant products, such as cereal grains, each have individual
characteristics with respect to hygroscopicity, volume changes
during moisture sorption, chemical reactivity, particle shape and
size, surface area and properties, and physical structure. The
precise determination of bound water is complicated by the varying
characteristics of these natural products.

A. Classification

Cereal grains and oilseeds are characterized by their proportions of primary constituents: (a) flour (for wheat, rye, oats,
barley, corn, rice, and millet), (b) protein (peas, lupines, and
beans), and, (c) oil (rape-seed, mustard, poppy, and sunflower).

Corn can be used to illustrate the problems involved in the
determination of the sorption isotherms for cereal grains.

B. Composition

Corn is a heterogeneous material, the main components of which
are endosperm - ∿85% of the total, embryo - ∿10% of the total, and
pericarp - ∿5% of the total. Figure 1 shows that the physical
composition of a kernel of corn is independent of moisture content.
During drying, the components are increased proportionately with
decreasing moisture content of the whole grain (Fig. 2). The
larger range of dispersion of the pericarp is explained by the
large surface area of this component and the difficulties involved in separating components and determining moisture.

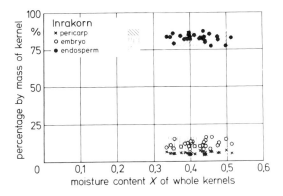

Fig. 1. Weight percentages of corn kernel components.

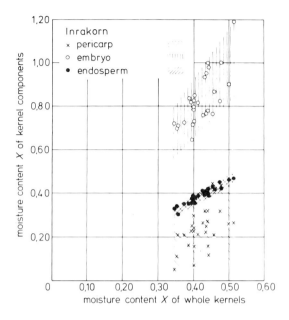

Fig. 2. Relationship between moisture content X of corn kernel components and moisture content X of whole kernels.

C. Release of Moisture

Release of moisture from kernels does not take place through the entire surface because of the waxy coat called the cuticule. The cuticule is impermeable to moisture if it has not been damaged. As a consequence, the water escapes from the kernels through conduction systems used for storage. From the interior of the kernel, water moves by osmosis and diffusion (if there are no cracks) until it has passed through the aleurone layer. According to Wolf (1952), for the major part, water first enters the kernel through the pericarp. The inner cellular spaces of the tubelike cells and cross-cells, as well as the dotted cells of the mesocarp, permit capillary water transport.

Water escapes outward from the kernel, for the most part, by capillary movement through the conducting vessels (such as the tracheas or tracheides) and, to a lesser extent, through

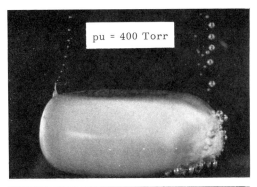

VARIETY:
    INRAKORN
PRE-TEATMENT:
    NATURALLY MOIST
HUMIDITY RATIO:
    U = 33,7 %
ATMOSPHERIC
PRESSURE:

    p = 721,5 Torr

*Fig. 3. Moisture released from corn components in oil sub-
jected to partial vacuum.*

the stigma rope extension. This means that the major part of
the water escapes at the kernel's connection to the spindle.
Figure 3 shows that the application of a vacuum to the grain in
an oil bath enables one to visualize the process. Figure 4 shows

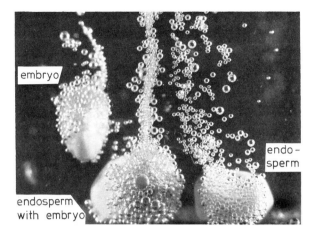

Fig. 4. Moisture released from corn components in oil under vacuum.

the various components of kernel, without the pericarp exposed to a vacuum in an oil bath. Water vapor escapes from nearly all parts of the surfaces.

II. PUBLISHED SORPTION ISOTHERMS FOR CORN

Many equilibrium moisture curves for the kernel and its components have been published since 1921: Bailey (1921), Coleman and Fellows (1925), Brockington *et al.* (1949), Sprenger (1953), Thomson and Shedd (1954), Hubbard *et al.* (1957), Hall and Rodriguez-Arias (1958), Sebestyen (1958), Haynes (1961), Poersch (1963), Shelef and Mohsenin (1966), Chung and Pfost (1967), Gerzhoj (1967), Shove (1971), Ballschmieter (1967), Anonymous, see Heiss (1968), Golik (1968), van Twisk (1969), Dengler (1971), Helmbrecht (unpublished), Galambos (1972), CNEEMA, see Srour (1972), Berry and Dickerson (1973), Gustafson and Hall (1974), Kumar (1974), Anonymous (1974), and Multon and Bizot (1976); see Table I.

TABLE I. Published Moisture Sorption Isotherms for Corn and Its Components

year	author	country	material	type of SIT	temperature °F	°C
1921	Bailey,C.H.		kernels	sorption	77	25
1925	Coleman,D.A. Fellows,H.C.	USA	kernels	sorption		25-28
1949	Brockington,S.F. Dorin,H.C. Howerton,H.K.		kernels	sorption	80	26,7
1953	Sprenger,J	NL	kernels	desorption		12-25
1954	Thompson,H.J. Shedd,C.K.		kernels	+desorption sorption	20 32 50 70	-6,7 0 10 21,1
1957	Hubbard,J.E. Earle,F.R. Senti,F.R.		kernels	desorption sorption		25 30 35
1958	Hall,C.W. Rodriguez-Arias,J.H.	USA	kernels	desorption	40 60 86 100 122 140	4,4 15 30 37,8 50 60
	Sebestyen,E.J.		corn	desorption	0++ 70 140++	-17,8 21,1 60
1961	Haynes,B.C.		kernels	sorption	30 60 90 120	- 1,1 15,6 32,2 48,9
1963	Poersch,W.		pericarp fodder	desorption		25 40 60 80
				sorption		40
1966	Shelef,L. Mohsenin,N.N.	USA	kernels embryo endosperm	desorption sorption	74	23,3
	Chung,D.S. Pfost,H.S		kernels	desorption sorption		22 50
			perikarp gluten embryo			25 50
	Gerzhoj,A.P	SU	corn	desorption		0 20 25 30 50

+ average of desorption and sorption
++assumption : extrapolated figures

Table Ia

year	author	country	material	type of SIT	temperature °F	°C
1967	Shove, G.C.	USA	kernels	desorption	30 40 50	- 1,1 4,4 10
	Ballschmieter, H.M.B.	ZA	corn meal corn farina product	sorption		30
1968	anonym	GB	corn meal	sorption		25
	Golik, M.G.	SU	kernels cob teguments filaments floral-integuments	desorption	unknown	
1969	van Twisk, P.	ZA	corn meal	desorption sorption		30
1971	Dengler, W.	D	kernels, ground	desorption sorption		40
1972	Helmbrecht, G.		corn	desorption		15
	Galambos, J.	H	corn	desorption		5 15 30 38 50 60
1973	CNEEMA	F	corn	desorption		5 15++ 20 25 35++ 80
	Berry, M.R. Dickerson, R.W.		corn, coarse ground	sorption		25 45 65
1974	Gustafson, R.J. Hall, G.E.	USA	kernels kernels, halved	desorption	50 90 120 155	10 32,2 48,9 68,3
	Kumar, M.		kernels, ground kernels without embryo, ground embryo, ground	sorption	78 122	25,6 50
	anonym	F	corn	desorption		15 35++
1976	Multon, J.L. Bizot, H.		corn, ground	desorption sorption		20

++ assumption : extrapolated figure

Table Ib

At the time of this survey, 27 reports were found, excluding
those for corn starch. These reports come from nine countries:
United States, Netherlands, Soviet Union, South Africa, Great
Britain, Federal Republic of Germany, Hungary, and France. They

deal with desorption and sorption isotherms in the temperature
range of -17° to 80 C. [In one case, no mention was made of
temperature (Golik, 1968)]. In 10 of the 27 articles, the results
are presented without any procedural details.

III. COMMENTARY ON PREVIOUS STUDIES

A comparative study has been made of both the desorption and adsorption isotherms. Considerable variation is apparent among the
individual curves. Based upon the desorption isotherms shown in
Fig. 5, and taking for example a relative humidity of ϕ_L = 0, 157,
the difference between the highest and lowest values of equilibrium
moisture content is 33%. As much as 29% difference is noted at
ϕ_L = 0.8 RH. Three of the five workers provided few details of
procedure (Sebestyen, 1958; Gerzhoj, 1967; CNEEMA, see Srour,
1972). The variations found are traceable to the different
methods and parameters used for the determination of equilibrium
moisture content.

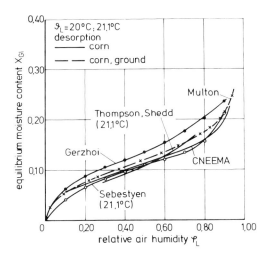

Fig. 5. Desorption isotherms for corn kernels.

IV. A DEFINITION FOR MOISTURE SORPTION ISOTHERMS

"A sorption isotherm is generally described as the relationship
between the total moisture content of the sorbent (material)
X_{GL} and the relative humidity of the surrounding air, at constant
temperature, when there is a state of thermodynamic equilibrium
between the sorbate (water) and the sorbent (material)."
This definition does not provide adequate information details
about the methods used in determining the isotherm. Therefore,
researchers have been forced to conduct their tests in an arbitrary
manner, and hence, there are major significant variations in
curves that have been published.

V. FACTORS INFLUENCING EQUILIBRIUM MOISTURE CONTENT

In order to achieve reproducible MSIs, it is necessary to establish all of the factors that influence equilibrium moisture
values (Tables II and III).

A. *Sorbent Factors Influencing Equilibrium Moisture in Corn Description of Sorbent*

Fifteen of the 27 published reports on corn isotherms present
descriptions of the materials used (Table IV). Twelve describe
the processing (preparation) of the material, eight specify the
variety (origin) of the corn, nine characterize the type of corn,
and seven indicate the crop year. Expressions such as "commercially available" were used by Coleman et al. (1925) and van Twisk
(1969). The method of processing the material is a very important
factor in determining sorption kinetics. There is a marked difference between corn types (e.g., dent versus flint) in quantitative and filling ratios between the horny and the floury endosperm

TABLE II. Factors Related to the Sorbent Affecting Moisture Equilibrium Values

description of sorbent	species condition stage of maturity variety (origine) processing grain type grain dimensions grain density crop year

composition of sorbent	mineral substances (ashes)
	organic substances:
	    raw protein
	    raw fat
	    raw fiber
	nitrogen-free extracts (N.F.E.): sugar
	                         starch

pre-treatment for desorption	storage
	before experiment : condition of sorbent
	                         temperature
	                         relative air humidity
	moistening : initial moisture content
	                         relative air humidity
	                         time of moistening
	                         water spraying
	moisture content before experiment

pre-treatment for sorption	method temperature time pressure
determination of dry matter	method temperature time pressure processing of sorbent quantity of sorbent

types. In addition, soft corn types contain predominantly floury starch, while sweet corn types have higher proportion of sugar in the endosperm.

B. *Composition of Sorbent*

Only one publication described the composition of the sorbent, and the description was incomplete (7% fat and 7% fiber) (van Twisk, 1969).

TABLE III. Measurement Factors Influencing Equilibrium Moisture Values

method of measuring	gravimetric : continuous
	                    discontinuous : static
	                                       dynamic
	manometric : continuous
	                    discontinuous : direct
	                                         indirect

method of air conditioning	water vapour pressure : regulating
	                                     measuring
	                                     determining

realisation of measuring	testing system condition of buoyant gas mass of specimen attaining of moisture equilibrium
tempering of experimental apparatus	test temperature variation of temperature time of tempering
execution of test procedures	differential integral
condition of moisture equilibrium	frequency of measuring definition of moisture equilibrium

C. Pretreatment for Desorption

Corn is obtained in its naturally moist state at harvest time. Since the test period for the determination of the sorption isotherms normally extends beyond harvest time, pretreatment is necessary. Initial moisture content was specified in only three of the previous studies and only five specified the method used for pretreatment and handling (Table V).

D. Pretreatment for Sorption

Eleven of the 27 studies specified pretreatments used prior to developing sorption isotherms (Table VI). Procedural differences can be seen in methods, temperatures, times, and pressures. Methods ranged from the use of natural drying to heated vacuum chambers. Temperatures ranged from ambient to 130°C. The length

TABLE IV. Characteristics of Corn Products Used in Developing Moisture Sorption Isotherms

author	material	processing	variety (origine)	grain type	crop year
Coleman, D A Fellows, H C	kernels, commercially available	whole		white dent yellow dent popcorn	
Brockington, S F Dorin, H C Howerton, H K	kernels	whole ground		yellow corn	1947
Thompson, H J Shedd, C K	kernels	whole		yellow dent	1938 1939
Hubbard, J E Earle, F R Senti, F R	kernels	whole ground	Schwenk 13 Dyar 444 Illinois 1277	yellow dent yellow dent white dent	1953 1954 1949
Hall, C W Rodriguez-Arias, J H	kernels			yellow dent	1954
Haynes, B C	kernels				
Poersch, W	pericarp (hulls from shelled kernels)		(German Maizena-Works, Krefeld)		
Shelef, L Mohsenin, N N	kernels embryo endosperm	whole	Pa 444	yellow dent	1964
Chung, D S Pfost, H B	kernels pericarp gluten embryo	whole	Dekalb 3×1 (Corn Products Company, USA)	dent	
Ballschmieter, H M B	corn meal corn farina product	special sieved			
van Twisk, P	corn meal, commercially available	unsifted sifted special sifted	(from various mills)		
Dengler, W	kernels	ground	Inrafruh	hybrid	1969
Berry, M R Dickerson, R W	kernels	coarse ground		yellow corn	
Gustafson, R J Hall, C E	kernels	whole halved	Dekalb XL-66		1971
Kumar, M	kernels kernels without embryo embryo	ground ground ground	WF9MST×H71 ×0·143RFXB 37RF (Robinson Hybrid Company, USA)		

of time ranged from 2 h to 7 days or more to obtain constant weight.   Those using vacuum heating did not report vacuum pressure.

E. Determination of Dry Matter

Only 14 studies reported how dry matter was determined.   Table VII summarizes the various procedures with respect to methodology, temperature, time, pressure, material processing, and the weight

TABLE V. Pretreatments Used for Desorption

author	storage before experiment			moistening				moisture content before experiment
	condition of material	temperature °C	relative air humidity	initial moisture content	relative air humidity	time h	spraying with water	
Hubbard, J.E. Earle, F.R. Senti, F.R.		5	0,6–0,64		0,97	~12		~0,299
Rodriguez–Arias, J.H. Hall, C.W.		4,4						0,254
Shelef, L. Mohsenin, N.N.				~0,13	0,97			
Chung, D.S. Pfost, H.B.					1,00	24		
Gustafson, R.J. Hall, G.E.	frozen							~0,400

TABLE VI. Pretreatment for Sorption

references	method	temperature °C	time h	pressure Pa
Coleman, D.A. Fellows, H.C.	heating chamber (water heated)			
Brockington, S.F. Dorin, H.C. Howerton, H.K.	natural drying			
Hubbard, J.E. Earle, F.R. Senti, F.R	vacuum heating chamber	72–76	72	
	drying above phosphorus pentoxide		to constant mass	
Haynes, B.C.	heating chamber	130	2	
Shelef, L Mohsenin, N.N.	vacuum chamber	room temperature		
Chung, D.S Pfost, H.B.	heating chamber	60	12	
Ballschmieter, H.M.B.	vacuum heating chamber	50		
van Twisk, P	vacuum heating chamber	50	72	
Dengler, W	evacuable sorption apparatus			0,00136
Berry, M.R. Dickerson, R.W.	heating chamber	~100	168	
Kumar, M.	heating chamber	75	72	

TABLE VII. *Determination of Dry Matter*

author	method	temperature °C	time h	pressure Pa	material	processing	mass g
Coleman, D.A. Fellows, H.C.	heating chamber (water heated)	100	to constant mass loss (120)		kernels	whole	
Brockington, S.F. Dorin, H.C. Howerton, H.K.	Brown-Duwel	190	0.33				100
	vacuum heating chamber, two stage	100 135	0.5 5		kernels	whole ground	20 2
Thompson, H.J. Shedd, C.K.	vacuum heating chamber	80	144	8 633	kernels	whole	~30
Hubbard, J.E. Earle, F.R. Senti, F.R.	heating chamber	130	1			ground	2
	heating chamber (water heated)		96		kernels		
Hall, C.W. Rodriguez-Arias, J.H.	heating chamber	100	72-80		kernels		
Haynes, B.C.	heating chamber	130	2		kernels	whole	
Poersch, W.	heating chamber	105-110	24		pericarp		
Shelef, L. Mohsenin, N.N.	heating chamber	103	72		kernels endosperm	whole	~2
					embryo		0.2-0.3
Chung, D.S. Ptost, H.S.	heating chamber single stage				gluten		
	heating chamber	50 130	24 2		kernels pericarp embryo	whole	
van Twisk, P.	vacuum heating chamber				corn meal	unsifted sifted special sifted	
Dengler, W.	evacuable sorption apparatus	105		0.00136	kernels	ground	
Berry, M.R. Dickerson, R.W.	heating chamber	100±15	48		kernels	coarse ground	
Gustafson, R.J. Hall, G.E.	heating chamber				kernels	whole halved	
Kumar, M.	heating chamber	103	72		kernels kernels without embryo embryo	ground	

used. The use of a heating chamber dominated. However, important differences can be observed: for example, water-heated, single-stage, two-stage, and combinations under vacuum. Corresponding temperatures and times ranged from 80°C to 190°C for 0.33 to 120 h. Furthermore, material processing varied. For a heterogeneous biological material such as whole corn, a sample of less than 2 g is unsatisfactory. In general, a heating chamber would be preferred since it gives reproducible results. Although the accuracy is lower, better reproducibility is achieved. In Germany the standard method for the determination of moisture content (DIN 10350) is not used by agricultural engineers who generally use a heating chamber at 105°C for 24 h. The development of this procedure took place between 1956 and 1968 and is presented in Table VIII.

TABLE VIII. Gravimetric Estimations of Moisture.

Temperature (°C)	Time (hr)	Material	Material
105	5	Lucerne Sugar beet leaves	Schneider, 1955
Storage above P_2O_5	4-6 (weeks)	Wheat Rape	Pichler, 1956
105	24	Potatoes	Görling, 1956
105	16	Cereal grains	Daiber-Kuhnke, 1956
105-110	24	Pericarp	Poersch, 1963
105	24	Grass	Tuncer, 1968

F. Measuring Procedure Factors Influencing Equilibrium Moisture

In order to be able to make reasonable comparisons, factors
that influence procedure variables in the determination of the
equilibrium moisture content must be properly characterized.

G. Method of Measuring

The method used for measuring moisture was specified in 15
studies (Table IX). The gravimetric rather than the manometric
method was predominantly used. Because of the biological nature
of the material, those methods that establish moisture equilibrium
in a short time are preferred. For example, in the case of the
discontinuous gravimetric methods, the dynamic, instead of the
static, method is preferred.

H. Method of Conditioning the Air

Fifteen of the 27 studies define, with respect to water vapor
pressure, the type of regulation used (Table X). Six indicated

TABLE IX. Methods of Vapor Pressure Measurement on Corn

author	method of measuring					
	gravimetric			manometric		
	contin-uous	discontinuous		contin-uous	discontinuous	
		static	dynamic		direct	indirect
Coleman, D.A. Fellows, H.C.			X			
Brockington, S.F. Dorin, H.C. Howerton, H.K.						X
Thompson, H.J. Shedd, C.K.		X				
Hubbard, J.E. Earle, F.R. Senti, F.R.		X		X		
Hall, C.W. Rodriguez-Arias, J.H.		X				
Haynes, B.C.				X		
Poersch, W.		X				
Shelef, L. Mohsenin, N.N.		X				
Chung, D.S. Pfost, H.S.		X				
Ballschmieter, H.M.B.			X[+]			
van Twisk, P.			X[+]			
Dengler, W.	X					
Berry, M.R. Dickerson, R.W.			X			
Gustafson, R.J. Hall, G.E.		X	X[+]			
Kumar, M.		X				

[+] turbulent air

how the water vapor pressure was measured, and 10 how it was determined.

I. Experimental Procedure

The way the experiments were conducted and the condition of
the carrier-gas are given in 15 of the studies (Table XI). Among
these, 12 indicated the quantity and the time at which equilibrium
moisture was attained. Advantages are offered by those methods

TABLE X. *Methods of Conditioning and Vapor Pressure Determination on Corn*

author	water vapour pressure		
	regulating	measuring	determining
Coleman,D A Fellows,H C	sulphuric acid solutions		Wilson,R E
Brockington,S F Dorin,H C Howerton,H K	quantity of water	electric hygrometer	water vapour tables of Weather Bureau, U S Department of Commerce
Thompson,H J Shedd,C K	satured salt solutions	psychometric	
Hubbard,J E Earle,F E Senti,F R	satured salt solutions		Carr,D S and Harris,B L O'Brien,F M
	Aminco-Dunmore-hygrometer	Tag-Heppenstall-hygrometer	Wexler,A and Hasegava,S Wink,W A and Sears,G R
Hall,C W Rodriguez-Arias,J H	satured salt solutions		
Haynes,B C	distilled water or drying at 43,3 °C (vacuum/no vacuum)	mercury manometer with cathetometer	
Poersch,W	sulphuric acid solutions	measuring of concentration	D'Ans,I and Lax,D E
Shelef,L Mohsenin,N N	satured salt solutions		Wexler,A and Hasegava,S Carr,D S and Harris,B L Wink,W A and Sears,G R
Chung,D S Pfost,H S	sulphuric acid solutions		Perry,J H
Ballschmieter H M B	sulphuric acid solutions		D'Ans,I and Lax,D E
	satured salt solutions		Carr,D S and Harris,B L O'Brien,F E M Wink,W A and Sears,G R
van Twisk,P	sulphuric acid solutions		Rockland,L B Richardson,G M and Malthus,R S O'Brien,F E M
	satured salt solutions		Carr,D S and Harris,B L Wexler,A and Hasegava,S Wink,W A and Sears,G R
Dengler,W	temperated quantity of water		
Berry,M R Dickerson,R W	spraying with distilled water	electric hygrometer	
Gustafson,R J Hall,G E	satured salt solutions		Wexler,A and Hasegava,S
Kumar,M	satured salt solutions		Wexler,A and Hasegava,S Carr,D S and Harris,B L Wink,W A and Sears,G R

that lead to the achievement of equilibrium moisture content in a short time.

TABLE XI. *Experimental Conditions and Measurement of Equilibrium Moisture*

author	testing system	condition of buoyant gas	mass of specimen g	attaining of moisture equilibrium		
				h	d	w
Coleman, D A Fellows, H C.	flow-through water and test containers	flowing	~60		6-8	
Brockington, S F Dorin, H C Howerton, H K	specimen container	motionless	~500	~2[*]		
Thompson, H J Shedd, C K	hygrostat	motionless	~100		209-446	
Hubbard, J E Earle, F R Senti, F R	evacuable desiccator	vacuum	2 (ground) 10 (whole kernels)		2-15	
	Aminco-Dunmore hygrometer	motionless	60	2-4[**]		
Hall, C W Rodriguez-Arias, J H	small chambers	motionless				>2
Haynes, B C	isotenoscope	vacuum	100	~7		
Poersch, W	evacuable desiccator	vacuum		>9		
Shelef, L Mohsenin, N N	evacuable desiccator	vacuum				1-3
Chung, D S Pfost, H B	evacuable hygrostats	vacuum	1,5-2,5		~5	
Ballschmieter, H M B	hygrostat with built-in agitator	turbulent	2 - 2,5			
van Twisk, P	hygrostat with built-in agitator	turbulent	2 - 2,5			
Dengler, W	electronic microbalance	motionless	~0,1	>14		
Berry, M R Dickerson, R W	air-conditioned test chamber	in motion	~20			
Gustafson, R J Hall, G E	hygrostat with or without built-in paddle agitator	in motion and motionless	30			
Kumar, M	desiccator	motionless				2-3

[*] previous assimilating for 16-24 h
[**] previous storage in moisture-proof container till moisture equilibrium reached

J. Tempering of the Experimental Apparatus

Nine studies report on the tempering of the experimental apparatus (Table XII). It can be seen that, whether by tempering with water or by air, the temperature variation remained about the same.

TABLE XII. Tempering Methods

author	temperature °C	variation of temperature °C	type of tempering
Coleman, D.A. Fellows, H.C.	25	+3	air tempered
Brockington, S.F. Dorin, H.C. Howerton, H.K.	26,7	±0,1	water tempered
Sprenger, J.	12-25		
Hubbard, J.E. Earle, F.R. Senti, F.R.	25 30 35	±0,15	water tempered
Hall, C.W. Rodriguez-Arias, J.H.	4,4 15 30 37,8 50 60	±0,6	
Poersch, W.	25 40	±0,3	air tempered
	60 80	±0,6	water tempered
Chung, D.S. Pfost, H.S.	22 25 50	±0,5	air tempered
Berry, M.R. Dickerson, R.W.	25 45 65	±0,5	thermostat
Gustafson, R.J. Hall, G.E.	10 32,2 48,9 68,3	±0,83	air tempered

K. Procedure for Preparing Isotherms

It is possible to distinguish between the integral, differential, and the integral-differential means (Fig. 6). Only a single sample, in the case of the differential method, is used for the determination of the sorption isotherm. Serious browning may result using the differential methods after long exposures at elevated temperatures.

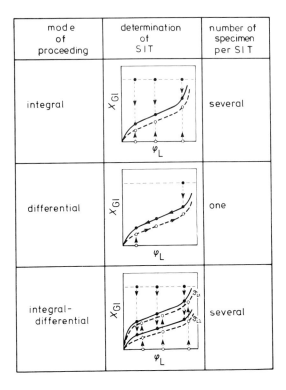

mode of proceeding	determination of SIT	number of specimen per SIT
integral		several
differential		one
integral-differential		several

Fig. 6. Procedures for Producing Sorption Isotherms

L. Execution of Test Procedures

Fourteen of the 27 studies report details of the test procedures (Table XIII) and in one study, Dengler (1971) gave a more
precise report. In some cases, the execution of the test procedures was not clearly stated.

M. Condition of Case at Moisture Equilibrium

The state of the material at equilibrium moisture was reported
in 11 of the studies (Table XIV). No standardized procedure was
used. According to the definition of equilibrium moisture, a constant weight is expected. The frequency of data collection varied
from one experimenter to another.

TABLE XIII. Procedures Employed for Preparing Sorption Isotherms

author	explanation of mode of proceeding	differential	integral
Coleman, D. A. Fellows, H.C.	—		X^+
Brockington, S.F. Dorin, H.C. Howerton, H.K.	—		X^+
Thompson, H.J. Shedd, C.K.	—		X^+
Hubbard, J.E. Earle, F.R. Senti, F.R.	—	X (mainly)	X^+
Haynes, B.C.	—	X^+	
Poersch, W.	—		X^+
Shelef, L. Mohsenin, N.N.	—		X^+
Chung, D.S. Pfost, H.S.	—		X^+
Ballschmieter, H.M.B	—		X^+
van Twisk, P.	—		X^+
Dengler, W.	X	X	
Berry, M.R. Dickerson, R.W.	—		X^+
Gustafson, R.J. Hall, G.E.	—		X^+
Kumar, M.	—	X^+	

[+] interpretated from the text

TABLE XIV. Conditions Employed for Determining Equilibrium Moisture Content

author	frequency of measuring	definition of moisture equilibrium
Coleman, D.A. Fellows, H.C.	daily	constant mass over 24 h
Brockington, S.F. Dorin, H C Howerton, H.K.	quarter-hourly	no change in relative air humidity
Thompson, H.J. Shedd, C.K.		constant mass
Hubbard, J.E. Earle, F.R Senti, F.R.		no change in relative air humidity over 1 h
Hall, C.W. Rodriguez-Arias, J.H.	periodic	constant mass
Haynes, B.C.	successive	no change in water vapour pressure
Shelef, L. Mohsenin, N.N.	periodic	constant mass
Ballschmieter, H.M.B.		constant mass
van Twisk, P.	weekly	constant mass
Dengler, W.	continuous	
Gustafson, R.J. Hall, G.E.	every three days	change in moisture content of < 0,001 over 3 days

M. Future Outlook

In order to arrive at uniform and comparable results, it will be necessary to agree upon and standardize the factors that influence equilibrium moisture curves.

References

Anonymous:Conservation du mais (1974). Nr. 112, *la documentation agricole BP.*
Bailey, C. H. (1921). Respiration of shelled corn. *Tech. Bull. 3*, The University of Minnesota Agricultural Experiment Station.
Ballschmieter, H. M. B. (1967). Die sorptionsisothermen von mehlen, mehlprodukten und anderen lebensmitteln bei 30°C. *Getreide und Mehl 10*, 118-120.

Berry, M. R., and Dickerson, R. W. (1973). Moisture Adsorption Isotherms for Selected Feeds and Ingredients. *Trans. ASAE,* 137-139.

Brockington, S. F., Dorin, H. C., and Howerton, H. K. (1949). Hygroscopic equilibria of whole kernel corn. *Cereal Chem. 26,* 166-173.

Chung, D. S., and Pfost, H. B. (1967). Adsorption and desorption of water vapor by cereal grains and their products. Part I: Heat and free energy changes of adsorption and desorption. *Trans. ASAE 10,* 549-551, 555.

Coleman, D. A., and Fellows, H. C. (1925). Hygroscopic moisture in cereal grains and flaxseed exposed to atmospheres of different relative humidity. *Cereal Chem. 2,* 275-287.

Daiber-Kuhnke, U. (1959). Das Feuchtigkeitsgleichgewicht von Luft und Getreide bei der Behältertrocknung. *Landtechnische Forschung 9,* 106-110.

Dengler, W. (1971). Praktische Bedeutung von Wasserdampfsorptions-isothermen. Messung von Sorptionsisothermen. *Chemie - anlagen + verfahren 4,* 68-75.

Galambos, J. (1972). *A Mezögazdasági termelési folyamatok gépesitése.* Mezögazdasági Kiado, Budapest.

Gerzhoj, A. P. *Grain Drying and Grain Driers* (origin Russian) (1967). Publishing-house EAR, Moscow.

Golik, M. G. (1968). *Storage and processing of corn ears and corn grains* (origin Russian). Publishing-house EAR, Moscow.

Görling, P. (1956). Untersuchungen zur Aufklärung des Trocken-verhaltens Pflanzlicher Stoffe. *VDI - Forschungsheft 458.* VDI - Verlag, Düsseldorf.

Gustafson, R. J., and Hall, G. E. (1974). Equilibrium Moisture Content of Shelled Corn From 50 to 155°F. *Trans. ASAE,* 120-124.

Hall, C. W., and Rodriguez-Arias, J. H. (1958). Equilibrium moisture content of shelled corn. *Agric. Eng. 39, 8,* 466-470.

Haynes, B. C. (1961). Vapor pressure determination of seed hygroscopicity. *Tech. Bull. Nr. 1229,* Agricultural Research Service United States Department of Agriculture in Cooperation with College Experiment Station, University of Georgia.

Heiss, R. (1968). *Haltbarkeit und Sorptionsverhalten wasserarmer Lebensmittel.* Springer-Verlag, Berlin, Heidelberg, New York.

Helmbrecht, G. Betrachtungen zur Getreidetrocknung. *MIAG - Speicher- und Umschlagstechnik,* 37-48.

Hubbard, J. E., Earle, F. R., and Senti, F. R. (1957). Moisture relations in wheat and corn. *Cereal Chem. 34, 11,* 422-432.

Kumar, M. (1974). Water vapour adsorption on whole corn flour, degermed corn flour and germ flour. *J. of Food Tech. 9,* 433-444.

Multon, J. L., and Bizot, H. (1976). Methods for studying water vapour sorption and water activity of materials. *Appl. Intersci. Pub.,* 149-183.

Neuber, E. E. Das Sorptionsverhalten von mais unter besonderer berücksichtigung der Einflußgrößen beim sorbens sowie beim

Vorgang des Messens. Dissertation der Universität Hohenheim (in preparation).

Neuber, E. E. (1976). Vorgehensweise beim Aufnehmen von Sorptionsisothermen für Körnerfrüchte. Arbeitssitzung des VDI-GVC (Gesellschaft für Verfahrenstechnik und Chemie-ingenieurwesen)-Unterausschusses PHYSIKALISCHE DATEN VON LEBENSMITTELN, Karlsruhe.

Neuber, E. E. (1977). Kritische betrachtungen z-m aufnehmen von sorptionsisothermen für körnerfruchte, dargestellt am feuchtegleichgewicht von körnermais. Internationale Tagung der VDI Fachgruppe LANDTECHNIK, Braunschweig.

Pichler, H. J. (1956). Sorptionsisothermen für Getreide und Raps. *Landtechnische Forschung 6, 2,* 47-52.

Poersch, W. (1963). Sorptionsisothermen und ihre ermittlung und auswertung. *Die Stärke 15, 11,* 405-412.

Schierbaum, F. (1960). Die hydration der stärke. *Die Stärke 12, 9,* 257-265.

Schneider, A. (1954). Untersuchungen über das charakteristische Trocknungsverhalten von Luzerne und Zuckerrubenblatt in Einzelschichten und durchströmten Schüttungen. Dissertation TH München.

Sebestyen, E. J. (1958). Aus theorie und praxis der getreidetrocknung. *Die Müllerei 41,* 558-559.

Shelef, L., and Mohsenin, N. N. (1966). Moisture relations in germ, endosperm and whole corn kernel. *Cereal Chem. 43, 5,* 347-353.

Shove, G. C. (1971). Die kühllagerung von mais und die kühltrocknung. *Die Muhle 33,* 483-484.

Sprenger, J. (1953). Einige aspekte der getreidetrocknungsanlagen, unter besonderer berücksichtigung ihrer verwendbarkeit für die verschiedenen arten landwirtschaftlicher pordukte. *Landwirtschaftliche Veröffentlichung der O.E.E.C.,* 27-34.

Srour, S. (1972). La conservation du mais par la ventilation. Le mais grain - récolte, réception, séchage, conservation, qualité. *Inst. Tech. Cereales Fourrages,* 56-66.

Thompson, H. J., and Shedd, C. K. (1954). Equilibrium moisture and heat of vaporization of shelled corn and wheat. *Agric. Eng. 35, 11,* 786-788.

Tuncer, J. K. (1968). Versuche zur Ermittlung des Trocknungsverhaltens einiger deutscher Futtergräser, Dissertation Universität Göttingen.

van Twisk, P. (1969). The sorption isotherms of maize meal. *J. Food Techn. 4,* 75-82.

Wolf, M. J., Buzan, C. L., MacMasters, M. M., and Rist, C. E. (1952). Structure of mature corn kernel, I-IV. *Cereal Chem. 29,* 321-382.

WATER BINDING ON STARCH:

NMR STUDIES ON NATIVE AND GELATINIZED STARCH

Hans T. Lechert

I. INTRODUCTION

The application of NMR spectroscopy to problems of the behavior of water in biological systems has its origin in the study of molecular structure and mobility (Shaw and Elsken, 1953).

For investigations of the binding of water especially, methods are used that give information on the mobility of the water protons (Bloembergen *et al.*, 1948; Resing, 1968; Pfeifer, 1972; Zimmerman and Brittin, 1957).

As the most powerful method for examining systems with very low water contents, the NMR pulse technique which gives the longitudinal and the transverse relaxation times T_1 and T_2 of the nuclear magnetization parallel and perpendicular to the applied magnetic field, must be considered.

T_1 and T_2 depend directly on the mobility of the molecules, which is usually measured by the so-called correlation time τ_c. $1/\tau_c$ can be regarded as the mean jump frequency of the motional process. The connection between T_1 and T_2 and τ_c is demonstrated in Fig. 1. The scale of the abscissa is proportional $1/T$, because generally an Arrhenius law can be expected for τ_c (Bloembergen *et al.*, 1948).

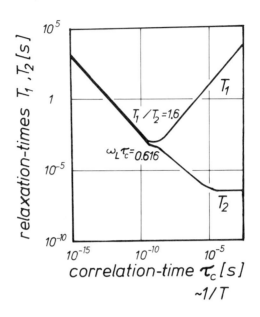

Fig. 1. Dependence of the nuclear relaxation times T_1 and T_2 on the correlation time τ_c, according to the theory of Bloembergen et al. (1948).

At the minimum of T_1, $\omega_L\tau_c = 0.616$ holds, where ω_L is the measuring frequency. From the slope of the high-temperature branch of the dependence of log T_1 on $1/T$, the activation energy of the process of the molecular mobility can be estimated in the case of an isotropic motion with only one τ_c. This situation is characterized by a value of $T_1/T_2 = 1.6$ at the T_1 minimum. For any deviation from the isotropic motion, T_1/T_2 increases. An increase is also observed if a distribution of τ_c is present, which occurs mostly for molecules in sorbed states (Resing, 1968; Pfeifer, 1972). In this case a mean activation energy can be obtained by measuring the T_1 minimum at different frequencies and calculating the respective τ_c from $\omega_L\tau_c = 0.616$. Often, from the magnetization decays two or more relaxation times can be observed, corresponding to regions with different molecular mobility.

The influence of exchange effects of nuclei between those regions and of different lifetimes have been thoroughly studied by Zimmermann and Brittin (1957), Woessner (1961), and Woessner and Zimmermann (1963).

II. RELAXATION TIMES OF WATER IN STARCHES

Figure 2 shows T_2 for different water contents in maize, rice, and potato starch. Up to about 15% water, only a single T_2 occurs. Above this water content two T_2 can be obtained from the decay of the transverse magnetization. The point of the appearance of the more mobile kind of water lies for all starches near the water content, for which, by sorption experiments, so-called capillary water can be detected (Lechert, 1976; Lechert and Hennig, 1976; Hennig and Lechert, 1976).

In contrast to this, only a single T_1 is observed over a large range of temperature for all starches. For water contents above 15%, T_1 is in no case below 80 msec, which means that the molecular exchange between the two regions, being responsible for the two T_2, must have a rate faster than T_1 and slower than T_2. Minima of T_1 lie for different water contents in the range between 220 and 260 K. The values of T_1/T_2 at the minima are larger than 50 and indicate broad distributions of τ_c. Because the potato starch shows the most distinct effects, more detailed experiments were carried out with this starch.

First, the frequency dependence of the minimum temperature of T_1 was measured. From τ_c, obtained by $\omega_L \tau_c = 0.616$, apparent activation energies between 30 and 50 kJ/mole can be estimated. The large spread of these values is due to the fact that the

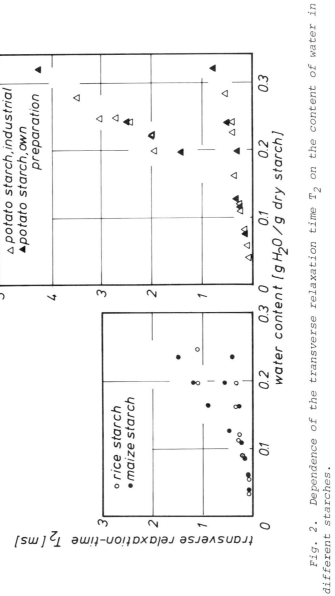

Fig. 2. Dependence of the transverse relaxation time T_2 on the content of water in different starches.

minima are rather flat and the values of the minimum temperature can be obtained only with rather low accuracy.

To get more insight into the exchange effects, the temperature dependence of T_2 has been measured for different water contents of the potato starch (Lechert, 1976). The results are demonstrated in Fig. 3. Striking in the different diagrams of Fig. 3 is the decrease of the longer T_2 with increasing temperature, which seems to be in contradiction with the behavior demonstrated in Fig. 1.

According to the theoretical consideration of Woessner and Zimmermann (1963), the apparent T_2', taken from the magnetization decays, depend in a quite complicated way on the exchange rates, the real T_2, and the probabilities of occupation of the different regions.

For slow exchange, T_2 as well as the probabilities of occupation can be taken undisturbed from the magnetization decays, because then the decay is simply the sum of the two exponentials belonging to the respective regions, weighted by the relative number of water molecules in these regions.

For the conditions of the measurements, demonstrated in Fig. 3, the mentioned relations can be simplified to

$$\frac{1}{T_2'\,_a} = \frac{1}{T_{2a}} + \frac{1}{\tau_a} \tag{1}$$

where T'_{2a} is the time constant taken from the decay for the region, denoted with a; T_{2a} is the respective real relaxation time of the proton in rhis region; and τ_a is the lifetime of a molecule herein. $1/\tau_a$ is then the exchange rate.

Sufficiently above the maxima observed in Fig. 3 for T_2 of the more mobile component, this T_2 is quite long and the relaxation rate $1/T_2$ cam be neglected. The apparent relaxation rate $1/T'_{2a}$ is then equal to the exchange rate.

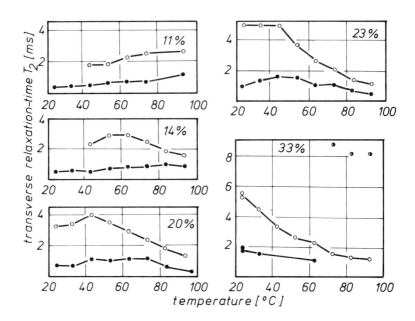

Fig. 3. Temperature behavior of the transverse relaxation
time T_2 of potato starch for different water contents.

Sufficiently above the maxima observed in Fig. 3 for T_2 of the
more mobile component, this T_2 is quite long and the relaxation
rate $1/T_2$ can be neglected. The apparent relaxation rate $1/T'_{2a}$
is then equal to the exchange rate.

The data of the 20% sample have been analyzed quite carefully
according to the theory of Woessner and Zimmermann (1963). The
values, obtained for the life-times lie at 4.5 msec at 43°C and
at 1.3 msec at 93°C for the region with the longer T_2. The
percentage of the more tightly bound water is at room temperature
17% and decreases at higher temperatures to about 12%.

The activation energy of the exchange is 31 kJ/mole for 20%,
28 kJ/mole for 23% and 26 kJ/mole for 29% water in the sample,
respectively. This is fairly below the value of about 50 kJ/mole
usually observed in hydrogen-bonded systems.

For the sample with 33% water, a further kind of water can be
observed, which has an increased mobility. In comparative calorimetric experiments, this water proved to be freezable and must be
regarded as "free" in terms of the definition of Kuprianoff (1958).

As further information on the mobility mechanism of the sorbed
water molecules, the linear self-diffusion coefficient can be
obtained from measurements of the magnetization decays in inho-
mogeneous fields (Steijskal and Tanner, 1965; Tanner, 1970).

In Table I the results of measurements of this kind on the
more mobile component of sorbed water are demonstrated for native
potato starch and some water contents and temperatures. From
Table I a distinct dependence of the self-diffusion coefficient
on the water content can be observed. For low water contents
because of technical reasons, only the values for higher temperatures are measurable.

For linear self-diffusion, the temperature dependence gives
54 ± 4 kJ/mole activation energies for the motion of the
water molecules, which are more exact than those obtained from
the T_1 minima. These values are distinctly higher than the values
obtained for starch gels, which lie with 20 kJ/mole near the
values of pure water (Basler and Lechert, 1974), and slightly
below the values obtained for water in ice.

III. ANISOTROPIC MOBILITY OF WATER

In some potato starch samples a careful analysis of the
magnetization decays showed shapes that could not be split into
exponentials. One of these decay shapes is given in Fig. 4 in
comparison to the respective wide-line spectrum. Thorough studies
of this phenomenon showed that these decays result from an anisotropic motion of the water molecules in the mobile phase of the
bound water, causing a splitting of the line of the proton resonance (Hennig and Scholz, 1976; Hennig, 1977).

TABLE I. Self-Diffusion Coefficient D_S of Water in the Grains
of Native Potato Starch at Different Water Contents and Temperatures

Water content (%)	Temperature (^{o}C)	Self-diffusion coefficient D_S (10^{-7} cm^2 sec^{-1})
13.4	50	22
	60	25
	80	27
20	30	8.0
	40	9.7
26	10	3.2
	30	10
	50	52
30.5	10	5.5
	30	14
	50	100

For the special case of the molecular structure of the water,
proton resonance as well as deuteron resonance can be used to get
information on the above-mentioned motional process. For proton
resonance, water must be regarded as a system of two spins
coupled by the interaction of its magnetic dipoles. This system adopts three energy levels in a magnetic field, causing a
split line, the components of which have the distance

$$\Delta\omega = \frac{3\gamma^2 n}{} (3 \cos^2\Theta - 1) \tag{2}$$

where γ is the gyromagnetic ratio of the protons, r their distance in the water molecule, and Θ the angle between the proton-
proton vector and the applied magnetic field. The deuteron has
spin 1 and its energy is split into three levels, too, if a system with D_2O or with HDO is brought into a magnetic field.

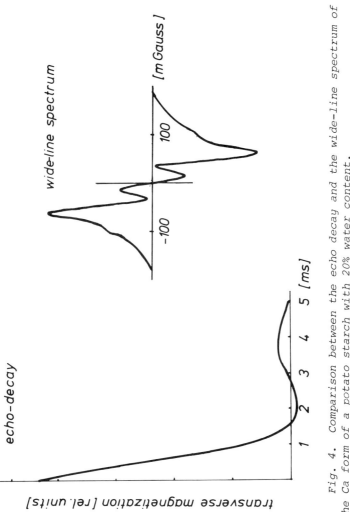

Fig. 4. Comparison between the echo decay and the wide-line spectrum of the Ca form of a potato starch with 20% water content.

The cause of this splitting is given, however, by the interaction of the quadrupole moment Q of the deuteron with the electric field gradient q in the field of the oxygen ion.   The resulting spectrum is, again, a doublet with a distance of its
components given by an expression quite similar to that of the
dipole-dipole interaction, where θ is in this case defined as the
angle between the direction of the OD bond with the applied magnetic field:

$$\Delta\omega = \frac{3}{4} \ \frac{e^2 qQ(3 \cos^2\theta - 1)}{h} \qquad\qquad (3)$$

For a reorientating water molecule in both expressions the
angular term can be split into

$$(3 \cos^2\theta - 1) = (3 \cos^2\theta' - 1) \ (3 \cos^2\theta'' - 1)/2 \qquad (4)$$

where θ is for both types of interaction the angle between the
applied magnetic field and the preferred axis of reorientation.

It contains the structural information on the reorientation process.   The term with θ'' describes the time behavior of the re-
orienting molecule.   For powders and possibly fibers the first
term has to be integrated over all the angles θ' that this axis
of reorientation can assume.   From this integration a two-peak
distribution results, which is measured in the spectrum.   The
second term has to be time-averaged.   The average vanishes, if
the motion is isotropic and can generally be taken as a measure
of the anisotropy of the motion, regardless of the slight difference in the definition of θ'' in Eqs. (2) and (3).

    If now an exchange of water molecules occurs, the energy
levels of a proton, left at the molecule are changed for the
dipole-dipole interaction, but not for the quadrupole interaction
of the deuteron.

    The occurrence of the split line is therefore quite sensitive
against exchange effects in the case of protons.   The exchange
can be, on the other hand, preferably studied by the proton
resonance.   The anisotropic motion itself is studied better with
deuteron resonance.

    From Fig. 4 a value of the splitting of the proton resonance
of 250 sec^{-1} can be estimated.   This corresponds to the fact
that the anisotropy of the motion must be retained at least 4 msec.
Regarding the common relation for the mean square of the diffusion
length

$$x^2 \;=\; 2D_s\Delta t \tag{5}$$

the self-diffusion coefficient D_s for the sample with 20% water
gives with $\Delta t = 4$ msec a diffusion length of about 1 μm over
which the anisotropy must be retained.   This corresponds to about
100 of the crystallites radially arranged in the granules of the
native potato starch.

    In the following, some applications of the behavior of the
deuterium resonance and the anisotropic mobility shall be dis-

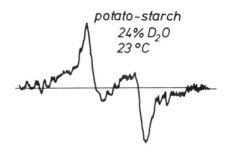

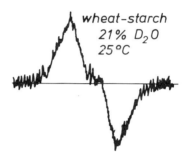

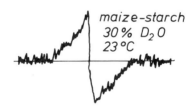

Fig. 5. Deuteron resonance spectra of deuterated potato, wheat, and maize starch.

cussed. The respective deuterated starch samples have been prepared by a treatment with heavy water. A careful treatment leads
to a complete exchange of the water and the OH groups of the
starch against deuterons, but not of the protons of the CH groups.

In Fig. 5 the deuteron spectra of three starches are demonstrated for room temperature and heavy water contents near 25%. As
must be expected from the effects demonstrated in Fig. 4, the

potato starch shows a distinct splitting of the deuteron resonance, wheat starch shows a splitting, but rather broad components of the single lines, and maize starch shows no splitting at all.

Indications of the presence of splitted lines have been observed also in samples of dried beans, lentils, and peas, swollen in D_2O. Because the effects are most distinct in the potato starch, more detailed measurements were carried out on this starch. An explanation of the effects demonstrated in Fig. 5 will be given in connection with the results of these measurements.

Figure 6 shows the temperature behavior for two samples with D_2O contents of 24 and 39% corresponding to water contents of about 21 and 34%, respectively. The sample with 24% D_2O should contain no "free" water, whereas the sample with 39% D_2O is in the range where freezable water can be detected. At 0^oC for both samples the spectra have broadened lines, indicating a decreasing mobility of the molecules at lower temperatures. For the 24% sample, above about 50^oC a decrease in the splitting can be observed, which is due to a loss of the anisotropy of the motional process. The splitting is present up to about 120^oC and collapses then to a broad line (see Fig. 6). The loss of anisotropy is most probably caused by a rearrangement of the starch structure, consisting of an expanding that reduces the strength of the interaction, causing the ordering of the molecules. Comparable effects have been observed in collagen fibres (Dehl and Hoeve, 1969). The most striking difference of the two samples in Fig. 6 is, however, the occurrence of the narrow line in the center of the spectrum. It can be seen that this line grows rapidly in intensity between 0° and 5^oC, which is near the melting point of the pure D_2O at 3.8^oC, and remains then constant over a large range of temperature.

Obviously, this line corresponds to the "free" water, discussed in connection with the results of the temperature behavior of T_2, demonstrated in Fig. 3. Raising the temperature above

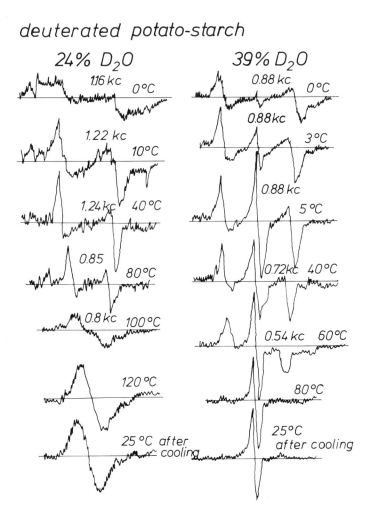

Fig. 6. Deuteron resonance spectra of deuterated potato starch at different water contents and temperatures.

80°C, the splitted line vanishes and cannot be restored by cooling the sample back to room temperature. Obviously, even the small amount of "free" water present in the sample destroys the structures responsible for the detectability of the aniso-

tropic motion. The most reliable explanation of this phenomenon
is that a partial disorder of the crystallites of the starch
granule is caused, so that the water molecules lose their
orientation on their way from one crystallite to another and
the short time the molecule is in an ordered state is not sufficient for detection with deuteron resonance.

This may also be an explanation for the lack of splitting
in maize starch. Maize starch contains a small amount of protein between the polysaccharide crystallites, which may have a
similar effect.

In deuterated B-type amylose, the crystallinity of which was
better than the above-discussed potato starch, no splitting could
be detected, which agrees with the suggestion that not only the
crystallinity but the whole granular structure is responsible for
the observed effects.

The sensitivity reaction of the deuteron spectrum offers
a lot of possibilities of investigations on changes in this
granular structure under the influence of different technical
processes. In the following, some examples are reported in short.

In Fig. 7 the deuteron spectra of a native and a weakly
cross-linked sample with 24% heavy water are compared for different temperatures. First, it can be seen that the cross-linking
does not influence the structural elements responsible for the
anisotropic motion and for the detectability conditions. For
native starch the splitting collapses at $120^{\circ}C$, as has been
mentioned above, and a broad line is left, which is not restored
by cooling back to room temperature.

For the cross-linked starch, at $120^{\circ}C$ a second line appears,
indicating a second more mobile kind of water. Cooling down,
a very broad line is superimposed by a line distinctly narrower
than in the case of the native starch. In both samples
some kind of extension of the structure occurs, allowing
the molecules an isotropic reorientation after the heat

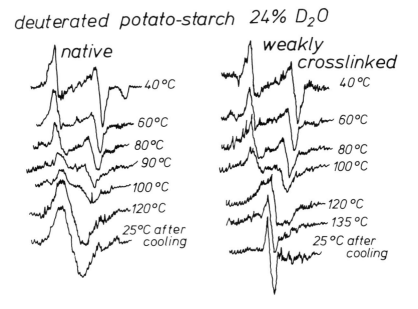

Fig. 7. Deuteron resonance spectra of deuterated samples of native and weakly cross-linked potato starch with 24% D_2O at different temperatures.

treatment. The cross-linked starch can obviously incorporate only a limited amount of water, whereas the rest is
expelled and sorbed at the surface of the grain in a region with
better molecular mobility.

It should be stressed at this point, that the steady collaps-
ing of the splitting must be distinguished from the vanishing
of the split line at 80°C in the presence of free water. The
first effect consists of the noted extension of the structure,
as it has been observed also in collagen fibers, the sudden
breakdown in the presence of free water in a disorder of the
crystallites preventing the detectability after the heat treatment.

Now, the question arises as to which effect drying and rehydration have on the two starch samples. For this experiment,

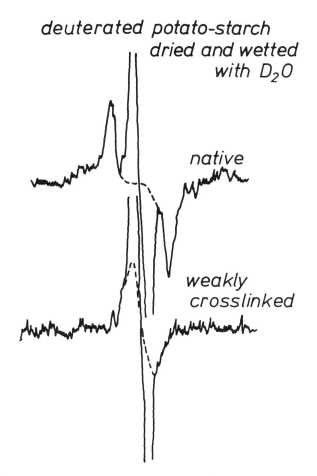

Fig. 8. Deuteron resonance spectra of deuterated samples of native and weakly cross-linked potato starch, which have been dried at 135°C and dispersed in excess of D_2O. The narrow line in the center of the spectrum belongs to the free D_2O.

the samples were heated to 135°C for 12 hrs and then water was added. The spectrum showed no change compared with the split spectra in Fig. 5, but only a narrow line in the case of the cross-linked starch, as can be seen in Fig. 8.

In another series of experiments the influence of heat-moisture treatment has been tested. Samples of deuterated starch were

mixed with an amount of D_2O giving mixtures of 1 g of water per
1 g of dry starch. starting at $50^{\circ}C$ these samples were heated
from a few degrees to higher temperatures. After at times one
day heat treatment, the deuteron resonance has been measured.
A change of the spectrum can be observed only above $70^{\circ}C$, giving
a decrease of the relative amount of the molecules in the anisotropic state of mobility. The splitting remains constant. Above
$80^{\circ}C$ the splitting collapses, as has been demonstrated in Fig. 5.
X-Ray diffraction patterns of the samples showed an increasing
amount of the C form with decreasing intensity of the splitted
line.

IV.   SWELLING, RETROGRADATION, AND FREEZE-THAW BEHAVIOR

As already mentioned, the OH protons and the water of the
starch can be easily replaced by deuterons without exchanging
the CH protons. In the most interesting problems concerned with
swelling, water is present in large excess. The influence of
the swelling process on the protons of the water is therefore
difficult to measure. Measuring, however, the proton resonance
of the CH protons of a starch, deuterated in the described
manner and swollen in D_2O, the swelling can be studied on the
high polymer itself, because the T_2 of the high polymer in the
swollen state is some orders of magnitude larger than that in
the solidlike state.

Figure 9 shows a so-called Carr-Purcell experiment at different temperatures in a mixture with only 3% potato starch. The
portion of the mobile starch chains, given by the relative intensity of the envelope of the slowly decaying peaks, grows near
the temperature of swelling with increasing rate, until at $96^{\circ}C$

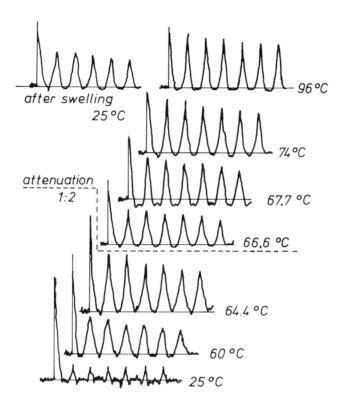

*Fig. 9. Magnetization decays of the CH protons of a
deuterated potato starch in the course of swelling in D_2O. The
figure in the upper left corner results after cooling the sample
from 96°C to room temperature.*

only mobile starch chains are present. The fast decay, just
at the beginning of the experiments, indicates the solidlike
part of starch in the system has disappeared.

Cooling back to room temperature and leaving overnight, the
signal of the solid is partly restored. This method seems to be
highly useful for studies of retrogradation and, for instance,
freeze-thaw phenomena, the investigation of the molecular processes of which are quite difficult.

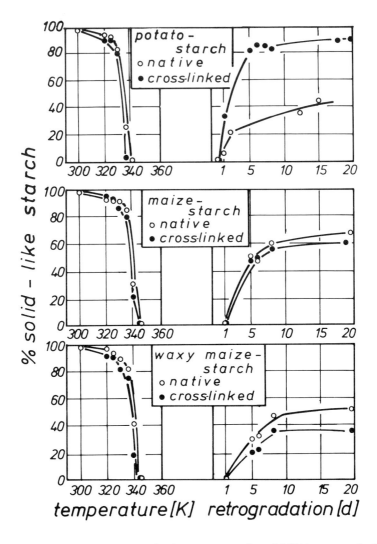

Fig. 10. Dependence of the amount of solidlike starch for the swelling and the retrogradation of potato, maize, and waxy maize starch in the native and weakly cross-linked form.

In Fig. 10, the amount of the solidlike component is determined for the swelling and the retrogradation process of native
and cross-linked potato, maize, and waxy maize starch. It can
be seen that the course of the swelling process does not differ

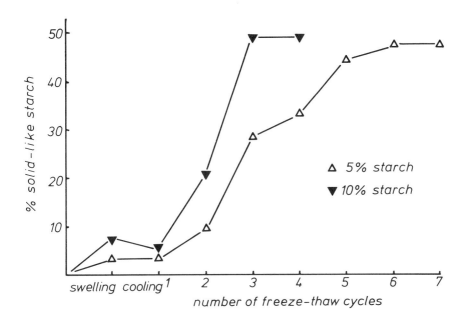

Fig. 11. Dependence of the amounts of solidlike starch in
two potato starch gels on the number of freeze-thaw cycles.

for all starches. The difference in the temperatures where the
solidlike component vanishes correspond roughly to experiences
that can be obtained from other experiments. Following this,
the swelling point of the potato starch lies slightly below
those of the two maize starch samples. The exact position of the
swelling temperature is, however, dependent on the rate of
heating, as shown in an earlier paper.

For the retrogradation experiments the most striking effect
is the big difference of the rate and the final state of the
retrogradation of the native and the cross-linked potato starch.
For all starches the final state is reached within 5-7 days.
The behavior of the other starches coincides roughly with that
from other experiments.

In Fig. 11 some experiments on the freeze-thaw behavior of potato starch gels with 5% and 10% dry starch are demonstrated. The samples had been swollen and left at room temperature for one day. Then the gels were cooled to -20°C with a defined rate. After thawing the amount of solid was determined. Repeating this procedure several times, this amount of solid increases as demonstrated in Fig. 11.

These experiments show that NMR methods can give detailed information on molecular processes in biopolymers, if well defined substances are employed.

ACKNOWLEDGMENTS

The author wishes to thank the Arbeitsgemeinschaft Industrieller Forschungsvereinigungen and the Forschungskreis der Ernährungsindustrie for support of this work.

REFERENCES

Basler, W. D., and Lechert, H. (1974). *Die Stärke-Starch 26,* 39.
Bloembergen, N., Purcell, E. M., and Pound, R. V. (1948). *Phys. Rev. 73,* 679.
Dehl, R. E., and Hoeve, C. A. J. (1969). *J. Chem. Phys. 50,* 3245.
Hennig, H. J. (1977). *Die Stärke-Starch 29,* 1.
Hennig, H. J., and Lechert, H. (1974). *Die Stärke-Starch 26,* 232.
Hennig, H. J., and Lechert, H. (1976). *Z. Naturforsch. 31a,* 306.
Hennig, H. J., and Lechert, H. (1977). *J. Colloid Interface Sci. 62,* 199.
Hennig, H. J., and Scholz, P., (1976). *Z. Phys. Chem. NF 103,* 31.
Hennig, H. J., Lechert, H., and Krische, B. (1975). *Die Stärke-Starch 27,* 151.
Kuprianoff, J. (1958). *Soc. Chem. Ind. (London)*
Lechert, H., and Hennig, H. J. (1976). ACS Symp. Ser. "Magnetic Resonance in Colloid and Interface Science," No. 34, P. 328.
Pfeifer, H., (1972). "NMR-Basic Principles and Progress," Vol. 7. Berlin.

Resing, H. A. (1968). *Adv. Mol. Relax. Processes 1,* 109.

Shaw, T., and Elsken, R. J. (1953). *J. Chem. Phys. 21,* 565.

Steijkskal, E. O., and Tanner, J. E. (1965). *J. Chem. Phys. 42,* 288.

Tanner, J. E. (1970). *J. Chem. Phys. 52,* 2523.

Woessner, D. E. (1961). *J. Chem. Phys. 35,* 41.

Woessner, D. E., and Zimmerman, J. R. (1963). *J. Phys. Chem. 67,* 1530.

Zimmerman, J. R., and Brittin, W. E. (1957). *J. Phys. Chem. 61,* 1328.

PULSED NMR
AND
STATE OF WATER IN FOODS

Nobuya Nagashima and Ei-ichiro Suzuki

I. INTRODUCTION

It has long been recognized that the water in food exists
both bound and free. However, a universally accepted definition
of bound water does not exist (Fennema, 1976). Generally,
bound water includes (1) water of hydration (2) water in micro-
capillaries or entrapped in macrocapillaries, and (3) adsorbed
water on a solid surface. It is difficult to measure the
amount of various types of water separately.

There are many reports (e.g., Bratton *et al.*, 1965; Toledo
et al., 1968; Hazelwood *et al.*, 1969; Kuntz *et al.*, 1969; Kuntz,
1971; Duckworth *et al.*, 1973; Leung *et al.*, 1976) concerning some
applications of wide-line or pulsed NMR for investigation of
water in biological systems or foods. As an example of distinction of state of water, Belton *et al.* (1972) inferred from
the transverse relaxation time the existence of three fractions
of water.

Using pulsed NMR, we measured the variation of free induction decay (FID) amplitude with decreasing temperature, and
obtained the variation of liquid or unfrozen water content as a
function of temperature, for which we used here the term
freezing curve.

The present study represents an attempt to distinguish the state of water in foods from freezing curves. We have found that the shape of the freezing curve reflects more closely the water state related to the properties of foods. The details of variation over the range of the rapid decrease between 0° and $-10^{\circ}C$ represent the slight difference of the state of water, and some quantitative data for the state of water can be obtained.

II. METHOD AND MATERIALS

A. NMR Measurements

The pulsed NMR spectrometer used for this work was a Bruker minispec p20 (20 MHZ for observation of ^{1}H) equipped with a temperature control system. The sample temperature was directly measured by a copper-constantan thermocouple inserted in the sample tube. Sample tubes had 7 mm OD.

For observations of FID, a repeated 90° pulse was applied, where the time interval between the sequential pulses is a few seconds. Unfrozen water content as a function of temperature was determined from the sequential FID amplitude measurements at about 40 μsec after a 90° pulse with decreasing temperature. The usual cooling rate was approximately $2 \pm 1^{\circ}/min$. Figure 1 shows FID curves appearing on CRT at several temperatures, and the variation of this amplitude with decreasing temperature is automatically recorded on the X-Y recorder as a freezing curve, as shown in Fig. 2. This automatic recording has not only made measurement easy, but also increased qualitative and quantitative information about specimens. It has been possible to obtain details of variations in the sudden decreases in signal between 0° and $-10^{\circ}C$.

Free Induction Decay

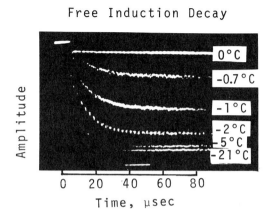

Fig. 1. FID curves on CRT at several temperatures.

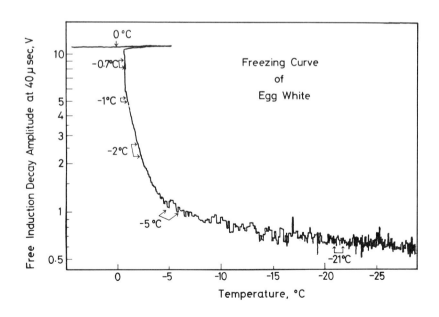

Fig. 2. Freezing curve of egg white recorded automatically.

B. Materials and Sample Preparation

The materials used were commercially available. Total
moisture content was determined by weighing after drying to
constant weight in an oven at 115°C. Gelatinzation was usually
taken at 100°C for 2 min in the sample tube sealed with parafilm
to prevent volatilization of water. Retrogradation for starch
was run at 5°C for about 35 hr.

III.   RESULTS AND DISCUSSIONS

A. Freezing curve

The freezing curves of several foods are shown in Fig. 3,
and in this figure the FID amplitude is converted to the unfrozen water content in grams of water per gram dry matter with
calibration of Curie's rule.

These curves usually show rapid decrease in unfrozen water
content between 0° and -10°C, followed by a gradual decrease at
lower temperature. The rapid and then the slow decrease in signal correspond to the freezing of free water and the gradual
freezing of bound water or difficult to freeze water, respectively.

The shape and temperature of the sudden drop portion of the
freezing curve are influenced delicately by the state of water.
The authors would like to emphasize that the freezing curve shows
not only the amount of unfrozen water at some temperature, but
also the force or strength of water binding to some extent.

A schematic freezing curve is shown in Fig. 4. We use here
tentatively the term "weakly bound water" (WBW) for the unfrozen
water between 0° and -20°C and "tightly bound water" (TBW) for
the unfrozen water below -20°C.

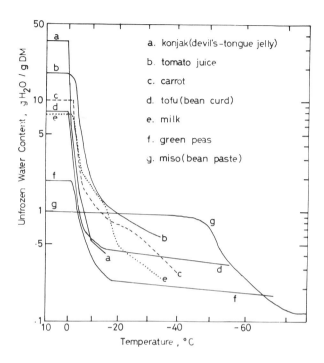

Fig. 3. Freezing curves of various foods.   The FID amplitudes were converted to the unfrozen water content.

B.   Water in Gel

Both carbohydrates and proteins are used to form gels in foods. It is believed that gel of starch or gelatin holds a large amount of water, but the amount of TBW to macromolecules as hydrated water is very small and a major proportion of the water is weakly entrapped, like nearly free water between a network of macromolecules.

As shown in the freezing curves of gelatin gel and starch gel (Figs. 5 and 6), it seems that water bound to the macro- molecule as a hydrated water and WBW like nearly free water correspond to the unfrozen water below -20°C and between 0° and -10°C, respectively.

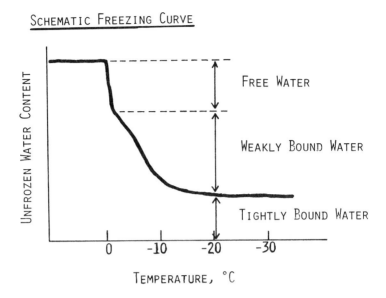

Fig. 4. Tentative definition of "weakly bound water" and "tightly bound water" on freezing curve.

Figure 7 shows the variation of TBW content (unfrozen water at -20°C) with concentration for gelatinized and retrograded potato starch. Generally, the amount of TBW in the retrograded state was larger than that of the gelatinized state, but the amount of WBW decreased with retrogradation, as shown in Fig. 8. This suggests that the increased association or partial crystallization of the starch molecules on the process of retrogradation acts to increase TBW. The water-filled regions between starch molecules become smaller by this association and water is forced out, with a resulting decrease of WBW.

Another important feature of the freezing curve is that the portion between O and -10°C relates to the progression of gelatinization of starch. The effect of gelatinzation temperature on freezing curves of 30% aqueous suspensions of potato starch is shown in Fig. 9. As a result, the higher the gelatin-

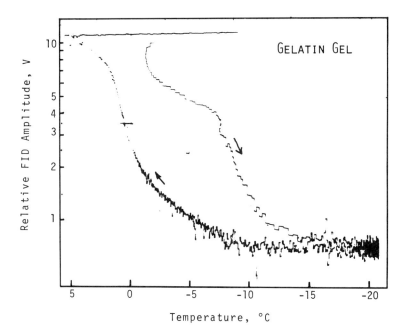

Fig. 5.   Freezing curve of gelatin gel.   Concentration 10%, gelatinized at 70°C.

ization temperature, the lower the freezing temperature at which
a rapid decrease in unfrozen water content is recorded. This
fact indicates that WBW content increased or freezing point decreased with the progression of gelatinization, and the gel
formation is not sufficient at 70°C and becomes complete at 99°C,
after the collapse of granules. These results correspond well
to the amylograph curve (the record of the viscosity change), that
is, the viscosity does not change until 60°C, when a rapid increase in viscosity occurs with the swelling of the granules and
reaches the highest viscosity at about 70°C. Consequently the
decrease occurs with the collapse and fragmentation of granules
until about 95°C.

Figure 10 shows the effect of amylose content on the freezing point for corn starch gelatinized at 99°C. As a result,

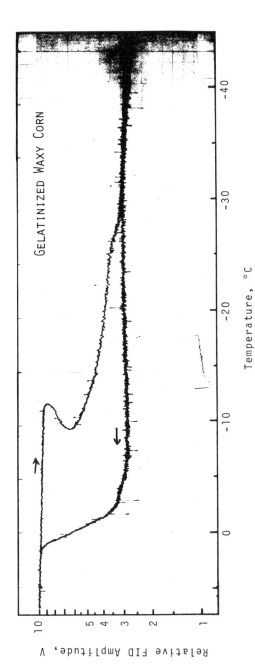

Fig. 6. Freezing curve of gelatinized waxy corn.

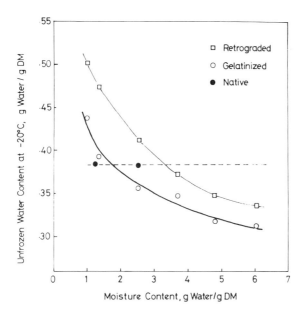

Fig. 7.   Variation of TBW with concentration for gelatinized
and retrograded potato starch.

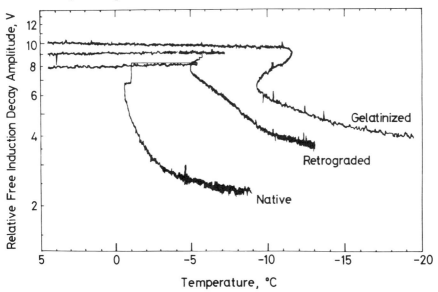

Fig. 8.   Freezing curves of native, gelatinized, and retrograded waxy corn starch.   Concentration is 50%.

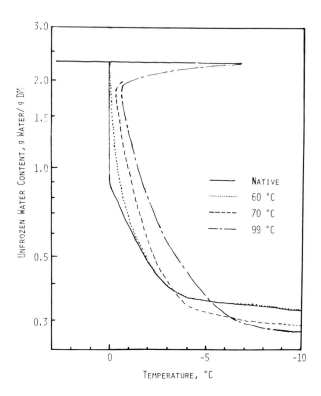

Fig. 9. Effect of gelatination temperature on freezing curve of 30% potato starch.

the higher the amylose content, the higher the freezing point, and these correspond to the fact that the higher the amylose content, the more difficult to gelatinize.

As comminuted animal muscle products, there are various sausages prepared from meat, and kamaboko and chikuwa (boiled fish paste) from fish fillets, the latter foods being popular in Japan. In these comminuted meat products, the process of the comminution and blending after the addition of sodium chloride (2.5-4.0%) is generally contained in their manufacture process. This fact indicates that the salting contributes to a cohesive network of coagulated proteins and to form stable meat emulsions.

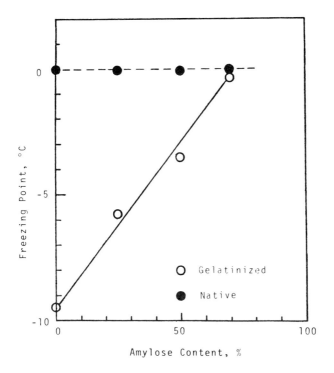

*Fig. 10. Effect of amylose content on freezing point of
corn starch gelatinized at 99°C.*

The effect of the salting on the freezing curves is indicated
in Fig. 11 for ground meat (beef) and in Fig. 12 for ground
fish fillet (Alaska pollack). Both TBW and WBW increased with
salting, and this fact indicates the increases not only in the
hydrated water content but also in the water-holding capacity,
which is an important property of these foods. This salting
effect appeared also on soy protein, as shown in Fig. 13.

In the manufacturing processes of kamaboko, fish fillet is
gelatinized by storage at low temperature (2°-5°C) overnight,
a process called *suwari* in Japanese, which means "settling."
In the case of the gelatinization at low temperature for ground
fish fillet, the amount of TBW almost did not change, but the

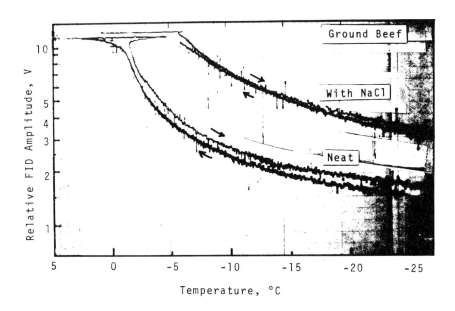

Fig. 11. Effect of salting on freezing curve of ground beef.

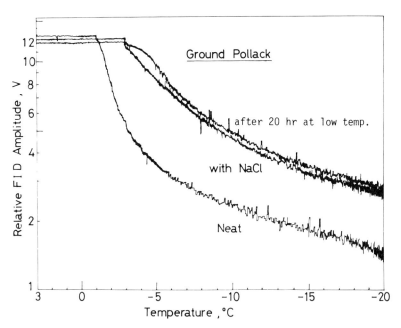

Fig. 12. Freezing curves of ground Alaska pollack with and
without NaCl, gelatinized at low temperature.

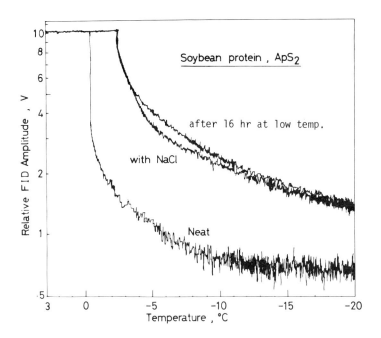

Fig. 13. Freezing curves of soy protein with and without NaCl, gelatinized at low temperature.

variation of WBW with time showed the sudden increase at about 15 hr as shown in Fig. 14. This suggests that the gel formation may progress after a some induction period. In the gelatinization at low temperature for soy protein, the amount of WBW also increased, but to a lesser degree than that of fish fillet. Moreover, the shape of the freezing curve near the freezing point was different from that of fish fillet, as shown in Figs. 12 and 13, and these indicate that the mechanisms of gel network formation are different in these proteins.

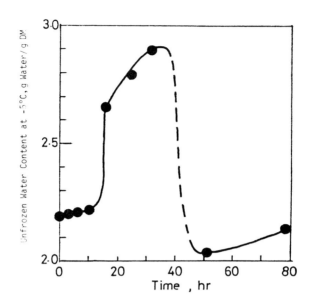

Fig. 14. Variation of WBW content with storage time at
5°C for a ground pollack.

C. Hysteresis

The freezing curves for many materials show hysteresis, as
is expected, since the gel or tissue structure is broken during
freezing.   In homogeneous material like a sucrose solution,
hysteresis in the freezing curve is small (Fig. 15), but when
supercooling occurs a small hysteresis 100p appears between
-20° and -30°C; this may be due to the association of sugar
molecules.

Hysteresis is clearly shown when the gel structure is
destroyed or altered by freezing, as is shown for gelatin gel
or starch gel in Figs. 5 and 6.

Hysteresis of the ground beef is much smaller than that of
native beef (Fig. 16).

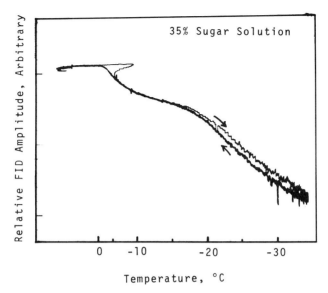

Fig. 15. *Hysteresis curve of freezing of 35% sugar solution.*

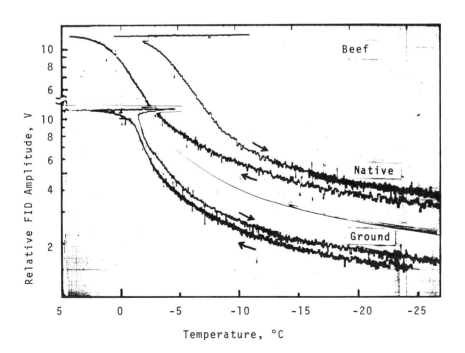

Fig. 16. *Hysteresis curves of freezing of native and ground beef.*

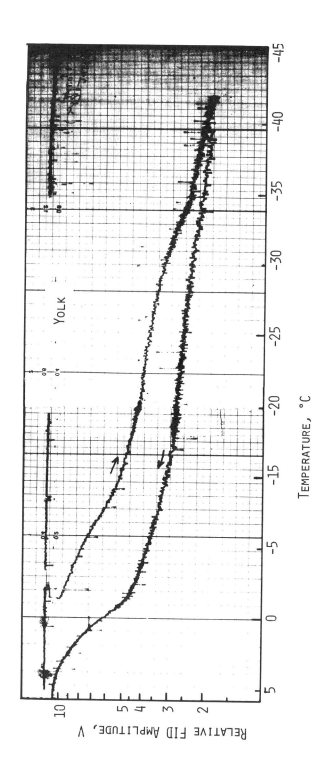

Fig. 17. Hysteresis curve of freezing of egg yolk.

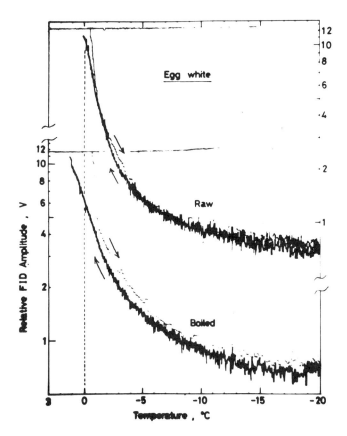

Fig. 18.   Hysteresis curves of freezing of raw and boiled egg white.

The freezing curve of yolk has a very large hysteresis as shown in Fig. 17.   Yolk seems to be homogeneous visually, but it has been reported by Powrie (1976) that yolk is a dispersion containing a variety of particles distributed uniformly in a protein solution, and the viscosity of the thawed yolk is much higher than that of native yolk.   Meanwhile, egg white did not show such a hysteresis, but the hysteresis of the boiled egg white showed the gel formation (Fig. 18).

IV.   CONCLUSIONS

The measurement of the freezing curve may be one of the useful methods for assessment of the state of water in food, which is related to food quality.

REFERENCES

Belton, P. S., Jackson, R. R., and Packer, K. J. (1972). *Biochim. Biophys. Acta 286,* 16.
Bratton, C. B., Hopkins, A. L., and Weinberg, J. W. (1965). *Science 147,* 738.
Duckworth, R. B., and Kelly, C. E. (1973). *J. Food Technol. 8,* 105.
Fennema, O. R. (1976). *In* "Principles of Food Science," Part 1, Food Chemistry (O. R. Fennema, ed.), p. 28.   Marcel Dekker, New York and Basel.
Hazlewood, C. F., Nicols, B. L., and Chamberlain, N. F. (1969). *Nature 222,* 747.
Kuntz, I. D. (1971). *J. Am. Chem. Soc. 93,* 514.
Kuntz, I. D., Brassfield, T. S., Law, G. D., and Purcell, G. V. (1969). *Science 163,* 1329.
Leung, H. K., Steinberg, M. P., Wei, L. S., and Nelson, A. I. (1976). *J. Food Sci. 41,* 297.
Powrie, W. D. (1976).   "Principles of Food Science," Part 1, Food Chemistry (O. R. Fennema, ed.), Chap. 15, Marcel Dekker, New York and Basel.
Toledo, R., Steinberg, M. P., and Nelson, A. I. (1968). *J. Food Sci. 33,* 315.

DETERMINATION OF BINDING ENERGY
FOR THE THREE FRACTIONS OF BOUND WATER

Soewarno T. Soekarto
M. P. Steinberg

I. INTRODUCTION

One of the peculiarities associated with bound water is its
high binding energy. However, reports on binding energy are con-
fusing. First, some authors (Kunzt and Kauzmann, 1974; Bull, 1974;
Othmer and Sawyer, 1943) reported the energy as all heat associated
with adsorption of water molecules to the solid, including heat of
liquefaction, while others (Berendsen, 1975; Brunauer et al.,
1938) reported the energy as the difference between total heat of
adsorption and heat of liquefaction. Second, the energy was reported as that involved in only the most tightly bound water or
"monolayer" (Brunauer et al., 1938), while others considered it
involved in all bound water, including that beyond the "monolayer"
(Fish, 1957; Chung and Pfost, 1967). Determinations of binding
energy are commonly classified into thermodynamic and calorimetric
methods. Calorimetric methods (Schrenk et al., 1949) are based
on the heat evolved or released when water is added to a dry,
vacuumized solid.

Based on the Boltzmann distribution law, Brunauer et al.
(1938) derived an equation that related a constant to binding
energy. This energy value is particularly related to the most
tightly bound water.

The most common method for determining water binding energy

is also based on thermodynamics, especially the Clausius-Clapeyron
equation. Based on this equation, Othmer and Sawyer (1943) calculated binding energy by plotting $\ln p$ vs. $\ln p_0$ at different
temperatures for a given moisture content. This procedure was
applied to grain products by Thompson and Shedd (1954) and Rod-
rigues-Arias et $al.$ (1963), to food substances by Fish (1957)
and Labuza (1968), and to soil by Zettlemoyer et $al.$ (1967).
This method requires sorption data from at least three different
temperatures. However, by integration of the Clausius-Clapeyron
equation, binding energy can be calculated from sorption data at
a minimum of two temperatures; this procedure was widely employed
(Fish, 1957; Becker and Sallans, 1956; Stitt, 1957; Chung and
Pfost, 1967; Kuntz and Kauzmann, 1974). Kruger and Helcke (1967)
measured binding enerby by combining NMR relaxation and thermodynamic principles. A modification of the Clausius-Clapeyron equation
has been applied to determine heat of absorption by Dole and
McLaren (1947). Bull (1944) calculated heat of absorption as the
difference between the free energy changes at two temperatures.

The concept of different fractions of bound water suggests
that the binding energy characterizes each fraction of bound
water. The objectives of this chapter are: to determine composite binding energies at different levels of bound water; and
to characterize binding energies of the three fractions of bound
water, called primary (PRI), secondary (SEC), and tertiary water
(TER).

II. BINDING ENERGY

Binding energy ΔH_B is a thermodynamic term defined as the difference between the heat of absorption of the water by the solid
and the heat of condensation of water vapor at the same temperature. Thermodynamic principles were applied to water sorption
concepts in order to study binding energy. Equilibrium sorption

data were taken from food ingredients, dextrinized tapioca (TAP), and sucrose powder. The equilibrations were carried out in desiccators with saturated salt solutions according to Rockland (1960) and sulfuric acid solutions according to Robinson and Stokes (1959) to cover a wide range of relative humidity (RH), from 5 to 97%.

Figure 1 represents a classical water sorption isotherm of dextrinized tapioca. The isotherm could be subdivided into the Langmuir, the lower, and the asymptotic, the upper parts. Binding energy was attained by deduction from the Langmuir energy equation for the PRI and Clausius-Clapeyron equation for SEC and TER fractions.

A. Binding Energy for Primary Bound Water

Brunauer *et al*. (1938) derived the following equation that related to binding energy:

$$c = K \exp(\Delta H_1 - \Delta H_L)/RT \tag{1}$$

where c is a constant related to binding energy, ΔH_1, ΔH_L the heats of absorption of monolayer and of condensation of water vapor, respectively ($\Delta H_L = 10{,}530$ cal/mol at 22°C; Weast, 1972), R the gas constant (1.987 cal/mol $^\circ$K), T absolute temperature ($^\circ$K), and K a constant.

The binding energy ΔH_B may be defined as $\Delta H_1 - \Delta H_L$. Then Eq. (1) becomes

$$c = K \exp(\Delta H_B/RT) \tag{2}$$

and

$$\ln c = \ln K + (\Delta H_B/R)(1/T) \tag{3}$$

Thus, an Arrhenius plot of $\ln c$ against $1/T$ should give a straight line with slope $= \Delta H_B/R$. Temperature and c values for SF, DF, and TAP were obtained from experiment and solution for c was obtained

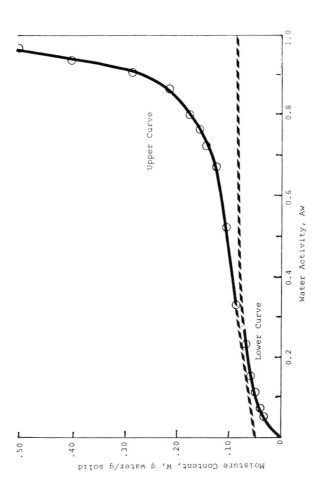

Fig. 1. *Water sorption data for dextrinized tapioca at 21.7°C divided into two isotherms.*

from the Langmuir equation applied to sorption data.   These values
were plotted according to Eq. (3).   The slope of each line is
given in Table I; the correlation coefficients were 0.98 or above,
showing a good correlation.   The proportionality constant K in
Eqs. (2) and (3) was found to be close to unity by Brunauer et al.
(1938) in case of gases such as N_2 on silica gel at low (190°C)
temperatures; however, we obtained large values, such as 430 in
the case of DF.

Each slope was multiplied by R (1.987 cal/mol °K) to obtain
binding energy for primary bound water.   The negative values of
the binding energy thus obtained (Table I) mean that this binding
is exothermic.   They show that ΔH_B of SF is lower than that of
DF.   When ΔH_B for SF is corrected for fat content (23.2% dry basis),
the result, -1410 cal/mol °K ΔH_2O, is still lower than that of DF,
-1678 cal/mol.   This supports the idea that some component of the

TABLE I.   c Values of Langmuir Isotherm and Binding Energy
of PRI

Products	Temp. (°C)	c values	Slope of $\ln c$ vs. $1/T$	r	Binding energy (cal/mol H_2O)
Soy flour (SF)	1.1	19.290	−545.3	−0.999	−1083
	21.7	22.125			
	37.0	24.295			
Defatted soy flour (DF)	1.1	20.090	−844.3	−0.984	−1678
	21.7	23.305			
	37.0	28.989			
Dextrinized tapioca (TAP)[a]	1.1	9.108	−731.0	−0.979	−1452
	21.7	11.567			
	37.0	12.251			
Dextrinized tapioca (TAP)[b]	1.1	11.104	−721.6	−0.988	−1434
	21.7	12.790			
	40.0	15.459			

[a] The raw tapioca was obtained from Ginza & Co., Champaign, IL.
[b] The cassava root was purchased and the raw tapioca was isolated in the laboratory.

remaining fat in the DF contribute strongly to the binding of
water (Soekarto, 1978). The DF, at 60.16% protein, show a primary
water binding energy here close to the heat of absorption of some
proteins as reported by Bull (1944) and Berendsen (1974); their
values fell in the range 1.30-1.75 kcal/mol.

B. *Binding Energy for Secondary (SEC) and Tertiary (TER) Bound
   Water*

Water molecules sorbed (i.e., bound) on a solid surface set
up an equilibrium with water vapor over the solid as follows:

$$\text{bound water} \rightleftharpoons \text{free water} \rightleftharpoons \text{water vapor}$$

Thermodynamics of such an equilibrium is expressed by the Gibbs
equation for the free energy and by the Clausius-Clapeyron equation for binding energy (Fish, 1957; Nash, 1970):

$$\frac{-\Delta H_B}{T^2} = R \left[\frac{\partial \ln a_w}{\partial T} \right]_{P,W} \tag{4}$$

Equation (4) is the Clausius-Clapeyron equation at fixed pressure (P) and moisture content (W) values.

Considering ΔH_B as the average binding energy over the working
temperature range and the P and W values fixed, integration in
Eq. (4) results in a working expression:

$$\ln a_w = \frac{\Delta H_B}{R} (1/T) + \text{const} \tag{5}$$

Thus, a plot of $\ln a_w$ vs. $1/T$ at a given moisture content should
result in a straight line with slope $\Delta H_B/R$. The first step taken
to apply Eq. (5) was to obtain corresponding W vs. a_w data at
each of three temperatures. In order to enable more accurate
interpolation, the data were plotted as $\log(1-a_w)$ vs. W. At a
given moisture content, a_w was determined for each temperature
and plotted as $\ln a_w$ vs. $1/T$. The slope of a regression line
$(r > 0.95)$ was determined from the data. This slope was multiplied

by $R = 1.987$ cal/mol water $^\circ$K to obtain ΔH_B at the given moisture
content.

Then, ΔH_B was plotted against corresponding moisture content
as in Fig. 2 for TAP. As moisture increased, ΔH_B decreased abruptly with a slight curvature to a breakpoint at 18.0%. This
moisture level was considered to be a moisture level at which SEC
ends, called W_{SEC}, and then follows a slower, linear decrease in

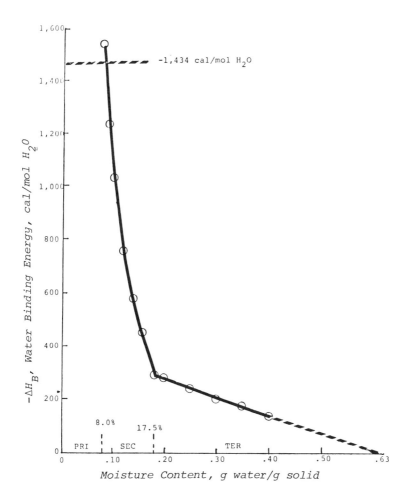

Fig. 2. Average energy of water binding by dextrinized
tapioca at each moisture content.

ΔH_B. The latter line was extrapolated to the abscissa to obtain the moisture content at $\Delta H_B = 0$, which should give the transition point between bound water and free water, 63.0% dry basis. This moisture level was considered a moisture level at which TER ended, called W_{TER}.

Also shown in Fig. 3 is a horizontal line at ΔH_B of 1434 cal/mol. This is the binding energy for PRI from Table I. At this ΔH_B, the moisture content was 8.0%, a moisture level at which PRI ended. Thus, by combining the Langmuir and Clausius-Clapeyron concepts, we can both determine energy of water binding at each moisture content and quantitatively demarcate the three fractions of bound water as shown in Fig. 2.

Figure 3 was constructed to show the relation between ΔH_B and a_w. First, the moisture isotherm was plotted. At a given moisture content, a_w was read from the isotherm and the corresponding ΔH_B was taken from Fig. 2, so that ΔH_B could also be plotted against a_w in Fig. 3. This shows a linear relation for SEC. The energy for PRI was taken from Table I and plotted as a horizontal line in Fig. 3. The intersection of this line with the energy line for SEC gives the upper limit of PRI. This point apparently coincides with the inflection point of the isotherm. The upper limit for SEC is taken as the lower end of the straight line. Apparently this coincides with the point where the isotherm starts its rapid rise.

C. Characteristic Binding Energy

Our hypothesis is that each fraction is characterized by a constant binding energy. According to this concept, each increment of PRI is bound to the solid by the same energy; similarly for SEC and TER. This binding energy will vary with the fraction and substrate; however, within a given fraction it is the same. For instance, if SEC ranges from 8 to 18%, the increment of water at 10% is bound with the same characteristic energy as the increment at 16%.

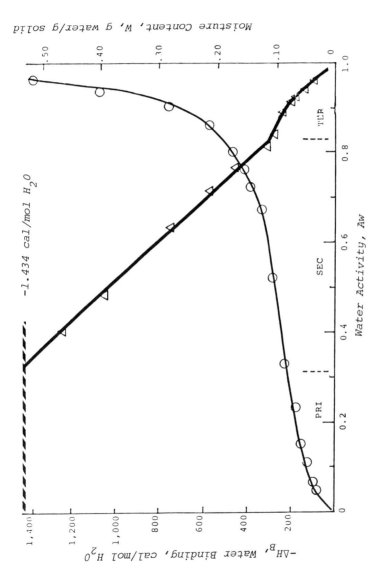

Fig. 3. Variation of water binding energy of dextrinized tapioca and moisture content with water activity at 21.7°C.

That ΔH_B is constant over the whole range of PRI has already been shown by Eq. (1) and discussion above, and calculated values were given in Table I. There remains to show that the same is true for SEC and TER:

$$W = W_{PRI} + W_r \qquad (6)$$

where W is total moisture, W_{PRI} PRI capacity, and W_r all the water in addition to W_{PRI}.

We note that ΔH_B as obtained from Eq. (5) at moisture level above W_{PRI} is the energy at a given moisture content averaged over all the water present or a composite bound water at a given moisture level. A total energy balance stating that the whole is equal to the sum of its parts gives

$$\Delta H_B W = \Delta H_{PRI} W_{PRI} + \Delta H_r W_r \qquad (7)$$

where ΔH_{PRI} is the characteristic binding energy for PRI, and ΔH_r the characteristic binding energy for water above PRI. Combining Eqs. (6) and (7),

$$\Delta H_B W = \Delta H_{PRI} W_{PRI} + \Delta H_r (W - W_{PRI}) \qquad (8)$$

Equation (8) is linear in $\Delta H_B W$ vs. $W - W_{PRI}$ with intercept $\Delta H_{PRI} W_{PRI}$ and slope ΔH_r, the energy value sought.

Since W_{PRI} is a constant here, $\Delta H_B W$ was plotted against W as shown in Fig. 4 for TAP. Both SF and DF also gave a broken line with a V shape as shown in Fig. 5.

The descending line in Figs. 4 and 5 represents SEC and the ascending line gives TER. Also, the slope of each line gives the corresponding ΔH_r, which is defined as the characteristic binding energy for SEC, ΔH_{SEC}, and the characteristic binding energy for TER, ΔH_{TER}, respectively. These energy values are given in Table II with ΔH_{PRI} taken from Table I.

Each flour in Table II showed a reversal of sign; ΔH_{PRI} was negative, ΔH_{SEC} was positive, and ΔH_{TER} was negative. This made it appear that, although binding of PRI and TER are exothermic as

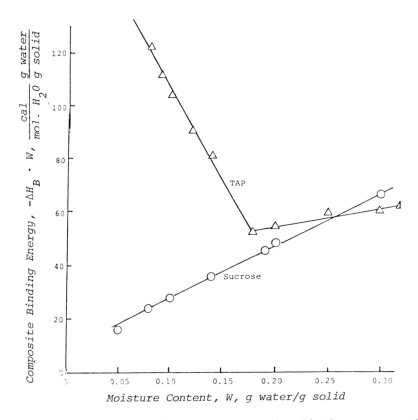

Fig. 4. Relationship between composite binding energy and moisture content for TAP and sucrose at 21.7°C.

TABLE II. Characteristic Binding Energy (cal/mol) of PRI, SEC, and TER

Materials	PRI (HPR)	SEC (HSEC)	TFR (HTER)
Full fat soy flour	−1083	+265	−73
Defatted soy flour	−1678	+549	−139
Dextrinized tapioca	−1434	+642	−63
Sucrose	−−	−−	−173

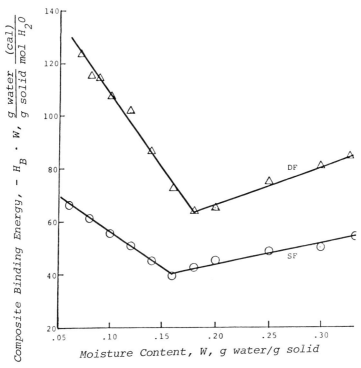

Fig. 5. Relationship between composite binding energy and moisture content for SF and DF at 21.7°C.

expected, binding of SEC is endothermic. Such is not the case, as shown in Fig. 6. Binding of SEC is also exothermic, but the positive value for ΔH_{SEC} means the heat of absorption is slightly less than the heat of condensation of pure water. This is still surprising but can be explained as follows: Sorption of SEC is characterized by a curve that rises slowly at low a_w but at an accelerated rate as a_w increases. This was classified as type III isotherm by Brunauer (1943) and Zettlemoyer et al. (1976). According to them, this type of isotherm involves a heat of absorption lower than or equal to but never higher than the heat of condensation of pure water. This statement is corroborated by our finding of a positive ΔH_{SEC}

ΔH_{TER} was slightly negative (Table II and Fig. 6), indicating that heat of absorption of TER was slightly higher than heat of

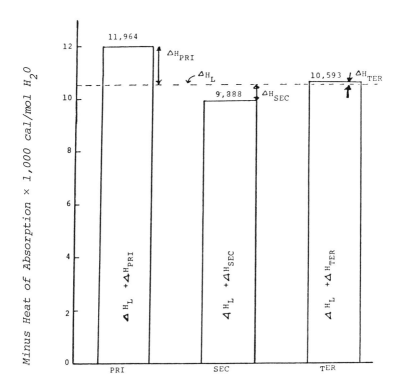

Fig. 6.   Schematic representation of characteristic binding energy of primary, secondary, and tertiary bound water on dextrinized tapioca as it relates to heat of condensation of water vapor H_L.

condensation of water and therefore higher than heat of absorption of SEC.   In the TER region, processes other than water absorption on a solid take place.   Examples are gelation, capillary absorption, and dissolution of solutes (Rockland, 1960).   These processes involve varying degrees of water attraction and of a_w lowering (Labuza, 1975).   Therefore, they would also involve varying levels of binding energy; thus ΔH_{TER} actually was a composite energy of all processes involved in the TER region, including dissolution of solutes as well as binding by macromolecules. This may be illustrated by analysis of sorption data for the soluble substance crystalline sucrose.   As shown in Fig. 4, only an ascending line characteristic of TER was obtained.   The value of ΔH_{TER} for sucrose was -173 cal/mol. This indicated that the characteristic heat of absorption for sucrose in the TER region was slightly higher than heat of condensation of pure water. In this region, the sucrose was no longer in a solid but in the liquid form, indicating that water association was no longer an absorption but rather dissolution process. This finding confirmed the negative value of ΔH_{TER} shown by DF, SF, and TAP.

References

AOAC (1970). "Official Methods of Analysis." 11th Ed. Assoc. Official Agr. Chemists. Washington, D.C.

Becker, H. A., and Sallans, H. R. (1956). *Cereal Chem. 33,* 79.

Berendsen, H. J. C. (1975). Specific interactions of water with biopolymers. *In* "Water, A Comprehensive Treatise," Vol. 5 (F. Frank, ed.). Plenum Press, New York.

Brunauer, S. (1943). "The Adsorption of Gases and Vapors," Vol. 1. Princeton Univ. Press, Princeton, New Jersey.

Brunauer, S., Emmett, P. H., and Teller, E. (1938). *J. Am. Chem. Soc. 60,* 309.

Bull, H. B. (1944). *J. Am. Chem. Soc. 66,* 1499.

Chung, D. S., and Pfost, H. B. (1967). *Trans. Am. Soc. Agri. Eng. 10,* 549.

Dole, M., and McLaren, A. D. (1947). *J. Am. Chem. Soc. 69,* 657.

Fish, B. P. (1957). Diffusion and thermodynamics of water in potato starch gel. *In* "Fundamental Aspects of the Dehydration of Foodstuffs." *Soc. Chem. Indr. 14.*

Greenwood, C. T. (1967). *Adv. Carbo. Chem. 22,* 483.

Kruger, G. J., and Helcke, G. A. (1967). Proton relaxation of water adsorbed on protein. *In* "Magnetic Resonance and Relaxation" (R. Blinc, ed.). North-Holland Publ., Co., Amsterdam.

Kuntz, I. D. Jr., and Kauzmann, W. (1974). Hydration of proteins and polypeptides. *Adv. Prot. Chem. 239.*

Labuza, T. P. (1968). *Food Technol. 22,* 263.

Labuza, T. P. (1975). Sorption phenomena in foods: Theoretical and practical aspects. *In* "Theory, Determination and Control of Physical Properties of Food Materials: (C. K. Rha, ed.). Reidel, Dordrecht.

Nash, L. K. (1970). "Elements of Chemical Thermodynamics." Addison Wesley, London.

Othmer, D. F., and Sawyer, F. G. (1943). *Ind. Eng. Chem. 35,* 1269.

Robinson, R. A., and Stokes, R. H. (1959). "Electrolyte Solution." Butterworths, London.

Rockland, L. B. (1960). *Anal. Chem. 32,* 1375.

Rodriguez-Arias, J. H., Hall, C. W., and Bakker-Arkema, F. W. (1963). *Cereal Chem. 40,* 676.

Schrenk, W. G., Andrew, A. C., and King, H. H. (1949). *Cereal Chem. 26*, 51.

Simatos, D., Faire, M., Bojour, E., and Couach, M. (1975). Differential thermal analysis and differential scanning calorimetry in the study of water in foods. *In* "Water Relations of Foods" (R. B. Duckworth, ed.). Academic Press, London.

Soekarto, S. T. (1978). Water Relations in Food Constituents and Their Application to the Development of a High Protein, Intermediate Moisture, Soybean Food. Ph.D. Thesis, Univ. of Illinois, Urbana.

Stitt, F. (1957). Moisture equilibrium and the determination of water content of dehydrated foods. *In* "Fundamental Aspect of Dehydration of Foodstuffs." *Soc. Chem. Ind. 14.*

Thompson, H. J., and Shedd, C. K. (1954). *Agri. Eng. 35*, 786.

Weast, R. C. (1972). "Handbook of Chemistry and Physics." The Chemical Rubber Co., Cleveland, Ohio.

Zettlemoyer, A. C., Micale, F. J., and Klier, K. (1976). Adsorption of water on well-characterized solid surfaces. *In* "Water, A Comprehensive Treatise," (F. Frank, ed.), Vol. 5. Plenum Press, London.

HYDRATION AROUND HYDROPHOBIC GROUPS

Hajime Noguchi

I. INTRODUCTION

When a macromolecule is dissolved in water three kinds of hydration occur: electrostrictional hydration around ionic groups, hydrogen-bonded hydration around polar groups, and hydrophobic hydration (the formation of icebergs) around hydrophobic groups. In the course of the study on the solution properties of ionic dextran derivatives (Noguchi *et al.*, 1973; Gekko and Noguchi, 1974, 1975), we were interested in the hydration around hydrophobic groups in the molecules. The clarification of the structure and role of the hydration should contribute to an understanding of the characteristic properties of the ionic polysaccharides and some biological phenomena such as their interaction with proteins (Gekko and Noguchi, 1978; Gekko *et al.*, 1978).

Water of hydration is considered to have a different structure than with bulk water (normal water). Therefore it is necessary to take into account the structure of bulk water in discussing the structure changes due to the hydration. Although extensive research on water structure has been conducted and several comprehensive reviews have been published (Berendsen, 1967; Eisenberg and Kauzmann, 1969; Franks, 1972), there has been no definite conclusion on the structure of normal water. At present it is understood that water is a highly structural liquid, with hydrogen bonds linking individual molecules to each other, although the actual arrangement of each molecule is not known. The structure

of normal water will be modified by any solute dissolved in the
water (Tanford, 1973).

II. AMOUNT OF HYDROPHOBIC HYDRATION

A. Estimation from the Partial Molal Volume of Ionic Dextran Derivatives

The partial molal volumes of sulfopropyl dextran and dextran
sulfate, which have various degrees of substitution, were determined with an ordinary pycnometer at 25°C (Gekko and Noguchi,
1974). The results are shown in Fig. 1. Supposing that the difference between their partial molal volumes at the degree of substitution 1.0 gives the partial molal volume of propylene group,
this volume is estimated to be 42.5 ml/mole, corresponding to
14.2 ml/mole of methylene.

If we could estimate the intrinsic volume V_{int}^{0} of methylene
in aqueous solution, then we can get the volume change due to the
hydrophobic hydration around the group ΔV_{ϕ} as follows:

$$\Delta V_{\phi} = V_{int}^{0} - V^{0} \tag{1}$$

where V^{0} is the partial molal volume of methylene. The V_{int}^{0} is
made up of two terms, the van der Waals volume V_{w}^{0}, and the void
volume (packing volume) V_{v}^{0}.

$$V_{int}^{0} = V_{w}^{0} + V_{v}^{0} \tag{2}$$

Bernal (1960) has proposed that simple liquids correspond in
structure to a randomly closed-packed array of equal size hard
spheres, and obtained the result that the packing density, that
is, $V^{0}/\left(V_{w}^{0} + V_{v}^{0}\right)$, becomes 0.634 for random close packing (Bernal
and Mason, 1960). Alder (1955) has shown that a 1:1 mixture of
two kinds of spheres, which have volumes in a ratio of as much as
5:1, have the same density for random packing as spheres of equal

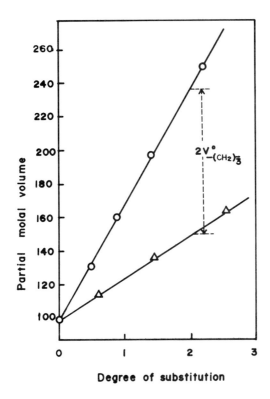

Fig. 1. The relation between partial molal volume and degree of substitution for dextran sulfate (△) and sulfopropyl dextran (○) at 25°C.

size. Therefore, difference in size of solute and water molecules may have only a minor effect on packing densities. The intrinsic volume of $-CH_2-$ in liquid state is obtained from the difference between the intrinsic volumes of homologous series of alcohol. Here, the intrinsic volume is obtained by dividing molecular weight of alcohol by its density. The intrinsic volume of methylene in liquid state thus obtained is 16.5 ml/mole (Gekko and Noguchi, 1974), and the intrinsic volume in aqueous solution should become the same value from the above explanation.

Therefore, from Eq. (1), the difference $14.2 - 16.5 = -2.3$ ml/mole corresponds to the volume decrease due to the hydrophobic

hydration around methylene group in aqueous solution. However,
it may remain to be confirmed whether the difference between the
partial molal volume of sulfopropyl dextran and that of dextran
sulfate gives the correct partial molal volume of propylene group.

On the other hand, the partial molal volume of $-CH_2-$ was
reported to be 15.9 ml, which was obtained from the partial molal
volumes of homologous n-alkanols (Nakajima et $al.$, 1975) and organic salts (Desnoyers and Arel, 1969) in aqueous solutions. In
this case the volume decrease due to the hydrophobic hydration
becomes $15.9 - 16.5 = -0.6$ ml/mole. Therefore, judging from both
above results we can conclude that the volume change accompanying
the hydrophobic hydration around the $-CH_2-$ group is $-1 \sim -2$ ml/
mole, which is much smaller than the value that has been accepted
for a long time (Kauzmann, 1959).

B. Estimation from the Partial Molal Volume of Surfactants

It is well known that the micelle formation of surfactant is
due to the hydrophobic bonds between the nonpolar chains of the
surfactant. Corkill et $al.$ (1967) found that the dependence of the
partial molal volume below and above the critical micelle concentration upon alkyl chain lengths was approximately linear in the
series studied and the difference between the partial molal volume
of the methylene group in the singly dispersed state and in the
micellar state was about $+1.5$ ml/mole on average. The micelle formation of surfactant involves the transfer of hydrocarbon from an
aqueous to a nonpolar environment, and the observation that micelle
formation is accompanied by an increase in the partial molal volume
can be attributed to the elimination of hydrophobic hydration
around the hydrocarbon chain of the surfactant. In order to derive
the amount of hydrophobic hydration around the methylene group from
their results, it is again necessary to take into account the intrinsic volume of the methylene group in the micellar and the singly
dispersed states.

Kresheck (1975) has reported that all the results obtained to date indicate that the micelle interior is generally fluid, although there are several indications that it is less fluid than a compatible hydrocarbon solvent of the same chain lengths as the surfactant at the same temperature. Harada *et al.* (1975) measured the partial molal adiabatic compressibility of dodecyl-hexaoxyethylene glycol monoether in aqueous solution and found that the compressibility of the dodecyl group in the micellar state was very close to that of *n*-dodecane at 25°C. They concluded that the micelle interior resembled liquid hydrocarbon. Therefore, it is reasonable to consider that the packing density of alkyl group in the micelle interior takes the same value as that in aqueous solution. Thus the difference of about +1.5 ml/mole is just the volume increment due to the dissolution of the hydrophobic hydration around the $-CH_2-$ group, that is, a volume decrease of about -1.5 ml/mole appears, accompanying the formation of hydrophobic hydration around methylene group.

III. ADIABATIC COMPRESSIBILITY OF HYDROPHOBIC HYDRATION

The adiabatic compressibility $\beta = -(1/V)(\partial V/\partial P)$ of solution is calculated from the speed of sound (u) in the solution and the solution density (d) using the following Laplace equation:

$$\beta = 1/u^2 d \tag{3}$$

The speed of sound is measured very accurately with a "sing-around" pulse velocimeter (Noguchi and Yang, 1971).

By analyzing the data on the adiabatic compressibilities of sulfopropyl dextran and dextran sulfate solutions, we obtained the result that the adiabatic compressibility of bulk water, β_w, is larger than that of water of hydrophobic hydration around propylene group, β_ϕ (Gekko and Noguchi, 1974). The result that the water of hydrophobic hydration becomes less compressible than bulk

water is consistent with Conway's (1966) proposition. Nakajima
et al. (1975) also reached the same conclusion on the basis of the
results of their research on the partial molal adiabatic compressibilities of n-alkanols and α,ω-alkane diols in dilute aqueous solutions. However it is very difficult to determine the absolute
value of β_ω at the present.

It has been recognized that the adiabatic compressibility of
water of hydrogen-bonded hydration, β_p, is nearly equal to that of
ice ($\beta_{ice} = 18\times10^{-12} cm^2/dyne$) (Shiio, 1958) and the electrostrictional water has a very small compressibility, β_e, being zero at
the first approximation (Yasunaga and Sasaki, 1951; Conway and
Verrall, 1966; Millero et al., 1978). Hence we can propose that
the relative order of increasing adiabatic compressibility of the
water participating in the hydration would be $\beta_e < \beta_p \sim \beta_\phi < \beta_w$.
But the order of β_p and β_ϕ remains undetermined. Determination of
the β_ϕ value, if possible, would help to clarify the structure of
water around the hydrophobic groups.

IV. TEMPERATURE DEPENDENCE OF HYDROPHOBIC HYDRATION

The temperature dependence of the hydrogen-bonded hydration
was investigated on various saccharides (Shiio, 1958) and dextran
(Nomura and Miyahara, 1964). Typical data are shown in Fig. 2
with our results (Noguchi, unpublished results). We have recon-
firmed the phenomena with the measurements on dilute saccharide
solutions (0.1-0.5%) since the previous measurements were carried
out on concentrate solutions (5-20%). However, the concentration
dependence of the hydration was not observed as seen in the
figure. The hydration around polar groups decreases quickly with
increasing temperature.

Ihnat et al. (1968) determined the amounts of hydrations of
D-glucose and tetrahydropyran-2-carbinol (THPA) as a function of
temperature (from 5° to 70°C) from the measurements on their

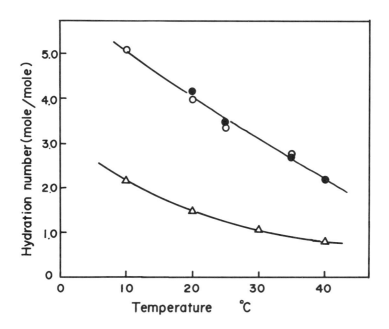

Fig. 2. Hydration number of D-glucose and dextran as a func-tion of temperature. ●, *Glucose (Shiio, 1958);* ▲, *dextran (Nomura and Miyahara, 1964).*

intrinsic viscosities. According to their results, the hydropho-bic hydration of THPA melts off a little more rapidly with in-creasing temperature than the hydrogen-bonded hydration of glucose.

We have tried to estimate the hydration around the hydrophobic groups in amino acids as a function of temperature (H. Noguchi, unpublished results). If M grams of solute having ionic and hy-drophobic groups is dissolved in V_0' ml of water, some water, say v_0' ml, will be attached to the solute and compressed to v_2' ml v_{2e}' around the ionic groups and $v_{2\phi}'$ around the hydrophobic groups-that is, $v_2' = v_{2e}' + v_{2\phi}'$, and there will result a solution of volume V' ml given by

$$V' = V_0' + V_1' - v_0' + v_{2e}' + v_{2\phi}' \qquad (4)$$

in which $V_1' = M/d_1$ is the volume of solute in V' ml solution, and d_1 the true density in solution where the hydration effect is not taken into account. Differentiating (4) with respect to the pressure P, and letting β_1 and β represent the adiabatic compressibility of solute and solution, respectively, we obtain the relation

$$V'\beta = V_0'\beta_w + V_1'\beta_1 - v_0'\beta_w + v_{2e}'\beta_e + v_{2\phi}'\beta_\phi$$

where β_w, β_e, and β_ϕ are same as previously defined. For a 1 ml solution,

$$\beta = V_0\beta_w + V_1\beta_1 - v_0\beta_w + v_{2e}\beta_e + v_{2\phi}\beta_\phi \qquad (5)$$

where $V_0 = V_0'/V'$, $V_1 = V_1'/V'$, $v_0 = v_0'/V'$, $v_{2e} = v_{2e}'/V'$, $v_{2\phi} = v_{2\phi}'/V'$, and there exists a relation

$$V_0 = (d-c)/d_0$$

in which d is the density of solution, c the concentration of solute in g/ml solution, and d_0 the bulk water density.

From Eq. (5) we obtain the formula

$$\Delta \equiv \beta/\beta_w - V_0 = V_1\beta_1/\beta_w - v_0 + v_{2e}\beta_e/\beta_w + v_{2\phi}\beta_\phi/\beta_w$$

and per gram of solute,

$$\Delta/c \equiv (\beta/\beta_w - V_0)/c = V_1\beta_1/\beta_w c - (v_0 - v_{2e}\beta_e/\beta_w - v_{2\phi}\beta_\phi/\beta_w)/c$$

$$(6)$$

Neglecting the compressibility of the solute

$$\lim_{c \to 0} \Delta/c = -(v_0 - v_{2e}\beta_e/\beta_w - v_{2\phi}\beta_\phi/\beta_w)/c$$

As β_e is assumed to be zero, if we measure $\lim_{c \to 0}\Delta/c$ for glycine and leucine,

$$\lim_{c \to 0} \Delta^G/c = -\left(v_0^G - v_{2\phi}^G\beta_\phi/\beta_w\right)/c \qquad (7)$$

$$\lim_{c \to 0} \Delta^L/c = -\left(v_0^L - v_{2\phi}^L\beta_\phi/\beta_w\right)/c \qquad (8)$$

The amount of hydration around ionic groups is same for glycine
and leucine because their pK_1' and pK_2' values are the same. Then
the difference δv_0 between the total amounts of hydration around
leucine and glycine becomes $\delta v_0 \equiv v_0^L - v_0^G = v_{2p}^L - v_{2\phi}^G$, provided
that the water of hydrophobic hydration is not so strongly com-
pressed. Then, subtracting Eq. (7) from Eq. (8),

$$\lim_{c \to 0} \Delta^L/c - \lim_{c \to 0} \Delta^G/c = -\delta v_0 (1 - \beta_\phi/\beta_w)/c \qquad (9)$$

When the value of β_ϕ is known, the amount of the hydration
around $-CH_2CH(CH_3)_2$ can be calculated from Eq. (9). At the
present this is impossible, but if we measure the values of the
left-hand side of Eq. (9) at different temperatures, we can obtain the relative amount of hydration as a function of temperature.
The values of $\phi v_0 (1 - \beta_\phi/\beta_w)/c$ thus obtained are given in Fig. 3
as a function of temperature. The hydration around the
$-CH_2CH(CH_3)_2$ group of leucine decreases more rapidly with increasing temperature compared with the case of THPA.

If we assume $\beta_\phi = \beta_w/2$, the amount of hydrophobic hydration
becomes 40 ml/mole at 20°C, corresponding to about 2 mole/mole in
hydration number, and the value seems to be appropriate one
compared with that of THPA. Recently Millero *et al.* (1978) determined the number of water molecules hydrated to amino acids at
25°C by the measurements on their partial molal volumes and adiabatic compressibilities. However, they computed the hydration
number based on the assumption that water molecules are attracted
only to ionic groups in amino acids, and so the approach masked
contributions from other types of interactions such as hydrophobic
hydration.

In the case of glycine, we can obtain the amount of electrostrictional hydration from Eq. (7) neglecting the second term on
the right-hand side. The results are given in Fig. 4 along with
the data of Goto and Isemura (1964), which were determined by the
speed of sound measurements on a 2% solution. As seen from the

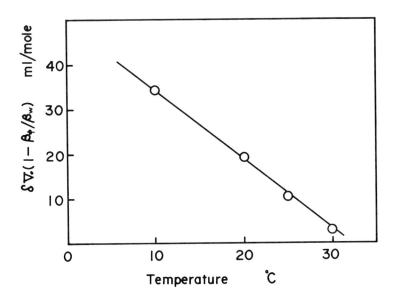

Fig. 3. Relative amount of hydrophobic hydration around leucine as a function of temperature.

figure, the temperature dependence of the electrostricted hydration is much smaller compared with the case of hydrogen-bonded hydration or hydrophobic hydration.

V. CONCLUSION

We have elucidated the characteristic properties of the hydration around hydrophobic groups and obtained the following results:

(1) The volume change due to the hydrophobic hydration around methylene group is $-1 \sim -2$ ml/mole, which means that there are not so many bound water molecules around the group. On the other hand, the hydration numbers around THPA were reported to be about 2 mole/mole at $20°C$. Therefore it is reasonable to suppose that

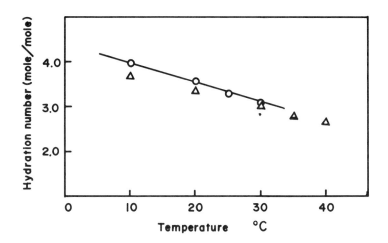

*Fig. 4. Hydration number of glycine as a function of tem-
perature   , Goto and Isemura (1964).*

an icelike block such as an "iceberg" does not exist around the
hydrophobic groups in aqueous solution.

(2) The diabatic compressibility of the water of hydrophobic
hydration is larger than that of the electrostricted water and
smaller than that of normal water.

(3) The water molecules participating in the hydration are
released more rapidly with increasing temperature in the order
electrostrictional hydration < hydrogen-bonded hydration < hydrophobic hydration.

Water molecules are attracted to biopolymers such as polysaccharides or proteins according to these hydration modes, and
as a whole the amount of hydration around the biopolymers at near
0°C should be more than twice that at room temperature. It has
been interpreted that unfreezable water around the biopolymers
below the freezing point corresponds to the water of hydration by
reason that the amount of the unfreezable water is almost equal

to the amount of hydration water at room temperature. However, it
is necessary to reexamine the interpretation since there exists
such a strong temperature dependence of the hydration.

Acknowledgments

The author would like to express his appreciation to Dr. K.
Gekko for his collaboration in this work. Special thanks are due
to Meito Sangyo Company, Nagoya, for supplying the dextran derivatives.

References

Alder, B. J. (1955). *J. Chem. Phys. 23*, 263.
Berendsen, H. J. C. (1967). *In* "Theoretical and Experimental Bio-
    physics" (A. Cole, ed.), Vol. 1, p. 1. Marcel Dekker, New
    York.
Bernal, J. D. (1960). *Nature 185*, 68.
Bernal, J. D., and Mason, J. (1960). *Nature 188*, 908.
Conway, B. E., and Verrall, R. E. (1966). *J. Phys. Chem. 70*,
    3952.
Corkill, J. M., Goodman, J. F., and Walker, T. (1967). *Trans.
    Faraday Soc. 63*, 768.
Desnoyers, J. E., and Arel, M. (1969). *Can. J. Chem. 47*, 547.
Eisenberg, D., and Kauzmann, W. (1969). "The Structure and
    Properties of Water." Oxford Univ. Press, London.
Franks, F., ed. (1972). "Water--A Comprehensive Treatise," Vol. 1.
    Plenum Press, New York.
Gekko, K., and Noguchi, H. (1974). *Macromolecules 7*, 225.
Gekko, K., and Noguchi, H. (1975). *Biopolymers 14*, 2555.
Gekko, K., and Noguchi, H. (1978). *J. Agri. Food Chem. 26*, 1409.
Gekko, K., Harada, H., and Noguchi, H. (1978). *Agri. Biol. Chem.
    42*, 1385.
Goto, S., and Isemura, T. (1964). *Bull. Chem. Soc. Japan 37*,
    1697.
Harada, S., Nakajima, T., Komatsu, T., and Nakagawa, T. (1975).
    Chem. Lett. (Chem. Soc. Japan), 725.
Ihnat, M., Szabo, A., and Goring, D. A. I. (1968). *J. Chem. Soc.
    A*, 1500.
Kauzmann, W. (1959). *Adv. Prot. Chem. 14*, 1.
Kresheck, G. C. (1975). *In* "Water--A Comprehensive Treatise"
    (F. Franks, ed.), Vol. 4, p. 95. Plenum Press, New York.
Millero, F. J., Surdo, A. L., and Shin, C. (1978). *J. Phys. Chem.
    82*, 784.

Nakajima, T., Komatsu, T., and Nakagawa, T. (1975). *Bull. Chem. Soc. Japan 48*, 783-788.

Noguchi, H., and Yang, J. T. (1971). *Biopolymers 10*, 2569.

Noguchi, H., Gekko, K., and Makino, S. (1973). *Macromolecules 6*, 438.

Nomura, H., and Miyahara, Y. (1964). *J. Appl. Polymer Sci. 8*, 1643.

Shiio, H. (1958). *J. Am. Chem. Soc. 80*, 70.

Tanford, C. (1973). "The Hydrophobic Effect," p. 19. Wiley, New York.

Yasunaga, T., and Sasaki, T. (1951). *J. Chem. Soc. Japan, Pure Chem. Ser. 72*, 366.

SOLUTE MOBILITY IN RELATION

TO WATER CONTENT AND WATER ACTIVITY

R. B. Duckworth

Behavior of solutes in partially hydrated absorbent systems
and the molecular mobility of simple soluble substances bear
directly on the important questions of the rates of chemical
change and of microbiological activity in foods. Mobility is,
of course, a term that may be used to describe a whole range of
degrees of molecular freedom. At the most restricted level, for
example, it is known that water in extremely small amounts can
facilitate intramolecular conformational changes in other substances (Chirgadze and Ovsepyan, 1972) and such limited mobility
could in turn influence the rates of monomolecular reactions in
very dry systems. On a different level, progress of chemical
reactions involving more than one reactant depends on all or all
but one of these reactants being sufficiently mobile at the
molecular level to move to the vicinity of the other(s). Substrates of enzymatic reactions must be capable of diffusing to
the active sites of the enzyme molecules or alternatively the
latter must be capable of approaching the substrate molecules.
Similarly, if microorganisms are to obtain the nutrients necessary
for their continued activity and growth, then these nutrients
must become accessible in the vicinity of the cell surfaces of the
microorganisms.

There are various ways in which solute behavior may be approached experimentally. In the simplest case, physical techniques such as infrared spectroscopy and x-ray analysis have been used to detect intramolecular conformational changes when initially dry films of pure test materials are progressively hydrated by stepwise exposure to atmospheres of higher relative humidities (Chirgadze and Ovsepyan. 1972). It is possible also to study more complex systems directly, looking for visible evidence of solute movement by using tracer techniques (Ducksworth, 1962; Ducksworth and Smith, 1962, 1963).

Alternatively, the measurement of various colligative properties may be utilized after a test substance has been added to and dispersed through the aqueous phase of a test material. This experimental approach is based on the assumption that an added soluble material will become uniformly distributed through all the water in the system that is free to act as a solvent, with the added proviso that no spurious surface effects are present that might interfere with this uniformity of distribution. Having added a known amount of test substance, its final concentration may be measured directly, or the effect of the addition on some known concentration-dependent property of solutions, such as the depression of freezing point or the reduction of water vapor pressure may then be measured. In cases where it is possible to express or otherwise extract a colloid-free solution from the test material, some convenient naturally occurring solute may be used as the "indicator" substance and its concentration determined both in the extract and in the residue, thus obviating the need to make any addition.

The simplest system for which this kind of experimental approach is possible is a three-component one in which two phases are represented. Inevitably, therefore, uncertainties arise with respect to possible interactions between solid components, surface effects, etc. which could influence behavior. Moreover, in attempting to extend this approach to even more complex systems such as food materials, these uncertainties are further compounded and increased.

Since the earliest measurements of Newton and Gortner (1922) on plant colloids, many authors have utilized variants of this approach to obtain values for the nonsolvent water present in various systems (Barthel *et al.*, 1928; Sandberg *et al.*, 1930; Newton and Cook, 1930; Barkas, 1932; Dumanskii, 1933; Brooks, 1934; Skovholt and Bailey, 1935; McDowell and Dolby, 1936; Vail-and Bailey, 1940; Eilers, 1945; Eilers and Laboret, 1946; Moquot, 1947; Wolowinskaya, 1953; Guilbot and Lindenberg, 1960; Lindenberg *et al.*, 1963; Gari-bobo, 1964; Dang-Vu-Bien, 1965; Kayser, 1968; Elford, 1970; Ball and Breese, 1970; Guerts, 1972; Walstra, 1973; Guerts *et al.*, 1974). Among food materials so investigated are cereals (Newton and Cook, 1930; Skovholt and Bailey, 1935; Vail and Bailey, 1940), beef and pork (raw and cooked) (Vail and Bailey, 1940), potato, (Dumanski, 1933), yeast (Guilbot and Lindenberg, 1960; Gari-bobo, 1964), and if one counts frog's legs among the culinary delicacies, frog leg muscle (Brooks, 1934). Most of this work, however, relates to dairy products, in particular, cheese (Barthel *et al.*, 1928; Sandberg *et al.*, 1930; Mc-Dowall and Dolby, 1936; Eilers, 1945; Wolowinskaya, 1953; Keyser, 1968; Guerts, 1972; Wolstra, 1973; Guerts *et al.*, 1974). Some pure macromolecular food constituents have also been studied in this way, for example, (Lindenberg *et al.*, 1963; Dong-Vu-Bien, 1965) egg albumin (Bull and Breese, 1970), gelatin (Bari-bobo, 1964) and casein (Guerts, 1972; Walstra, 1973; Guerts *et al.*, 1974).

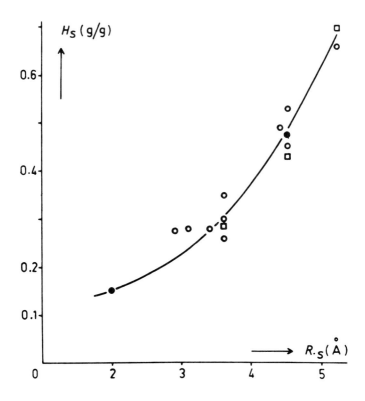

Fig. 1. Amount of nonsolvent water (Hs) in paracasein as
found for solutes of different molecular radius (Rs) (Walstra,
1973).

A number of workers have shown that the amount of nonsolvent
water in a given system differs according to the particular
solute under consideration (Eilers and Labout, 1946; Moquot, 1947;
Dang-Vu-Bien, 1965; Bull and Breese, 1970; Guerts, 1972; Walstra,
1973; Guerts et al., 1971). The discussion of this phenomenon
has centered largely round the situation in dairy products.
Figure 1 illustrates results obtained at Wageningen by Walstra
and Guerts and their colleagues for various indicator solutes in
hydrated paracasein. In particular, for the sugars fructose,
lactose, and raffinose, the corresponding values for nonsolvent

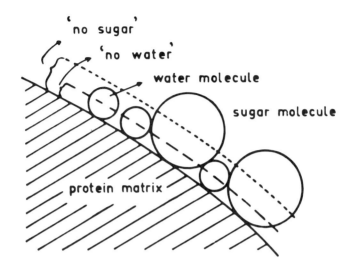

Fig. 2. A certain space around the protein surface is accessible to water molecules but not to those of sugar. Water present within this space is nonsolvent for the sugar. (Geurts et al., 1974).

water are, respectively, 28.7, 42.8, and 69.6 per 100g of protein.

The explanation proferred for these differences in behavior by different solutes is that of steric exclusion (Walstra, 1973; Geurts *et al.*, 1974). It is postulated that surface films of water and water-filled capillaries of cross-sectional dimensions less than the diameter of the solute molecules concerned cannot accommodate such molecules and the water in such films cannot therefore act as a solvent for the particular solute. An illus-tration of this effect, again taken from Guerts *et al.* (1974) is shown in Fig. 2. An interesting conclusion arrived at in this work is that the water nonsolvent for sodium chloride is much less than for the sugars examined and this is in accord with the fact that the radii of Na^+ and Cl^- ions are comparable with that of the water molecule itself and much smaller than those of the sugars. Indeed, in some cases negative values were obtained for water nonsolvent for the Cl^- ion, indicating a net adsorption of ions by the protein. However, from sodium determinations values of between about 7.5 and 15 g per 100 g protein were estimated for the water nonsolvent for sodium chloride in hydrated para-caseinate.

During the period between 1963 and 1971, Gal (1975) and his colleagues at Berne carried out a very thorough, detailed, and elegant series of investigations into the system casein-sodium chloride-water, the results of which he summarized in 1974. Gal was able to show that sodium chloride may be present in four different forms in this system, i.e. crystalline salt, amorphous salt, salt bound to the protein, and dissolved salt. He prepared a phase diagram for the casein-sodium chloride-water system with water activity as a parameter.

Crystalline sodium chloride was measured by x-ray diffraction and the presence of solvent water for sodium chloride was inferred when changes in the degree of crystallization (necessitating translational motions of the salt ions) occurred during the approach to equilibrium. On this basis, solvent water could be said to be present a water activity >0.3, which for pure casein would correspond to a water content of some 7-8 g per 100 g of dry casein, a level agreeing with the lowest value reported by Guerts *et al.* (1974).

A further interesting contribution to the discussion of the solvent properties of absorbed water was made by Duprat and Guilbot (1975). This concerned the water available as a solvent for alcohol in the specific case of starch, in particular granular starch. In this case, the solute, being volatile, could be introduced through the vapor phase. It was found that in granular starch water capable of acting as a solvent for ethanol begins to accumulate above a water activity of 0.1, when the most active water sorption sites in the starch are saturated. However, because of the restrictions imposed by the structure of the starch granule, which strictly limit the extent to which it can expand and therefore the total quantity of water that it can absorb, alcohol becomes excluded as maximum uptake of absorbent is approached, presumably because of sheer lack of space. Otherwise the water present could presumably be quite capable of acting as a solvent for the alcohol. In gelatinized starch, ethanol was absorbed directly by the absorbent from zero water activity and it was not possible to distinguish a fraction of nonsolvent water. This serves a useful further reminder that structural considerations can be important in determining whether water is free to exert its solvent properties.

Since 1970 we have been applying an experimental procedure involving the use of wide-line nuclear magnetic resonance in an attempt to further elucidate some of the outstanding questions in this area. Briefly, the instrument used (Newport Quantity Analyzer) is tuned to induced resonance in hydrogen nuclei and measures the energy absorbed from a radio frequency (RF) field when this resonance occurs. The gate of the instrument is set so that only the sharp signal due to resonance of protons in the liquid state is measured. As a result, neither the hydrogen atoms of the solid in a sample nor those of the most firmly bound water molecules, which behave in effect as part of the solid, contribute to the signal. However, when molecules of a hydrogen-containing solute dissolve in the water of a sample these do contribute to the liquid-state signal. A means is therefore provided of detecting the solution of an added test solute and therefore of measuring nonsolvent water.

In the first place experiments were carried out on a range of suitalbe test solutes in a number of model absorbents. The experimental procedure used for the main part in this work was to admix a given weight (normally 0.1 g) of finely divided test

solute into each of a series of samples (usually 1 g) of the
absorbent hydrated to progressively higher levels, to allow time
(several days) for equilibration, and then to measure the resonance signals in comparison with those from control samples of
absorbent hydrated to the same levels as the test samples. The
solutes used were selected for the main part because of their
relatively high hydrogen content, low molecular weight, and high
solubility, and included a number of common food constituents
or additives. They had of course, also to be solid at the
temperature of testing (25°C).

Determinations were made in each case of the level of hydration of the absorbent at which an incremental contribution
to the resonance signal was obtained in the presence of the test
solute. These have been referred to as mobilization points but
may also be regarded as the limits of nonsolvent water for the
systems concerned.

One of the earliest generalizations possible from the results of this work, in agreement with the conclusions of some
previous workers referred to above, was that the mobilization
point for a given system is peculiar to that system, different
test solutes being mobilized in a given absorbent at different
levels of hydration (Duckworth, 1972; Duckworth and Kelly, 1973;
Duckworth et al., 1975). However, attempts to explain the
pattern of results in terms of molecular radii and steric exclusion have been singularly unsuccessful. While it is generally
true that of the solutes tested those with lower molecular weights
and smaller molecules have lower mobilization points there are
notable exceptions that invalidate the general application of
such a rule. For example, among the sugars, sucrose and glucose
have closely similar mobilization points (23 and 22 g/100 g
dry starch, respectively) whereas fructose has one substantially
lower (-16.5) in gelatinized potato starch. Again, glycine, with
molecular weight 75.07, becomes mobilized in agar at -41 g/100 g

dry agar, whereas citric acid (molecular weight 192.12) is mobilized at 28 g/100 g. On the other hand, all the substances with
especially low mobilization points have small highly polar
molecules each with groupings capable of forming firm hydrogen
bonds. These groups will interact strongly with water molecules
and with polar sites on the absorbent surfaces to which the water
may become bonded, and such molecules will therefore be particularly effective in disrupting existing intermolecular associations
in a hydrated absorbent system. Again, it is noteworthy that
the determined mobilization points for these especially mobile
substances lie close to the BET monomolecular layer values for
the respective absorbents calculated from sorption data.

Another example of such "competition" for water is the
formation of crystalline hydrates of added test substances by the
withdrawal of water from the absorbent. This phenomenon has been
found to occur with several of the substances examined and, since
water in crystals does not contribute to the liquid-state resonance
signal, this results in a reduction of the signal given by the
test sample below that of the control (Duckworth et al., 1975).

When the various soluble substances tested are ranked in
order of increasing mobilization points, it is found that the
order is the same for each different absorbent. Although the
agreement is not always as close as in the examples cited by
Duckworth (1972), it is generally the case that the mobilization
points for a given solute in different absorbents, though differing substantially in terms of actual water content, correspond
to similar levels of water activity in each of the different systems concerned.

A specific case where comparison is possible between our
results and those of Guerts and his co-workers for dairy products
is that of sucrose in hydrated casein. The values reported by
Guerts et al. (1974) for the water nonsolvent for sucrose (or
sucrose + lactose) in cheese (expressed as grams of water per

100 g protein) range from 24.3 to 41.6. The mobilization point
for sucrose in pure hydrated Hammerstein casein determined in my
laboratory lies between 22 and 23.5 g water per 100 g casein
(Duckworth et al., unpublished results). It should be added that
Guert's lowest value relates to a sample containing a substantial
amount of salt that according to that author would cause a reduction in the determined value.

The pattern of mobilization of a given solute in fact is
usually altered when a second soluble component is added to the
system (Duckworth et al., 1975). The change usually takes the
form of a reduction in the level of hydration that will support
solution, resulting either in the mobilization of both solutes
at the lower of the two points determined for the two components
individually or in a reduction of the mobilization point for the
mixture even below this level. An example of this is illustrated
in Fig. 3 for the case of γ-aminobutyric acid and glucose. It
should be added, however, that for all of the many combinations
of solutes investigated (Duckworth et al., unpublished results)
no cases have been recorded of the lowering of a combined mobilization point below a level of hydration corresponding roughly
to the calculated BET monomolecular layer value for the given
absorbent. The implications of this pattern of results for more
complex food systems containing numerous low-molecular weight
soluble constituents is, of course, that some mobility of the
soluble solids (the precise pattern depending on the spectrum and
solubility relations of the components) is to be expected upwards
from around the monomolecular layer moisture levels for the
supporting macromolecular constituents. Such a picture is in
general accord with the results of the earlier investigations
using radiotracer techniques (Duckworth, 1962; Duckworth and Smith,
1962, 1963), which showed that diffusibility of simple solutes is
detectable in dried food materials at water contents slightly in

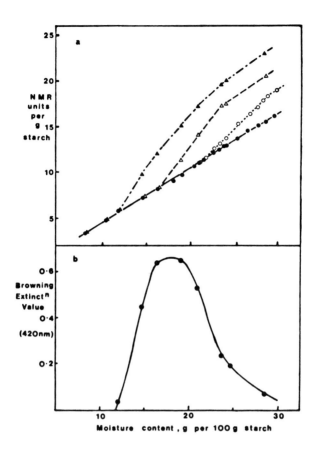

*Fig. 3. Relationship between reactant mobilization and re-
action rate. (a) Mobilization of γ-amino butyric acid, glucose,
and a mixture of the two in hydrated gelatinized potato starch.
Control starch, O; + γ-aminobutyric acid, Δ; + glucose hydrate,
O; + γ-aminobutyric acid; + glucose hydrate, Δ. (b) Browning
extinction values of water extracts of the samples used in (a)
after seven days at 25°C (From Duckworth, Allison and Clapperton,
1976).*

excess of the respective monomolecular layer values. These

results also help to highlight one of the problems associated with the application of wide-line nuclear magnetic resonance to the determination of moisture in low- and intermediate-moisture foods.

Returning to the data illustrated in Fig. 3, this example also serves to illustrate the practical significance of this sort of information about solute mobility in relation to the area of food stability and rates of chemical spoilage. In this instance close agreement is shown between the mobilization of a mixture of nonenzymic browning reactants, γ-aminobutyric acid and glucose, as demonstrated by the NMR technique, and the rapid build-up in the rate of the resulting reaction. The peak in the reaction rate curve corresponds to the hydration level at which the reactants have been completely mobilized, above which the subsequent dilution of the reactants causes the rate of reaction to fall away. We reported previously similar effects of mobilization of ascorbic acid on the rate of loss of this vitamin in dehydrated cabbage and in potato starch gel (Duckworth et al., 1975; Seow, 1975).

The behavior of electrolytes is of special interest because of their almost universal presence in foods, their major influence on water structure, their possible interactions with food macromolecules, and their relatively strong direct effects in solution on water activity. In their studies of the solvent properties of water in cheese, Guerts et al., (1974) found that the amount of water nonsolvent for sugars was substantially reduced in the presence of sodium chloride, the more so as the concentration of salt was increased, whereas the water nonsolvent for salt itself remained fairly constant between 10 and 15 g per 100 g protein irrespective of salt concentration. This they argue is consistent with the concept of the steric exclusion of solute molecules, the reduction in water nonsolvent for sugar resulting from a decrease in surface area of the protein caused by shrinkage at the higher salt concentrations.

We have also found that sugars become mobilized at lower levels of hydration in the presence of sodium chloride. This has been shown to be the case with sucrose in hydrated casein and in gelatinized potato starch (Baird, 1976) and for glucose in starch and in agar (Duckworth et al., 1975). The higher the concentration of the salt, the greater is the reduction in the amount of absorbed water rendered unavailable as a solvent for sucrose (Baird, 1976) (see Fig. 4) a pattern that is again in agreement with that found by Guerts et al. (1974). The chlorides of lithium and potassium and the acetates of lithium, sodium, and potassium have also been shown to produce similar effects (see Table I).

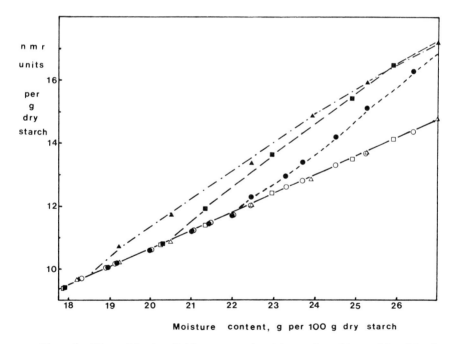

Fig. 4. The effect of the concentration of sodium chloride in hydrated starch gel on the mobilization of sucrose (25°C). Control starch, O; starch + sucrose, ●; control starch with 4% NaCl, ; starch with 4% NaCl + sucrose, ; control starch with 8% NaCl, Δ; starch with 8% NaCl + sucrose, ▲ (after Baird, 1976).

The order of the effect for the chlorides in each case follows the lyotropic series, which is closely related to the ionic radii of the cations. Bull and Breese (1970) have pointed out that whereas the unhydrated radii of these cations are in the order K+ > Na+ > Li+, the order for the hydrated ions is reversed. These same authors were able to show, using an isopiestic method, that the alkali metal chlorides reduce the amount of water bound as nonsolvent water in egg albumin, the larger the hydrated radius of the cation the greater being the reduction. Thus, lithium chloride produced the greatest effect followed by sodium chloride and finally potassium chloride.

The behavior of the acetates of the alkali metals examined in the present work is interesting because here the order of the effect was found to be reveresed (Table I). Bull and Breese (1970) comment that it appears from their results that in contrast to the effect of the cations, the larger the monovalent anion, the greater is the tendency to bind to the protein and the greater is its dehydrating effect upon the protein. This, however, would not explain the greater influence of potassium and sodium acetates than of lithium acetate in reducing the amount of water nonsolvent for glucose in the agar system and this difference can only be attributed for the present to unexplained individual variations in behavior of the salts concerned.

The results of experiments with such relatively simple model systems help to provide answers to some basic questions and allow predictions to be made with respect to the behavior of soluble constituents in more complex materials. Detailed patterns of behavior may, however, differ according to the particular experimental procedures used. For example, when the test solute and the absorbent are initially separate, as when the solute is admixed with samples of an absorbent already hydrated to appropriate levels, the results may be regarded as reflecting the properties of water sorbed on the pure absorbent and even then the detailed pattern of behavior may depend on the form in which

TABLE I. Effects of Alkali Metal Salts on the Mobilization Points of Sugars in Hydrated Polysaccharide Absorbents

Absorbent	Test solute	Mobilization points in absorbents containing added salts (grams of water/100 g dry absorbents)				
		Control	LiCl	NaCl	KCl	Level of salt addition
Gelatined potato starch	Sucrose	22.25	14.0	18.5	18.5	8 g per 100 g dry absorbent
	Glucose	25.5	16.25	18.25	19.5	0.154 moles per 100 g dry absorbent
Agar	Glucose	36.5	25.3	28.0	30.0	0.154 moles per 100 g dry absorbent

Absorbent	Test solute	Mobilization points in absorbents containing added salts (grams of water/100 g dry absorbents)				
		Control	Li acetate	Na acetate	K acetate	Level of salt addition
Agar	Glucose	36.5	25.2	22.0	19.5	0.154 moles per 100 g dry absorbents

the test solute is introduced, i.e., whether crystalline or
amorphous, anhydrous, or in the form of a hydrate (Duckworth *et*
al., unpublished results, 1975). This is particularly true of
substances that form crystalline hydrates.

A different situation arises when, as is more usually the
case in manufacturing practice, soluble substances are finely
distributed through an absorbent before the latter is dried.
Here, full scope exists for interactions between macromolecular
components and any small solute molecules present and the water-
binding properties of the absorbent may therefore be altered
(Bull and Briese, 1970).

Mobilization of solutes can be detected in such cases by
comparing the resonance signals from hydrated samples of absorbent containing the solute, with comparable samples of pure
absorbent. When the solute is mobilized, an incremental increase
in signal is found relative to that for the pure absorbent and
this increase, as before, is roughly proportional in size at its
maximum to the number of hydrogen protons contained in the solute
present. While it is to be expected that the absorbent/solute
interactions mentioned above might have some effect on the solvent
properties of the absorbed water, for almost all of the absorbent/
solute combinations so far examined using each of the two different methods of sample preparation described, patterns of
mobilization determined by the two different procedures have
shown good agreement.

It has already been suggested that water activity, more so
than water content itself, is a useful indicator of the ability
of absorbed water to support solute mobility in that it reflects
the condition of the water present rather than its sheer amount.
At the same time, the very presence of a solute in the system,
as we have seen, in turn affects the condition of the water and
through this the water activity. At higher moisture levels, any
soluble substance dissolving in the free water of a good material
acts through Raoult's law to reduce the activity of the water,

and in practice the reduction in the majority of cases is substantially greater than the theoretical. As a result, the water
sorption isotherm for a system containing a soluble additive runs,
in the higher moisture range, *above* the isotherm for the absorbent
alone (see Fig. 5). Generally, over the lower moisture range the
additive has the effect of reducing the amount of water held by
the absorbent at any given water activity and therefore causes
the isotherm to run *below* that for the absorbent alone. At some
intermediate point, the two isotherms intersect and it can be
shown that, for moderate levels of additive, the point of intersection is little affected by the proportion of added solute in
the system. This zone of intersection is, however, different and
apparently characteristic for each particular solute.

A reasonable inference from this pattern of behavior is that
the intersection point of the isotherms in such a case coincides
with the mobilization point or limit of nonsolvent water for the
solute concerned and in the large majority of cases so far examined a close agreement has indeed been found between the mobilization point determined by the NMR method and the intersection
point of water sorption isotherms for the pure absorbent itself
and for the absorbent with the respective solute added to it
(see Fig. 6).

The methods developed for studying the patterns of mobilization of water-soluble additives by measuring the NMR of hydrated
systems may therefore also be used as simple and relatively rapid
methods for determining the water content/water activity levels
above which given hydrogen-containing substances will bring about
reductions in the water activity of a particular absorbent system.
Alternatively, spot checks may quickly be made at a particular
level of hydration to establish whether or not A_w reduction will
result from the introduction of a range of selected additives.

Finally, in attempting to summarize present knowledge concerning the solvent properties of water in foods at low and inter-

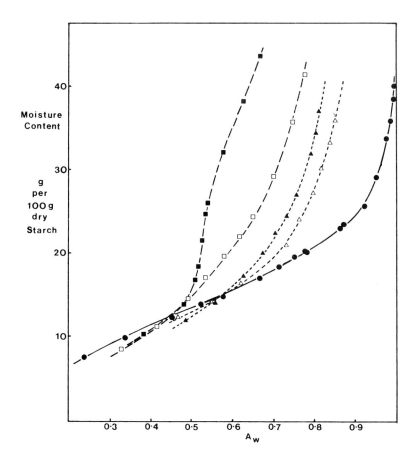

Fig. 5. Water sorption isotherms for gelatinized potato
starch and for similar starch-containing added solutes in dif-
ferent proportions. Control starch, ●; starch + 1-% Ch₃COONₐ,
Δ; starch + 15% CH₃COONₐ, ▲; starch + 20% HCOONa, ▢ ; starch +
30% HCOONa, ■ (all percentages on a dry starch basis.)

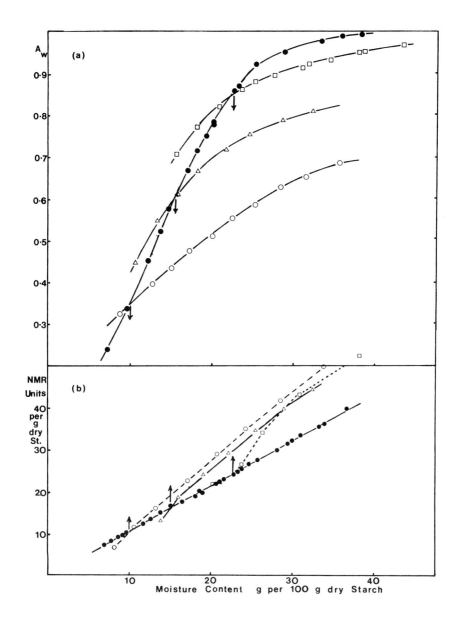

mediate levels of hydration the following general conclusions may
be drawn:

1. The first very firmly bound layer of absorbed water molecules in food and food constituents is effectively devoid of solvent properties, although some degree of intermolecular mobility
may be possible within this moisture range.

2. The bulk of the water in excess of this monomolecular layer
is capable of supporting the mobility of whole solute molecules
of small size, especially if these are highly polar in nature.

3. The solution of particular solutes in absorbed water is
facilitated in the presence of other small polar molecules of
small size, especially if these are highly polar in nature.

3. The solution of particular solutes in absorbed water is
facilitated in the presence of other small polar molecules and
of ions, electrolytes with ions of relatively small radius being
especially effective in reducing the amounts of nonsolvent water.

4. Because of (2) and (3), solution processes will normally
occur in biological materials and other multicomponent food
systems at all levels of hydration above that corresponding
roughly to a monomolecular layer of absorbed water.

5. Structural and steric factors may, however, play some part
in determining the levels of water that are nonsolvent for particular soluble components, according to the molecular size and
shape.

*Fig. 6. Relationships between the intersection points of
water sorption isotherms for gelatinized potato starch with or
without added solutes and the mobilization points, determined
by NMR, of the respective solutes in hydrated starch gel. Control starch, ●; starch + glucose (10%), □ ; starch + COONa (10%)
and glucose (10%), Δ; starch + HCOONa (10%) CH$_3$COOK (10%) and
C$_2$H$_5$COONa (10%), O.*

6. At levels above the mobilization point for a particular solute, the addition of this substance will result, through the effect of Raoult's law, in a consequent reduction in water activity at any given moisture content.

7. The mobilization of substances capable of taking part in reactions leading to deterioration in quality results in increased rates of reaction and may cause rapid spoilage.

ACKNOWLEDGMENTS

The author wishes to acknowledge that assistance with experimental work described in this chapter was provided by Mrs. Joy Y. Allison, B.Sc., Mr. S. Baird, B.Sc., and Miss H. A. Anne Clapperton, B.Sc., in the Department of Food Sciences and Nutrition, University of Strathclyde, Glasgow, Scotland, and by Miss Angelina Bainiah and Mr. J. K. Attah in the Department of Nutrition and Food Science, University of Ghana, Legon, Ghana.

Acknowledgment is also due to Dr. T. J. Geurts and Dr. P. Walstra of the Department of Dairy Technology of the Agricultural University of Wageningen, Netherlands, and the editors of the Netherland Milk and Dairy Journal and of Kolloid Zeitung for permission to reproduce Figs. 1 and 2, and to Applied Science Publishers, Ltd. for permission to reproduce Figs. 3-5.

REFERENCES

Baird, S. (1976). B. Sc. Thesis, Univ. of Strathclyde, Glasgow.
Barkas, W. (1932). *Nature (London) 130,* 699.
Barthel, C., Sandberg, E., and Haglund, E. (1928). *Lait 8, 285,* 762, 891.
Brooks, J. (1934). *J. Gen. Physiol. 17,* 783.
Bull, H. B., and Breese, K. (1970). *Arch. Bioch. Biophys. 137,* 299.
Chirgadze, Y. N., and Ovsepyan, A. M. (1972). *Biopolymers 11,* 2179.

Dang-Vu-Bien (1965). "Contribution a l'étude de l'absorption preferentialle de l'eau par la cellulose a partir des solutions ioniques et non ioniques," Thesis, Faculty of Science, Paris.

Duckworth, R. B. (1963). In "Recent Advances in Food Science," Vol. 2 (J. Hawthorn and J. M. Leitch, eds.), p. 46. Butterworths, London.

Duckworth, R. B. (1972). Proc. Inst. Food Sci. Technol. 5, 60.

Duckworth, R. B., and Kelly, C. E. (1973). J. Food Technol. 8, 105.

Duckworth, R. B., and Smith, G. M. (1962). In "Recent Advances in Food Science," Vol. 3 (J. M. Leitch, and D. N. Rhodes, eds.), p. 230. Butterworths, London.

Duckworth, R. B., and Smith, G. M. (1963). Proc. Nutr. Soc. 22, 182.

Duckworth, R. B., Allison, J. Y., and Clapperton, J. A., (1975). In "Intermediate Moisture Foods" (Davies, R., Birch, G. G., and Parker, K. H., eds.), p. 89. Applied Science Publ. London.

Dumanski, A. V. (1933). Kolloid Z. 65, 178.

Duprat, F., and Guilbot, A. (1975). In "Water Relations of Food" (R. B. Duckworth, ed.). Academic Press, London.

Eilers, H. (1945). Versl. Landbouwk Onderz. 50, 1001.

Eilers, H., and Labout, J. W. A. (1946). Proc. Symp. Fibrous Proteins, Leeds, 1946, p. 30.

Elford, B. C. (1970). Nature (London) 227, 282.

Gal, S. (1975). In "Water Relations of Foods" (R. B. Duckworth, ed.), p. 183. Academic Press, London.

Gari-bobo, C. (1964). "Contribution a l'étude des propriétés solventes de l'eau dellulaire (levure, hématies) et tissuaire (muscle) ainsi que de l'eau d'hydration des protéines en solution (hémoglobine) et a l'état de gel (gélatine)," Thesis, Faculty of Science, Paris.

Guerts, T. J. (1972). Versl. Landbouwk Onderz., 777.

Guerts, T. J., Walstra, P., and Mulder, H. (1974). Neth. Milk Dairy J. 28, 46.

Guilbot, A., and Lindenberg, A. B. (1960). Biochim. Biophys. Acta 39, 389.

Hill, A. V. (1930). Proc. Roy. Soc. A 106, 477.

Keyser, W. L. (1968). Diss. Abstr. Sect. B 29(2), 422.

Lindenberg, A. B., Dang-Vu-Bien, and Castan-Recherioq, E. (1963). Nature (London) 200, 358.

McDowall, F. H., and Dolby, R. M. (1936). J. Dairy Res. 7, 156.

Moquot, G. (1947). Lait 27, 576.

Newton, R., and Cook, W. H. (1930). Can. J. Res. 3, 560.

Newton, R., and Gortner, R. A. (1922). Bot. Gazette 74, 442.

Sandberg, E., Haglund, E., and Barthel, C. (1930). Lait 10, 1.

Seow, C. C. (1975). Ph.D. Thesis, Univ. of Strathclyde,
    Glasgow.
Skovholt, O., and Bailey, C. H. (1935). *Cereal Chem. 12,*
    321.
Vail, G. E. and Bailey, C. H. (1940). *Cereal Chem. 17,* 397.
Walstra, P. (1973). *Kolloid Z. Z. Polymere 251,* 603.
Wolowinskaya, X. (1953). Quoted from Kuprianoff, J. (1958),
    in "Fundamental Aspects of the Dehydration of Foods," p. 14.
    S.C.I., London.
Wong, H. Y. (1972). M.Sc. Thesis, Univ. of Strathclyde,
    Glasgow.

USE OF ELECTRON SPIN RESONANCE FOR THE STUDY OF SOLUTE MOBILITY
IN RELATION TO MOISTURE CONTENT IN MODEL FOOD SYSTEMS

D. *Simatos*
M. *Le Meste*
D. *Petroff*
B. *Halphen*

I. INTRODUCTION

Water activity is not the only criterion determining the stability of low-moisture systems. Mobility of molecules, or of reacting groups, may not be necessarily correlated with water activity (Loncin et al., 1965; Acker, 1969; Labuza et al., 1970;
Schneeberger et al., 1978). Solute mobility should be a more
direct criterion for the stability of dehydrated foods.

Studies on mobility of solutes in low- and intermediate-moisture food systems are scarce. By using autoradiography, Duckworth
and Smith (1963) demonstrated mobility of several solutes in plant
materials with moisture contents a little higher than those corresponding to the BET monolayer.

Karel (1975) pointed out that "in general the amount of nonsolvent water measured by various techniques seems to be about
2-3 times larger than the BET monolayer. But some solute mobility
can be imparted by amounts of water corresponding to the monolayer
value," as is indicated by acceleration of some chemical or enzymatic reactions, release of volatiles entrapped in freeze-dried
foods, etc.

In this chapter, the use of electron spin resonance (ESR) for
the study of solute mobility is discussed. Free radicals, which

were created by irradiation in dehydrated products, have been used as test substances for mobility studies (Simatos, 1965; Rockland, 1969; Guex, 1975). The recombination rate of these radicals increased with the moisture content of the material. The same was demonstrated with free radicals produced by oxidation of lipids (Karel, 1975). A limitation of the method, however, is that the molecular structure of these species cannot be easily controlled, and most often is not exactly known.

On the contrary, stable free radicals, having a well-known chemical structure, may be used in model systems. Such free radicals, especially nitroxides, are currently being used to study mobility characteristics in biological membranes or around macromolecules (Berliner, 1976). The present work was devoted to testing the capability of the recent spin probe technique for the problem of solute mobility in dehydrated food systems.

II. THE SPIN PROBE ESR TECHNIQUE

The basic principle of the ESR technique is summarized briefly (Fig. 1). The sample is submitted to a magnetic field H and to a microwave. If the molecule contains an unpaired electron, its energy level is split into two levels by the magnetic field, following the relationship

$$\Delta E = g\beta H$$

The molecules are distributed between these two levels following the Boltzmann law. Providing the energy ΔE through the microwave may increase the population in the upper level, if the following condition is fulfilled:

$$\Delta E = h\nu$$

The ESR spectrum is the derivative curve of the energy absorption when the magnetic field is increased.

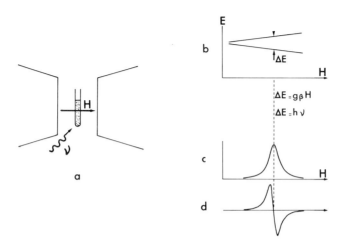

Fig. 1. For an ESR experiment, the sample is exposed to a
magnetic field H, and a microwave of frequency ν. (b) Splitting
of the energy level of an unpaired electron by the magnetic field,
(c) record of the energy absorption, (d) derivative curve of the
energy absorption.

The first of the spectrum parameters is the g value, which is
obtained from the value of the resonance field. The g value
deviates from 2, which is the expected value for the free electron,
the amount of the deviation depending on the interaction between
the spin and the orbital motion of the electron. The g value is
of little use in the present type of work. More interesting parameters can be derived from the ESR spectra of nitroxide free radicals (Rozantsev, 1970; Berliner, 1976).

Nitroxide-free radicals (Fig. 2), which are used as probes or
labels, are stabilized by means of four methyl groups. The chemical groups R_1R_2 are chosen so as to design the probe with the desired molecular properties. In some studies, the nitroxide-free
radicals are fixed on a macromolecule through a covalent bond
(labeling studies). In other studies, they are dispersed in the
system as probes. The analysis of the ESR spectrum provides information on the mobility of the probe, and also on the polarity
of the environment.

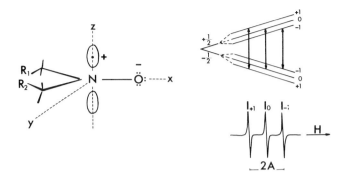

Fig. 2. Nitroxide free radical and the ESR spectrum.

The unpaired electron is largely localized on a 2p Π orbital of the nitrogen atom. As a result of the interaction of the magnetic moment of the nitrogen nucleus with the spin magnetic moment of the electron, the energy levels are divided into three sublevels. Permitted transitions between these levels give rise to a three-line spectrum. The hyperfine splitting A is the distance in gauss between two lines. The A value is sensitive to the polarity of the medium, and of the probe.

A more important property of nitroxide spin probes is the fact that the spectrum shape is very sensitive to the probe mobility. This results from the anisotropy of magnetic parameters of the nitroxide radical. Figure 3 (b,c,d) shows spectra for a rigidly oriented nitroxide radical obtained with the substance included in a monocrystal. Different spectra are obtained according to the probe axis being oriented parallel to the magnetic field.

These spectra reveal the anisotropy of the g parameter and the especially large anisotropy for the A parameter. Most often, however, the parameters for the first two orientations (x and y axis parallel to H) are not very different: the molecule has an axial symmetry. The g and A values, which are observed when the z axis is parallel to H, are indicated as $g_{\parallel}$ and $A_{\parallel}$. The g and A parameters, which are observed when the x or y axis is parallel to H, are almost equal, and are indicated as $g_{\perp}$ and $A_{\perp}$.

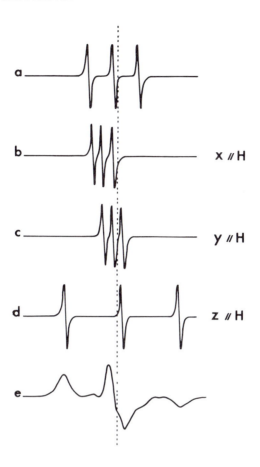

Fig. 3. ESR spectra of a nitroxide free radical. (a) In a solution of low viscosity; (b,c,d) rigidly oriented nitroxide in a monocrystal, with the magnetic field along each axis of the probe; (e) randomly oriented nitroxides in a rigid glass.

The spectrum observed when the probe is dispersed in a diluted solution with a low viscosity is shown in Fig. 3a. As a consequence of the fast molecular reorientation, there is an averaging process; the positions of the spectrum in the field (g value) and the hyperfine splitting are determined by the average values of the g and A tensors.

The so-called powder spectrum is obtained when the probe is randomly oriented and practically without motion, for instance,

when it is dispersed in a polycrystalline material or in a vitreous
solid.   The principal axes of the radical assume all possible angles
relative to the magnetic field.   The ESR spectrum is expected to be
the superposition of all the spectra corresponding to all possible
orientations and to be spread over the entire field range determined
by the g and A anisotropy.   Rather well resolved lines are, however,
observed (Fig. 3e).

The distribution of the absorption intensity is actually not
uniform in this range (Fig. 4).   Let us consider a system with an

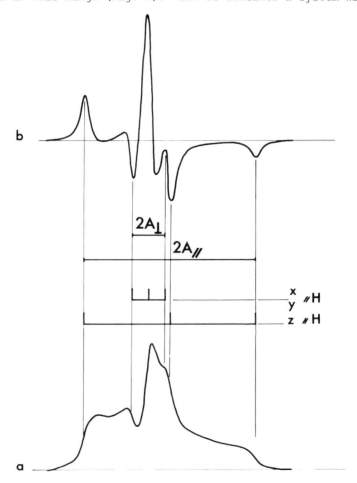

*Fig. 4.   Calculated ESR spectrum for randomly oriented ni-
troxides in the absence of motion.   (top) Absorption graph,
(bottom) derivative curve (from Leniart, 1975).*

axially symmetric g value, the symmetry axis being the z axis.
As a consequence of the random orientation, there will be many
more molecules with the z axis nearly perpendicular to the direction of the magnetic field than there are with the z axis parallel
to H.   Therefore, there will be more spins absorbing at the resonance field determined by $g_\perp$ than those absorbing in the field re-
gions determined by $g_\|$.   The shape of the spectral envelope arising
from g and A anisotropy has been computed and it shows a reasonable
agreement with experimental spectra.   Figure 4 shows the characteristic spectrum computed (and observed) with nitroxide radicals, for
which the g anisotropy is small, and A_{zz} much larger than A_{xx} and
A_{yy}.

Most important for our purpose are spectra that are intermediate between the solution spectrum and the powder spectrum, and
that correspond to correlation times larger than 10^{-8} sec.   Correlation time may be defined as the length of time over which the
molecules persist in a given orientation.   With the present usual
ESR techniques, it is possible to detect motion for probes having
a correlation time shorter than 10^{-6} sec.

In this "slow tumbling region," the line shapes are found to
be sensitive to the details of the molecular reorientation process: rate of tumbling, symmetric or asymmetric rotation, oscillations about a molecular axis, type of reorientation process:
Brownian rotational diffusion or other.

If the magnetic parameters of the probe are known, it is possible to calculate the theoretical spectra following various
models for the reorientation process and thus, by comparison with
the experimental spectra, to obtain information on the probe motion.

III. METHODS

    The nitroxide probes that have been used in this study are
shown in Fig. 5.   They have been chosen among the commercially
available ones (Varian, Palo Alto, California), so as to get different physicochemical characteristics.   From the molecular struc-

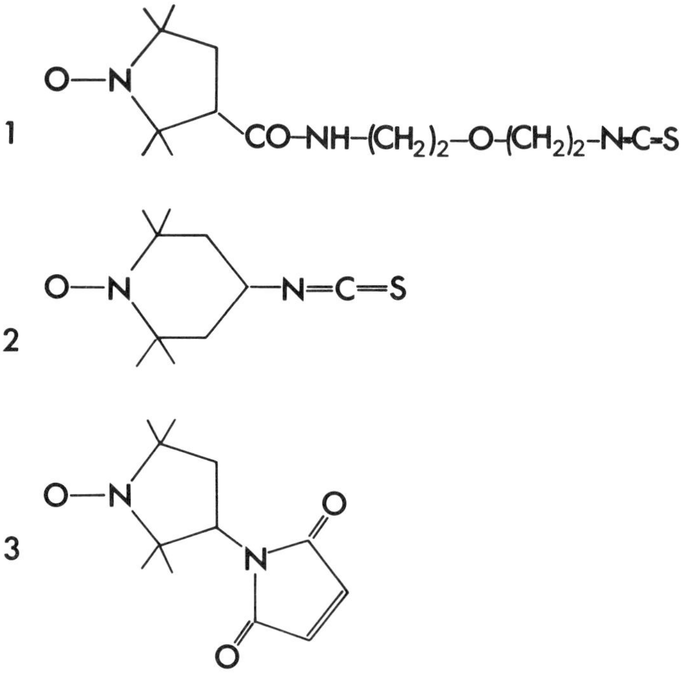

Fig. 5.   Structure of the nitroxide-free radicals.   (1) 3-
{[2-(2-Isothiocyanatoethoxy)ethyl]carbamoyl}-2,2,5,5-tetramethyl-
1-pyrrolidinyloxyl; (2) 4-isothiocyanato-2,2,6,6-tetramethylpiperi-
dinooxyl; (3) 3-maleimido-2,2,5,5-tetramethyl-1-pyrrolidinyloxyl.

tures, probe 2 has probably a more polar character than probes 1 and 3; (polar groups of probe 1 being probably engaged in an intramolecular H bond) this higher polarity may contribute to the higher value for the hyperfine splitting constant (A_0 in Fig. 5). The larger molecular dimension may explain the higher value of the correlation time for probe 1 (τ_c in Fig. 5).

The procedure for the preparation of samples was as follows:

(1) The nitroxide probe was dissolved in gelatin or dextran solutions (dry matter content was 2% w.w for gelatine, 10% w.w for dextran; concentration of probe was between 1 and 2 mg/100 g solution).

(2) The solutions were freeze-dried.

(3) The dry material was rehydrated to various water activities by water vapor sorption, by equilibrium with sulfuric acid solution.

(4) Samples were submitted to ESR when they had attained a constant weight. The material was introduced into quartz tubes 4 mm in internal diameter, which were closed with Teflon tape.

Water contents were determined on controls by drying at $105°C$ for 48 hr.

In most experiments, gelatin (OR-NOIR-Rousselof) solutions were freeze-dried without aging. In a preliminary experiment with matured gelatin (24 hr at $20°C$), however, results seemed very similar. The dextran (T40; Pharmacia) had a mean molecular weight of about 40,000.

Figure 6 shows the sorption isotherms that have been obtained for the polymers. For both, the sorption process revealed a structure reorganization similar to crystallization. In atmospheres of high relative humidities (higher than 80% for gelatin or 92% for dextran) the samples showed an increase in weight, then a decrease, exactly as is observed with amorphous sugars. Reorganization processes have been reported for dextran (Gal, 1967) and for collagen (Esipova and Chirgadze, 1969; Nomura, 1977;

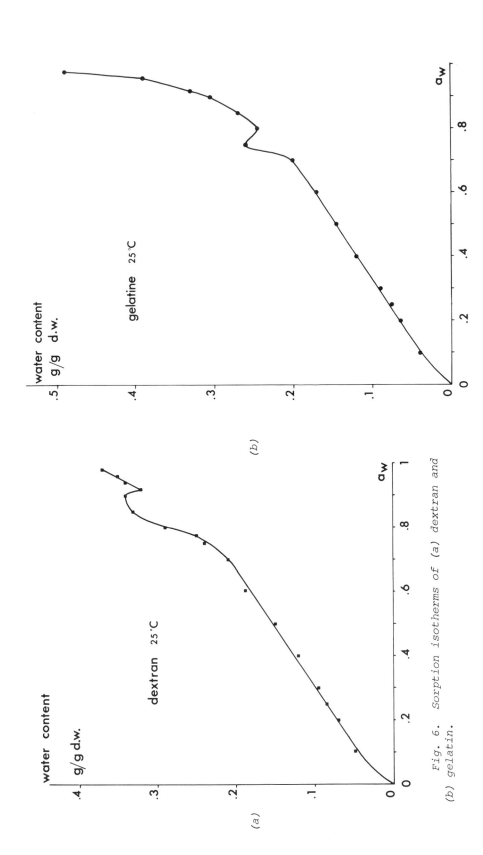

Fig. 6. Sorption isotherms of (a) dextran and (b) gelatin.

Nagamura and Woodward, 1977). Dextran T40 has been shown to have a very high proportion of one-six bonds, and thus a relatively linear molecule, which should make crystallization easier (Aizawa, 1976).

The BET values derived from the isotherms were 0.081 g/g for dextran and 0.087 g/g for gelatin. The ESR spectrometer was E9 (Varian, Palo Alto, California). ESR spectra were recorded at room temperature; standard recording conditions were microwave power 20 mW, scan rate and time constant adjusted so that distortion of spectra was avoided.

IV. RESULTS

A. *Probe 1 in Dextran*

Figure 7 shows the ESR spectra that were obtained for probe 1 dispersed in dextran. Considering samples with increasing water activities, an evolution from a typical powder spectrum to a solution spectrum can be observed. From the lowest a_w to ~ 0.75, a progressive change in the central part of the spectrum occurs. For $a_w > 0.75$, the narrow lines of the solution spectrum appear, superimposed on the powder spectrum. With increasing a_w, the solution spectrum progressively grows up, and the powder spectrum vanishes.

In dextran with $a_w > 0.75$, there are thus two populations of probes: One corresponds to the powder spectrum, consisting of probes that have only slow tumbling, e.g., reorientation correlation time higher than 10^{-8} sec. They may be called immobilized probes. The second population, corresponding to the narrow lines, consists of probes with correlation time in the range 10^{-9} or 10^{-10} sec. They may be called mobile probes. When water activity is raised above 0.74, the population of mobile probes increases, and the population of immobilized ones decreases.

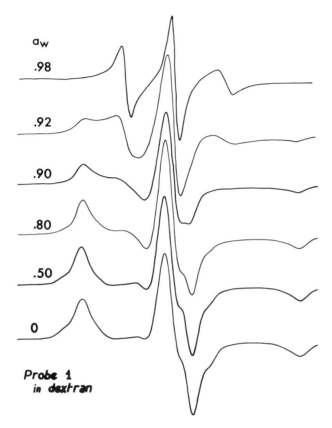

Fig. 7.   ESR spectra of probe 1 in dextran at different water activities.

The different steps of this evolution can be more accurately determined by measuring some spectrum parameters (Fig. 8):

1.   The amplitudes of the I and M lines, which respectively increase with the numbers of immobilized and mobile probes.

2.   The width of the central line (ΔH), which is known to decrease when the mobility of the probe increases.

3.   The parameters $2A_{max}$ and $2A_{min}$, which are approximately equal respectively to the parameters $2A_{\parallel}$ and $2A_{\perp}$. It has been shown by simulation of spectra that the values of $2A_{max}$ and $2A_{min}$ respectively decrease and increase when the mobility of the probe increases.

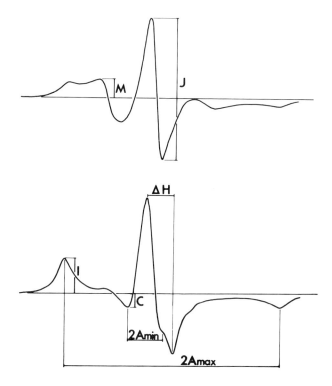

Fig. 8.   Spectrum parameters.

4. The amplitude in C, which can also be related to the mobility of the probes (see discussion).

Figure 9 shows the variations of the ratios C/J, I/J, M/J, as
a function of a_w. Amplitudes of I, M, and C lines have been
divided by the amplitude of the medium line, because these ratios
do not depend so much on the density of the sample or the recording
conditions. From these graphs, three ranges of a_w can be distinguished:

1. a_w < 0.75: The ratio C/J increases linearly with a_w.
2.   0.75 < a_w < 0.90: A more important increase of C/J is observed, simultaneously with an increase of M/J.

3. a_w > 0.90: The increase of the ratios C/J and M/J becomes
even more important.

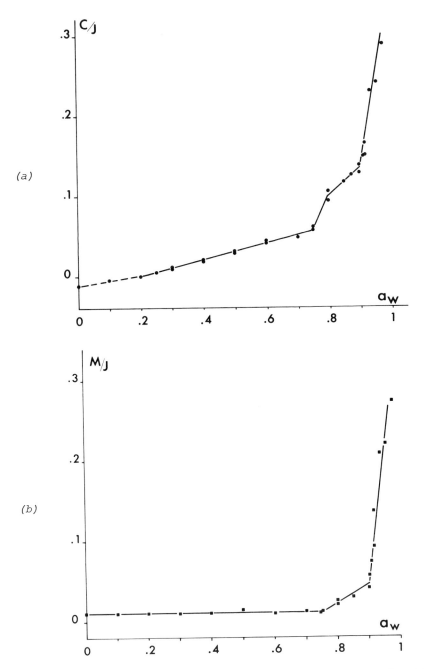

Fig. 9. *Evolution of the spectrum parameters as a function of water activity: probe 1 in dextran.*

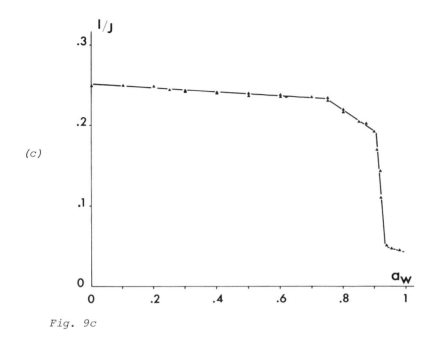

Fig. 9c

The variations of the linewidth ΔH and of the parameters A_{max} and A_{min} also make evident the same a_w ranges: ΔH and A_{max} decrease slowly for a_w increasing from 0.75 to 0.90 and decrease more abruptly for a_w increasing above w.90. A_{min} shows the reverse variations.

An approximate value of the correlation time for the immobilized probes can be calculated by the method of Goldman et al. (1972), which is based on the decrease of $2A_{max}$ below the maximum value corresponding to the limit powder spectrum. This calculation indicated that the correlation time, still close to 10^{-6} sec for a_w = 0.75, decreased to about 10^{-7} sec for a_w = 0.90. For the mobile probes, the correlation time can be approximated from the following equation, if it is in the range 10^{-9}-10^{-11} sec (Lassman, 1973):

$$\tau_c = 6.65\times10^{-10} \, \Delta H(+1) \left[\frac{I(+1)}{I(-1)}\right]^{\frac{1}{2}} - 1$$

For the "mobile" probes 1 in dextran, the calculation gives

$$a_w = 0.92, \quad \tau_c = 3.2 \times 10^{-9} \text{ sec}$$

$$a_w = 0.98, \quad \tau_c = 1.3 \times 10^{-9} \text{ sec}$$

B. *Comparison of Different Probes*

Figure 10 shows the variations in the ratios *C/J* and *M/J* that were observed for the different probes dispersed in dextran. The parameter *C* was not easily measured for probe 2, especially at

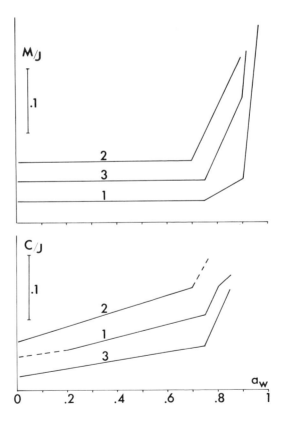

Fig. 10. Evolution of the spectrum parameters as a function of water activity: probes 1-3 in dextran. Curves are displaced vertically to facilitate comparison.

high moisture contents, because lines were broadened by dipolar interactions. The increase in C/J in range 1, however, appeared to be more important for probe 2 than for probes 1 and 3. The increase of M/J (range 2) was observed to begin for lower a_w and to be more important, with probe 2 than with probes 1 and 3.

In gelatin (Fig. 11) the same differences between probes 1 and 2 were observed, but differences in behavior of probes 1 and 3 became evident: for probe 3 an increase of ratio C/J in ranges 1 and 2 was less important; an increase of ratio M/J only began at $a_w = 0.90$.

In conclusion, the mobilization processes occurred at lower a_w, and increased more rapidly with a_w for probe 2 than for probe 1, and for probe 1 than for probe 3.

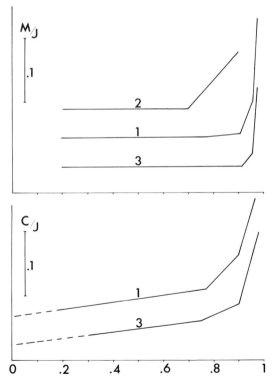

Fig. 11. *Evolution of the spectrum parameters as a function of water activity: probes 1-3 in gelatin. Curves are displaced vertically to facilitate comparison.*

C. *Comparison of Probe Behavior in Different Polymers*

As shown in Figs. 10 and 11, the variations of ESR parameters for the three types of probes made evident the same ranges of a_w for both gelatin and dextran. The amplitude of the variations, however, were different. The increase of the C/J and of the M/J ratios was less important for gelatin than for dextran. The proportion of mobile probes at the upper limit of range 2 ($a_w = 0.90$) for probe 1 was only 6% in gelatin, as opposed to 15% in dextran. As already noticed, mobility in gelatin was especially low for probe 3.

D. *Reduction of the Probe by Ascorbic Acid*

Experiments have been designed to demonstrate the connection of the mobility of solutes, as studied by the spin probe ESR technique, with the possibility for this solute to participate in a chemical reaction.

Nitroxide radicals can be reduced by ascorbic acid. Solutions of dextran containing either probe 1 or ascorbic acid were freeze-dried. The dry materials were mixed and then rehydrated to different increasing a_w successively. Sorption was effected in a moisture-saturated atmosphere from which air had been evacuated, in order to perform a rapid sorption process (less than 1 hr). The a_w of the sample were evaluated from the features of the ESR spectra obtained from the control sample without ascorbic acid.

The importance of the probe reduction was evaluated from the decrease of the ESR lines' amplitude (Fig. 12). No reduction of probe was observed in these conditions for $a_w < 0.75$. A limited reduction occurred for a_w in range 2, and the reduction was very important for $a_w > 0.90$.

When samples were stored with constant water content for 24 hr, no further change of ESR amplitudes was observed. Diffusion of the probe and/or the ascorbic acid molecules, and the oxydation-

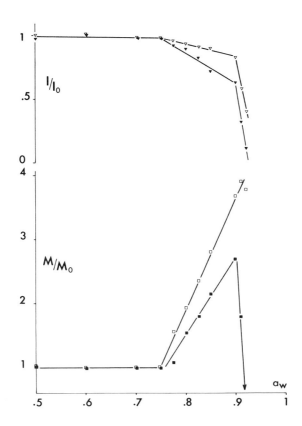

Fig. 12. Influence on the ESR spectrum of the reduction of the probe by ascorbic acid. ▽ *,* □ *, Probe 1 in dextran;* ▼ *,* ■ *, probe 1 + ascorbic acid, in dextran. The figures represent the ratio of the lines amplitude after water sorption to the amplitude before sorption.*

reduction reaction appeared to occur very rapidly, e.g., during the mixing and rehydration stages, and to stop at a definite level, which depended on the water activity of the sample.

V. DISCUSSION

A. *Influence of Water Activity on the ESR Spectrum*

The above results are consistent with the data published by
Nagamura and Woodward (1977). These authors used the spin probe
technique to investigate molecular motion in collagen as a function of temperature. The ESR parameters used were the hyperfine
splitting and the ratio of amplitude of narrow and broad lines.
The ESR spectra were interpreted as indicating the appearance of
mobile probes, for temperatures and moisture contents above certain levels.

Some methodological problems that occur when samples with different water contents are submitted to ESR must be discussed, especially when quantitative studies are concerned. Comparison of
the amplitudes of an ESR line in successive samples with different
water contents was difficult:

1. The weight of the ESR sample being small (a few milligrams),
it was difficult to know accurately the weight of the sample in the
spectrometer cavity.

2. It has been shown (Sarna-Lukiewisz, 1971) that, as a consequence of its high dielectric constant, the water contained in the
sample may induce an increase, or a decrease, of the amplitude of
the ESR signal. The influence of water on the signal obtained from
samples with different water contents is thus not predictable.

To overcome these problems, the ratios of the amplitudes of different lines (I, M, C) to the amplitude of the medium one (J) have been
determined. An experiment has been designed, however, to observe
the individual variation of each line: one sample (probe 1 in dextran T40) was rehydrated in the ESR tube to different a_w successively. The amplitude of the medium line J was observed to be constant
for all a_w < 0.90, and to decrease for a_w increasing above this
value. Line a was constant for a_w < 0.75, decreased for a_w rising

from 0.75 to 0.90, and decreased more rapidly for a_w > 0.90. Line
M underwent the reverse variations. The amplitude in C increased
slowly for a_w rising from 0 to 0.75, and more rapidly for a_w >
0.75.

In conclusion, the ratios I/J, M/J, C/J are not representative
of the amplitudes of the lines I, M, C. These ratios, however,
may be used to define different a_w ranges corresponding to variations in the mobility of the probes.

Another side effect of water on ESR spectra must be taken into
account. It is well known that, in order to avoid saturation
phenomena and anomalous ESR spectra, it is necessary to use a microwave power below a critical level. The specific problem here
is that the saturation risk is more important for paramagnetic
species with large correlation times. The larger the mobility of
the probe, the higher saturating power may be. It was thus necessary to wonder whether the increase of line M or of the amplitude
in C was dependent on the saturation effects.

To answer this question, ESR spectra for samples with various
a_w have been recorded with increasing microwave power between 1
and 60 mW. It could be observed that the saturation effects were
already noticeable with a power of about 10 mW for solution spectra, and about 5 mW for powder spectra. A microwave power of 20
mW could be considered, however, as a satisfactory compromise between saturation effects and sensitivity (for spectra recorded at
20°C).

However, spectra recorded with a microwave power well below
the saturation level (1 mW) showed an increase of the parameter C
(ranges 1 and 2) and of the amplitude of line M (range 2) with increasing a_w. This increase was small for a power of 1 mW, but was
amplified for a power of 20 mW. It thus seems that the increase
of parameter C with increasing a_w was partly the direct consequence
of increasing mobility of the probe, but above all it was the result of a decreasing saturation effect, which was itself the consequence of the increasing rate of motion of the probe.

To summarize, the ESR spectra of spin probes permit three ranges of a_w to be defined:

1. The probes have only very slow tumbling (correlation time higher than 10^{-6} sec). The rate of motion, however, increases with increasing a_w in this range, as indicated by the increasing amplitude in C.

2. Most probes still have low mobility similar to that of range I (correlation time between 10^{-6} and 10^{-7} sec). But a limited fraction of probes have much higher mobility (correlation time between 10^{-8} and 3×10^{-9} sec).

3. All probes enter progressively the more mobile class.

The more mobile probes that appear in range 2, and give rise to an ESR solution spectrum may be called "dissolved probes." Range 2 thus represents the appearance of "solvent water" for this particular species. The tumbling rate of the probes, however, still remains rather low as compared to their tumbling rate in diluted solution: correlation time is $1.3 \ 10^{-9}$ sec for probe 1 in dextran with $a_w = 0.98$; it is 5×10^{-11} sec in the 2% dextran solution (not significantly different from the value in pure water).

It may be noticed that the ESR behavior, in the range of strucural reorganization of the polymer, is related to a_w and not to water content: dextran samples with $a_w = 0.92$, although they have a water content lower than samples with $a_w = 0.90$, always have more mobile probes.

B. *NMR Relaxation Times for Water*

The proton NMR relaxation times have been determined, in order to correlate the spin probe behavior with properties of water. Relaxation times T_1 and T_2 have been determined by means of pulsed NMR. The frequency of the spectrometer was 20 MHz (Minispec P20, Brucker).

The spin-lattice relaxation time (T_1) has been measured only

for gelatin samples. The spin-lattice relaxation process exhibited
a single-phase behavior. The value of T_1 was minimum for samples
with a_w = 0.75.

The spin-spin relaxation process, on the contrary, appeared to
be resolvable in two or even three components for both gelatin and
dextran. Figure 13 shows the T_2 values as a function of water content, because this representation seemed to be more coherent in the
a_w range 0.80-0.96. The discrepancy between the proton relaxation
behavior and that of the spin probe is worth further considerations.
It is probable, however, that the reorganization that takes place
in the polymer is connected with a complex behavior of water.

If the smallest values of T_2 (lower than 50 μsec) are neglected,
three ranges of moisture contents may be distinguished in the case
of dextran:

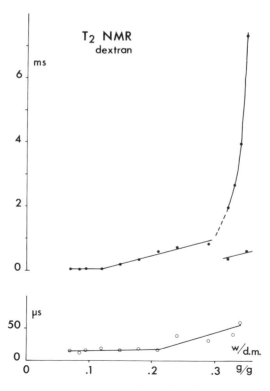

Fig. 13. Proton NMR relaxation time (T_2) for dextran.

1. Up to 0.12 g/g, T_2 levels at about 60 µsec;

2. From 0.12 to 0.29 g/g, T_2 increases from 200 to about 800 msec;

3. Above 0.30 g/g, two values of T_2 may be determined: one close to 500 msec, the other one increasing up to 7 sec.

With gelatin, the spin-spin relaxation behavior of the proton was found to be very similar. The three ranges are limited by moisture contents: 0.15 and 0.26 g/g.

These results are very similar to the data reported by Lechert (in this volume). Tables I and II summarize the data that have been obtained for dextran and gelatin from ESR and NMR studies.

The various moisture ranges that have been defined by NMR do not correspond exactly to those that have been defined by ESR. This fact does not seem surprising since it is quite probable that ESR results depend on the type of probe used as solute. For both gelatin and dextran, however, the moisture content that corresponds to the beginning of the rapid increase of T_2 is in the range of moisture contents corresponding to the beginning of the probes' "solubilization."

TABLE I. *Data Obtained with Dextran for Proton NMR Relaxation Time* (T_2) *and ESR*

a_w	T_2 NMR (sec)	ESR		
		1	2	3
0				
0.20	$< 8 \times 10^{-5}$			
0.30				
0.40				
0.50				
0.60	$2 - 8 \times 10^{-4}$			
0.70				
0.75				
0.80				
0.85				
0.90	$2 - 8 \times 10^{-3}$			
0.95				
1				

TABLE II. Data Obtained with Gelatin for Proton NMR Relaxation Time (T_2) and ESR

a_w		Mobilization points[a]	T_2 NMR (sec)	ESR 1	2	3
0						
0.20						
0.30	BET		$\sim 1.8 \times 10^{-4}$			
0.40	Monolayer					
0.50		Urea				
0.60						
0.70			$2\text{-}5 \times 10^{-4}$			
0.75						
0.80						
0.85		Sucrose				
0.90	Unfreezable	Glucose	$1\text{-}3 \times 10^{-3}$			
0.95	water					
1	limit[a]					

[a] Data from Duckworth and Smith (1963).

C. Mobility of the Probes and Physicochemical Characteristics of the System

For probe 2, mobile probes appeared for lower water activities than for probes 1 and 3, both in dextran and in gelatin. Moreover, mobilization developed more rapidly when a_w was raised. This behavior can be attributed to the more polar character of probe 2.

Mobilization processes of probe 3 in gelatin seemed to be more difficult than for probe 1 in the same polymer. Since this difference was not observed in dextran, it should not be connected with a difference in polarity of the probe. The low mobility of probe 3 in gelatin may be the result of physical or chemical interactions of the probe with the polymer.

It may be noticed that interactions of the probes with the polymers could be only weak, noncovalent interactions. Comparison of the ESR spectra that were obtained with the initial solutions with the same ones after dialysis (48 hr) showed that a maximum of 9% of the probes were covalently fixed on gelatin and none on dextran.

A great variety of spin probes can be designed. This possibility may be used to study the influence on mobility of small solutes of their molecular properties: molecular size, polarity, electric charge, etc. It should be noticed, however, that experiments with probe 2 were difficult because ESR lines were broadened. This is a consequence of the polar character of this probe, which results in important dipole interactions between probes, or between probe and water molecules.

VI. CONCLUSION

The ESR spectra of spin probes make it possible to identify three ranges of water activities. The intermediate range corresponds to the appearance of solvent water for the particular substance used as probe. In this medium range, only a limited number of probes are "dissolved"; most of them still have a slow tumbling rate, as in the dry material.

The three ranges of water activities, as defined from the tumbling rate of the probe, also correspond to different levels in the ability of the probe to be reduced by ascorbic acid. The moisture level corresponding to the appearance of solvent water, as defined by ESR, is close to the moisture level above which a rapid increase of the spin-spin NMR relaxation time is observed.

In the lower moisture range, ESR makes evident some kind of mobility of the probe, increasing linearly with water activity. Using the techniques of spectra simulation should permit to get more detailed qualitative information on the mobilization processes of the probes.

The mobility level of spin probes results from several parameters:

1. Water activity: there seem to be characteristic ranges

of a_w for one probe, even if dispersed in different macromolecular systems.

2. Characteristic of the probe: size, polarity. In the present study, polarity appeared to be more important than size of the molecule.

3. Properties of the macromolecule, and/or interactions with the probe.

By using various spin probes with adequately chosen molecular properties, it should be possible to obtain information on the mobility of small solutes, and thus to help understand their ability to participate in chemical reactions.

Acknowledgments

NMR experiments were performed at the Institut für Lebensmittelverfahrenstechnik (Universität Karlsruhe). We thank Doctor Weisser and Professor Director Loncin for their kind cooperation.

References

Acker, L. W. (1969). *Food Technol. 23,* 1257-1270.
Aizawa, M. (1976). *Bull. Chem. Soc. Japan 49,* 2061-2065.
Berliner, L. J. (1976). "Spin Labeling, Theory and Applications." Academic Press, New York.
Duckworth, R. B., and Smith, G. M. (1963). *Recent Adv. Food Sci. 3,* 230-238.
Esipova, N. G., and Chirgadze, Y. N. (1969). *In* "Water in Biological Systems" (L. P. Kayushin, ed.), pp. 42-50. Consultants Bureau.
Gal, S. (1967). "Die Methodik du Wasserkampf-Sorptionmessungen," p. 117. Springer-Verlag, Berlin.
Goldman, S. A., Bruno, G. V., and Freed, J. H. (1972). *J. Phys. Chem. 76,* 1858-1860.
Guex, M. M. (1975). *In* Freeze-drying and Advanced Food Technology" (S. A. Goldblith, L. Rey, and W. W. Rothmayr, eds.), pp. 413-420. Academic Press, New York.
Karel, M. (1975). *In* "Water Relations of Foods" (R. B. Duckworth, ed.), pp. 435-453, 639-657. Academic Press, New York.
Labuza, T. P., Tannenbaum, S. R., and Karel, M. (1970). *Food Technol. 24,* 35-42.

Lassman, G. (1973). *In* Elektron Spin Resonanz und anderen spek-
    troskopik Methoden in Biologie und Medezine" ⁾(J. Wyard, ed.),
    Akademie-Verlag Berlin, pp. 381-403.
Leniart, D. S. (1975). Varian ESR Workshop, Zurich.
Loncin, M., Jacqmain, D., Provost, A. M., Lenges, J. P., and
    Bimbenet, J. J. (1965). *C.R. Acad. Sci.*, 3208-3211.
Nagamura, T., and Woodward, A. E. (1977). *Biopolymers 16,* 907-
    919.
Nomura, S. (1977). *Biopolymers 16,* 231-246.
Rockland, L. B. (1969). *Food Technol. 23,* 1241-1251.
Rozantsev, E. G. (1970). "Free Nitroxyl Radicals." Plenum Press,
    New York.
Sarna-Lukiewisz, T. S. (1971). *Cytochem. 9,* 203-216.
Schneeberger, R., Voilley, A., and Weisser, H. (1978). *Int. J.
    Refrig. 1,* 201-206.
Simatos, D. (1965). Ph.D. Thesis.

A THERMODYNAMIC MODEL FOR TERPENE-WATER SOLUTIONS

Marc LeMaguer

I. INTRODUCTION

Recent studies, involving the retention of flavoring organic
compounds during the freeze-drying of aqueous solutions, have
shown that the degree of volatile retention is dependent on the
solubility of the flavoring compound (Flink and Gejl-Hansen,
1972; Flink et al., 1973; Massaldi and King, 1974a; Smyrl and
LeMaguer, 1978). Similar studies in the area of evaporation-
drying have also indicated the usefulness of this parameter (King
and Massaldi, 1974b; Menting et al., 1970a; Thijssen, 1972).

Although qualitative observations (Flink and Gejl-Hansen,
1972) have shown that reaching the solubility limit of the organic
phase led to the formation of droplets, the lack of information on
actual quantitative values for these same solubilities has limited
its possible use in practical calculations.

The prediction of solubilities in complex aqueous solutions is
a difficult task. Still, it is quite important to know not only
the solubility limit, but also the behavior of the system with the
nature and composition of the phases that appear when one cools,
heats, or removes water from actual liquid foods. This knowledge
is necessary to determine the true composition of the food after
such operations as freezing, freeze-drying, evaporation, distillation, and membrane separation. Equilibrium values as a function
of temperature, pressure, and composition are used in conjunction

with transport equations describing heat and mass transfer within
the food to calculate the actual retention level of the components
of interest.

This chapter represents an attempt to adapt a thermodynamic
model to the system terpene-water with which our recent work
(Smyrl and LeMaguer, 1978) has been concerned.

II. THERMODYNAMIC MODEL FOR SPARINGLY SOLUBLE COMPOUNDS IN WATER

The thermodynamic modeling of nonideal liquid mixtures is
based on semitheoretical equations. The basis of development of
all the new models is the excess Gibbs energy function. Depending
on the hypothesis on which it is constructed one obtains some of
the more well-known equations, such as the Van Laar, Wilson, and
athermal Flory-Huggins equations. The following equations indicate
some of the most important relationships:

$$G^E = H^E - TS^E \tag{1}$$

$$\left(\frac{\partial G^E/T}{\partial T}\right)_{P,x} = -\frac{H^E}{T^2} \tag{2}$$

$$\left(\frac{\partial G^E}{\partial T}\right)_{P,x} = -S^E \tag{3}$$

Therefore, from the knowledge of G^E at different temperatures and
composition, one can predict the excess enthalpy and excess entropy.

This, of course, applies also to the partial excess functions
where we have, for example,

$$\bar{G}_i^E = \left(\frac{\partial n_T G_E}{\partial n_i}\right)_{P,T,n_{j \neq 1}} = RT \ln \gamma_i \tag{4}$$

$$\left(\frac{\partial \overline{G}^E_i/RT}{\partial T}\right)_{P,x} = -\frac{\overline{H}^E_i}{T^2} \tag{5}$$

$$\left(\frac{\partial \overline{G}^E_i}{\partial T}\right)_{P,x} = -\overline{S}^E_i \tag{6}$$

Equations (4)-(6) are useful to describe at infinite dilution the effect of the solute on the water and to predict some of the most practical properties such as solubilities and activity coefficients.

A. Basis for the Model

We present here a very rapid introduction to the basic ideas behind the universal quasi-chemical equation. A more detailed discussion is offered by Renon and Prausnitz (1968) and Abrams and Prausnitz (1975). When nonelectrolyte liquids are mixed at constant temperature and pressure there is very little volume change. Although small volume changes can have a significant effect on the entropy and enthalpy of mixing they approximately cancel in the excess Gibbs energy. This allows for the replacement of the excess Gibbs energy at constant temperature and pressure by the excess Helmoltz energy of mixing at constant temperature and volume. This substitution permits the development of a theory based on the lattice structure of a liquid as developed originally by Guggenheim (1952). With the use of the local-composition concept introduced by Wilson (1964) molecules of different size and shape can be accepted in the previous model. The excess Gibbs energy can be considered to be made up of two terms, one combinatorial in nature, which depends essentially on the difference in size and shape of the molecules, the second residual due to energetic interactions:

$$G^E = G^E(\text{comb}) + G^E(\text{res}) \tag{7}$$

where

$$\frac{G^E}{RT}(\text{comb}) = x_1 \ln \frac{\Phi_1}{x_1} + x_2 \ln \frac{\phi_1}{x_2}$$

$$+ \frac{Z}{2} q_1 x_1 \ln \frac{\theta_1}{\phi_1} + q_2 x_2 \ln \frac{\theta_2}{\phi_2}$$

$$\frac{G^E}{RT}(\text{res}) = -q_1 x_1 \ln(\theta_1 + \theta_2 \tau_{21}) - q_2 x_2 \ln(\theta_2 + \theta_1 \tau_{12}) \qquad (9)$$

with

$$\theta_1 = \frac{q_1 x_1}{q_1 x_1 + q_2 x_2} \ , \quad \theta_2 = \frac{q_2 x_2}{q_1 x_1 + q_2 x_2} \qquad (10)$$

$$\phi_1 = \frac{r_1 x_1}{r_1 x_1 + r_2 x_2} \ , \quad \phi_2 = \frac{r_2 x_2}{r_1 x_1 + r_2 x_2} \qquad (11)$$

$$\tau_{12} = \exp - \frac{U_{12} - U_{22}}{RT} \ , \quad \tau_{21} = \exp - \frac{U_{21} - U_{11}}{RT} \qquad (12)$$

Equations (10) and (11) contain pure-component structural parameters r_1, r_2, q_1, q_2. These are evaluated from bond angles and
bond distances as described later. As can be seen from Eq. (8)
the combinatorial term depends only on the structural parameters
and only the residual term requires the two additional empirical
parameters τ_{12}, τ_{21} to be determined from experimental data. The
significance of the energy of interaction represented by the U_{ij}
terms in Eq. (12) can be best visualized using Scott's (1956) two-
liquid theory of binary mixtures as represented in Fig. 1. U_{11}
and U_{21} represent interaction energies between a 1-1 and a 1-2
pair of molecules, where molecule 1 occupies the center of the
cell indicated on Fig. 1.

From Eqs. (4) and (7)-(11) one obtains for γ_1 in a binary
system

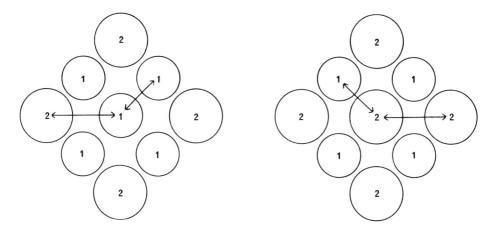

MOLECULE 1 AT CENTRE   MOLECULE 2 AT CENTRE

Fig. 1.   Scott's model for a binary liquid mixture.

$$\ln \gamma_1 = \ln \phi_1/x_1 + (Z/2)q_1 \ln \theta_1/\phi_1 + \phi_2(\ell_1 - r_1\ell_2/r_2)$$

$$-q_1 \ln(\theta_1 + \theta_2 \tau_{21}) + \theta_2 q_1 \left[\frac{\tau_{21}}{\theta_1 + \theta_2 \tau_{21}} - \frac{\tau_{12}}{\theta_2 + \theta_1 \tau_{12}} \right] \tag{13}$$

where

$$\ell_1 = \frac{Z}{2}(r_1 - q_1) - (r_1 - 1)$$

$$\ell_2 = \frac{Z}{2}(r_2 - q_2) - (r_2 - 1) \tag{14}$$

For component 2, γ_2 can be found by interchanging subscripts 1 and 2.

As shown by Abrams and Prausnitz (1975) these equations can be extended to three or more components. We refer to their paper for the explicit expressions of the excess Gibbs energy and activity coefficient in these cases.

B. Determination of Parameters

1. Structural Parameters

From the chemical structure of the solute (Fig. 2) and using
the technique described by Abrams and Prausnitz (1975) or Fredenslund *et al.* (1977), which calls for the Van der Waals group
volumes and surface areas V_k and A_k given by Bondi (1968), one
determines the values of r_i and q_i as indicated in the appendix.
Using this technique the values calculated for carvone, pulegone,
and piperitone are given in Table I where the value for water is
also given. As can be seen, the three compounds are very similar
in nature in terms of their contribution to the combinatorial excess Gibbs energy. It is expected in these conditions that differences, if they appear, would be at the level of the interaction
with the water molecule. Therefore, we have to obtain the information necessary to the determination of the energy of interaction.

2. Residual Parameters τ_{12}, τ_{21}

The determination of these parameters will obviously depend
on the existence of experimental values for the system. Among
these we have vapor pressure, solubility, heat of mixing, activity

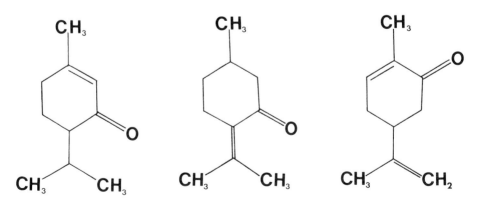

piperitone   pulegone   carvone

*Fig. 2.   Chemical structure of piperitone, pulegone, and
carvone.*

TABLE I. Structural Parameters r_i, q_i

	r_i	q_i	m_i
Water	0.92	1.40	18
Carvone	6.38	5.15	150
Pulegone	6.61	5.63	152
Piperitone	6.39	5.01	151

coefficients with temperature and composition, and any other relevant data pertaining to the system. We should realize here that the system to be solved is usually overdetermined in the sense that we have more data points than parameters to determine. In this case the solution has to be of the least-squares type, in which one minimizes an objective function. Suppose, for example, that we have measured at constant temperature the variation of the total pressure in the gas phase with the composition of the liquid phase. In the case of a binary system we know T_1, x_1, x_2, and P_{exp}. We know the structural parameters that have been previously calculated. From Eq. (12) and its equivalent for γ_2, if we give initial values to τ_{12} and τ_{21} we can calculate the activity coefficients. From these and

$$P = P_1^0 x_1 \gamma_1 + P_2^0 x_2 \gamma_2 \tag{15}$$

where P_1^0 and P_2^0 are the pure vapor pressure of components 1 and 2, one obtains the first approximation for P_{cal}. The objective function used in this case will be

$$\sum_{i=1}^{n} \left(P_{cal}^i - P_{ex}^i \right)^2 \tag{16}$$

where n is the number of experimental values. Using a numerical technique to find the optimum of a function of two variables, the best values of τ_{12} and τ_{21} will be obtained.

From this simple case it is easy to see that at constant temperature one could choose a criterion such as

$$\sum_{i=1}^{n_1} \left| P_{cal} - P_{exp} \right|^2 + \sum_{i=1}^{n_2} \left| x^s_{cal} - x^s_{exp} \right|^2$$

$$+ \sum_{i=1}^{n_3} \left| H_{cal} - H_{exp} \right|^2 + \sum_{i=1}^{n_4} \left| \gamma_{cal} - \gamma_{exp} \right|^2 \qquad (17)$$

The advantage of such a method, provided the data are thermodynamically consistent, is that we are using all the information
available at the same time. With modern methods of computation
very effective programs exist to treat such problems.

In our case the data treatment was done for constant-tempera-
ture experiments. The reason for this is that the interaction
parameters $U_{21} - U_{11}$ and U_{22} are usually temperature dependent.
An a priori temperature variation could have been introduced in
the model, but it is more reasonable to determine a set of τ_{12}
and τ_{21} at each temperature and then determine their variation
with temperature.

The precision of the measurements should also be reflected in
the choice of the optimization criterion by the introduction of
weighing factors inversely proportional to the standard deviation
for example. This gives more weight to the more precise measurements and still utilizes all the available information.

C. Application to Terpene-Water Solutions: Experimental

1. Materials and Chemicals

Carvone (K and K), piperitone (ICN), and pulegone (Fluka)
were used as received. Analysis by gas chromatography revealed
these compounds to be at least 90% pure. All solutions were prepared with distilled water.

2. Equipment

Saturated solutions were prepared in a Labline environmental chamber (Labline, Inc.) and temperatures are accurate to ±0.2°. A Vortex-Genie (Scientific Industries, Springfield, Massachusetts) was used to mix the aqueous solution with the added organic species. Separation of the saturated aqueous layer from excess organic was accomplished on a Janetzki T5 centrifuge (Heinz Janetzki Kg.), which was equipped with a fixed-angle rotor. Ultraviolet absorption measurements were conducted on a Unicam SP1800 ultraviolet spectrometer (Pye Unicam). The instrument was operated on the fixed-wavelength mode using a bandwidth of 1.2 mm and a slit width of 0.4 mm. Quartz cells (Canlab) were used for reference and sample during ultraviolet analysis.

For the measurement of solubility of water in pure carvone a vapor pressure technique was used, which provided a monitoring of the total vapor pressure along with a knowledge of the composition of the gas and liquid phase.

3. Preparation of Standard Curves

Standard solutions of the organic species in water were prepared so that the range of absorbances for the standard solutions fell between 0.4 and 1.4. It has been found that absorbances falling within this range are the most reliable for double-beam spectrometers. The standard curve for carvone (Fig. 3) illustrates the validity of the Beer-Lambert law for absorbances between 0 and 1.2.

4. Preparation of Standard Solutions and Solubility Determinations

All saturated solutions were prepared in a Labline environmental chamber (±0.2°C). Conical 12-ml centrifuge tubes with screw tops were filled with 8.0 ml of aqueous solution. Approximately 0.1 ml of the organic species were added. The contents of the centrifuge tubes were mixed on the vortex mixer for 8 min.

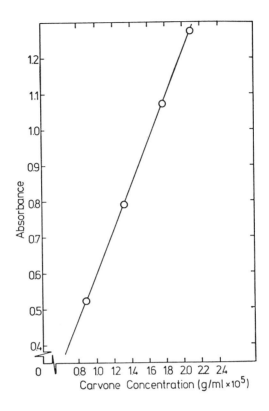

Fig. 3.   Standard curve for carvone-water solutions.

After mixing the tubes were centrifuged for as long as was required to effect separation between excess organic and the aqueous
phase.   In some cases centrifugation times of several hours were
required.   The centrifugation was carried out at the temperature
at which the saturation was performed.   When centrifugation was
completed, as evidenced by a totally clear aqueous phase, a 1-ml
graduated pipette with a Teflon extension was used to remove aliquots from the aqueous phase.   To ensure that none of the less
dense organic phase entered the pipette, a gentle blowing action
was applied as the Teflon tip passed through the organic layer.
After the 1-ml aliquot had been withdrawn from the centrifuge
tube, the Teflon tips were removed and the contents of the pipette
were emptied into a suitable volumetric flask and diluted with

water. The absorbance of the diluted saturated solution was measured and the concentration of the organic species was determined from the previously obtained standard curves.

D. Application to Terpene-Water Solutions: Results

1. Experimental

Table II presents the experimental solubilities obtained using the previously described technique. As can be seen, the reproducibility is quite good. Only for the carvone-water system were pressure measurements conducted. These values formed the base from which the residual interaction parameters τ_{12} and τ_{21} were determined.

2. Adjustment of Model

As indicated previously, the determination of the best values of τ_{12} and τ_{21} for the model is done for each temperature independently. Because of the nature of the experimental data, solubilities, and variation of vapor pressures with composition in the more complete case, it is necessary to take the following steps.

a. *Solubility calculations.* For a given set of values τ_{12}, τ_{21} solve the system of nonlinear equations

$$(x_w \gamma_w)_1 = (x_w \gamma_w)_2 \tag{18}$$

TABLE II. *Experimental Solubility Values for Carvone, Pulegone, and Piperitone (Mole Fraction of Terpene)*

T, °C	Carvone	Pulegone	Piperitone
4	0.000211		
10	0.000204 ± 2%	0.000191 ± 0.2%	0.000359 ± 0.3%
20	0.000185 ± 2%	0.000164 ± 0.6%	0.000303 ± 0.3%
30	0.000194 ± 0.6%	0.000149 ± 2.6%	0.000283 ± 0.5%
40	0.000199		

$$(x_t \gamma_t)_1 = (x_t \gamma_t)_2 \tag{19}$$

$$(x_w + x_t)_1 = 1 \tag{20}$$
$$(x_w + x_t)_2 = 1$$

The results are then compared to the experimental values and τ_{12} and τ_{21} changed to minimize the sum of squares of residuals.

   b. *Vapor pressure.* For a given set of τ_{12}, τ_{21} and a value of the mole fraction of water, for example, one calculates the activity coefficient from Eq. (13) and then the pressures from

$$P_w = P_w^0 x_w \gamma_w \tag{21}$$

$$P_t = P_t^0 x_t \gamma_t \tag{22}$$

$$P_{tot} = P_t + P_w \tag{23}$$

Again, the sum of squares of residuals between experimental and calculated pressure will be minimized.

   More refinements can be introduced in the procedure when information is known on the precision of the experimental measurements. Table III shows the values of τ_{12} and τ_{21} obtained for carvone, pulegone, and piperitone in water at various temperatures along with the variation on the parameters when applicable. From these the energies of interaction were calculated from Eq. (12). They are presented in Table IV. One should point out at this stage that it is usually important to have information available on both phases of the system. The penalty, otherwise, can be of an excellent adjustment on the water side and a very poor one on the terpene side. This is the reason why we conducted the vapor pressure measurements on the pure terpene phase. The values for carvone at 10 and 20°C in Table III represent the average of all the parameters obtained and do not offer a variation much greater than the ones for pulegone and piperitone, which were obtained from solubility data alone. The precision of the adjustment is,

TABLE III. Interaction Parameters τ_{12}, τ_{21} Variation with Temperature (Best Values for Carvone, Pulegone, Piperitone)

T (K)	Carvone		Pulegone		Piperitone	
	τ_{12}	τ_{21}	τ_{12}	τ_{21}	τ_{12}	τ_{21}
277	0.716	0.182				
283	0.712 ± 0.002	0.181 ± 0.004	0.690 ± 0.002	0.143 ± 0.002	0.672 ± 0.003	0.441 ± 0.002
293	0.703 ± 0.002	0.173 ± 0.004	0.678 ± 0.002	0.131 ± 0	0.657 ± 0.001	0.432 ± 0.002
303	0.707 ± 0.002	0.178 ± 0.002	0.671 ± 0.002	0.125 ± 0.002	0.651 ± 0.002	0.426 ± 0.002
313	0.711	0.177				

TABLE IV. Interaction Parameters U_{12}, U_{21} (kJ/kmole) Variation with Temperature

T (K)	Carvone		Pulegone		Piperitone	
	$U_{12}-U_{22}$	$U_{21}-U_{11}$	$U_{12}-U_{22}$	$U_{21}-U_{11}$	$U_{12}-U_{22}$	$U_{21}-U_{11}$
277	770	3926				
286	800 ± 8	4029 ± 53	1148 ± 2	4857 ± 11	936 ± 18	2202 ± 9
293	859 ± 6	4277 ± 26	1221 ± 2	5227 ± 0	1023 ± 3	2321 ± 7
303	874 ± 4	4355 ± 9	1280 ± 7	5521 ± 54	1081 ± 3	2422 ± 4
313	888	4509				

of course, contained in the sum of squares of residuals. This can be used as a guide as to the goodness of the fit and allows for the determination of confidence intervals on the parameters.

3. Variation of Parameters with Temperature

In order to increase the prediction capability of the model, the parameters $U_{12} - U_{22}$ and $U_{21} - U_{11}$ were fitted with polynomials in T:

$$U_{12} - U_{22} = A_{12} + B_{12}T + C_{12}T^2 \tag{24}$$

$$U_{21} - U_{11} = A_{21} + B_{21}T + C_{21}T^2 \tag{25}$$

The results are shown in Figs. 4 and 5, and the coefficients are given in Table V. In view of the behavior of carvone, which shows a definite curvature in the temperature variation, it was felt that the second-degree polynomial was a better choice as long as we did not extrapolate the results too far out of the range of experimental measurement. It can be assumed that the validity of the proposed correlation will be valid in the range of temperature $-10°$ to $+50°C$.

4. Predictions from the Model

a. Vapor pressures. Figure 6 shows results obtained for the carvone-water system at $10°$ and $20°C$. The average errors for these two temperatures were, respectively, 1.6 and 1.15%. In this case solubilities of carvone in water were simultaneously used with the pressure measurements to obtain the best values of τ_{12} and τ_{21}.

b. Solubilities. The parameters obtained from Eqs. (24) and (25) can be introduced in Eqs. (18), (19), and (20) and these in turn solved to obtain the relevant solubilities in each phase. Results are presented for carvone, piperitone, and pulegone in the water phase in Fig. 7, and for both phases in Table VI.

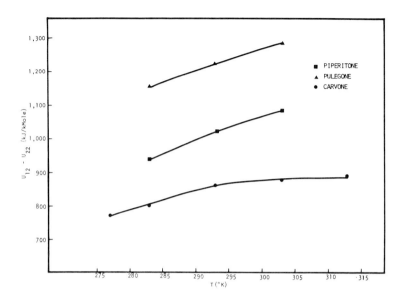

Fig. 4. Variation of $U_{12}-U_{22}$ with temperature. Carvone-, pulegone-, piperitone-water system.

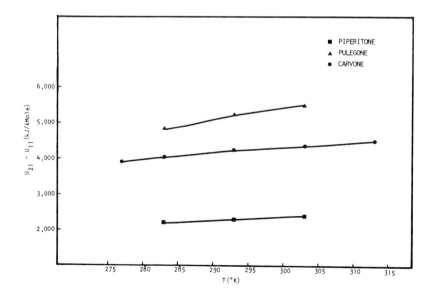

Fig. 5. Variation of $U_{21}-U_{11}$ with temperature. Carvone-, pulegone-, piperitone-water system.

TABLE V. Coefficients for Carvone, Pulegone, and Piperitone in Eqs. (24) and (25)

	A_{12}	B_{12}	C_{12}	A_{21}	B_{21}	C_{21}
Carvone	-10,900.7	76.286	-0.1234	-7.626.5	64.952	-0.0838
Pulegone	-7.498.8	51.019	-0.0757	-37,283.9	254.95	-0.3782
Piperitone	-18,151.8	123.046	-0.1966	-9.374.9	66.858	-0.0951

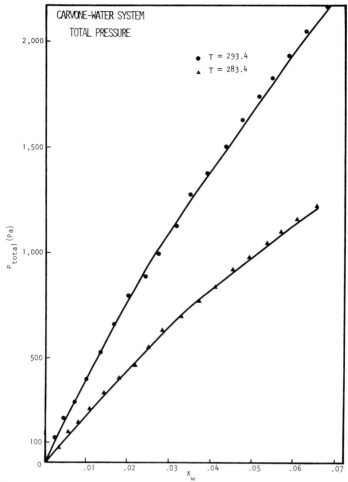

Fig. 6. Predicted and experimental (● , ▲) pressures in the system water-carvone at 20°C.

c. *Activity coefficient, Gibbs energy, enthalpy and entropy at infinite dilution.* From the model, the behavior of the terpenes in the water phase at infinite dilution can be obtained. In order to study interactions in the water phase two types of thermodynamic parameters are usually considered. Hydration properties corresponding to the process, pure solute (ideal vapor) → solute (solution), and solution properties, pure solute (liquid) → solute (solution). Subscripts h and s will be used respectively.

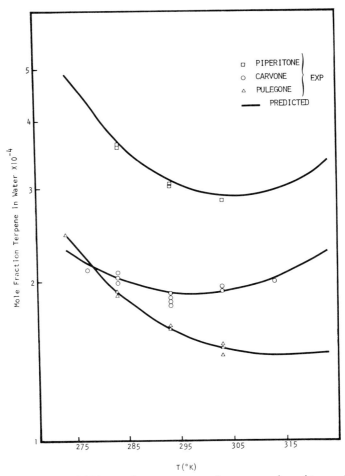

Fig. 7. Solubility of carvone, pulegone, piperitone in water. Variation with temperature.

Tables VII-IX represent these properties for carvone, pulegone, and piperitone. The activity coefficients at infinite dilution are presented in Table X. They have been calculated from Eqs. (4)-(6) along with supplementary Eqs. (26) and (27) (Cabani *et al.* 1971a,b; Franks *et al.*, 1970):

$$\Delta G_h = \Delta G_s + RT \ln P_t^0 \tag{26}$$

$$\Delta H_h = \Delta H_s - \Delta H_v \tag{27}$$

TABLE VI.[a] Results of Solubility for Carvone, Piperitone, and Pulegone

T, °C	Carvone		Pulegone		Piperitone	
	$(x_t)_w$	$(x_w)_t$	$(x_t)_w$	$(x_w)_t$	$(x_t)_w$	$(x_w)_t$
273	0.000230	0.074476	0.000245	0.075775	0.000491	0.235609
283	0.000203	0.070654	0.000193	0.061153	0.000370	0.213948
293	0.000192	0.068841	0.000164	0.052840	0.000312	0.201625
303	0.000192	0.068676	0.000150	0.048533	0.000291	0.196199
313	0.000204	0.069990	0.000145	0.047093	0.000298	0.196448
323	0.000226	0.072756	0.000148	0.048024	0.000332	0.201933

[a] $(x_t)_w$ mole fraction terpene in water phase, $(x_w)_t$ mole fraction water in terpene phase.

TABLE VII. Thermodynamic Properties at Infinite Dilution: Carvone

T (°K)	ΔG_h (kJ/kmole)	ΔG_s (kJ/kmole)	ΔH_h (kJ/kmole)	ΔH_s (kJ/kmole)	ΔS_h (kJ/kmole K)	ΔS_s (kJ/kmole K)
273	20,239	18,913	-71,399	-10,401	-335	-107
283	23,357	19,899	-66,695	-6,213	-318	-92
293	26,324	20,743	-61,803	-1,845	-301	-77
303	29,139	21,444	-56,786	2,641	-283	-62
313	31,803	22,002	-51,693	7,196	-267	-47
323	34,316	22,418	-46,560	11,782	-250	-32

TABLE VIII. Thermodynamic Properties at Infinite Dilution: Pulegone

T (°K)	ΔG_h (kJ/kmole)	ΔG_s (kJ/kmole)	ΔH_h (kJ/kmole)	ΔH_s (kJ/kmole)	ΔS_h (kJ/kmole K)	ΔS_s (kJ/kmole K)
273	17,228	19,451	-78,329	-18,396	-350	-138
283	20,681	20,041	-73,099	-13,688	-331	-119
293	24,582	21,149	-68,016	-9,134	-316	-103
303	28,265	22,108	-63,078	-4,734	-301	-88
313	31,742	22,926	-58,284	-486	-287	-75
323	35,018	23,606	-53,633	3,610	-274	-62

TABLE IX. Thermodynamic Properties at Infinite Dilution: Piperitone

T (°K)	ΔG_h	ΔG_s	ΔH_h	ΔH_s	ΔS_h	ΔS_s
273	19,407	16,851	-83,686	-22,488	-377	-144
283	22,706	18,173	-76,535	-15,855	-350	-120
293	25,762	19,253	-69,049	-8,895	-323	-96
303	28,578	20,093	-61,338	-1,717	-296	-71
313	31,155	20,694	-53,482	5,597	-270	-48
323	33,495	21,058	-45,549	12,980	-244	-25

TABLE X. *Activity Coefficients at Infinite Dilution*

T ($^{\circ}K$)	Carvone γ_t^{∞}	Pulegone γ_t^{∞}	Piperitone γ_t^{∞}
273	4135	3875	1669
283	4685	4978	2251
293	4964	5868	2695
303	4951	6448	2898
313	4675	6671	2831
323	4202	6544	2535

where P_t^0 is the saturation pressure of the pure terpene at temperature T and ΔH_v the latent heat of evaporation of pure terpene at T.

III. DISCUSSION

The results obtained from the model show that it is capable of producing experimental data. The precision is, of course, a function of the validity of the experimental data and the optimizing factor chosen to adjust the parameters. As compared to similar adjustments performed on numerous systems (Abrams and Prausnitz, 1975) the present one is particularly extreme because of the very low solubility of the compounds involved. This poses problems at the experimental and the optimization levels. The experimental technique used here (Smyrl, 1976) proved to be accurate to about 2% or better in most cases as indicated by Table II. This has only limited consequences on the variability of the parameters as shown in Tables III and IV. Consequently, the curves representing the variations of $U_{12}-U_{22}$ and $U_{21}-U_{11}$ can be fitted to a good approximation by a second degree in temperature polynomial. This added constraint on the original values of the parameters seems to have little effect for carvone and pulegone and a slightly greater one on piperitone as revealed by Fig. 6.

This is probably due to the fact that the piperitone is more
soluble in water than the others, but also that water is much
more soluble in the terpene phase as revealed by Table VI, with
up to 20% of water in the piperitone phase. Because the model
was adjusted without any information on the piperitone phase, it
is probable that a distortion results that moves the solubility
curve toward higher values in this case.

As mentioned earlier, another difficulty is the computation
of the parameters from the optimizing procedure. The usual procedures for the search of the optimum of a function of two or
more variables had to be modified to take into account the extremely low values of the solubilities. Special tests and procedures had to be used in order to converge on proper values for
the parameters.

The parameters so obtained can now be used to predict solubilities and relative volatilities for the simple binaries that
were studied. We can also postulate that the solutions of car-
vone-piperitone, carvone-pulegone, and pulegone-piperitone behave
ideally. This is probably a reasonable hypothesis to formulate
since they are molecularly alike and as shown in Table I their
structural parameters are very close to each other. On this
basis Eqs. (13) and (14) extended to multicomponent systems
(Abrams and Prausnitz, 1975) can be used to predict the behavior
of water and the three terpenes simultaneously. This is equivalent
to fixing for the new model the energies of interactions between
terpenes $U_{12}-U_{22}$ and $U_{21}-U_{11}$ to zero. This is equivalent to assuming that terpene molecules do not differentiate between each
other. Only water has specific interactions with each of them.

From the model one can also draw conclusions related to the
activity of the terpene when the two phases are in equilibrium.
It is usually assumed because of the extremely high dilution of
the terpene in the water phase that this activity is close to one.
From this one equates the activity coefficient to the reciprocal
of the solubility expressed in mole fraction. As can be seen from

TABLE XI. *Activity of the Terpene Phase at Equilibrium*

T	Carvone	Pulegone	Piperitone
273	0.934	0.933	0.791
283	0.937	0.946	0.809
293	0.939	0.953	0.820
303	0.939	0.957	0.825
313	0.938	0.958	0.825
323	0.936	0.957	0.820

Table XI the more water there is in the terpene phase the lower the activity of the terpene compound and the greater the error on the activity coefficient.

We now turn to the information we can gain from thermodynamic parameters at infinite dilution. Values of ΔH_h, tabulated in Tables VII-IX for piperitone, pulegone, and carvone, respectively, show that the process of transferring one mole of solute from the ideal vapor to a solution at infinite dilution (mole fraction of solute at saturation is approximately 10^{-4}) is an exothermic process, i.e., heat is evolved. The overall magnitude of ΔH_h is determined from the contributions of two mechanisms. A positive heat (endothermic) corresponds to the heat required to form a cavity for the solute molecule to occupy. A negative heat (exothermic) results from the formation of a more icelike water structure, i.e., icebergs. Obviously, the formation of a more ordered form of water is the main contribution to ΔH_h. It is interesting to note that ΔH_h values for cyclic ethers (Cabani *et al.*, 1971b; Franks *et al.*, 1970), dialkylamines (Frank and Watson, 1969), and the three compounds of this study all fall within the range -50000 to -70000 kJ kmole^{-1}.

Tables VII-IX indicate that the respective ΔH_h values for piperitone, pulegone, and carvone become less negative with an increase in temperature. Several previous studies have shown similar results. Gross *et al.* (1939) noted that ΔH_h for many aliphatic ketones became less negative with increased temperatures

in the range 0-50 C. ΔH_h for cyclic ethers (Franks et al., 1970)
and dialkylamines (Franks and Watson, 1969) were found to exhibit
exactly the same tendency in pure water. Again the model of the
iceberg may be used to rationalize the less negative ΔH_h values
at higher temperatures. More positive ΔH_h values indicate that at
higher temperatures ΔC_p values are higher as well. A larger value
of ΔC_p at higher temperatures is a direct consequence of the
"melting" of a greater number of iceberg species at the higher
temperature.

Table VII illustrates that at a temperature very near 20°C,
ΔH_s for carvone has a value of zero. Here $\Delta H_s = 0$, the negative
heat due to increased ordering in water exactly balances the
positive heat of cavity formation. According to Bohon and Claussen (1951) the temperature at which $\Delta H_s = 0$ corresponds to the
point of minimum solubility of the solute in water.

ΔG_h values for piperitone (Table IX), pulegone (Table VIII),
and carvone (Table VII) in water appear to be very similar in magnitude. In most cases the values of ΔG_h are seen to increase with
an increase in temperature. Similar observations were made for
aliphatic ketones (Gross et al., 1939). As shown in Tables VII-
IX, values of ΔG_s are invariably lower than ΔG_h. The relative
magnitudes of ΔG_h and ΔG_s are determined by the units of p^0 in
the expression $\Delta G_h = \Delta G_s + RT \ln p^0$. In this study, p^0 was expressed in terms of Pascals; hence the $RT \ln p^0$ term is always positive. Cabani et al. (1971a) expressed p^0 values in terms of atmospheres; hence the calculated ΔG_h values were less than ΔG_s values.

The partial molar entropies of hydration and solution, ΔS_h and
ΔS_s, respectively, are of considerable interest since their values
reflect the degree of ordering or structure within the solution.
ΔS_h values for piperitone (Table IX), pulegone (Table VIII), and
carvone (Table VII) in water are seen to be highly negative. Such
highly negative values of ΔS_h point out that the solubilization of
piperitone, pulegone, or carvone brings about a less random or
more ordered state within the solution.

An alternative, more thermodynamically based explanation for the observed solubilities of piperitone, pulegone, and carvone is derived from considering the iceberg model for water. The guest molecule or solute occupies a cavity or partial cavity within the cluster. The ability of the guest molecule to occupy a cavity is dependent on the size of the guest molecule and it has been shown that there exists an optimum molecular diameter for maximum solubility (Franks, 1973). If it can be assumed that in piperitone, pulegone, and carvone the carbonyl-water interactions are similar (the carbonyl has two proton acceptor sites) then the difference in the solubilities of these compounds may be due to the difference in the molecular diameter of the molecules. It is evident that pulegone and carvone will have different diameters than piperitone, since the former compounds both possess unsaturated isopropyl groups outside the six-membered ring, whereas piperitone does not. The presence of this unsaturated isopropyl group may render the molecular structure incompatible with the structure of the cavity; hence carvone and pulegone exhibit lower water solubilities. Piperitone, on the other hand, may not be sterically hindered to such a great extent so as to prevent inclusion into a water cavity.

Piperitone (Table IX) and pulegone (Table VIII) display decreased solubilities in water with higher temperatures. This behavior (Gross et al., 1939; DeSantis et al., 1976; Rice et al., 1976) is followed by very many liquid organic compounds in aqueous solution. An explanation for a decreased solubility with an increase in temperature can be found, once again, by considering the iceberg model of water. The $(H_2O)_b$ species, which exists in equilibrium with $(H_2O)_d$, has many characteristics that have been linked to ice. An increase in temperature has the effect of decreasing the number of $(H_2O)_b$ species, i.e., decreasing the number of host species for the solute. Thus the consequence is lower solubilities at higher temperatures.

Of the three compounds studied, carvone is unique in that it exhibits a solubility minimum between 10° and 30°C. Bohon and Claussen (1951) noted a similar solubility minimum for aromatic hydrocarbons in water and correlated this solubility minimum with the temperature at which the heat of solution became equal to zero. Tables VIII and IX show that ΔH_s values for piperitone and pulegone will similarly equal zero at a temperature between 30° and 50°C.

IV. CONCLUSION

We have adapted successfully the original thermodynamic model of Abrams and Prausnitz. The systems considered are highly non-ideal and represent a very severe test of the capability of the original model. This opens up for liquid foods an important area of application. The calculation of parameters needed for the design of operations based on concentration and distillation is now possible in a wide temperature range with the need of a minimum number of experimental data. It is conceivable that solutes other than the flavoring compounds considered could be added to the model. Solutes such as carbohydrates and proteins will be studied since the model can a priori accept polymer type molecules. Apart from the practical applications mentioned and taking into account the basic limitations of the original model one can also gain insight into the structure of the solution. Water-solute and solute-solute interactions can be estimated and lead to a better understanding of the effect of different compounds in aqueous solutions.

Acknowledgment

The author is grateful to the National Research Council of Canada for financial support of this work.

SYMBOLS

A, B, C	constants, Eqs. (24) and (25)
G	Gibbs free energy
H	enthalpy
ℓ	parameter, Eq. (14)
n	number of moles
P	pressure
q	surface structural parameters (Appendix)
r	volume structural parameter (Appendix)
R	gas constant
S	entropy
T	temperature
U	interaction energy
x	mole fraction
z	lattice coordination number

Greek Letters

γ	activity coefficient
θ	surface fraction
ϕ	volume fraction
τ	parameter, Eq. (12)

Subscripts

i, j	components i and j
cal	calculated
exp	experimental
w	water
t	terpene
s	solution
h	hydration
v	vaporization

Superscripts

E   Excess

0   pure component

——   partial molar quantity

APPENDIX

The structural parameters r and q are, respectively, the Van der Waals volume and area of the molecules relative to those of a standard segment:

$$r_i = V_{wi}/V_{ws}, \quad q_i = A_{wi}/A_{ws}$$

$$A_{ws} = 2.5 \times 10^9 \; cm^2/mole, \quad V_{ws} = 15.17 \; cm^3/mole$$

V_{wi} and A_{wi} are obtained from Bondi (1968), where contributions to V_{wi} and A_{wi} are given for each bond of the molecule and these contributions added.

References

Abrams, D. S., and Prausnitz, J. M. (1975). *Am. Inst. Chem. Eng. J. 21*, 116.
Bohon, R. L., and Claussen, W. F. (1951). *J. Am. Chem. Soc. 73*, 1571.
Bondi, A. (1968). *In* "Physical Properties of Molecular Crystals, Liquids and Glasses," Ch 14. Wiley, New York.
Cabani, S., Conti, G., and Lepori, L. (1971a). *Trans. Faraday Soc. 67*, 1933.
Cabani, S., Conti, G., and Lepori, L. (1971b). *Trans. Faraday Soc. 67*, 1943.
DeSantis, R., Marrelli, L., and Muscetta, P. N. (1976). *J. Chem. Eng. Data 21*, 324.
Flink, J. M., and Gejl-Hansen, F. (1972). *J. Agri. Food Chem. 20*, 691.
Flink, J. M., Gejl-Hansen, F., and Karel, M. (1973). *J. Food Sci. 38*, 1174.
Frank, H. S., and Evans, M. W. (1945). *J. Chem. Phys. 13*, 507.
Franks, F. (1973). In "Water: A Comprehensive Treatise" (F. Franks, ed.), Chap. 1, Vol. 3. Plenum Press, New York.

Franks, F., and Watson, B. (1969). *Trans. Faraday Soc. 65,* 2339.
Franks, F., Quickenden, M. A. J., Reid, D. S., and Watson, B. (1970). *Trans. Faraday Soc. 66,* 582.
Fredenslund, A., Gmehling, J., Michelsen, M. L., Rasmussen, P., and Prausnitz, J. M. (1977). *IEC Process Design Devel. 16,* 450.
Gross, P., Rintelen, J. C., and Saylor, J. H. (1939). *J. Phys. Chem. 43,* 197.
Guggenheim, E. A. (1952). "Mixtures." Clarendon Press, Oxford.
King, C. J., and Massaldi, H. A. (1974a). Paper presented at *4th Int. Congr. Food Sci. Technol., Madrid, Spain.*
Massaldi, H. A., and King, C. J. (1974b). *J. Food Sci. 39,* 438.
Menting, L. C., Hoogstad, B., Thijssen, H. A. C. (1970a). *J. Food Technol. 5,* 111.
Menting, L. C., Hoogstad, B., and Thijssen, H. A. C. (1970b). *J. Food Technol. 5,* 127.
Renon, H., and Prausnitz, J. M. (1968). *Am. Inst. Chem. Eng. J. 14,* 135.
Rice, P. A., Gale, R. P., and Barduhn, H. J. (1976). *J. Chem. Eng. Data 21,* 204.
Smyrl, T. (1976). Ph.D. Thesis. University of Alberta.
Smyrl, T. G., and LeMaguer, M. (1978). *J. Food Process Eng. 2.*
Thijssen, H. A. C. (1972). *Proc. 3rd Nordic Aroma Symp. June 6-8, Finland.*
Wilson, G. M. (1964). *J. Am. Chem. Soc. 86,* 127.

HYDRATION OF
BIOLOGICAL MEMBRANES

Allan S. Schneider

I. INTRODUCTION

The extraordinary growth in biological membrane research
over the past decade is an indication of the recognition membranes have received for their central role in a host of key
cellular processes (Finean *et al.*, 1974). Biomembranes are now
known to regulate ion and nutrient transport, cellular energy
production, the conduction and transmission of nerve impulses,
the sending and receiving of hormonal messages, exo- and endo-
cytosis, key elements of the immune response, etc. The details
of membrane structure are beginning to reveal themselves in
descriptions such as the fluid mosaic model (Singer and Nicolson,
1972), and will no doubt play a definitive role in our understanding of these cellular processes. Most recent structural
studies of membranes have emphasized the dynamic properties of
membrane lipids and proteins. There is, however, a third major
component of biological membranes, which has more or less been
ignored and which is likely to be important in stabilizing the
bilayer structure (Finean, 1969; Finean *et al.*, 1966; Ladbrooke
and Chapman, 1969; Ladbrooke *et al.*, 1968), in transport of

aqueous solutes (Hays, 1976; Hays *et al.*, 1971, Kuiper, 1972),
in the lateral and transverse mobility of membrane lipids and
proteins (Finch and Schneider, 1975), and in membrane fusion
(LeNeveau *et al.*, 1976). This component is the membrane-bound
water or membrane hydration and is the subject of the present
chapter. In what follows we review the field of membrane hydration with emphasis on recent findings from our own laboratory.
We confine ourselves to the hydration of natural biological
membranes and only occasionally refer to the massive literature
on artificial model lipid membranes when directly relevant to
the argument.

II. HYDRATION OF BIOLOGICAL MEMBRANES

    Although biomembrane hydration has been postulated to be
important in various aspects of membrane physiology, there have
been relatively few direct studies of the amount and properties
of membrane bound water. Some of the earliest investigations
in this field were those pioneered by F. O. Schmitt and co-
workers in the 1940's on red cell membranes (Waugh, 1950; Waugh
and Schmitt, 1943) and myelin (Schmitt *et al.*, 1941. Reflectivity measurements using the analytical leptoscope were performed on red cell membranes and their thickness determined as
a function of drying. These measurements yielded a rough estimate of the membrane water content as 25% of the hydrated
membrane. Another early approach to estimating the water content of membranes was offered by Ponder (1954). He measured the
hematocrit volume of rehydrated red cell membranes and by this
somewhat crude method derived a hydration value of approximately
65% of the wet membranes. Subsequently, more comprehensive
studies of membrane hydration involved the use of X-ray diffraction, calorimetry, direct sorption isotherms, and a variety of
spectroscopic techniques. These are discussed below.

A. *X-Ray Diffraction*

Schmitt *et al.* (1941) measured the thickness of the water layer between myelin membranes by X-ray diffraction. By determining the amount of water required to produce the same spacing in isolated nerve lipids, they were able to estimate a water content of at least 30% by weight of the myelin membrane. A rather interesting and extensive approach to the problem of membrane hydration was presented by Finean and co-workers using X-ray diffraction techniques (Finean, 1957, 1969; Finean *et al.*, 1966; Finean and Millington, 1957). In their earlier studies they determined the effects of ionic strength and drying kinetics on the low-angle diffraction patterns of peripheral nerve myelin. They noted an expansion rather than contraction of the layer spacing in hypertonic solution and took this as evidence that the water present between the myelin layers was not free water. From the kinetics of the changes in the diffraction pattern accompanying drying, they suggested that the myelin sheath contains 40-50% water and that the greater part of this water is bound or organized around the polar groups of the myelin lipids and proteins. Further X-ray diffraction studies by Finean *et al.* (1966; Finean, 1969) on myelin and red cell membranes during slow drying gave accurate determinations of the extent of the membrane-bound water and, more importantly, demonstrated that it was required for the structural integrity of the membrane bilayer. It was found that upon drying the membranes, the low-angle X-ray diffraction characteristic of the bilayer structure undergoes an abrupt change at 10-20% hydration for red cell membranes and 30-40% hydration for myelin membranes (Finean *et al.*, 1966). Furthermore, the high-angle diffraction pattern characteristic of liquid water could no longer be detected at hydration levels that coincide with the change in the low-angle pattern (Finean, 1969). Thus the 20-30%

hydration level clearly marks a membrane-bound water component
with properties different from liquid water and which is required
for the structural integrity of the membrane.

B. *Calorimetry*

The results of Finean were independently confirmed and
extended by Ladbrooke, Chapman, and co-workers using calorimetric
methods.   Expanding on their earlier studies of phospholipids and
cholosterol (Ladbrooke and Chapman, 1969) these authors discovered
for myelin membranes below the 20% hydration level that the
liquid water-ice thermal transition disappears and suppressed
endothermic transitions of the cholesterol and phospholipid
appear (Ladbrooke *et al*. 1968).   The differential thermal analysis (DTA) heating curves they obtained for myelin at different
hydration levels are shown in Fig. 1. Curve (a) shows the
ice-water transition, but no lipid transitions at a 30% by weight
water content.   In curve (b) (15% water) the ice peak disappears
and an endothermic transition appears at $35^{O}C$, which they
demonstrate to be due to a transition of crystalline cholesterol.
Further drying produces a second endotherm at $55^{O}C$ due to a
phospholipid transition.   These results lead to a clear determination of membrane-bound water, in terms of nonfreezable water,
as 20% by weight of the hydrated membrane.   They also demonstrate that this amount of bound water is required to maintain
the organization of the lipid layers in myelin and mixing of
cholesterol and phospholids.

C. *Sorption Isotherms*

There has been only one direct study of water binding to
biological membranes, that by Schneider and Schneider (1970).

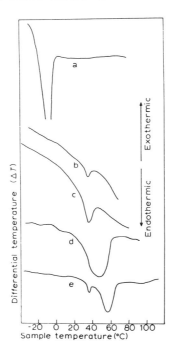

Fig. 1.   Dehydration of myelin.   Differential thermal analysis heating curves for samples of myelin previously equilibrated at different relative humidities.   Approximate final water content (wt.%): (a) 30, (b) 15, (c) 10, (d) 5, (e) 3.   From Ladbrooke et al. (1968).

Water vapor sorption isotherms on human red cell membranes were measured at 20° and 4°C, and the results were used to determine the coverage of primary sites, the saturation hydration, and the heats and entropies of binding. A typical membrane adsorption and desorption isotherm at 20°C is shown in Fig. 2. The familiar sigmoid-shaped curve characteristic of many multilayer sorbing systems, e.g., proteins and phospholipids, is apparent and the existence of hysteresis in the desorption-adsorption cycles is noteworthy. The latter may reflect reversible, hydration-dependent, structural changes in the membrane or possibly capillary condensation. The shape of the hysteresis loop varied slightly with different membrane preparations. The 4°C isotherm gave significantly greater water sorption than the 20°C isotherm.

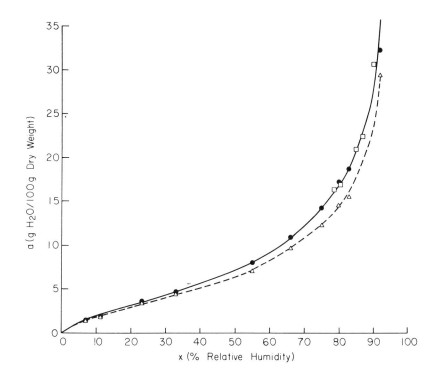

Fig. 2. Sorption isotherms of water on human red cell mem-
branes at 20°C. Upper curve, desorption; lower curve, adsorp-
tion. From Schneider and Schneider (1972).

The sorption data gave a good fit to the BET equation (Brun-
auer-Emmet-Teller) which is plotted in Fig. 3. The BET analysis
led to a value for the monolayer hydration coverage a_1 of 4-5 g
H_2O /100 g dry membrane. Although the concept of a monolayer on
a complex heterogeneous system such as a membrane or a protein
is at best only a rough first approximation, it has been commonly
used. From the monolayer hydration, Schneider and Schneider
(1970) estimated the area that would be covered by the most
tightly bound water and compared it with the surface area of the

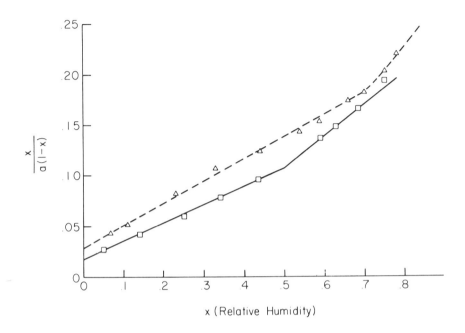

*Fig. 3. BET plots using desorption isotherms at 20°C
(upper line) and 4°C (lower line). a = amount adsorbed in g
$H_2O/100$ g dry membrane; x = water activity (relative humidity
x .01). From Schneider and Schneider (1972).*

membrane. They found that the amount of water bound in the
primary layer is only enough to cover one surface of the membrane. This is an interesting result in that it suggests that
the primary binding sites are not densely and uniformly spaced
over both surfaces of the membrane, but may instead exist at
separated discrete locations, which may be bridged by interconnected, hydrogen-bonded water lattices at low hydrations.
There is also the possibility that many of the polar sites at

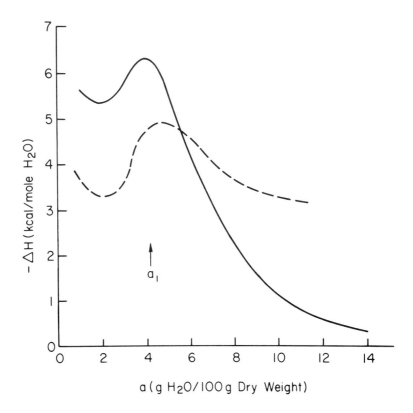

Fig. 4. Heats of water sorption on red cell membranes.
Differential or isostearic heat (——) and integral heat
(- - -) as a function of hydration. Reference state for $\Delta H=0$ is
liquid water at 20°C. a_1 BET monolayer coverage. Schneider and
Schneider (1972).

the membrane surface interact with each other and are thus
shielded from the water at low hydration levels. The saturation
hydration was also determined (hydration at water activity of 1)
and found to be 70-80 g H_2O/100 g dry membrane or about 41-44%
of the hydrated membrane. This would include a large component
of loosely bound water.

The BET parameter, C, which is a measure of the strength of
binding to the primary binding sites, was found to be 7-8 at

20^0C and 9-11 at 4^0C (Schneider and Schneider, 1970). These
values are typical of water binding of phospholipids and proteins.
A better approach to the energetics of binding is to determine
the differential and integral heats and entropies of water binding by thermodynamic analysis of the 20 and 4 C sorption isotherms.
The results for the differential (isostearic) and integral heats
of water sorption are shown in Fig. 4. The reference state
($\Delta H=0$) is liquid water, and the heats shown in Fig. 4 represent
binding energies above that of the liquid state. The differential
heat (solid curve) is particularly interesting in that it represents the energies for water molecules binding at a particular
hydration level, in contrast to the integral heat, which is the
average binding energy of all water molecules already bound at
that level. In Fig. 4 we note an extremum near the BET monolayer
value (^{a}1 arrow) and then a dceline in the differential heat so
that above 0.15-20 g H_2O/g dry membrane the energetics of water
vapor bonding are close to that of condensation into liquid water.
These results are interesting in that they allow a classification
of the total bound water determined from the sorption isotherms
into several categories according to their binding energies:

1. Hydrations of 0.04-0.05 g H_2O/g dry membrane would mark
the water bound to the primary hydration sites (~monolayer) and
have binding energies greater than 5 kcal/mole above the latent
heat of condensation (~10.4 kcal/mole). (For comparison we note
that water molecules in ice are held with energies only 1.5 kcal/
mole above liquid water.

2. Hydrations up to ~0.20 g/g dry membrane would make up
most of the tightly bound water and would have binding energies
ranging from 0 to 5 kcal/mole above liquid water.

3. Hydrations ranging from 0.2-0.7 g H_2O/g dry membrane
would represent loosely associated water or "boundary" water,
with binding energies little different from condensation of vapor
into liquid water.

TABLE I. Sorption, BET, and Thermodynamic Values for Water Bound to Membranes and Membrane Components[a]

Adsorbent	g H_2O/100 g dry weight				ΔH_{a_l} (kcal/mole)	ΔS_{a_l} (e.u.)
	a_l	$a_{90\%}$	a_s	c		
Red cell membranes[b]	4	25-30	70-80	7-8.3	6	19
Globular proteins[c]	6-7	30-40	45-65	9-14	3-6.6	17-18
Lipids (lecithin)[d]	4.5-5.6	20-30	25-44	7.7-10.7	1.8-2.4	10-13
Polysaccharides[e]	5-11	25-50	-	10	6-8	12-22

[a] a_l, monolayer hydration; $a_{90\%}$, a_s, water sorbed at 90% and 100% (saturation) relative humidity; c, BET energy parameter; ΔH_{a_l}, differential or isosteric heat of water binding at monolayer hydration, ΔS_{a_l}, differential entropy of water binding at monolayer hydration; liquid water at 20° is reference state for ΔH_{a_l} and ΔS_{a_l}. The latent heat and entropy of condensation (10.4 kcal/mole and 28 e.u.) should be added to obtain the vapor reference state values.

[b] Schneider and Schneider, 1972.

[c] Altman and Benson, 1960; Bull, 1944; Eley and Leslie, 1964; Foss and Reyerson, 1958; Killion et al., 1970; Kuntz and Kauzmann, 1974; Katchman and McLaren, 1951; Dole and McLaren, 1947.

[d] Elworthy, 1961, 1962.

[e] Bettelheim and Volman, 1957; Bettelheim et al., 1970.

The binding and thermodynamic data for red cell membranes are summarized in Table I and compared with equivalent values for proteins and phospholipids. The monolayer hydration (a_1) of membranes is seen to be somewhat lower than that of lipids (Elworthy, 1961, 1962), proteins (Altman and Benson, 1960; Bull, 1944; Eley and Leslie, 1964; Foss and Reyerson, 1958; Killion et al., 1970; Kuntz and Kauzmann, 1974; Katchman and McLaren, 1951; Dole and McLaren, 1947), or polysaccharies (Bettelheim et al., 1970; Bettelheim and Volman, 1957), and this is likely due to shielding of polar sites in the membrane by their mutual interaction. The saturation hydration of membranes is considerably higher than phospholipids and close to or slightly higher than globular proteins. This may be related to the membranes' surface charge or to swelling or capillarity of the membrane matrix. The differential heats and entropies of water binding to the membrane at monolayer hydration are similar to those for globular proteins (Eley and Leslie, 1964; Foss and Reyerson, 1958; Kuntz and Kauzmann, 1974; Dole and McLaren, 1947) polysaccharides (Bettelheim et al., 1970, Bettelheim and Volman, 1957) and somewhat higher than for phospholipids (lecithin) (Elworthy, 1962). This might suggest involvement of glycoprotein in the hydration of primary sites on the membrane, although additional data would be required to confirm this.

III.   SPECTROSCOPIC STUDIES

The number of spectroscopic studies of biomembrane hydration is relatively few and these have been aimed at both a characterization of the bound water (infrared absorption, NMR, extrinsic fluoresence) and a determination of the effects of hydration on membrane structure (circular dichroism, infrared absorption).

A. *Circular Dichroism*

Along with their sorption isotherm study, Schneider and
Schneider (1970) measured the circular dichroism of red cell
membranes during dehydration at relative humidities of 92 and
0%. Circular dichorism (CD) is probably the most reliable and
precise measure of protein secondary structure (other than
X-ray diffraction) and the CD spectra of the alpha helix, beta-
sheet, and random coil are distinct and well defined. CD spectra
of red cell membranes in suspension indicate a relatively large
amount of alpha helix for the membrane protein (~45%) (Schneider
et al., 1970; Gitter-Amir *et al.*, 1976). CD spectra of red
cell membrane films as a function of hydration are shown in
Fig. 5. The spectrum clearly resembles that of the alpha helix
and upon dehydration from 92 to 0% relative humidity, there was
little or no change, indicating a stable membrane protein
secondary structure independent of membrane hydration. In contrast, a similar change in membrane hydration is known to dis-
rupt the ordered lipid bilayer X-ray pattern (Finean, 1969;
Finean *et al.*, 1966) and induce a separation of cholesterol and
phospholipids (Ladbrooke and Chapman, 1969; Ladbrooke *et al.*,
1968). That such a major change in membrane lipids could occur
without effect on membrane protein conformation is consistent
with a central concept of membrane structure, namely, that the
bulk of membrane lipids exist in domains separate from membrane
proteins.

B. *Extrinsic Fluorescence*

Extrinsic flurescent probes have been used extensively over
the past decade to explore the properties of various regiospecific
compartments of biomembranes (Radda and Vanderkooi, 1972; Azzi,
1975). Perhaps the most widely used and best characterized is
the dye molecule 1-anilino-8-naphtalenesulfonate (ANS). It is

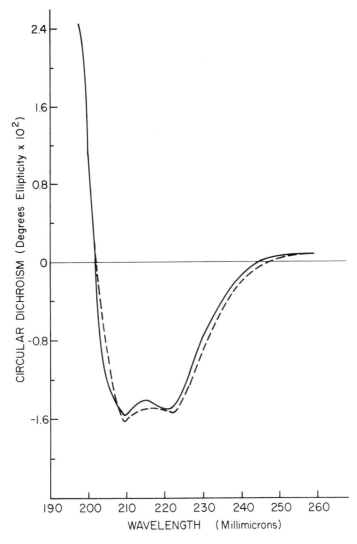

Fig. 5. *Circular dichroism of red cell membranes as a function of hydration: 92% relative humidity (——); 0% relative humidity (---). From Schneider and Schneider (1972).*

generally thought to bind to membranes at the bilayer-aqueous interface. ANS is also known to be sensitive to both the polarity and rigidity of its solvent microenvironment through a mechanism

of excited-state solvent relaxation (Brand and Gohlke, 1972).
Polar fluid solvents such as liquid water quench the fluorescence
and induce a red shift in emission peak while less polar (ethanol)
or viscous (ice) solvent environments produce an intense fluorescence and blue shifted emission peak. The fluorescence spectrum of ANS may thus provide a useful probe of hydration effects
on the polarity and microviscosity of the membrane-aqueous interface. Schneider et al. (1979) have measured the fluorescence
emission spectra of ANS bound to red cell membranes as a function of membrane hydration (relative humidity) and their results
are shown in Fig. 6. A 2½-fold increase in fluorescence intensity, accompanied by a 10 nm blue shift in emission peak
wavelength, occurs with membrane dehydration between 95% and 0%
rh. These changes are characterstic of an increasing rigidity
and/or decreasing polarity of the microenvironment of ANS at the
membrane-aqueous interface in the region of the lipid polar head
groups. Similar spectral changes have been observed for ANS
and related dyes in model phospholipid membranes with dehydration
(Galik-Krzywicki et al., 1970; Wells, 1974). In these studies
it was further determined by X-ray methods that the area per lipid
head group increases and, by fluorescence polarization that there
is increased molecular motion at the lipid-water interface with
increasing hydration.

C. Infrared Spectroscopy

Infrared absorption spectroscopy has been used in exploring
membrane-bound water and hydration effects on other membrane
components (Schneider et al., 1979). Infrared absorption is
particularly sensitive to hydrogen bond strength and vibrational
motion of water and membrane protein, lipid, and polysaccharide
structural groups. Schneider et al. (1979) have measured the

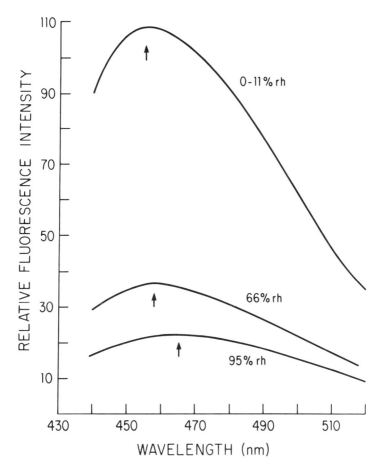

Fig. 6. Fluorescence emission spectra of ANS bound to red cell membranes as a function of hydration. rh, relative humidity. Excitation wavelength=350nm. Arrows indicate location of emission peak. From Schneider et al. (1979).

IR spectra of red cell membranes at various hydration levels and in several different IR spectral regions. Figure 7 shows the IR spectra from 3800 to 400 cm^{-1} for wet (95% rh) and dried (0% rh) red cell membranes that have been equilibrated against H_2O (Fig. 7a) and D_2O (Fig. 7b). The spectra of liquid H_2O and

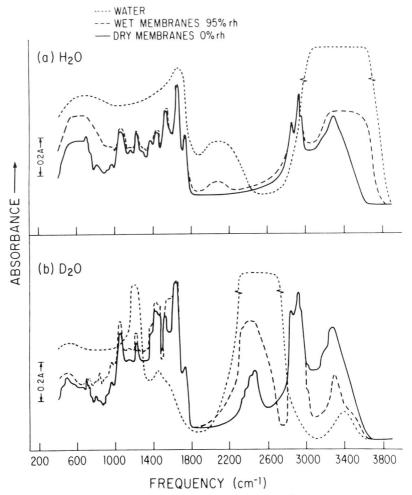

Fig. 7. Infrared spectra (4000 - 400 cm⁻¹) of red cell
membranes at 95 and 0% relative humidity. (a) Membranes equilibrated against H_2O; (b) membranes equilibrated against D_2O.
Spectra of liquid water shown for comparison. From Schneider
et al. (1979).

D_2O are shown for comparison. A number of dried membrane spectral bands can be identified in Fig. 7a (Chapman, 1965; Bessette,
1975; Dobriner et al., 1953): Amide A (~3300 cm⁻¹), CH stretching
(3100-2800 cm⁻¹), ester carbonyl stretching (1738 cm⁻¹), amide I

(1650 cm^{-1}), amide II (1540 cm^{-1}), P=O stretching (1230 cm^{-1}),
POC aliphatic stretching (1080-1060 cm^{-1}), and cholesterol vinyl
CH bending (834-800 cm^{-1}). The H_2O IR bands overlapping some of
the membrane bands include the OH stretching modes (3400 cm^{-1}
broad band) and the HOH bending mode (1640 cm^{-1}). D_2O and the
deuterated membrane specra show the corresponding shifted bands.
Dehydration effects are obvious in the OH and OD stretching
regions (3200-3600 cm^{-1}, Fig. 7a; and 2300-2700 cm^{-1} Fig. 7b)
and in the 600-1000 cm^{-1} region of Fig. 7b. The D_2O equilibrated film spectra can be useful in that there is less masking
of membrane groups by the strong H_2O absorptions and the un-
deuterated, non-exchangeable, structural groups can be more
easily explored.

Figure 8 shows dehydration difference spectra in the OH
(Fig. 8a) and OD (Fig. 8b) stretching regions for red cell
membranes equilibrated between various hydration levels (relative
humidities) (Schneider et al., 1979). Three difference bands
are apparent with frequencies 3550, 3500, and 3200 cm^{-1} for H_2O
hydrated membranes and with corresponding frequencies 2650,
2540-2500, and 2350 cm^{-1} for D_2O equilibrated membranes. The
3550 (2650) cm^{-1} difference band is most apparent at high hydration levels and may represent removal of loosely bound water
having weaker hydrogen bonds than ordinary liquid water. The
3500 (2540-2500) cm^{-1} difference band may also be due to removal
of bound water somewhat more tightly bound than that corresponding to the 3550 cm^{-1} since it is more prominent at lower
hydrations. Similar blue-shifted (relative to liquid water)
dehydration difference bands have been observed for stratum
corneum (2480 cm^{-1}, OD) (Hanson and Yellin, 1972), polyproline
(3490 cm^{-1}, OH) (Kuntz and Kauzmann, 1974), molecular sieves
(3500 cm^{-1}, OH) (Hino, 1977), and phospholipid micelles (3530
cm^{-1}, OH)(Wells, 1974), and have been assigned to a bound-water
component.

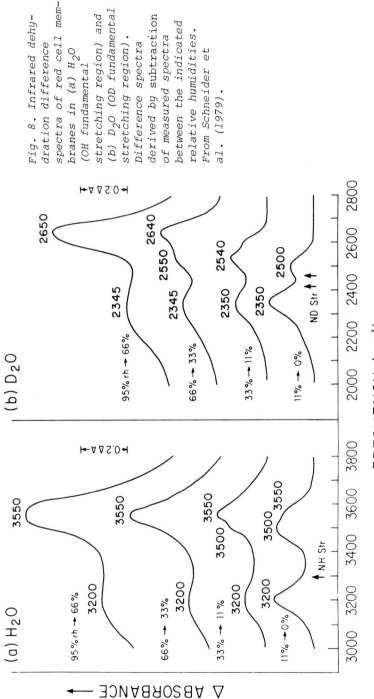

Fig. 8. Infrared dehydration difference spectra of red cell membranes in (a) H_2O (OH fundamental stretching region) and (b) D_2O (OD fundamental stretching region). Difference spectra derived by subtraction of measured spectra between the indicated relative humidities. From Schneider et al. (1979).

The difference band observed at 3200 (2340) cm^{-1} appears
at quite low hydrations, and several assignments are possible.
Protein structural changes may alter NH hydrogen bond strengths
and these may be manifested in amide A and B bands in this
spectral region. The cholesterol hydroxyl, which normally
absorbs at 3400 cm^{-1}, has been shown to shift to 3250 cm^{-1}
upon interaction with phospholipids (Zull et al., 1968). The
observed 3200 cm^{-1} difference band in Fig. 8 may thus have a
contribution due to lipid separation and loss of cholesterol-
phospholipid interaction with dehydration. This interaction
is known to require water (Ladbrooke and Chapman, 1969; Ladbrooke
et al., 1968). There is also the possibility of involvement of
OH groups from saccharide units of membrane glycoproteins.
Finally, there is a good chance that the 3200 cm^{-1} difference
band represents a class of the most tightly bound water. This
difference band occurs most prominently at very low hydrations,
which would correspond to the primary binding sites. Its low
frequency (3200 cm^{-1} H_2O, 2350 cm^{-1} D_2O) is similar to the OH
stretch in ice, which indicates stronger hydrogen bond strengths
than liquid water, and would be consistent with the high heats of
binding found from thermodynamic analysis of sorption isotherms
(Schneider and Schneider, 1970). Extensive dehydration leads
to complete loss of this band from D_2O equilibrated films, making
its assignment to ND vibrations unlikely (Schneider et al., 1979).
A number of other workers have also assigned dehydration difference bands at similar frequencies to a tightly bound-water component on proteins (Buontempo et al., 1976), stratum corneum
(lipids and proteins) (Hansen and Hellin, 1972), polyelectrolyte
membranes (Zundel, 1969), and molecular sieves (Hino, 1977).

A lower-frequency region of the IR spectrum of red cell membranes has been measured over a range of relative humidities in
D_2O (Schneider et al., 1979). The spectra are shown in Fig. 9
and the region covered (900-650 cm^{-1}) has contributions from CH

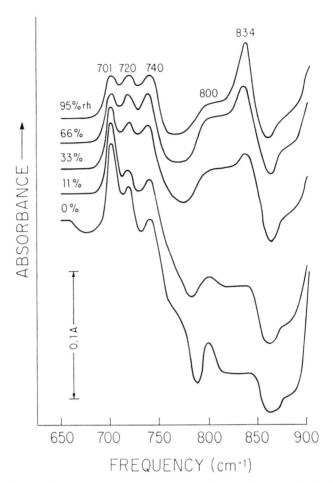

Fig. 9.   Infrared spectra of red cell membranes in D_2O in the region 650-900 cm^{-1} as a function of membrane hydration. rh, relative humidity.   From Schneider et al. (1979).

bending and NH deformation modes.   The 720 cm^{-1} bond has been assigned to the CH rocking vibration of the lipid fatty chains (Chapman *et al.*, 1968) and is insensitive to membrane hydration. The 701 and 740 cm^{-1} bands may be due to amide V vibrations of membrane proteins that are inaccessible to deuteration.   The hydration-dependent changes in these bands may signify changes in the hydrogen bonding of the NH groups.   Among the more

intereting IR changes with dehydration are those in the 800 and
834 cm^{-1} bands, which may be due to the vinyl CH bending vibration of cholesterol (Chapman, 1965; Dabriner et al., 1953). If
cholesterol and phospholipids separate with dehydration in the
red cell membrane, it is conceivable that these IR changes may
be monitoring this event.

D. NMR

Nuclear magnetic resonance methods have been used extensively
to study the dynamic properties of water in tissues and protein
solutions (Kuntz and Kauzmann, 1974) and these methods have
begun to be applied to the problem of membrane-bound water
(Finch and Schneider, 1975; Clifford et al., 1968). Clifford
et al. (1968) measured NMR linewidths and relaxation times of
red cell membrane-water systems by broad line and pulse techniques. Spin-lattice (T_1) and spin-spin (T_2) relaxation times
were determined for samples containing 92, 50, and 8 wt.% water.
They found their data to be consistent with a model in which
water molecules exist in two phases, one with restricted motion
designated as membrane-bound water and the other with motions
similar to ordinary free water. The water molecules are assumed
to exchange rapidly between the two phases and the measured NMR
relaxation times thus represents a weighted average between these
phases. Clifford et al. concluded that the bound water represents
about 30% of the hydrated membrane and is mostly associated
with the membrane proteins (Clifford et al., 1968). They also
concluded that the bound water is very different from bulk liquid
water and that it exists in a wide range of states, some of which
are very tightly bound and comparable to water in ice or solid
hydrates. This is consistent with the high heats of binding
determined from sorption isotherms (Fig. 4) and the infrared

dehydration difference band (Fig. 8) which has a frequency
near that of ice (3200 cm^{-1}).

In an effort to obtain more detailed information about the
dynamic state of the membrane-bound water, Finch and Schneider
(1975) have measured the NMR spin-lattice relaxation time in
both the laboratory (T_1) and rotating frame ($T_{1\rho}$) of reference
for aqueous suspensions of red cell membranes. The T_1 results
were consistent with those of Clifford et al. (1968) and the
$T_{1\rho}$ measurements were more sensitive to motion of the bound
water. The data support a model in which water molecules are
exchanging rapidly between a bound phase with restricted motion
and a free phase with dynamic properties similar to liquid water
(Finch and Schneider, 1975). Possible relaxation mechanisms
for the bound water were explored and support was found for an
intermolecular interaction modulated by translational motion
similar to that found by Burnett and Harmon for glycerol (Burnett and Harmon, 1972). These authors have shown that the intermolecular mechanism gives rise to a linear dependence of relaxation rate ($1/T_{1\rho}$) on the square root of Larmor frequency in the
rotating frame, W_1 with a slope which is a function of the
translational diffusion constant (Burnett and Harmon, 1972):

$$\frac{d(1/T_{1\rho})}{d\sqrt{W_1}} = \frac{\sqrt{2}\ \gamma^4 h^2 N}{80\ \pi D^{3/2}}$$

where γ is the gyromagnetic ratio for protons, h is Plank's
constant, N is the number of proton spins per cm^3 and D is the
translational diffusion constant. From a plot of relaxation rate
($1/T_{1\rho}$) in the bound phase vs. $W_1^{1/2}$, it is possible to derive
the diffusion constant of membrane-bound water. Such a plot was
prepared and the resulting diffusion constant of the bound water
was determined to be 2×10^{-9} cm^2/sec (Finch and Schneider, 1975).

This value is four orders of magnitude slower than the self-diffusion constant of liquid water ($D \simeq 10^{-5}$ cm^2/sec), and represents an effective or average self-diffusion constant for a bound-water component representing 30 wt.% of the membrane. There no doubt will be a range of mobilities within the bound phase, with some molecules having greater and lesser values than the average. The value of 2×10^{-9}cm^2/sec for the bound-water diffusion constant is provocatively close to the rates of lateral diffusion of membrane lipids ($10^{-7} - 10^{-9}$cm^2/sec) and proteins (10^{-9}-10^{-12} cm^2/sec), and within half an order of magnitude of the self-diffusion constant of ice. The bound-water diffusion constant may also be compared with the rate of passive diffusional transport of water across the red cell surface membrane. If no assumptions are made about membrane pores, and the total area of the cell surface is taken as the area for aqueous transport, then it is possible to derive a value for the rate of water diffusional transport from red cell permeability coefficients by simply multiplying by the thickness of the transport barrier (Finch and Schneider, 1975; Stein, 1967). This rate turns out to be $\sim 10^{-9}$ cm^2/sec (Finch and Schneider, 1975). The closeness of the membrane-bound water self-diffusion constant to both the rates of lateral diffusion of membrane proteins and lipids and to the derived rates of aqueous diffusional transport across the membrane suggests several possible roles of bound water in membrane physiology. It is conceivable that the membrane hydration layers provide anchorage sites that influence the lateral mobility of membrane lipids and proteins and restricts their transverse ("flip-flop") mobility. The slowly diffusing bound water is likely to coat all polar membrane surfaces and could also affect the rate-limiting barrier to aqueous diffusional transport across the membrane.

IV.   ROLE OF BOUND WATER IN BIOLOGICAL MEMBRANE STRUCTURE

A potpourri of information has been presented above about
biological membrane hydration, which has been obtained by diverse
methods. It may be summarized in terms of (a) the bound-water
content of membranes, (b) the properties of the membrane-bound
water, and (c) its effects on membrane structure.

Although the different methods of measuring membrane water
content tend to define bound water in different terms, there
is remarkable agreement among the various techniques suggesting
that 20-40 wt. % of the hydrated membrane consists of bound
water. Half or more of this water is tightly bound, with binding
energies comparable to and greater than water molecules in ice.
These numbers derive from X-ray diffraction (Finean, 1969;
Finean et al., 1966; Schmitt et al., 1941), calorimetry (Ladbrooke and Chapman, 1969, Ladbrooke et al., 1968), sorption
isotherms (Schneider and Schneider, 1970), NMR (Finch and
Schneider, 1975; Clifford et al., 1968), and reflectivity
measurements (Waugh, 1950; Waugh and Schmitt, 1940).

The properties of membrane-bound water have been found to be
distinctly different from ordinary free liquid water. High-
angle X-ray diffraction patterns typical of liquid water can no
longer be discerned in membranes containing less than 20-30% water
(Finean, 1969). Both NMR (Clifford et al., 1968) and calorimetry
(Ladbrooke and Chapman, 1969; Ladbrooke et al., 1968) indicate
that this amount of water is nonfreezable (Fig. 1) presumably
because it cannot crystallize into an ice lattice. Yet the
bound water appears to have a highly restricted motion relative
to liquid water as indicated by NMR (Finch and Schneider, 1975;
Clifford et al., 1948) and suggested by extrinsic fluorescence
spectroscopy (Fig. 6). The bound water may in some respects
resemble amorphous ice in that it does not crystallize (Fig. 1),
its self-diffusion constant is near that of ice (Fig. 10), its

binding energies are comparable to and greater than water molecules in ice (Fig. 4), and it may have an infrared vibrational
frequency close to that of ice (Fig. 8). Thus the bound water
may be considered "icelike," not in the sense of having a
crystal structure, but rather because it is very tightly bound
and has restricted motion relative to liquid water.

   The effects of hydration on membrane structure are beginning
to be understood. Both the X-ray (Finean, 1969; Finean et al.,
1966) and calorimetry (Ladbrooke and Chapman, 1969; Ladbrooke
et al., 1968) studies indicate that approximately 20% hydration
levels are required for the membrane bilayer structure to be
preserved and for the mixing of cholesterol and phospholipids.
Below this level of hydration changes can also be seen in the
infrared spectra of membrane cholesterol and amide N-H, when
observed in D_2O (Fig. 9). However comparable changes in membrane
protein secondary structure were not observed in H_2O by circular
dichroism (Fig. 5) and this may be due to differences between
H_2O and D_2O effects during dehydration and/or the CD studies not
covering high enough hydration levels. The similarity in diffusion rates for bound water and laterally mobile membrane proteins and lipids suggests an influence of one on the other (Finch
and Schneider, 1975). Yet at present it is hard to know whether
the hydration is affecting protein and lipid lateral mobility or
whether the mobilities observed for the bound water are the
result of their being attached to and moving with the proteins
and lipids. The evidence coming from fluorescence emission
spectra of the probe ANS bound to red cell membranes would
suggest removal of the loosely bound water results in an increased rigidity and/or decrease polarity of the membrane -
aqueous interface. This may simply be telling us that part of
the probe molecule senses the more mobile, loosely bound water,
which when removed would leave a more rigid environment.

The investigation of membrane-bound water is still in its infancy. The handful of studies that have been reported reveal that bound water is a major membrane component, required for the structural integrity of the membrane bilayer, and influencing the dynamic properties of membrane lipids and proteins. The properties of the bound water itself have only begun to be explored. Further characterization of membrane-bound water should allow its incorporation into current models of membrane structure and give insight into the relevance of membrane hydration to cell surface function.

ACKNOWLEDGMENTS

It is a pleasure to thank my past and present collaborators, Mary-Jane T. Schneider, Edward D. Finch, C. Russell Middaugh, and Mary D. Oldewurtel, for their stimulating ideas and experimental expertise. I would also like to express appreciation to Professor Martin Sonenberg for support, encouragement, and interesting discussions. I am in special debt to my mentor the late Professor Aharon Katzir Katchalsky in whose department at the Weizmann Institute our studies of membrane hydration began and whose extraordinary intelligence, wit, and enthusiasm inspired our work. Part of the research described in this chapter was done during the tenure of an Established Investigatorship of the American Heart Association. The work was supported in part by grants from the National Science Foundation (PCM 76-04079) and National Institutes of Health (CA 15773, CA 08748, AM 18759).

REFERENCES

Altman, R. L., and Benson, S. W. (1960). *J. Phys. Chem. 64,* 851.

Azzi, A. (1975). *Quart. Rev. Biophys. 8,* 237.

Bessette, F. (1975). *Can. J. Spectrosc. 20,* 126.

Bettelheim, F. A., and Volman, D. H. (1957). *J. Polymer Sci. 24,* 445.

Bettelheim, F. A., Block, A., and Kaufman, L. M. (1970). *Biopolymers 9,* 1531.

Brand, L., and Gohlke, J. R. (1972). *Ann. Rev. Biochem. 41,* 843.

Bull, H. B. (1944). *J. Am. Chem. Soc. 66,* 1499.

Buontempo, U., Careri, G., and Fasella, P. (1972). *Biopolymers 11,* 519.

Chapman, D., Kamat, V. B., and Levene, R. J. (1968). *Science 160,* 314.

Chapman, D. (1965). "The Structure of Lipids by Spectroscopic and X-Ray Techniques." Wiley, New York.

Clifford, J., Pethica, B. A., and Smith, E. G. (1968). *In* "Membrane Models and the Formation of Biological Membranes" (L. Bolis and B. A. Pethica, eds.), pp. 19-42, North-Holland Publ. Co., Amsterdam.

Dobriner, C., Katzenellenbogen, E. R., and Jones, R. N. (1953). "Infrared Absorption Spectra of Steroids." p. 41. Wiley (Interscience), New York.

Dole, M., and McLaren, A. D. (1947). *J. Am. Chem. Soc. 69,* 651.

Eley, D. D., and Leslie, R. B. (1964). *Adv. Chem. Phys. 7,* 238.

Elworthy, P. H. (1961). *J. Chem. Soc. 1961,* 5385.

Elworthy, P. H. (1962). *J. Chem. Soc. 1962,* 4897.

Finch, E. D., and Schneider, A. S. (1975). *Biochim. Biophys. Acta 406,* 146.

Finean, J. B. (1957). *J. Biochem. Biophys. Cytol. 3,* 95.

Finean, J. B. (1969). *Quart. Rev. Biophys. 2,* 1.

Finean, J. B., and Millington, P. F. (1957). *J. Biophys. Biochem. Cytol. 3,* 89.

Finean, J. B., Coleman, R., Green, W. G., and Limbrick, A. R. (1966). *J. Cell. Sci. 21,* 287.

Finean, J. B., Coleman, R., and Michell, R. H. (1974). "Membranes and Their Cellular Functions." Wiley, New York.

Foss, J. G., and Reyerson, L. H. (1958). *J. Phys. Chem. 62,* 1214.

Gitter-Amir, A., Rosenheck, K., and Schneider, A. S. (1976). *Biochemistry 15,* 3131.

Gulik-Krzywicki, T., Schechter, E., Iwatsubo, M., Ranck, J. L., and Luzzati, V. (1970). *Biochim. Biophys. Acta 219,* 1.

Hansen, J. R., and Yellin, W. (1972). *In* "Water Structure at the Water-Polymer Interface" (H. H. G. Jellinek), pp. 19-28. Plenum Press, New York.

Hays, R. M. (1976). *Kidney Int. 9*, 223.

Hays, R. M., Franki, N., and Soberman, R. (1971). *J. Clin. Invest. 50*, 1016.

Hino, M. (1977). *Bull. Chem. Soc. Japan 50*, 574.

Katchman, B., and McLaren, A. D. (1951). *J. Am. Chem. Soc. 73*, 2124.

Killion, P. J., Reyerson, L. H., and Cameron, B. F. (1970). *J. Colloid Interface Sci. 34*, 495.

Kuiper, P. J. C. (1972). *Ann. Rev. Plant Physiol. 23*, 157.

Kuntz, I. D., and Kauzmann, W. (1974). *Adv. Prot. Chem. 28*, 239.

Ladbrooke, B. D., and Chapman, D. (1969). *Chem. Phys. Lipids 3*, 304.

Ladbrooke, B. D., Jenkinson, T. J., Kamat, V. B., and Chapman, D. (1968). *Biochim. Biophys. Acta 164*, 101.

LeNeveau, D. M., Rand, R. P., and Parsegian, V. A. (1976). *Nature 259*, 601.

Ponder, E. (1954). *Nature 173*, 1139.

Radda, G. K., and Vanderkooi, J. (1972). *Biochim. Biophys. Acta 265*, 509.

Schmitt, F. O., Bear, R. S., and Palmer, K. J. (1941). *J. Cell. Comp. Physiol. 18*, 33.

Schneider, A. S., Schneider, M.-J. T., and Rosenheck, K. (1970). *Proc. Natl. Acad. Sci. 66*, 793.

Schneider, A. S., Middaugh, C. R., and Oldewurtel, M. D. (1979). *J. Supramol. Struct.*, in press.

Schneider, M.-J. T., and Schneider, A. S. (1970). *J. Membrane Biol. 9*, 127.

Singer, S. J., and Nicolson, G. (1972). *Science 175*, 720.

Stein, W. D. (1967). "The Movement of Molecules Across Membranes." Academic Press, New York.

Waugh, D. F. (1950). *Ann. N. Y. Acad. Sci. 50*, 835.

Wells, M. A. (1974). *Biochemistry 13*, 4937.

Zull, J. E., Greanoff, S., and Adam, H. K. (1968). *Biochemistry 7*, 4172.

Zundel, G. (1969). "Hydration and Intermolecular Interaction: Infrared Investigations with Polyelectrolyte Membranes." Academic Press, New York.

STRUCTURES OF WATER

IN AQUEOUS SYSTEMS

Werner A. P. Luck

I. INTRODUCTION

Food science, like modern biology and medicine, assumes two
different states of water in cells: bound and free. How can
we be sure these assumptions are correct? To answer this
question we need detailed knowledge of the structure of water
in aqueous systems.

II. STRUCTURE OF LIQUID WATER

The simplest model of nonpolar liquids is a disturbed lattice-
like arrangement of molecules with spheric dispersion forces and
in addition some temperature-dependent hole defects (Ditter and
Luck, 1971). In polar molecules with H bonds we have to correct
this model, recognizing the angular dependence of H bonds. The
H bond interactions are usually bigger than dispersion interactions — in the case of OH-O interactions about 3.5-5 kcal/mole
- and therefore strongly influence the liquid structure. The
maximum number of interactions that occurs when the angle β,
between the proton axis and the axis of lone-pair electrons, has

value zero (Luck, 1976b). This energy minimum determines the ice tridymite structure with a coordination number 4 and six-membered rings of six H_2O molecules. X-Ray results (Luck, 1976a) and the spectroscopically determined C_{2v} symmetry (Luck, 1974, 1976c) demonstrate that the disturbed icelike structure with its energy minimum is dominant in liquid water. Overtone IR spectra (Luck, 1963, 1965a, 1966, 1974, 1976c; Luck and Ditter, 1969) give us a fine method for estimating the angular defects in liquid water counting the non-H-bonded OH groups. IR studies of alcohols (Mecke, 1960; Pimentel and McClellan, 19XX; Luck and Ditter, 1968), amines (Wolff, 1976), oximes (Luck, 1961a), and so on make it possible to determine the free OH groups in water (Luck, 1963, 1966; Luck and Ditter, 1969).

Figure 1 shows that at the melting point of ice there is a jump of free OH of about 10%. It then slowly increases under saturation conditions until the critical temperature $T_c = 374^{\circ}C$. The result of Fig. 1 is in agreement with calorimetric properties of water (Luck, 1974, 1976, 1967), but it contradicts some older theories of water properties. Figure 1 shows that we could neglect the content of free monomeric water. We have to stress the difference between free OH groups and monomeric molecules with two free OH. If we try to calculate theoretically the details of the liquid-water IR spectrum with a two-state model of two different bands corresponding to the free and the bound OH groups, we see discrepancies (Luck, 1967a, 1969). We have to assume, therefore, a third band with a maximum between these two bands.

Matrix spectroscopy (Luck, 1973a) suggests that in the diluted systems H_2O/solid Ar and N_2, some preferred H bond angles β induce favored arrangements of cyclic dimers, trimers, and tetramers (Luck, 1965b, 1967b, and 1973a).

Figure 2 gives the frequency shifts $\Delta\nu$ of the OH stretching band as function of β. $\Delta\nu$ is the difference between the band

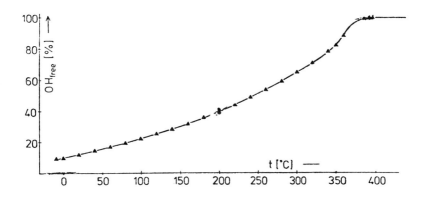

Fig. 1. Content of "free" non-H-bonded OH groups in liquid water under saturation conditions.

maxima of monomeric water and the H-bonded ones, and $\Delta\nu_0$ corresponds to the shift of the ice band. Figure 2 illustrates a phenomenon of fundamental importance to all intermolecular structures in biochemistry (Luck, 1965b, 1967b, 1968). For example, in DNA or in the α helix all H bonds are adjusted at β = O. The region 100° <β< 110° corresponds to an antiparallel position of two OH groups. H bond spectroscopy has proved the validity of the so-called Badger-Bauer rule, which states that $\Delta\nu$ of X H vibrations is directly proportional to the H bond interaction energy ΔH per bond. In the antiparallel position of two OH groups two groups are engaged, which means $\Delta H \approx 2\,\Delta\nu$. Because $\Delta\nu$ in this position of one OH group is about 0.45 $\Delta\nu$ (ice), this ΔH is not far from the icelike linear H bond. In consequence, the linear icelike bonds and antiparallel positions of two OH groups are energetically favored. The $\Delta\nu$ of this antiparallel arrangement is the same medium band of liquid water that computer

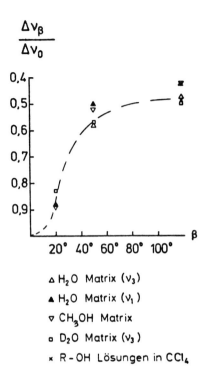

Fig. 2. Dependence of the H bond interaction energy (deter-mined by the frequency shift) as a function of the H bond angle β.

analysis has claimed. Therefore, we can assume the unknown angle distribution of liquid water can be approximated fairly well be a three-step function: (1) free non-H-bonded OH, (2) closed linear H bonds, (3) an intermediate antiparallel arrangement with two energetically unfavored H bonds.

Figure 2 also reveals the strong cooperative mechanicsm in liquid water, which means that the energy to install defects is a function of the defect concentration. If we assume a hole defect in the ice lattice separating one H_2O from a six-membered ring, the remaining five molecules will form five H bonds with

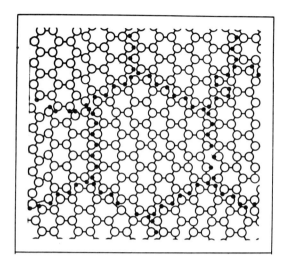

Fig. 3. Simplified model of liquid water: open circles,
oxygen atoms; connecting lines, H-bonded protons, and closed
circles, non-H-bonded protons. The defects of non-H-bonded OH
groups are not statistically distributed because of the cooperative mechanism.

with $\beta \approx 10^{\circ}$. Every one of these five molecules is in an energetically unfavored position with a ΔH loss adjusted by the corresponding $\Delta \nu$ ($\beta \approx 10^{\circ}$) of Fig. 2. A further transfer of heat to the
system would probably induce a second defect in a disturbed five-
membered ring rather than in the undisturbed region.

The consequence of the cooperativity is that in liquid water
we have to assume that the angle defects are not statistically
distributed, but are more or less concentrated in defect areas.
This induces the simplified cluster model of liquid water (Fig. 3).
In a strongly idealized model, Fig. 1 allows us to estimate the
extension of the H-bonded network (cluster) with the result that
at 0°C in liquid water the extension of the H-bonded system may
be several hundred molecules, at 50°C about 100, and at the
boiling point about 40 (Luck, 1976c, Luck and Ditter, 1969).

The relaxation time in liquid water of about 5 X 10^{-12} sec, determined by dielectric measurements, may be the lifetime of H bonds (Hasted). In the idealized cluster model, the free OH trap a "free" lone electron pair, binding a new H bond, which produces about 4 kcal/mole. This energy may open a neighboring H bond. The defect areas therefore flicker very fast.

III. AQUEOUS SOLUTIONS

This flickering cluster model is in agreement with the viscosity of liquid water and its activation energy of 5 kcal/mole, which is the bond energy of the surface molecules of the clusters (Luck, 1976c). In a similar way we could give a simplified model of the ability of water to form gels with small additions. If we assume that solutes with smaller interactions than H_2O/H_2O are assembled in the defect areas of the water cluster, a fairly small amount would be able to stop the flickering mechanism of the clusters. The result is a gel with a high viscosity. We can understand then why some dyes form gels in concentrations as low as 5 x 10^{-4} moles per mole of H_2O.

IV. THE "HYDROPHOBIC" BOND

We know from the existence of the so-called gas hydrates of water with hydrophobic guests like CH_4, SO_2, and Cl_2 that in these cases water forms five-membered plain rings (Luck, 1970b). Twenty water molecules in this arrangement give a pentagonicosaeder (Fig. 4), which forms a much bigger hole than the ice lattice. The diameter of these holes has the size of about one benzene molecule. The cause of this structure determined by X rays in solid gas hydrates may be that no protons and no lone-pair elec-

Fig. 4. Model of the water structure in solid-gas hydrates. The planar water rings form a hole the size of a benzene molecule.

trons lie inside these holes; all are in the "surface" of the pentagonicosaeder. In this structure water turns its most hydrophobic back to the inside of the hole.

We can expect that water orients similarly to neighboring bigger hydrophobic groups of solved organic molecules. Since we need 20 water molecules for one guest molecule, this arrangement can only work in dilute solutions. At higher concentrations hydrophobic groups interact. This is observed in micelles above the so-called critical concentration (Luck, 1960). In such cases hydrophobic molecules try to form a spheric hydrophobic micelle nucleus. This has the consequence that the interaction area water/hydrophobic groups becomes a minimum and the reduction of H bonds/volume unit becomes a minimum. The so-called hydrophobic bond is only a sum of energy minima of H bonds and dispersion forces. The use of the term "hydrophobic bond" had induced a lot of misunderstanding. This "bond" is due to strong

H bonds of water and not to the weaker dispersion interactions of
the hydrophobic group. A better nomenclature might be "hydrophobic
aggregation." Comparing Figs. 4 and 2 we can see during the
gas hydrate structure that the H bond angles β have to change
from zero, requiring energy.

The so-called iceberg structure of water around hydrophobic
groups was introduced by Franck and Evans (1945) as a consequence
of the entropy change while dissolving hydrophobic molecules in
water. But entropy change does not necessarily mean more icelike
structures. Hertz (1964) has reviewed all known spectral data
and concludes that there are no indications of more H bonds in
so-called icebergs. Our view is that the five-membered rings in
gas-hydrate-like structures are planes, which have a different
entropy than the icelike six-membered rings that we expect to
dominate in the liquid-water structure. The number of arrangements of a planar ring is smaller than a nonplanar ring. This
could include an entropy effect without changing the number of
H bonds. We suggest that instead of "iceberg" we use the term
"entropy-preferred arrangements". The weakening of H bonds
by gas hydrate formation could also be demonstrated by our IR
method (Luck, 1979b).

V. HYDRATION OF ORGANIC MOLECULES

The above examples have been idealized. Normally in chemistry
of aqueous systems we have organic molecules with different groups:
hydrophilic and hydrophobic. Very good models for such systems
are polyethers with one hydrophobic endgroup, for example,

$$C_8H_{27} - O - (O-CH_2-CH_2)_n -OH$$

It has mixed groups on the right side: H bond acceptors and small
hydrophobic $-(CH_2)_2$ groups and a big hydrophobic group at the

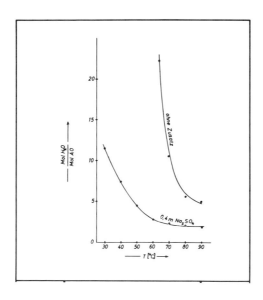

Fig. 5. Water content of the organic phase in the two-phase system of aqueous $H_{17}C_8$ *–   0   – (OCH_2CH_2-)_9OH   (PIOP 9).*

"left" side. These systems are dissolved as monomers in dilute solutions (Luck, 1960, 1970b) and as micelles at higher concentrations. The solubility is due to the ether-oxygen atoms and to entropy-preferred arrangements around the hydrophobic groups. With increasing temperature we reach a turbidity point T_K and above it a two-phase formation.

The H bonds are too weak above T_K and the resulting phase separation is the same as formation (Kryut, 1958; Lauffer, 1975). Our simple model system enables us to study details of coacervate formation. In Fig. 5 we give the water content of the organic phase of C_8H_{17} – ⟨O⟩ – (OCH$_2$-CH$_2$)$_9$-OH above T_K. In water T_K is 64°C and the water content is about 23 H$_2$O/atom of oxygen. With increasing T the water content decreases rapidly to a limit value of two H$_2$O/oxygen atoms. If we add structure-breaking ions like

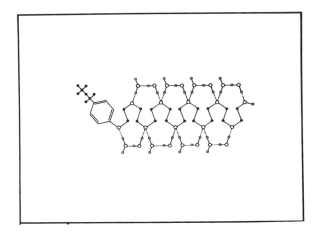

Fig. 6. Model of the dihydrate of PIOP 9 with β = 0.

Na_2SO_4, T_K and the water content in the organic phase decrease
(Fig. 5). In this case the limit of the dihydrate is distinct.
This dihydrate has special properties (Luck, 1964): it has a
viscosity maximum of the velocity of sound and particular X-ray
structure (Boedeker, 1941; Rösch, 1957, 1963), which demonstrates
a neighboring of different $-(CH_2)_2$-groups. With molecular models
of this type or helix structure, we have found that two water
molecules can form an H bond bridge from an ether-oxygen-atom
to the second-nearest ether-oxygen with H bond angles β = 0
(Fig. 6). We think this is a good example of the possibility
that organic molecules can have primary hydrate layers similar
to ions. Higher water content up to about 20 moles/mole (Fig. 5)
is described as secondary hydrates.

VI.   ELECTROLYTE SOLUTIONS

The Debye-Hückel theory provides a simplified model of electrolyte solutions accenting the long-range Coulomb forces of ions

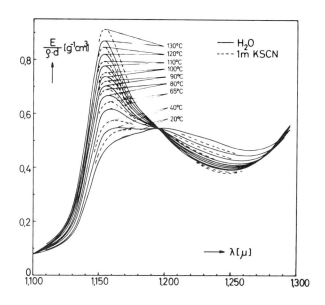

*Fig. 7.   One overtone water band at different temperatures
(full lines) and with addition of 1 M KSCN (dashed lines).   The
salt additions change the water spectra like a temperature increase.*

and neglecting the detailed water structure. The influence of
water is taken into account only by its macroscopic dielectric
constant and by its viscosity. Colloid chemists recognized early
the importance of the Hofmeister or lyotropic ion series (Luck
1965), which can not be understood by the simplified model of
long-range Coulomb forces. IR spectra can be a useful tool to
recognize the cause of the Hofmeister ion series. Figure 7 shows
the change of the water spectra by salt additions (Luck, 1976c,
1974; Kleeberg and Luck, 1978). The salt spectra behave as if
the temperature of pure water were increased about 5°-8°C. That
effect may be caused by a decrease in H bonds. We could say that
water has become more hydrophilic by the solute KSCN additions.
Such solutes are called structure breakers because their effect

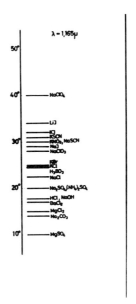

Fig. 8. "Structure temperatures" T_{str} of different salts determined at 20°C and anion concentrations of 1 m at 1.165 μm.

is similar to breaking the H bond structure by T increase and induce salting-in effects. Some salts change the water spectra as if T were decreased. As a result, water becomes less hydrophilic. Such solutions have a salting-out effect for other solutes. To refine this rough model more quantitatively we can describe the H-bonding state of electrolyte solutions by defining the so-called structure temperature T_{str}, the temperature of pure water with a similar IR water spectrum (Luck 1965c), 1973b). For effects on other solutes, the state of the strongly bonded primary hydration shell around the ions is less important than the disturbance of the water structure in the secondary transition phase between the primary hydrates and the normal water structure. We prefer to determine the T_{str} values from the frequency region of the non-H-bonded (free) OH groups (Luck, 1965c, 1973b). Figure 8 gives T_{str} of different solutions with 1 mol anions at 20°C,

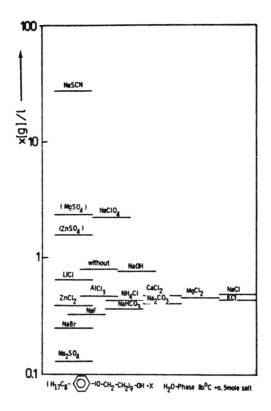

Fig. 9. Content of PIOP 9 in the aqueous phase above the turbidity temperature at 80°C without and with 0.5 mol added salts.

measured in the region of free OH at 1.165 µm. This series of T_{str} is identical with the Hofmeister ion series, and represents series of the disturbed water structure. In all cases in which salt effects can be ordered in the Hofmeister ion series (Luck, 1976a; Kleeberg and Luck, 1978) we can assume that it is due to a change of water structure.

The change of water structure by electrolytes also influences the compositions of the different phases in coacervates. Figure 9 shows the composition of the water phase above T_K of C_8H_{17} - ⬡ - $(O-CH_2-CH_2-)_9OH$ independence of salt additions. The water

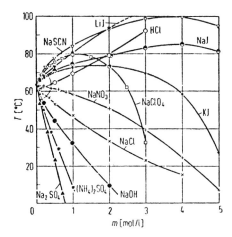

Fig. 10. Turbidity point of PIOP 9 as function of salt concentration.

activity in solutions is changed by salt content and therefore the water content of polymers (Luck, 1961b) in contact with such solutions is also affected. We would expect a change of the water content of food in contact with salt solutions too.

The Hofmeister ion series is valid at medium concentrations. At concentrations above 2 M, overlapping of different hydration spheres can occur. Figure 10 shows that KI salts-in the poly- ether below and salts-out above 3.5 M. In a simplified model one can assume that the hydration of the ions is stronger as the either hydration. Therefore, at high salt concentration the hydration of the ether groups is hindered. T_{str} of KI shows no anomalies, which supports the assumption that the maximum in Fig. 10 is caused by the rivalry of the hydration spheres of the ion and the polyether.

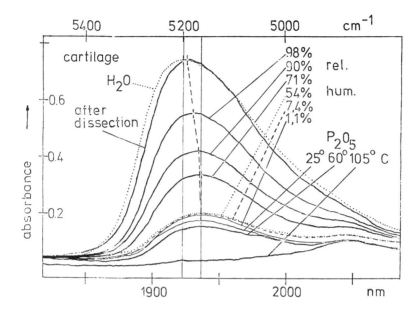

Fig. 11. H₂O combination band in cartilage samples as a function of relative humidity.

VII. HYDRATION OF BIOPOLYMERS*

In Fig. 11 are plotted the spectra of the water combination band in cartilage samples at different relative humidities. In comparison with the band of pure water (dashed line) the frequency of the maximum is shifted to higher wavelengths. In Fig. 12 we plot T_{str} derived from these spectra vs. relative humidity (rh). In the rh range of approximately 10-70% corresponding to the first hydration shell, water has a spectrum similar to that of supercooled water at about -55° as characterized by its T_{str}. This spectroscopic result may imply that the H bonds between water and collagen are in the average stronger than

Coauthor of this section is H. Kleeberg.

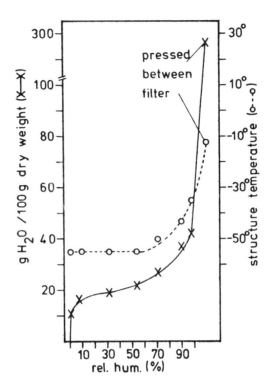

Fig. 12.   Position of the H_2O band maximum (given as structure temperature T_{str}; right scale) in collagen (tendon) as a
function of humidity; desorption isotherm (left scale).

those of liquid water but not as strong as an ice. The frequency shift is small but adequate to demonstrate that this
property of the hydrate water is really different from liquid
water. At higher humidities the maximum of the water band
shifts to the direction of pure water (Fig. 13).

     This experiment suggests that we recognize at least three
different water types in the collagen system: (1) primary hydrate,
(2) a transition state, and (3) liquid water under saturation
conditions. Dry cartilage is composed mainly of collagen and
chondroitin sulfate. In Fig. 14 are plotted the desorption iso-

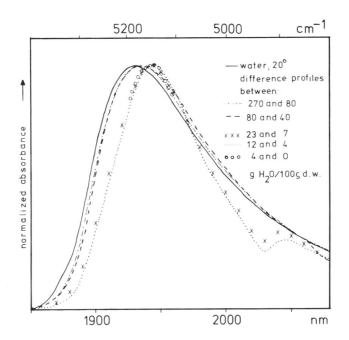

*Fig. 13. Differences between the spectra of gelatin at
different H_2O contents (270 g H_2O/100 g dry weight minus spectra
of 80 g H_2O/100 g dry weight, etc.)*

therms, positions of the maximum of the IR water combination
band, and the derived T_{str} for these materials. For all samples
T_{str} remains nearly constant below 54% rh.

    In this rh range H_2O is desorbed from the first hydration
sphere (first HS), i.e., H_2O that is H bonded to the polymer.
It is obvious that T_{str} of the first HS depends considerably on
the polymer to which it is attached, indicating the formation of
polymer-H_2O H bonds of, on average, a different H bond strength:
for the hydration of chondroitin sulfate, cartilage, and collagen,
T_{str} is 27, -15, and -55°, respectively. The H bond balance,
measured with our near — IR method, demonstrates that the state

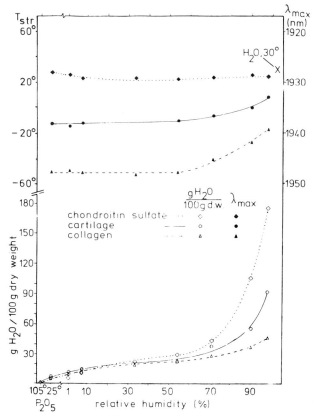

Fig. 14. Hydration of cartilage as a sum of effects of hydration of chondroitin sulfate and collagen. Upper part, positions of H₂O bands (30°C); lower part, desorption isotherms (23°C).

of water in cartilage corresponds to a sum effect of collagen and chondroitin sulfate. Collagen has a stronger H bond strength compared with liquid water.

In Fig. 15 are shown spectra of mixtures of H_2O with model substances. N-Ethylacetamide may be considered a model for protein and ethanol for polysaccharides. We find similar bands of the corresponding organic molecules (above 2000 nm in Fig. 15). The H_2O combination bands for amide and gelatin (Fig. 15 upper

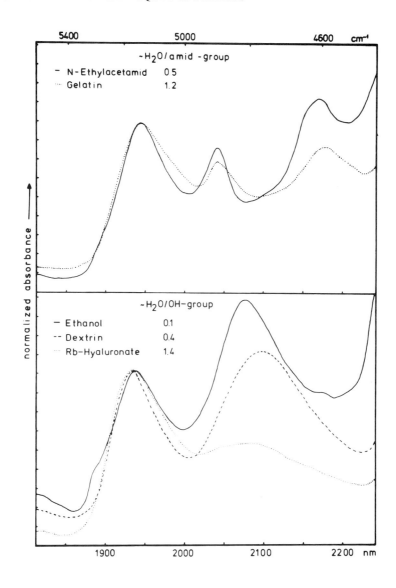

Fig. 15.   Spectra of different hydrophilic substances in water.

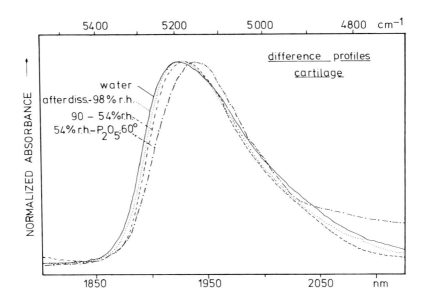

Fig. 16.   Difference spectra of H_2O in nasal cartilage at different humidities.

panel) are very similar, and so we can say that H_2O in these systems is bound to the -NHCO- groups or H bond acceptors of similar strength.

Comparing the band of H_2O in ethanol, dextrin, and rubidium-hyaluronate (the sodium and potassium salts give closely identical results) there are slight differences (Fig. 15 lower panel). Although large amounts of H_2O seem to be H bonded to the OH groups of these polysaccharides, a certain amount of H_2O is H bonded to weaker acceptors like -O- groups, for example.

At rh above 54% the maximum of the H_2O combination band shifts into the direction of the band of pure water at the experimental temperature (30°, Fig. 14).   The normalized differences between the spectra at different humidities are shown in Figs. 13 and 16 for gelatin and cartilage.   Both series of difference profiles are similar; this was not the case for the polysaccharides. Since the available sites of collagen capable of forming H bonds with H_2O are already mainly occupied by the first HS as indicated by calorimetry (Berlin et al., 1970; Haly and Snaith, 1971; Susi, Ard, and Carroll, 1971) we have to look for the reason why this H_2O does not give a difference profile like pure water. Calorimetric (Berlin et al., 1970; Haly and Snaith, 1971; Privalov and Mrevlishvili, 1967), NMR (Dehl, 1970; Migchelsen and Berendsen, 1973), and -IR (Kleeberg and Luck, 1977) experiments indicate that, in addition to the first HS there is a fraction of H_2O with properties slightly different from liquid water. We assume that this H_2O may form H bonds with the H_2O molecules of the first HS. These H bonds may have unfavorable angles since the first HS is not easily available or sterically hindered by the surrounding apolar residues.

If we remember that collagen comprises approximately 40% of the dry substance of bovine nasal cartilage (Mathews and Decker, 1977), the difference profiles in cartilage between 54 and 98% rh seem to be caused by H_2O that is H bonded to the first of the protein too (Fig. 16).

In cartilage the difference profiles between 54 and 98% rh are closely identical to one another, but clearly different from the band of pure water and the first HS. H_2O that gives rise to these different bands may be called the secondary hydration sphere (second HS).

Figure 17 shows that the IR H_2O combination bands of cartilage at 54 and 98% rh can be regarded as being composed of the respective bands of chondroitin sulfate (CS) and collagen. This indicates that the hydration of the pure polymers is not much changed if they are present in a complex system (Kleeberg and Luck, 1977). In gelatin the second HS amounts to not more than 250 g H_2O/100 g dry weight. T_{str} of this second HS is approximately 10°. The properties of this H_2O with respect to salt

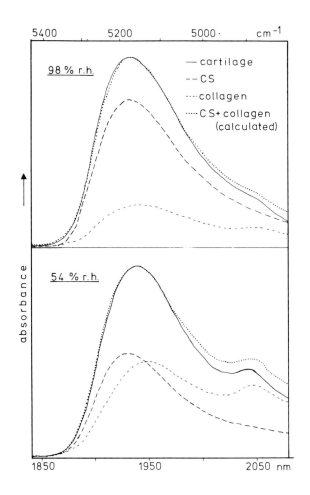

Fig. 17. *Cartilage spectra at 98 and 54% relative humidity are nearly the sum of the spectra of its main components (Mathews and Decker, 1977): chondroitin sulfate (CS) and collagen.*

solubility, for example, are changed. The difference profiles of H_2O in gelatin above 270 g H_2O/100 g dry weight are indistinguishable from the band of pure water. We conclude that the hydration of bovine nasal cartilage (360 g H_2O/100 g dry weight) can be

explained by the hydration of collagen and chondroitin sulfate—
its major components (Mathews and Decker, 1977). The first hydration sphere (20 g/100 g dry weight) is directly H bonded to
the polymers with T_{str} = -150°; the second hydration sphere
(approximately 75 g/100 g dry weight) seems to be caused by
collagen, T_{str} approximately 10°; the remaining 260 g H_2O/100 g
dry bovine nasal cartilage are bulk water.

VIII.   MECHANISM OF DESALINATION MEMBRANES

On the basis of Fig. 5 one could predict a simple desalination methods; separating the two phases of solution of $H_{17}C_8$ -
⬡⟨O⟩ - $(OCH_2CH_2)_9$ OH in 0.4 M Na_2SO_4 solution, after heating
the organic phase at 70°C, separation of 9 mol desalted water
per oxygen atom is expected. The amount of separation of 9 mol
H_2O could be established experimentally, but this water has a
salt concentration similar to the original concentration. We
learn by this experiment that the secondary hydration water has
a structure similar to liquid water, and therefore has the ability
to dissolve ions. One could try to prevent the formation of the
big secondary hydration sphere. This has been realized by mixed
polymers polyethylene/polypropylene (Luck, 1973c), which are not
water soluble and therefore have reduced hydration shell. With
these mixed polymers we had success in obtaining water with a
reduced salt content.

The working hypothesis is that a desalination membrane should
have a built-in steric hindrance to the formation of liquidlike
water layers.

Figure 18 seems to confirm this hypothesis. In cellulose
acetate desalination memberanes we found water spectra with a
frequency shift to shorter wavelengths in comparison with pure
water. This is valid up to high humidities. We can interpret

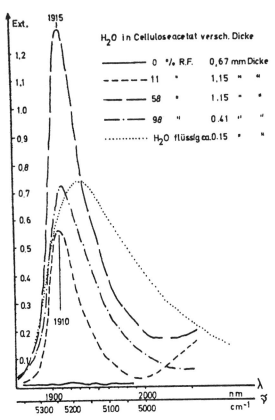

Fig. 18. *Water spectra in cellulose-acetate membranes in dependence on relative humidity. Maxima are at shorter wavelength than in liquid water (pointed).*

this by assuming that water in these membranes is more weakly H bonded to the membrane groups than in liquid water. Therefore, the solubility of ions in the membranes is reduced and ions are rejected by these membranes, in accordance with their position in the Hofmeister series. Figure 19 shows the ion reflection coefficient of cellulose acetate measured by Pusch (1974). Structure-making ions like $NaSO_4$ show more rejection than ions with positions nearer to the structure-breaking ions in the Hofmeister series. Ions that stabilize the water structure

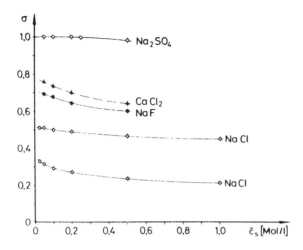

Fig. 19. Reflection coefficients of cellulose-acetate membranes as function of concentration C_s of different salts (Kluberg and Luck, 1977).

seem to need a larger water hydration sphere in polymers to become dissolved.   In addition, we have to take into account that
structure-making ions reduce the water content in fibers like
polyamide (Luck, 1961b) and structure-breaking ions like $NaClO_4$
increase the water content. The interaction of water with the
surface of other phases is similarly changed by ions in the
water phase. The force-area diagrams of monolayers of non-water-
soluble polyethers change with salt content of the aqueous sub-
phase corresponding the Hofmeister ion series (Luck and Shan,
1978).

IX.   MECHANISM OF BOUND WATER

     We conclude that the so-called free water in food is similar
to liquidlike water; bound water seems to be water bound to

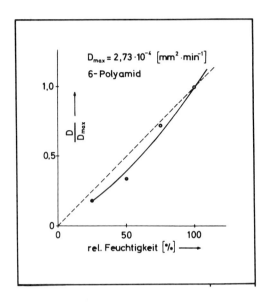

Fig. 20. Mean diffusion coefficient of water in 6-polyamide at 80°C as function of relative humidity (%), determined by the slope of the water amount-time curve (Pusch, 1974).

stronger H bond acceptors than in liquid water of bound with favored H bond angles. As shown in the membrane example, bound water has a reduced solubility for other compounds. In addition, the diffusion velocity is reduced in bound water (Luck, 1979). As an example, Fig. 20 shows the reduction of the apparent water diffusion coefficient in 6-polyamide (Nylon 6) with decreasing water content. This decreased mobility impedes drying processes. If the water content near the surface is reduced, the water diffusion in the surface layer is reduced. This effect may be important for protecting drying of insects in deserts and natural-ly for the water content of food, too. Parallel to the reduced water diffusion, a reduction of the diffusion of water-soluble solutes in polymers also observed (Luck, 1965d, 1979). We could show that the water activity is changed by dissolved ions in the diffusivity of other solutes (Luck, 1965e).

REFERENCES

Berendsen, H. J. C. (1967). *J. Chem. Phys. 36*, 3297-3305.
Berlin, E., Kliman, P. G., and Pallansch, M. J. (1970). *J. Coll. Interface Sci. 34*, 488-494.
Boedeker, K. (1941). *Kolloid Z. 94*, 161.
Dehl, R. E. (1970). *Science 170*, 738-739.
Ditter, W., and Luck, W. A. P. (1971). *Tetrahedron 27*, 201-220.
Frank, H. S., and Evans, M. W. (1945). *J. Chem. Phys. 13*, 507.
Haly, A. R., and Snaith, J. W. (1971). *Biopolymers 10*, 1681-1699.
Hasted, J. B. (19XX). *In* "Structure of Water and Aqueous Solutions" (W. A. P. Luck, ed.) p. 377. Verlag Chemie, Weinhem.
Hertz, H. G. (1964). *Ber Bunsenges. Phys. Chem. 68*, 907.
Hoeve, C. A. J., and Tata, A. S. (1978). *J. Phys. Chem. 82*, 1660-1663.
Kleeberg, H. (1980). Ph.D. Thesis, Univ. of Marburg, in preparation.
Kleeberg, H., and Luck, W. A. P. (1977). *Symp. Biol. Connective Tissue, Uppsala, Sweden,* 4-9 September.
Kleeberg, H., and Luck, W. A. P. (1978). *6th Mtg. Eur. Fed. Connective Tissue Clubs, (Creteil, France).* August 1978, *Coll. Intern. CNRS 287.*
Kruyt, H. R. (1958). "Culloid Science," Vol. I., Elsevier Amsterdam.
Lauffer, M. A. (1975). "Entropy Driven Processes," Springer Heidelberg.
Luck, W. A. P. (1960a). *Ang. Chem. 72*, 57.
Luck, W. A. P. (1960b). *3rd Int. Kongr. Grenzflächenaktive Stoffe, Köln,* Vol. I, p. 264.
Luck, W. A. P. (1961a). *Z. Elektrochem. 65*, 355-362.
Luck, W. A. P. (1961b). *Melliand 42*, 221.
Luck, W. A. P. (1963). *Ber. Bunsenges. Phys. Chem. 67*, 186-189.
Luck, W. A. P. (1964). *Fortschr. Chem. Forsch. 4*, 653.
Luck, W. A. P. (1965a). *Ber. Bunsenges. Phys. Chem. 69*, 626.
Luck, W. A. P. (1965b). *Naturwissenschaften 52*, 25, 49.
Luck, W. A. P. (1965c). *Ber. Bunsenges. Phys. Chem. 69*, 69.
Luck, W. A. P. (1965d). "100 Jahre BASF," p. 259.
Luck, W. A. P. (1965e). *Ber. Bunsenges. Phys. Chem. 69*, 255-264.
Luck, W. A. P. (1966). *Ber. Bunsenges. Phys. Chem. 70*, 1113.
Luck, W. A. P. (1967a). *Discuss. Faraday Soc. 43*, 115.
Luck, W. A. P. (1967b). *Naturwissenschaften 54*, 601.
Luck, W. A. P. (1968). *Naturwiss. Rundsch. 21*, 236.
Luck, W. A. P. (1970a). *Phys. Blatter 26*, 133.
Luck, W. A. P. (1970b). *J. Phys. Chem. 74*, 3687.

Luck, W. A. P. (1973a). *In* "Water: A Comprehensive Treatise"
    (F. Francks, ed.), Vol. II, pp. 225-320. Plenum, New York.
Luck, W. A. P. (1973b). *In Proc. 4th Int. Symp. Fresh Water
    from the Sea, Heidelberg, Sept. 1973, 4,* 531-538.
Luck, W. A. P. (1973c). Deutsches Patentment Offenlegungsschrift
    2141207.
Luck, W. A. P. (1974). "Structure of Water and Aqueous Solutions."
    Verlag Chemie, Weinheim.
Luck, W. A. P. (1976a). *Topics Curr. Chem. 64,* 113.
Luck, W. A. P. (1976b). *In* "The Hydrogen Bond" (P. Schuster,
    G. Zundel, and C. Sandorfy, eds.). Vol. II, pp. 527-562.
    North-Holland Publ. Co., Amsterdam.
Luck, W. A. P. (1976c). *In* "The Hydrogen Bond" (P. Schuster,
    G. Zundel, and C. Sandorfy, eds.), Vol. III, pp. 1369-2423.
    North-Holland Publ. Co., Amsterdam.
Luck, W. A. P. (1976d). *Naturwissenschaften 63,* 39.
Luck, W. A. P. (1979). *In* "Handbuch der Mikroskopie" Vol. VI,
    pp. 345-388.
Luck, W. A. P., and Ditter, W. (1968). *Ber. Bunsenges. Phys.
    Chem. 72,* 365.
Luck, W. A. P., and Ditter, W. (1969). *Z. Naturforsch. 24b,* 482.
Luck, W. A. P., and Shah, S. S. (1978). *Colloid and Polymer
    Sci.*
Mathews, M. B., and Decker, L. (1977). *Biochim. Biophys. Acta,
    497,* 151-159.
Mecke, R. (1960). *Wiss. Veröffentlichungen,* 1937-1960.
Migschelsen, C., and Berendsen, H. J. C. (1973). *J. Chem. Phys.
    59,* 296-305.
Mrevlishvili, G. M. (1976). *In* "L'Eau et les Systemes Biologiques"
    (C.N.R.S., ed.), No. 246, pp. 139-146. Paris, France.
Privalov, P. L., and Mrevlishvili, G. M. (1967). *Biophysics 12,*
    19-28.
Pusch, W. (1974). *In* "Structure of Water and Aqueous Solutions"
    (W. A. P. Luck, ed.), p. 551, Verlag Chemie, Weinheim.
Rösch, M. (1957). *Fette, Seifen, Anstriche 59,* 1, 745.
Röscj. ,? (1963). *Fette, Seifen, Anstriche 65,* 223.
Schröder, W. (1969). *Z. Naturforsch. 24b,* 500.
Susi, H., Ard, J. S., and Carroll, R. J. (1971). *Biopolymers 10,*
    1597-1604.
Troschin, A. S. (1958). "Das Problem der Zellpermeabilität."
    Fischer, Jena.
Wolff, H. (1976). *In* "The Hydrogen Bond" (P. Shuster, G. Zundel,
    and C. Sandorfy, eds.), Vol. III. p. 1259. North-Holland
    Publ. Co., Amsterdam.

Pimentel, G. C., and McClellan, A. L. (19XX). "The Hydrogen
    Bond." Freeman, San Francisco.

THEORETICAL STUDIES OF WATER

IN CARBOHYDRATES AND PROTEINS

Donald T. Warner

I. INTRODUCTION

The exact role that water serves in the structure and function of biopolymers from living things is still one of the deep
mysteries in our present scientific knowledge. Water is often
thought of as liquid medium in which macromolecules float around
without making any sensible or purposeful interaction. Yet as
we shall see from theoretical studies using molecular models,
many of these biopolymers have structural features that could
permit meaningful interactions with water. The studies to be
presented concern two kinds of polymers found in all plant and
animal tissues — proteins and carbohydrates. Since most of our
foods are of plant and animal origin, it is hoped that these
theoretical studies will be helpful to those interested in the
role of water in foods and food processing.

These studies began when I was assigned to research project
with the goal of synthesizing a series of polypeptides to be
screened for biological activity. As a preliminary to this
project, we decided to examine the molecular structure of some
relatively small known peptides using molecular models. We
hoped that such models would furnish clues about common geometrical
features that might be related to the activity of these peptides.

II.   SOME SPACE-FILLING PEPTIDE MODELS

The first molecular models we assembled were related to the
disulfide-stabilized ring sequences found in oxytocin and vaso-
pressin (Warner, 1961). Space-filling models were used for the
early work. The rings in these polypeptide hormones are composed
of six peptide residues and have a disulfide bond joining cys-
teine-1 and cysteine-6. This -S-S- bridge stabilizes the ring
and determines its size. For such small rings, the usually
accepted structural modes (α-helix and pleated sheet) cannot be
applied to the conformation of these molecules without notable
difficulty and significant distortion of the helix or sheet.
From our model studies another possible conformation for this
ring was achieved by placing all of the amide linkages of the
peptide backbone in a plane, with the oxygens of each =C=O
group pointing outward from a central point of the ring. In this
way the six peptide oxygens of the six amino acid residues form
the corners of a regular hexagon (Fig. 1). The surface of the
model illustrated here contains all of the principal atoms of
the polypeptide backbone. Arbitrarily this surface has been
designated the "hydrophilic" surface of the structure for this
conformation. The side chains attached to the α-carbon atoms
are sacrcely visible here. If the model is turned over (Fig. 2),
the "side chains" were used in this instance). This face of
the model is arbitrarily designated the "hydrophobic" surface,
although in some instances side chains will have polar groups
attached. For the sake of this discussion, however, we use the
terms "hydrophilic" and "hydrophobic" to designate the *peptide
backbone* surface and *side chain* surface, respectively.

The regular hexagonal pattern of peptide oxygens in this small
structure is, of course, an esthetically pleasing array. As we
shall see, this repeating pattern of hexagons can be applied to
more complicated molecules equally well. For example, there

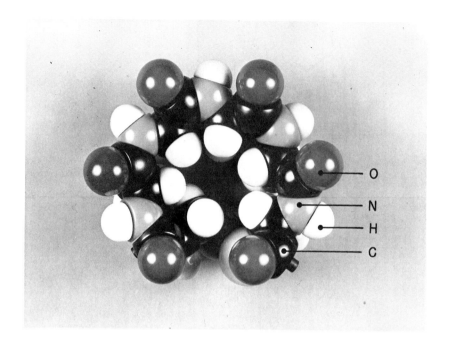

O
N
H
C

Fig. 1. Disulfide-stabilized cyclic hexapeptide (hydrophilic surface).

are a number of polypeptide antibiotics that are cyclic decapeptides and contain ten amino acid residues in a single continuous peptide backbone. These structures have no free amino
terminal or carboxyl terminal ends. With such structures the
peptide chain can be arranged so that the ten carbonyl oxygens
of the backbone form two contiguous hexagons, much in the same
manner as the ten carbon atoms in naphthalene form two fused
benzene rings. A space-filling molecular model of gramicidin
S, an example of this class of cyclic decapeptides, is shown
in Figs. 3 and 4. In a previous publication (Warner, 1961) we
discussed some additional features and possible advantages of

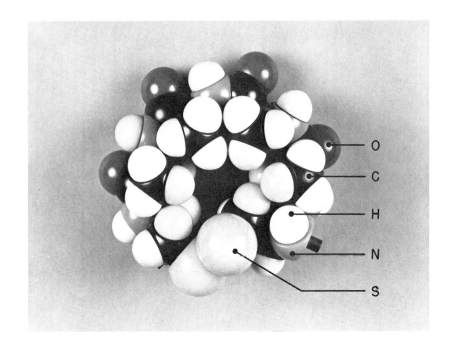

Fig. 2. Disulfide-stabilized cyclic hexapeptide (hydrophobic surface).

this conformation for gramicidin S, including the side-by-side
placement of the two D-phenylalanine aromatic rings as seen
in Fig. 4. However, for the purpose of the present discussion,
emphasis will be on the peptide carbonyl oxygens. In the
suggested molecular conformation, not only do they form the two
units of a hexagonal network but all of the oxygens in the
network are coplonar.

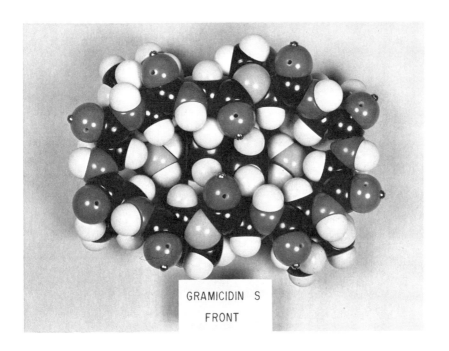

Fig. 3. Gramicidin S (hydrophilic surface).

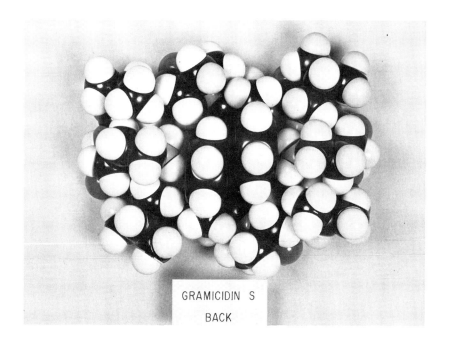

Fig. 4.   Gramicidin S (hydrophobic surface).

III.   WATER-PEPTIDE STUDIES WITH DREIDING STEREOMODELS

The regularity of such coplanar hexagonal networks took on
an added possible biological significance as a result of some
simultaneous studies we were making with ice lattice models. The
ice lattice studies were being made with Dreiding steromodels.
Stereo models emphasize the bonding directions and central
positions of the various atoms very precisely but do not emphasize the van der Waal's radius of each atom. Because of the

openness of the stereomodel structures it is much easier to see
the layers of atoms and measure their center-to-center distances.
It was noted that the distance between the peptide oxygens in
the proposed honeycomb network was about 4.8 Å and that this
distance was also about the same as the "second-neighbor" oxygen distance in the ice lattice. From an examination of the
Dreiding models of the ice lattice, it was immediately evident
that each of the coplanar layers of oxygen atoms in the ice
lattice is formed by a network of second-neighbor water molecules whose oxygens also comprise a regular hexagonal or honeycomb pattern. When the dimensions of the coplanar second-neighbor network in ice were compared with the coplanar hexagonal
network of peptide oxygens in the proposed polypeptide conformations, the two dimensions were nearly identical. Therefore,
such a second-neighbor water layer in the icelike lattice could
lie above a polypeptide layer in the proposed hexagonal conformation, and exact collinear hydrogen bonds could be formed
at each position to satisfy all of the bonding requirements of
the two partners. Such an interaction at each bonding site
could elegantly serve to stabilize the bound water in an icelike
lattice and simultaneously stabilize the polypeptide backbone
in the "hexagonal" conformation. The close fitting of the two
types of oxygen lattices is nicely demonstrated if a Dreiding
stereomodel of a cyclic hexapeptide (Fig. 5) is superimposed
on a Dreiding stereomodel of the ice lattice as in Fig. 6. It
is interesting to note that Wolfenden (1978) has made a strong
case for the association of water with peptide bonds, especially
with the =C=O group.

IV. "OPEN-CHAIN" PEPTIDES AND ADDITIONAL OBSERVATIONS ON WATER

It was of interest to determine if these possible water-
peptide interactions could exist in open-chain polypeptides having
no covalent cyclizing bonds, and in larger subunits of proteins.

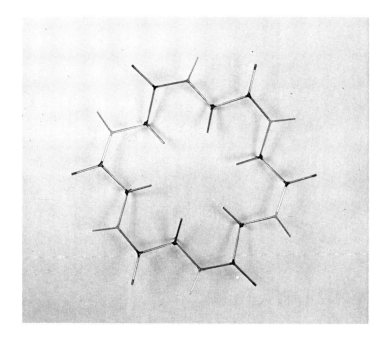

Fig. 5. Cyclic hexapeptide — Dreiding stereomodel.

To illustrate the feasibility of these interactions in open-
chain peptides, a model of glucagon (29 amino acids) is shown
from the hydrophilic side in Fig. 7. Beginning with the coil-
ing of the chain from the amino-terminal end, eight complete
hexagons and a ninth nearly completed hexagon are formed by
this 29 unit polypeptide. This model also contains only
trans amide bonds to satisfy a criticism by Dickerson (1965)
of some of our earlier models. Again for this model, a layer
of bond water in an icelike lattice can be superimposed on the
hexagonal peptide network, and each peptide oxygen can make an
exact collinear hydrogen bond with a matching water oxygen of
the overlying "second neighbor" icelike layer.

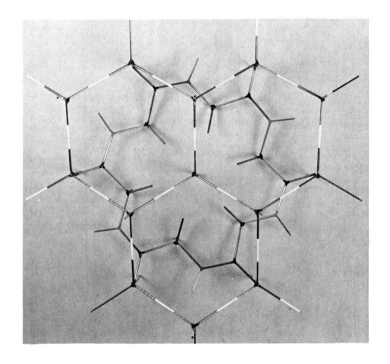

Fig. 6. Cyclic hexapeptide imposed on ice lattice.

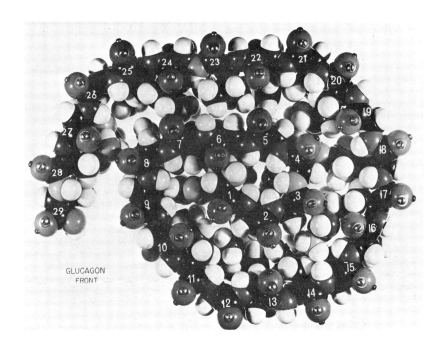

Fig. 7.   Glucagon (hydrophilic surface).

Two other points about these interacting hexagonal networks
should be made.   First, if one considers Fig. 6, where the cyclic
hexapeptide is superimposed on the ice model, it can be seen
that there is a centrally located second-neighbor position in
the ice that is not contacting a peptide oxygen.   However, if
such "open positions" are viewed in terms of the glucagon model
(Fig. 7), it is evident that these central points in the peptide
network are frequently occuped by polar or "hydrophilic" side
chains.   Because of their central location in the regular
openings of the hexagonal peptide framework, these side chains
could serve as contact points with the central second-neighbor
oxygens of the water network.   Perhaps in some instances an
"interstitial water" molecule, itself bonded to the polar side

chain, might be needed as a useful "space filler" between the
main water layer and the peptide layer. However, an equally
valid "bonding possibility" for such a centrally located interstitial water molecule would be the ever-present -N-H network of
the peptide conformation. Each hexagon of the peptide honeycomb
has at least two -N-H groups pointing into the central part of
the individual hexagons. These two -N-H groups could hydrogen-
bond (and center) a water molecule (for convenience again designated as interstitial water) between them. This centered
interstitial water would be at about the same level as the other
coplanar *peptide* oxygens, and could make contact with the
previously unbonded oxygen of the overlying second-neighbor
water lattice. If this alternative -N-H bonding mode for some
of the interstitial water is accepted as a theoretical possibility,
it means that the suggested hexagonal peptide conformation can
satisfy *not only* the hydrogen bonding requirements of every =C=O
but also of every -N-H group of the entire polypeptide backbone.
Furthermore it can accomplish this through a completely ordered
interaction with an icelike water layer. In addition, the -N-H
centered interstitial water would still lie above the available
polar side chains of the peptide honeycomb centers, and could
thus facilitate their participation in the very favorable proton
transfer pathway that the icelike bound water layer should provide.

A second point about the interacting hexagonal networks of
water and peptide is apparent if one examines the ice lattice
in more detail. Each coplanar second-neighbor water layer in
the ice lattice is completely bonded by a series of first-
neighbor bound water contacts to adjacent second neighbor water
layers above or below it. If one now isolates two such immediately adjacent layers for further study (Fig. 8), the following
points are clear. The two adjacent layers have three out of
four of the hydrogen-bonding possibilities of their individual

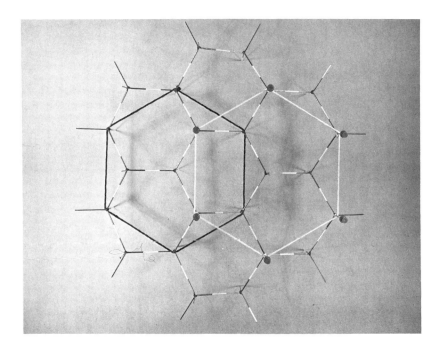

Fig. 8.   Ice lattice with outlined "second-neighbor" layers.

water oxygens occupied in interlayer bonding with each other.
However the remaining one out of four bonding possibilities
of each oxygen is available for bonding with an additional water
layer or, what is more important, for bonding with the oxygen
of a peptide. For example, in Fig. 8 the white string at the
right side of the figure joins six second-neighbor water oxygens
in the upper layer. These circumscribed oxygens have their
remaining bonding legs pointing perpendicularly above the plane
of the white string. These bonding points are indicated on the

figure by the six plastic pins pointing upward above the plane.
These six pin positions could be occupied by the six peptide
oxygens of a cyclic hexapeptide, such as illustrated in Fig. 5,
or else by one of the hexagons in the glucagon model (Fig. 7).
Also in Fig. 8, the black string on the left of the figure
circumscribes six second-neighbor oxygens in the adjacent lower
layer of the lattice. These second neihgbors have their remaining bonding positions pointing downward *below* the plane of
the black string. Here too, the six circumscribed oxygens of
this layer could make exact collinear hydrogen bonds with the
hydrophilic surface of a cyclic hexapeptide or any peptide
hexagon lying below that plane. It is thus immediately apparent
that the black string and white string layers (each already in
intimate and complete contact with each other) can use their
remaining bonding contacts to serve as an elegant cement between the hydrophilic faces of two peptide layers in the hexagonal conformation. Here again the icelike lattice could also
serve as a excellent proton transfer medium not only *within*
the same peptide layer but also *between* two such "cemented"
peptide or protein layers.

V.   PROTEIN MODELS WITH HEXAGONAL PAPERS

    In seeking to explore the hexagonal conformation of larger
protein subunits, the use of actual molecular models was too
cumbersome and expensive. One molecular model of cytochrome
C (104 amino acid residues) was made just to determine that
the honeycomb pattern of oxygens persisted throughout. For
most of the studies of protein subunits, however, hexagonally
ruled papers have been used. For convenience each of the
hexagonal units of the honeycomb paper pattern was drawn to the
dimensional scale of 1 cm on a side, so that for the sake of
measurement, 1 cm = 4.8 Å. By placing numbers at the hexagonal

corners corresponding to the sequence number of the amino acid
residue in the chain, the entire peptide or protein chain could
be laid out on the honeycomb network. The position of the various
side chains was indicated by a code, usually consisting of the
three letter symbol for the amino acid enclosed in a circle or
box. For example, Fig. 9 represents the structures of two calcitonins. The numbers at the corners indicate the sequence of
the chains and the symbols Met and S indicate the positions of
methionines and cysteines in the sequences. These paper models
represent a sequence about as large as that of glucagon (space-
filling model, Fig. 7). Using the hexagonal dimension of 1 cm =
4.8 $\overset{o}{A}$ the overall size of such models can be determined. Figure
9 illustrates some of the things that can be learned from these
models. For example, both porcine and human calcitonin have a
disulfide linkage between cystein-1 and cystein-7. However, the
porcine molecule has a Met-25 while the human calcitonin has a
Met-8. On the basis of primary sequence alone, these two calcitonins would appear to have widely different methionine placements.
However, if the two sequences are arranged in the hexagonal con-

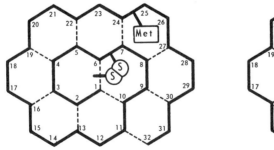
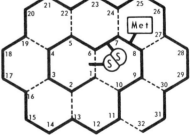

PORCINE   **HUMAN**

Fig. 9. Calcitonins.

formation with the paper models as in Fig. 9, it is clear that
the side chain -S-CH$_3$ groups of Met-8 and Met-25 can have
equivalent placements in the two models. In addition, the -S-S-
bridge between cysteine residues 1 and 7 can be readily formed
in both structures. Neither the α-helix nor the pleated sheet
conformations for these two sequences would permit formation of
the -S-S- bridges or permit equivalent positioning of the Met-8
and Met-25 residues. A wide variety of other protein subunits
have been studied with these paper models and in many instances
good correlations exist with known information about the molecules
(Warner, 1970b, 1974).

VI. WATER AND PROTEIN AGGREGATION

    Hexagonal paper models also have been used to study the
aggregation structures of proteins. The most detailed study was
made of tobacco mosaic virus (TMV) protein (Warner, 1964). Some
of the general forces that may be involved in the aggregation
of proteins and polypeptides were also elaborated in a review
article (Warner, 1967). Although the possible role of water in
the aggregation phenomenon will be emphasized, the role of other
forces will be mentioned to clarify the total procedure.
    It has already been stated that in the hexagonal conformation
of protein structure, the protein subunit is represented as a
flat disk with hydrophilic (peptide bond) and hydrophobic (side
chain) surfaces. The term "hydrophobic," as indicated before,
is an oversimplication in part since some of the amino acid side
chains have polar (or hydrophilic) parts in addition to the
hydrophobic segments. However, the aggregation of these disk
like subunits may be approached from the standpoint of attempting
to define mutually attracting surfaces.

In the case of layers of lipids, which also have hydrophilic (or polar) and hydrophobic surfaces, it is generally accepted that in an aqueous environment, the hydrophobic surfaces of two lipid layers will associate to yield a "lipid bilayer." In this bilayer all of the hydrophobic hydrocarbon chains are in the interior of the structure and the bilayer is left with all of its polar groups on the two outer surfaces. If the same thing happened with protein subunits in an aqueous environment then the hydrophobic surfaces of two subunits would associate, yielding a protein bilayer in which all of the amino acid side chains are meshed together in the *interior* and the hydrophilic groups exist at the two *exterior* surfaces of the protein bilayer. These bilayers can then be coated with monolayers of bound water arrayed in an ice-like lattice on the honeycomb network of peptide oxygens that comprise the two outer hydrophilic surfaces. Such "water-coated" bilayers of protein, under the proper conditions of temperature and pH, could conceivably associate with other water-coated protein bilayers, with the respective monolayers of water serving as an interfacing cement. Since the protein bilayers are bifunctional with respect to water, many bilayers could be cemented together into rodlike structures in a chain reaction. Such rodlike structures are well known in the aggregation of TMV protein.

The possible role of temperature in the aggregation of proteins may be theoretically approached as follows. Although the oxygen-oxygen distances in the *protein surface* (honeycomb network) may be relatively invariant with temperature since they are determined by covalent bonds, the second-neighbor oxygen distances in the *bound water* layer will vary somewhat with temperature. At a certain temperature the dimensions of the water and peptide oxygen networks will be identical, and the mutual interaction should produce an absolutely flat peptide surface. If the temperature of the medium is *increased* above that point, the

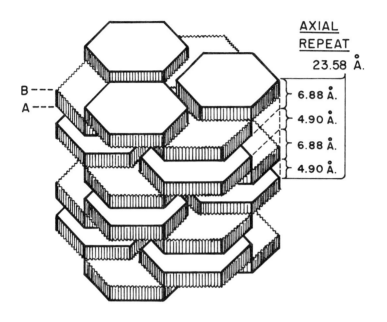

AXIAL
REPEAT
23.58 Å.

6.88 Å.

4.90 Å.

6.88 Å.

4.90 Å.

B

A

A —— HYDROPHILIC SURFACE
B ⌇⌇⌇ HYDROPHOBIC SURFACE

Fig. 10. Schematic TMV Rod.

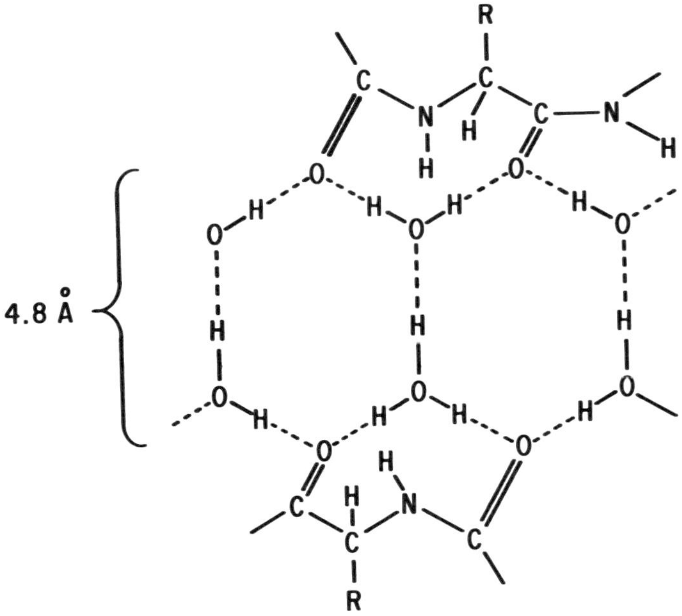

Fig. 11. Water bonded to keto form of peptide oxygens.

hexagonal second-neighbor dimensions of the water layer will increase. Theoretically, the dimensionally invariant protein layer
can attempt to adjust if the entire layer assumes a *convex* form.
In this convex form it can still maintain collinear hydrogen
bonding with its own overlying water hexagons, at least within
certain limits. Similarly, if the temperature of the medium
decreases, the second-neighbor water dimensions will decrease.
The invariant protein layer can then attempt to adjust to the
smaller hexagons of its water layer by assuming a *concave* form.
However, protein surfaces in either the concave or convex form
will have great difficulty in coming together with similar nonplanar surfaces to achieve strong binding forces between them.
For example, two concave hydrophilic protein surfaces, like two
saucers, will be able to touch only at the edges. With such a
limited contact between them, the monolayers of water on their
respective surfaces will adhere poorly. Furthermore, any possible
points of electrostatic attraction between these layers will be
weak since electrostatic forces are relatively short range. Only
when the protein layers are nearly perfectly flat will the
hydrophobic, hydrophilic, and electrostatic forces be able to
exert their maximum combined interlayer attraction. A schematic
representation of how these combined forces can operate in the
formation of the tobacco mosaic virus rod is shown in Fig. 10.
Further details of the rodlike formation of TMV are given in a
previous publication (Warner, 1964).

Properties of the interstitial bound water between the layers
of Fig. 10 provide interesting theoretical possibilities for
explaining biological processes such as muscle contraction, membrane permeability, and ion pumps. The principal mode of action
for these theoretical processes involves rearrangements within
the interstitial water, which can bring about different spacings
between the water-bonded protein layers. The mechanism of the
process is illustrated in Figs. 11 and 12. These rearragements

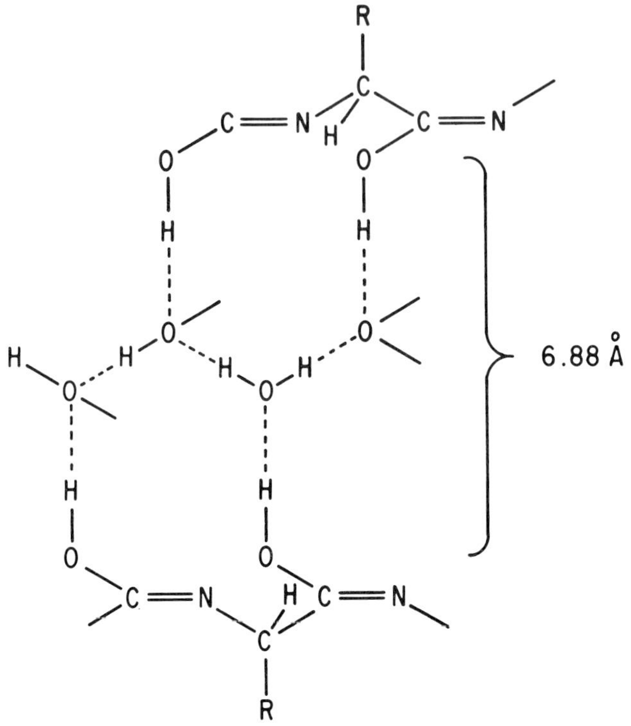

Fig. 12.   Water bonded to enol form of Peptide Oxygens.

could be activated by proton shifts into or out of the water
layer.   The use of this mechanism has been described in detail
for muscle contraction (Warner, 1970a) and advantages of this
model have been listed there.   Unlike the Huxley model of muscle
contraction (Huxley, 1965), which requires discontinuities
and the momentary complete separation of the active parts, this
"water model" of contraction never involves a separation of the
working components.

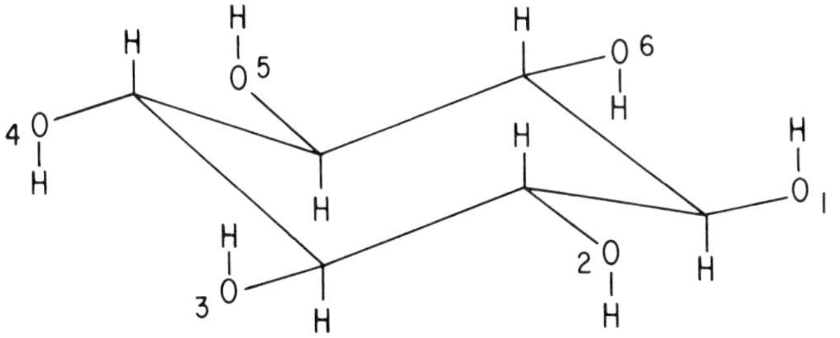

Fig. 13. Scyllo-Inositol.

VII. MODEL STUDIES OF WATER-CARBOHYDRATE INTERACTIONS

A. Inositol

In order to evaluate some possible interactions of polysaccharides with water in an icelike lattice, one of the inositol sugars, *scyllo*-inositol will first be considered. The
cyclohexane ring in this sugar has six equatorial hydroxyl
groups, one on each carbon of the ring, and six axial hydrogens
on these same carbons. A model of *scyllo*-inositol based on
Dreiding stereomodels is given in Fig. 13, using the chair
conformation of the cyclohexane ring. Such a model has a very
rigid configuration and in the chair form of the ring only the
hydroxyl hydrogens are capable of rotation. Oxygen atoms 1, 3,
and 5 are in a common plane and form an equilateral triangle
with sides of 4.8 Å. Similarly, oxygens 2, 4, and 6 are in
another plane and form another triangle of similar dimensions.
The 4.8 Å distance between these sugar oxygens is again equivalent to the second-neighbor oxygen distance in ice. The
possible ways in which *scyllo*-inositol can interact within an
ice lattice are illustrated in Fig. 14. It is readily seen that

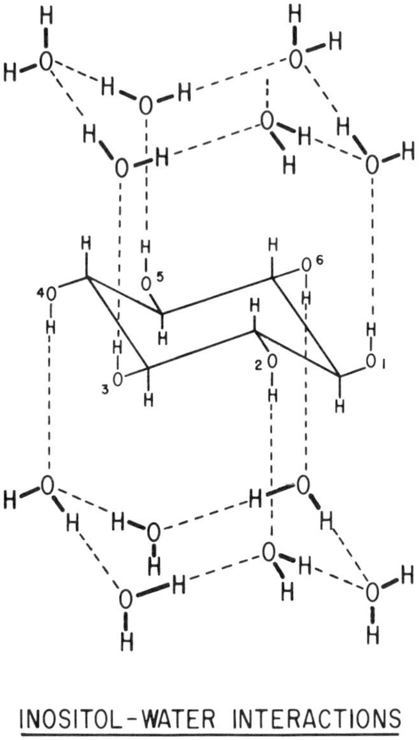

INOSITOL-WATER INTERACTIONS

Fig. 14. Scyllo-inositol inserted into layer of ice lattice.

oxygens 1, 3, and 5 can make precise collinear hydrogen bonds

with the ice layer lying above the sugar, while oxygens 2, 4, and

6 make similar contacts with an ice layer below. In this figure

it would seem that in many ways the six sugar oxygens are equi-

valent to six oxygens in an ice layer. However, there are some

differences. For example, the 1-3-5 and 2-4-6 oxygen planes

(Fig. 14) are not the same distance apart as the similar planes

in an ice lattice (Warner, 1962). However, if we employ an

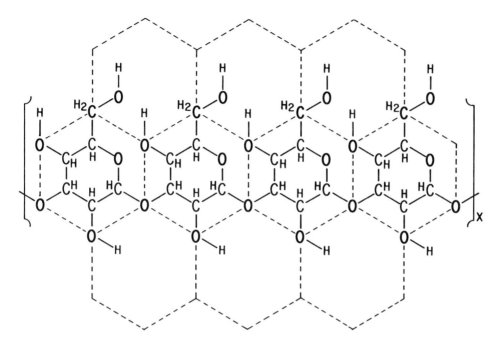

Fig. 15. β-1,3-Polyglucan in ice lattice.

H-O-H anagle of 104°45' (instead of the usual ice tetrahedral
angle of about 109°) as Vandenheuvel (1965) has suggested in
some of his water calculations, then the water layers are closer
together too and approach more nearly the interlayer distances
in the sugar. Other differences involve the permissible bonding
directions.

B. Polysaccharides

    Those polymers which contain pyranose sugars and have the
β-form of the glycosidic linkage between hexose units can be
arranged in a conformation that allows the entire polysaccharide
chain to be fitted into the ice lattice. An interesting example
is the β-1,3-polyglucan of Fig. 15. In this figure the dotted

lines give a two-dimensional picture of the ice lattice in which the hexagonal corners represent first-neighbor oxygens in the lattice. The β-1,3 structure fits smoothly into the confines of the lattice and five out of six of the water oxygen positions of each hexagon are occupied by sugar oxygen atoms. The remaining position is occupied by the carbon atom of the $-CH_2OH$ group comprising the C-6 of the pyranose unit. When the β-glycosidic linkage is in the 1,3 position between the hexoses, each of the $-CH_2OH$ groups points in the same direction away from the main polysaccharide chain. This particular β-1,3-polyglucan has been identified as the principal constituent of zymosan, a yeast cell wall fraction that initiates activation of the reticuloendothelial system to increase a variety of host defense mechanisms (Riggi and Di Luzio, 1961). This component was characterized by Hassid *et al.* (1941) who reported a molecular weight in the range of 6500 or about 40 glucose units. Such a structure could substitute for or bind with a considerable quantity of water in an icelike lattice.

Figure 16 shows another polysaccharide consisting of β-1, 4*N*-acetylglucosamine units, which bears the common name of chitin. Chitin is a major structural material in the invertebrate world (Rudall, 1955). It can be deacetylated chemically with strong alkali to produce polyglucosamine, commonly called chitosan. Chitosan has been sulfated on the amino and hydroxyl groups to produce a synthetic heparinoid (Doczi *et al.*, 1953; Wolfrom *et al.*, 1953; Warner and Coleman, 1958). In Fig. 16, units 1 and 2 are *N*-acetylglucosamines characteristic of chitin, and units 3 and 4 are deacetylated units as found in chitosan. It should be noted that, although the entire structure of this β-1,4 linkage also fits smoothly into the ice lattice, in this instance the $-CH_2OH$ groups of the C-6 position appear on alternating sides of the main backbone at neighboring hexose units (compared to "same-side" placement for β-1,3- glycosidic linkages).

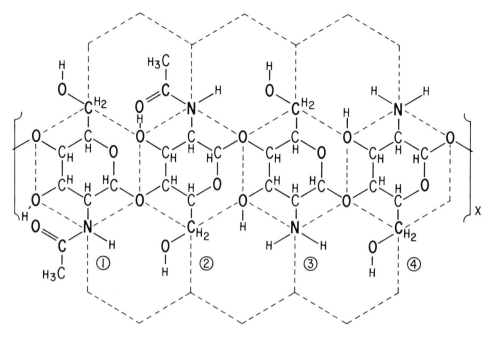

Fig. 16.   Chitin-chitosan in ice lattice.

    A polysaccharide of considerable biological importance as a
principal constitutent of synovial fluid and the vitreous humor
of the eye is hyaluronic acid. This component contains a repeating disaccharide unit consisting of N-acetylglucosamine in a
β-1,3 linkage to the next N-acetylglucosamine..(Marchessault
and Sundararajan (1976) give the chemical structure of this and
many other polysaccharides as verified in many cases by X-ray
data.) The possible fitting of hyaluronic acid into the ice
lattice is shown in a two-dimensional outline in Fig. 17, where
the position of the chain is indicated with atomic symbols, and
also in an excellent drawing (Fig. 18) of a Dreiding stereomodel
of hyaluronic acid superimposed on a Dreiding stereomodel of the
ice lattice. The outline of Fig. 17 and the drawing of Fig. 18
show identical portions of the polymer chain. The symbol M in

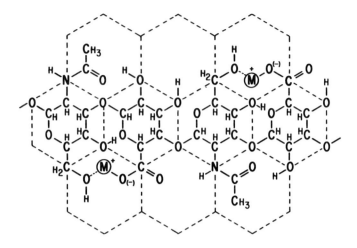

Fig. 17. Hyaluronic acid-schematic drawing in ice lattice.
M, Monovalent metal; ---, outline of "first-neighbor" oxygen
network of ice.

both figures indicates the possible binding site for a monovalent metal, ionically bound to the carboxyl group of the
glucuronic acid and coordinately involved with the 4-hydroxyl
and 6-hydroxyl of the *N*-acetylglucosamine residue. The 4-hydroxyl
is in the *equatorial* configuration on the pyranose ring, as may
be clearly seen in Fig. 18. It is interesting to note that the
polysaccharide chondroitin-4-sulfate has a structure quite
similar to hyaluronic acid except that *N*-acetylgalactosamine
replaces *N*-acetylglucosamine as one of the sugars and that its
4-hydroxyl group is sulfated (Marchessault and Sundarajan, 1976).
This 4-hydroxyl group is here attached to the pyranose in the
axial position, which in the model provides space for the sulfate ester group to sterically fit in and furnish an additional
ionic group adjacent to the glucuronate carboxyl. Consequently
in chondroitin-4-sulfate the symbol M could be a divalent metal

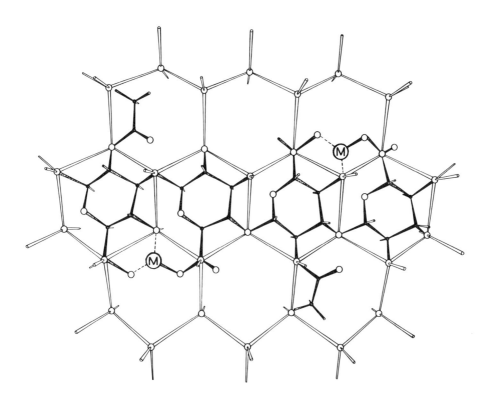

Fig. 18. Hyaluronic acid in ice lattice — drawing of Dreiding stereomodel.

ionically bound to $-SO_3^-$ and $-CO_2^-$ with an additional coordination to the 6-hydroxyl of the *N*-acetylgalactosamine moiety.

VIII.   A PROPOSED ROLE FOR GLYCEROL AND OTHER MONO- AND
        DISACCHARIDES IN SOME LIVING SYSTEMS

Glycerol is a three-carbon sugar that also fits into the ice pattern as a potential "water replacer". Of course, it can also serve as an energy source. Referring back to Fig. 13, the three carbon atoms and three hydroxyl groups of one glycerol molecule would occupy the same positions as carbons 1, 2, 3, and oxygens 1, 2,3 in the *scyllo*-inositol model. Therefore, two moles of glycerol could be as effective as one mole of *scyllo*-inositol or six molecules of water in cementing two protein (or peptide) hydrophilic surfaces together. Furthermore, since the oxygen distances in glycerol (or inositol) would not be expected to vary appreciably with temperature over a considerable range, these sugars might serve as "cements" and structure-stabilizing materials at temperatures where the fitting of water would not be suitable. If they can fulfill this role, perhaps it should not be surprising to find that insects attempt to accomodate to low temperatures by increasing the quantity of glycerol in their body fluids (Salt, 1961). Its presence may represent something more than a mere "antifreeze" effect. It could signify glycerol's ability to replace water over a wide temperature range as a structure stabilizing agent. Webb *et al.* (1963) have already shown that the death of bacterial cells or the inactivation of viruses during periods of partial desiccation results from loss of bound water, and their viability is preserved by adding inositol before the desiccation.

Part of the molecular accommodation of cold-adapted Antarctic fishes has been related to a unique glycoprotein containing large quantities of galactose and *N*-acetylgalactosamine in a disaccharide side chain bound to regularly spaced threonine residues on the protein (Feeney, 1974). Model studies show that this disaccharide also fits into an ice lattice structure.

Glycerol at 4 to 6 M concentration induces the formation of ring structures for "6S" tubulin even at $0^{O}C$ and also can produce microtubules from this usually inactive material at $35^{O}C$ (Erickson, 1971). These experiments again suggest the possibility that glycerol could be replacing water as a cementing material for proteins even at low temperature, where the assembly of these proteins would not oridinarily be maintained by the colder water alone.

Recently Di Paola and Belleau (1978) studied the effect of a series of monosaccharides on the isothermal denaturation (unfolding) of aqueous β-lactoglobulin by urea at 25°. Glycerol (2.4 M) decreased unfolding by as much as 48%. D-mannitol, D-sorbitol, and myo-inositol were also highly effective, while $meso$-erythritol and D-arabitol were somewhat less useful. Ethylene glycol, however, has only a very small effect on unfolding at concentrations where glycerol is very effective. From our model studies, this observation might be attributed to the fact that the 1,2-oxygens of ethylene glycol cannot substitute for the 4.8 Å second-neighbor spread of water, while the 1,3 oxygens of glycerol can. Di Paola and Belleau discuss a similar possible explanation for the glycerol phenomenon.

Although attention here has been directed primarily to model studies of water-protein and water-carboyhydrate interactions because of their probable biological significance, other workers have made equally interesting studies for other reasons. For example, Kabayama and Patterson (1958) also have noted certain similarities between the oxygen placements in an ice lattice and the oxygen positions in certain mono- and disaccharides. They too used comparative model drawings of the two structures to support their theoretical proposals on the thermodynamics of mutarotation in sugars. They even elaborate on the possibility that one of the pyranose rings in amylose (an α-linked disaccharide) may shift from the chair to the boat form so that the glycosidic linkage can assume an equatorial position and better accommodate the entire disaccharide to the water lattice. Such shifts could also be extremely important in a biological sense, and they require more study and thought.

IX.   CONCLUSIONS

The role of water in biology, life processes, and many
aspects of the food industry will continue to be an interesting
problem and challenge for many scientists for many years.   No
doubt the things brought out in this discussion will provide
more questions than answers.   No one who has taken, for example,
a viscous 1% solution of guar gum, gelled it in a beaker with
the addition of sodium borate, then literally poured out the
mass cleanly into the palm of the hand like a spongy nonwetting
cluster of frog's eggs can fail to be inquisitive about the
processes that are occurring at the molecular level in that
very dilute gel.   The open cooperation of scientists of many
disciplines and interests will be required to furnish the
answers to such questions.   It is our hope that the model
studies of water-protein and water-carbohydrate situations
that we have presented will provoke some new insights into the
problems that food processors and food chemists are continually
facing.

ACKNOWLEDGMENTS

I would like to thank Eugene Beals, Robert Simonds, and Jerry
Emrick of our Research Laboratory Services group for their help
in the preparation of drawings and photographs.   I also thank
the editors of Nature for permission to use Figs. 1, 2, and 13
and the Journal of the American Oil Chemists Society for the use
of Figs. 3, 4, 5, 6, and 7.   My gratitude to the art department
of the City of Hope National Medical Center for their detailed
drawing of the Sreiding stereomodel (Fig. 18).   Academic Press
has granted permission to use Figs. 9, 10, 11, and 12 from the
Journal of Theoretical Biology as well as Fig. 17 from their

book "Annual Reports in Medicinal Chemistry-1969, C. K. Cain, ed., Chapter 23 (copyrights by Academic Press, Inc.).

REFERENCES

Dickerson, R. E. (1965). *Nature 208*, 139.
Di Paola, G., and Belleau, B. (1978). *Canad. J. Chem. 56*, 848.
Doczi, J., Fischman, A., and King, J. A. (1953). *J. Am. Chem. Soc. 75*, 1512.
Erickson, H. P. (1971). *J. Supramol. Structure 2*, 393.
Feeney, R. E. (1974). *Am. Sci. 62*, 712.
Hassid, W. Z., Joslyn, M. A., and McCready, R. M. (1941). *J. Am. Chem. Soc. 63*, 295.
Huxley, H. E. (1965). *Sci. Am. 213(6)*, 18.
Kabayama, M. A., and Patterson, D. (1958). *Canad. J. Chem. 36*, 563.
Marchessault, R. H., and Sundararajan, P. R. (1976). *In* "Advances in Carbohydrate Chemistry and Biochemistry," Vol. 33 (R. S. Tipson and D. Horton, eds.), pp. 387-404. Academic Press, New York.
Riggi, S. J., and Di Luzio, N. R., (1961). *Am. J. Physiol. 200*, 297.
Rudall, K. M. (1955). *Symp. Soc. Exp. Biol. 9*, 49.
Salt, R. W. (1961). *Ann. Rev. Entomol. 6*, 55.
Vandenheuvel, F. A. (1965). *J. Am. Oil Chem. Soc. 42*, 481.
Warner, D. T. (1961). *Nature 190*, 120.
Warner, D. T. (1962). *Nature 196*, 1055.
Warner, D. T. (1964). *J. Theor. Biol. 6*, 118.
Warner, D. T. (1967). *J. Am. Oil Chem. Soc. 44*, 593.
Warner, D. T. (1970a). *J. Theor. Biol. 26*, 289.
Warner, D. T. (1970b). *J. Theor. Biol. 27*, 393.
Warner, D. T. (1974). *J. Theor. Biol. 46*, 329.
Warner, D. T., and Coleman, L. L. (1958). *J. Org. Chem. 23*, 1133.
Webb, S. J., Bather, R., and Hodges, R. W. (1963). *Canad. J. Microbiol. 9*, 87.
Wolfenden, R. (1978). *Biochemistry 17*, 201.
Wolfrom, M. E., Shen, T. M., and Summers, C. G. (1953). *J. Am. Chem. Soc. 75*, 1519.

HYDRATION OF MILK PROTEINS

Elliott Berlin

I. INTRODUCTION

The proteins of bovine milk exhibit unique water-binding
properties in addition to behavior common to most proteins. Hydration characteristics of purified whey proteins are generally
not atypical when compared with those of most animal and vegetable
proteins. Indeed many of the classic studies of the interactions
of proteins with water were conducted with β-lactoglobulin. In
contrast, the intramolecular association of the various casein
subunits into the casein micelle results in unique water-binding
properties. Differences in capacities for water were reported for
micellar casein and sodium caseinates (Berlin et al., 1968).
Similarly the effects of rennet coagulation on water binding by
caseins were studied (Ruegg et al., 1974). Factors determining
the special nature of water binding by casein include the hydro-
phobicity of the casein subunits, the loss of the polar glycomacropeptide upon coagulation, and the extent of phosphate hydration in the caseinates.

Though studies of water uptake by purified milk proteins are
of more than academic interest, the major practical concerns pertain to water binding by milk proteins in complex food systems,
including concentrated and dehydrated milk or whey products and
fabricated composites such as milk-citrus and whey-soy combinations (Berlin et al., 1973a). Research efforts have been focused

upon the role of milk protein hydration in these products and upon
the effects of water binding by low-molecular-weight components in
processing and upon protein stability. Such effects include the
hygroscopicity of amorphous lactose, the influence of lactose
crystallization on the physical characteristics of milk powders
and concentrates, and the loss of protein solubility in association
with milk salt hydration.

Pertinent research in this area encompasses work on both hydration and dehydration of milk components by means of appropriately
varied experimental approaches. In this chapter, the current
status of our understanding of certain phases of the interactions
of water with milk proteins is reviewed, with emphasis placed upon
results demonstrating water-induced changes in the proteins.
Studies drawn upon include: (1) measurements of water vapor
sorption and desorption isotherms and (2) calorimetric studies including measurements of "bound" or "unfreezable" water, energies
of vaporization of sorbed water, and heat capacities of protein-
water systems.

II. WATER VAPOR SORPTION

A. Isotherms and Isochrones

The water vapor sorption isotherm for dried milk proteins,
derived from equilibrium gravimetric sorption data as a function
of relative pressure, usually assumes the sigmoid shape of a type
II isotherm according to the classification of Brunauer et al.
(1940). As is apparent from the shape of the isotherm, different
chemical and physical processes occur as functions of relative
pressure. These processes include the entire span of events
between the initial interactions of water molecules with the surface of the dried protein and the ultimate aqueous dispersion or
dissolution of the protein on approaching the saturation pres-

sure. Events during H_2O sorption by a protein, which may be described as sorbate-induced effects upon the sorbent, include swelling of the solid matrix and conformational changes in the macromolecule to ultimately approach the conformation normally taken
by the protein when in solution.

Kinetic data also relate to these H_2O vapor-induced changes
in dried milk proteins. Generally, sorption rates are not uniform
over the entire isotherm, and equilibrium is reached very rapidly
at intermediate pressures, somewhat more slowly at low relative
pressures, and much more slowly as the saturation pressure is approached. This is demonstrated with the isochronal sorption data
in Fig. 1 for water vapor on EMPRO 80, an experimental whey protein concentrate. Each isochrone represents the millimoles of
water sorbed per gram protein as a function of water vapor pressure at a specified time. The isochrones form a family of curves
that converge toward the equilibrium isotherm (see Fig. 5). The
sorption data used to construct the isochrones (Fig. 1) were obtained by pumping the protein sample dry before it was exposed to
water vapor at each H_2O vapor pressure level. This contrasts with
the usual sorption experiment in which the sample is dried only
before sorption is measured at the first vapor pressure level and
then is exposed to higher vapor pressures after reaching sorption
equilibrium without intermediate drying.

Deitz and Turner (1971) developed the use of the adsorption
isochrone in the interpretation of data for water adsorption on
glass fibers. Deitz and Bitner (1973) pointed out the merit of
the isochrone concept for distinguishing between time-dependent
and concentration-dependent factors in a gas-surface reaction.
Sedlacek (1972) developed a theoretical equation for the isothermal isochrone that in the limiting case expresses the equilibrium
adsorption isotherm.

In Fig. 1, the isochrones corresponding to 10 min or more
resemble a type II isotherm, similar to the equilibrium isotherm
for the whey protein. However, the isochrones for shorter inter-

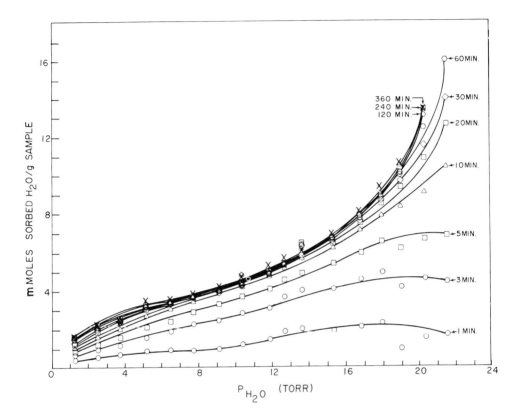

Fig. 1. *Sorption isochrones for water vapor on whey protein concentrate (EMPRO 80).* **X** , *Sorption points from the equilibrium sorption isotherm.*

vals assume different shapes. Apparently the isochrones for 1-5 min display a shift from a type I (Langmuir) to a type IV iso-therm. A limited pore volume usually is responsible for a type IV isotherm in physical adsorption. The curves in Fig. 1 suggest that the mechanism for water sorption requires a time-dependent expansion of the solid phase to permit the complete hydration of the protein.

B. *Sorption by the Caseins*

Casein is a major water-sorbing component in dried milk.
Comparison of water sorption by dehydrated milk components led
Berlin *et al.* (1968) to conclude that at low relative humidity
casein is the preferred sorption site in dried milk powder.

The extent of water sorption by casein is controlled by its
physical-chemical status. Data for water vapor sorption by lyophilized preparations of micellar casein and caseinate are shown
in Fig. 2 (Berlin *et al.*, 1968). The micellar casein was obtained
by ultracentrifugation from skim milk, washed several times in
distilled water, and freeze-dried. Acid casein was precipitated

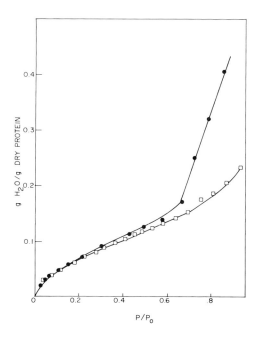

*Fig. 2. Sorption isotherms for H_2O vapor at 24.2 C on
micellar casein (□) and sodium caseinate (●) (Berlin et al.,
1968).*

at pH 4.7 from skim milk, washed with dilute acid, titrated to
neutral pH with NaOH, and the resulting sodium caseinate was lyophilized. Both casein forms yielded almost identical isotherms
at low to intermediate relative pressures (P/P_0 < 0.6), beyond
which the sodium caseinate sorbed far more water than the micellar
form. Increased H_2O sorption at these higher humidities may be
associated with the ions present in the caseinate preparation.
It is not uncommon for ionic materials to be very hygroscopic at
elevated P/P_0. Water sorption measurements with a synthetic
milk-salt mixture (Berlin et al., 1968) yielded a type III isotherm, with almost zero sorption up to 0.6 P_0, beyond which the
salts absorbed large amounts of water. The difference in water
sorption properties of the caseins is likely associated with the
nature of the ionic species rather than with the total ion content.
Ash values were similar for the two caseins, but calcium content
was 2.87% for the micellar casein and only 0.15% for the caseinate.
Indeed the alkali metal cations do adsorb greater amounts of water
than the alkaline earth cations.

   Ruegg and Blanc (1976) reported profound effects of pH on
water vapor sorption by micellar and acid-precipitated casein.
The acid-precipitated casein sorbed more water at high pH while
the micellar casein has minimal hydration at the higher pH values.
The micellar casein also revealed a minimum in water sorption in
the pH region 5.4-6. Ruegg and Blanc attributed the properties
of the micellar casein to the inorganic ions. Lowering or raising
the pH from neutrality destabilizes the casein micelles due to
release of bound cations and dissociation or dissolution of mineral
compounds. The subsequent increase in concentration of ionic substances is accompanied by an increase in water sorption capacity.
Ruegg and Blanc described the effect of reduction in pH from 7 to
5 as a decrease in protein molecule hydration with increased water
sorption by the calcium phosphate-caseinate system due to released
mineral salts.

   Processing of milk into finished dairy products often involves

destabilization of the casein micelle into subunits that differ in
hydrophobic amino acid content and hence possibly differ in solubility and water sorption capacity. Similarly, the loss of the
glycomacropeptide during rennet coagulation enhances the hydro-
phobicity of the residual casein. Ruegg et al. (1974) found little
difference between water vapor sorption–desorption isotherms for
rennin–coagulated and micellar caseins at 25°C. At low relative
pressures (to 0.35 P_0) the sorption isotherms for the two caseins
are superimposable; however, at intermediate to high water activities the native casein sorbed more water. As the saturation pressure was approached the curves came together again. The desorption
legs of the isotherms were not superimposable over the whole range,
and hysteresis was greater for native than for coagulated casein.
These authors attributed the decreased water sorption to the loss
of the hydrophilic glycomacropeptide during rennet coagulation and
concluded that the native form has the greater capacity for swelling.

The availability of genetic polymorphs of α_{S1}-casein permits
examination of water sorpticn by protein moieties that are identical in all respects except for the absence of a sequence of eight
hydrophobic amino acids from α_{S1}-casein A. Thompson et al.
(1969c) reported that α_{S1}-casein B is less solvated than α_{S1}-casein
A but micelles containing α_{S1}-casein A were less solvated, which
they attributed to increased calcium binding. More recently,
Dewan et al. (1974) reported no difference in solvation of micelles
from α_{S1}-casein A or pooled skim milks.

Sorption isotherms for water vapor on α_{S1}-casein A and α_{S1}-
casein C are compared in Fig. 3. At low relative pressures (up
to 0.25 P_0) the isotherms of the two α_{S1}-caseins are almost
exactly superimposable. At intermediate to higher relative pressures, α_{S1}-casein A, was more sorptive. Though not great, the difference in water sorption was clearly reproducible. These data
led Berlin et al. (1969) to conclude that the hydrophobic amino
acid segment in α_{S1}-casein C assumes a surface conformation, per-

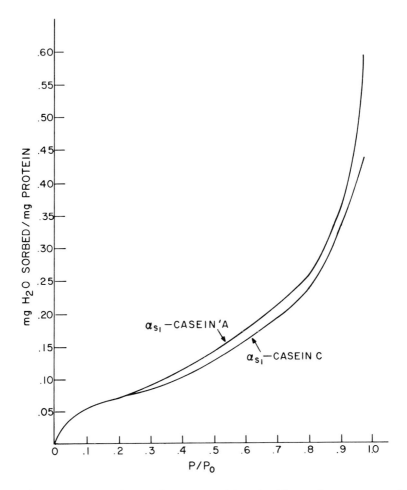

Fig. 3.   Comparison of H_2O sorption isotherm for α_{S1}-caseins A and C (Berlin et al., 1969).

mitting hydrophobic interactions between α_{S1}-casein C molecules and drastically reducing its solubility in water, as reported by Thompson *et al.* (1969b).

The two genetic polymorphs also differed in water vapor desorption properties.   Both variants exhibited reproducible hysteresis loops through two sorption-desorption cycles, but the loop was wider for the A variant.   Removal of sorbed water from the A variant after the first cycle required 7-10 days evacuation at 10^{-6} torr at ambient temperature to restore the protein to its

initial dry weight. In contrast, desorption from the C variant
was complete in a day or two. These data suggest that swelling
capacity is depressed in the more hydrophobic polymorph in con-
sonance with the conclusion of Ruegg et al. (1974) with respect
to rennin-coagulated casein.

III. SORPTION AND PROTEIN SWELLING AND CONFORMATION

   Extensive reference was made above to the concept that sorbed
water induced swelling of the protein. Several lines of experimental evidence from milk proteins demonstrate such effects of
sorbed water.

A. Sorption Hysteresis

   Hysteresis loops have often been observed in the sorption-
desorption isotherms for milk proteins (e.g., Berlin et al., 1969,
1970a; Ruegg et al., 1974; Rao and Das, 1968). Berlin et al.
(1969) reported fully reversible hysteresis loops through at least
two sorption-desorption cycles for genetic variants of α_{s1}-casein.
Rao and Das (1968) reported that hysteresis loops for casein decreased in size and finally disappeared in subsequent cycles.
They attributed this phenomenon to a gradual collapse of internal
cavities in the dried protein due to swelling and contraction
during the several sorption-desorption cycles.
   Physical adsorption of an inert gas on a rigid adsorbent may
also yield a type II isotherm with a hysteresis loop, in which
case the isotherm describes both multilayer adsorption on available
surfaces and capillary condensation within pores. It has been
tempting to apply physical adsorption theory to H_2O sorption on
proteins, utilizing the BET multilayer adsorption equation
(Brunauer et al., 1938). Treatment of data for N_2 adsorption at
-195°C and H_2O vapor sorption at 24.2°C on genetic variants of

α_{S1}-casein according to the isotherm (FHH) equation of Frenkel
(1946), Halsey (1948), and Hill (1952) clearly distinguished
between the adsorption processes involved. Data for nitrogen adsorption followed straight lines when plotted according to the
FHH equation, as expected for physical adsorption on a surface
with no unusual capillarity (Pierce, 1960). Data for water
sorption yielded a linear segment, $P/P_0 > 0.6$, with $r = 2.5$ in
the FHH isotherm equation:

$$\log \log (P_0/P) = r \log \theta$$

where θ corresponds to the fraction of surface coverage or any
variable proportional to it. The value 2.5 is consistent with
that observed by Halsey for H_2O adsorption. The major portion
of the H_2O isotherm ($< 0.6P_0$), however, deviated widely from
linearity. The curvature of the bulk of the isotherm results from
swelling and physical changes in the solid protein as a result of
the sorption process.

B. Desorption and Protein Contraction

When casein micelles were washed with water and dehydrated by
serial transfer through liquids of decreasing polarity before
final solvent removal under vacuum, the resulting dry material
had a specific surface area ten times higher than washed micelles
dried directly from aqueous systems by lyophilization (Berlin et
al., 1970a). Micelles dried by evaporation of water exhibited
surface areas of 3-5 m^2/g, determined by nitrogen adsorption at
-195°C, whereas micelles dried by solvent replacement through
methanol, pentane, hexane, or benzene yielded areas of 41.0,
32.0, 36.1, and 71.0 m^2/g, respectively. Subjecting the expanded
dry proteinaceous material to cyclic H_2O sorption and desorption
resulted in contraction and loss in surface area. The decrease
in surface area was dependent upon the amount of water vapor previously sorbed and desorbed, as shown in Fig. 4. Surface area

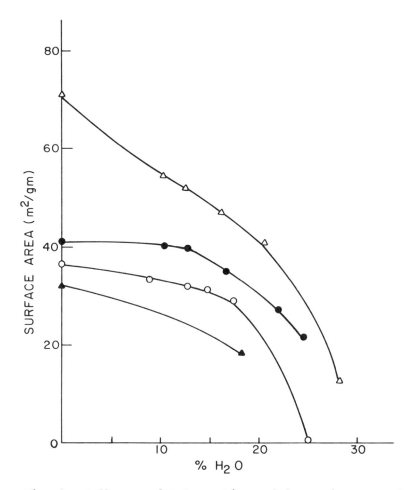

Fig. 4.   Influence of H_2O sorption and desorption on surface area.   The curves represent casein dried originally from ○ *, hexane;* ● *, methanol;* △ *, benzene; and* ▲ *, pentane (Berlin et al., 1970a).*

decreased at each H_2O sorption-desorption increment, but the loss became greater with an apparent change in slope in the curve at about 18% moisture. Apparently 18% moisture corresponds to a critical level of protein hydration, leaving the casein extensively swollen and likely assuming a conformation similar to that in aqueous solution. Desorption of H_2O from this state would then be a process similar to lyophilization of an aqueous protein solution or dispersion, hence the great loss in surface area due to collapse of the expanded macromolecular structure.

C. Calorimetric Evidence

Berlin et al. (1970b) observed a variation in ΔH_v, the enthalpy for vaporization of sorbed water from proteins, related to the amount of water sorbed. Enthalpy data were determined by differential scanning calorimetry with lyophilized preparations of micellar casein, β-lactoglobulin, bovine serum albumin, and calf-skin collagen at different moisture levels. Once a critical level, 0.18 g H_2O per g protein, of sorbed moisture was attained, ΔH_v was 80-125 cal/g higher than the heat of vaporization of pure water, reflecting a change in the energy through which water is bound to the protein. These results were unexpected as physical adsorption generally is accompanied by decreasing heat of adsorption that approaches the heat of liquefaction of the sorbate with increasing surface coverage. At very low humidity H_2O is hydrogen bonded to specific hydrophilic sites and the heat of adsorption usually is about 12 kcal/mole. However, as more water is sorbed the binding is weaker and the heat of adsorption decreases.

The DSC data were rationalized in terms of protein swelling and conformation change. The sorption of H_2O water leads to swelling of the solid such that surfaces available for H_2O sorption increase. At high P/P_0 the solid protein is sufficiently swollen to approach complete hydration, and the protein assumes a conformation resembling that in an aqueous solution or dispersion. Under these circumstances the sorbed water could form a quasi-solid structure as has been postulated for water in the vicinity of dissolved macromolecules. This could elevate ΔH_v values to levels observed when the sorbed water exceeded 0.18 g H_2O per g protein. Thus once a critical amount of H_2O is sorbed, the protein undergoes a change in conformation approaching its form in solution. In this regard casein and whey proteins exhibited identical properties that also resembled those of other very different types of proteins: BSA and collagen.

The differential scanning calorimeter has also been used to determine the heat capacity of proteins and sorbed water (Berlin *et al.*, 1972). The specific heat of β-lactoglobulin containing sorbed water was a linear function of moisture content according to the equation

$$C_p = 0.284 + 0.007 \times (g\ H_2O/g\ sample)$$

in calories/gram. In these experiments samples of the protein were adjusted to various moisture levels in humidostats, hermetically sealed in aluminum capsules at ambient temperature, cooled to 0°C in the DSC, and then heated to 25°C. Under the conditions of this experiment the heat capacity data are free of desorptive heat capacity effects. The linear relation cited above was determined by least-squares analysis of specific heats calculated for 12°C, the midpoint of the calorimeter scans. Actually any temperature in the range scanned would have yielded the same results.

The calculated apparent partial specific heat values were $\bar{C}_{p1} = 0.947 \pm 0.137$ cal/g/°C for the β-lactoglobulin sorbed water and $C_{p2} = 0.283 \pm 0.02$ cal/g/°C for β-lactoglobulin. The observed isosteric sorbate heat capacity value C_{p1} for the water is compatible with the notion that such water exists in an associated form involving multiple hydrogen bonding. The excess heat capacity of liquid water over that of ice or water vapor is sometimes called "structural" heat capacity (Berendsen, 1967) and is attributed to the thermal breakdown of the associated structures present in the liquid. Water molecules bound to isolated specific sites on a protein surface should not exhibit such a structural heat capacity contribution; hence the elevation of C_{p1} for the β-lactoglobulin sorbed water over the specific heat of ice or water vapor may be taken as evidence for the association of sorbed water into a structured hydration shell as would be found in an

aqueous protein solution. Kuntz and Kauzmann (1974) have suggested the importance of a "configurational" heat capacity term
for water-protein solutions.

IV. BOUND OR UNFREEZABLE WATER

   Bound water has often been defined as sorbent- or solute-
associated water molecules that differ thermodynamically from
ordinary water, hence the numerous studies of the thermodynamic
properties of protein solutions. Berlin *et al*. (1970b) used a
differential scanning calorimeter to measure the heat of fusion
of water associated with wet dispersions of milk proteins. The
apparent heat of fusion, based on total water, averaged 40 cal/g
instead of the usual 79.6 cal/g for pure water at $0°C$. The data
were understood in terms of the melting of only a fraction of the
total water; the heat of fusion of pure water was used to calculate the mass of unfrozen water per unit mass of protein on a dry
weight basis. The "nonfreezing water" was considered to be
"bound water," and did not freeze even when the samples were
cooled to $-70°C$. Observed values, in grams H_2O per gram protein,
were 0.55 for washed casein micelles, 0.55 for β-lactoglobulin,
0.50 for purified total whey protein, and 0.49 for bovine serum
albumin. In these experiments, repeated freezing and thawing
cycles of the samples did not alter the extent of water binding.
Thus the several different proteins bind water to an extent approximately equivalent to 50% of their dry weight, a value similar
to the quantity of water sorbed as P/P_0 approaches unity.

V. DAIRY PROCESSING AND MILK PROTEIN HYDRATION

A. *Protein Denaturation and Bound Water*

It is often suggested on theoretical grounds that denaturation
would increase the water-binding capacity of proteins, but few sub-
stantiating data are available. Kuntz and Brassfield (1971) and
Mrevlishvili and Privalov (1969) reported small increases in hydration of denatured proteins. Since dairy products are often
subjected to elevated temperatures during processing, it is important to consider the effects of thermal denaturation on milk protein hydration.

Berlin *et al.* (1973b) observed no changes caused by denaturation in the unfreezable water content of dispersions of whey protein concentrates. They examined a group of spray-dried whey
protein products containing 35-80% total protein with various
contents of insoluble or denatured protein. The protein concentrates were separated into high- and low-molecular-weight fractions
by suspending the powders in distilled water, dialyzing the suspensions against repeated changes of distilled water, and lyophiliz-
ing the dialyzable and nondialyzable fractions. Pellets of denatured or insoluble protein were isolated by suspending powdered
samples in water and centrifuging at high speed in a preparative
ultracentrifuge. Data for unfreezable water in dispersions of the
total protein and insoluble fractions of several powders are shown
in Table I. These data show that protein denaturation did not
consistently affect water binding in these whey protein products.

Thermal denaturation and water binding were also studied directly in the calorimeter. After measuring the amount of unfreezable water by heating from $-70°$ to $+20°C$ the protein sample, sealed
in an aluminum capsule, was heated rapidly ($80°C$/min) to $85°C$ *in
situ* in the calorimeter and held there for 1 hr to complete thermal
denaturation. The sample was then cooled and unfreezable water redetermined to assess the effects of denaturation on water binding.

TABLE I. Water Binding by Insoluble Protein Fractions of
Whey Protein Concentrates (WPC)[a]

Sample[b]	$\dfrac{g\ H_2O\ bound}{g\ total\ protein}$	Total insoluble protein (%)	$\dfrac{g\ H_2O\ bound}{g\ insoluble\ protein}$
EMPRO 80	0.45	29.9	0.49
WPC 75	0.48	14.3	0.45
WPC 50	0.52	48.6	0.56
WPC 35	0.49	41.0	0.47

[a]Berlin et al. (1973b).
[b]Number in sample designation refers to approximate protein
content of the WPC.

Bound water values were almost identical before and after heating
wet samples of pure whey protein and several concentrated whey
protein products. Thus loss in protein solubility through thermal
denaturation apparently has little or no effect on the capacity
of the protein to bind water against freezing.

Ruegg et al. (1974) measured the amount of unfreezable water
in micellar, rennin-treated, and isoelectric casein by differential
scanning calorimetry. They reported no differences in water binding in the various caseins. They also reported identical sorbed
water contents for micellar and rennin treated caseins at
$P/P_0 > 0.8$. These data are compatible with our conclusion that
unfreezable water content is comparable to water sorption at the
saturation pressure.

In contrast with these results, McDonough et al. (1974) reported improved water-holding properties for heat-denatured whey
protein concentrates. Heating a 10% solution of whey protein
concentrate, with 50% of the solids as protein, yielded a firm
gel with no water leakage. The water retained in the gels is not
only bound water but is primarily occluded water trapped within
a network of cellular protein filaments. Such occlusion is analogous to the casein pellet solvation reported by Thompson et al.
(1969a). Ruegg et al. (1974) concluded that the loss of water

accompanying syneresis in caseins results from contraction of the three-dimensional network formed by linked casein micelles and not from significant changes in the nature of hydration. Tarodo De La Fuente and Alais (1975) studied casein solvation cryoscopically by measuring nonsolvent water as water that would not dissolve sucrose. They stated that such water includes more than unfreezable water, conclusions that increases in nonsolvent water associated with growth in micelles and micelle clusters upon increasing the total solids in milk results from purely mechanical confinement within the clusters. However, they did report large increases in nonsolvent water upon heating. They defined these results as increased binding and more stable binding between protein clusters and water molecules as a result of thermal denaturation.

B. Nonprotein Components and Water Binding in Milk Protein Products

Briggs (1932) postulated a general concept that water binding by nonreacted components of mixed systems is additive. This idea was based on Briggs' (1931) earlier work with derivatives of a milk protein, sodium and calcium caseinates. Although this concept may not operate in all systems, Berlin et al. (1968, 1973a,b) observed additive water binding of the components in a number of milk-protein-containing systems including such diverse products as whey protein concentrates, dehydrated whey-soy beverage, and dried milk-orange drink. Measured water vapor sorption isotherms and isotherms calculated from data for water sorption by the system components were almost identical, except for the region corresponding to lactose crystallization. Even products containing the hydropholic colloid carboxymethyl-cellulose displayed additive H_2O sorption.

Sorption data for EMPRO 80, a whey protein concentrate, and its fractions obtained by dialysis are shown in Fig. 5. (Berlin et al., 1973b). These data indicate that the relative significance of the powder components varies with water activity. At low rela-

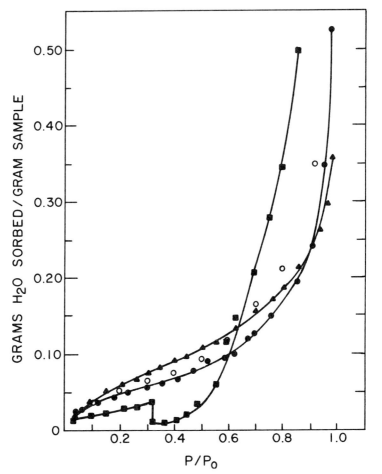

Fig. 5. Water vapor sorption at 24 C on dehydrated whey protein concentrate (EMPRO 80) and dialyzed fractions thereof. ● *, EMPRO 80;* ▲ *, nondialyzable fractions;* ■ *, dialysate;* ○ *, EMPRO 80, calculated from fractions (Berlin et al., 1973b).*

tive pressures sorption by the protein fraction dominated, but at $P/P_0 > 0.6$ sorption by the dialyzable fraction became more significant. In addition agreement may be noted between the observed isotherm and that calculated from sorption data of the fractions. Hence water sorption is additive in this system. Powders containing 73, 83, and 87% whey protein exhibited identical isotherms at low relative pressure, but above 0.4 P_0 water sorption was inversely related to protein content.

TABLE II. Water Binding by Fractions of Whey Protein Concentrate (WPC) Powders[a]

Sample	$\dfrac{g\ H_2O}{g\ protein}$	$\dfrac{g\ H_2O}{g\ dry\ dialysate}$	$g\ H_2O/g$ WPC powder	
			Calc.	Obs.
EMPRO 80	0.45	1.17	0.56	0.54
WPC 50	0.52	1.86	1.11	1.08
WPC 35	0.49	1.17	0.91	0.91

[a]Berlin et al., 1973b.

Unfreezable water in wet dispersions of whey proteins was additive as per the data in Table II (Berlin et al., 1973b). Individually, the dialyzable fraction bound much more water than the protein or nondialyzable fraction; however, water binding in the total protein product was additive. The observed bound water in the product was almost identical with that calculated from the water-binding capacity of the components. These results are not artifacts of dialysis and lyophilization, as physically recombining the separated fractions and redetermining the bound water yielded the same results as were obtained with the original powder.

Low-molecular-weight components, including amorphous lactose and minerals, assume an even greater role in dairy products containing much less protein such as dried milks and wheys. These materials have unusual influences on H_2O sorption and desorption isotherms that are related to lactose crystallization with its attendant change in product morphology and vaporization of lactic acid from wheys (Berlin and Anderson, 1975). San Jose et al. (1977) recently reported that water sorption by components is not additive in dried lactose hydrolyzed milk, as distinct from normal dried milk.

Detailed discussion of results with such low protein products is beyond the scope of this chapter. However, one type of experiment could be cited that relates both to protein swelling and conformation change and lactose crystallization. Berlin and Kliman

(1974) determined the specific heat of cheese whey as a function of moisture content. A linear relationship was maintained between specific heat and moisture content when dried whey solids were re-hydrated to moisture levels between 3 and 93% H_2O with the partial specific heat of sorbed water being 0.995 cal/g/°C. A change in slope was noted in the relation between specific heat and water content at 50% H_2O when the specific heat data were obtained with concentrated whey samples prepared by evaporating water from fluid whey. The apparent partial specific heat values for the water were 0.966 cal/g/°C above 50% and 1.203 cal/g/°C below 50% H_2O. Apparently the water is in a more structured form in concentrated systems provided the solids were initially fully hydrated. The results may be associated with changes in protein conformation that depend upon the presence of a critical moisture content, or they may reflect the effect of lactose crystallization on specific heat.

VI. CONCLUSION

Throughout the studies reviewed emphasis was placed upon the effects of sorbed or bound water on protein swelling and conforma-tion and upon the importance of nonprotein components in determining total bound water in preparation of milk protein. The presence of ionic material assumes a far greater role in controlling water bind-ing than the status of either the caseins or the whey proteins with regard to denaturation. Thus close attention should be given to pH and the presence of ionic species in the production of foods that contain milk protein and require a water-holding capacity.

References

Berendsen, H. J. C. (1967). *In* "Theoretical and Experimental Biophysics" (A. Cole, ed.), p. 26. Marcel Dekker, New York.
Berlin, E., and Anderson, B. A. (1975). *J. Dairy Sci. 58,* 25.
Berlin, E., and Kliman, P. G. (1974). *In* "Analytical Calorimetry," Vol. 3 (R. S. Porter and J. F. Johnson, eds.), p. 497. Plenum Press, New York.
Berlin, E., Anderson, B. A., and Pallansch, M. J. (1968). *J. Dairy Sci. 51,* 1912.
Berlin, E., Anderson, B. A., and Pallansch, M. J. (1969). *J. Phys. Chem. 73,* 303.
Berlin, E., Anderson, B. A., and Pallansch, M. J. (1970a). *J. Colloid Interface Sci. 33,* 312.
Berlin, E., Kliman, P. G., and Pallansch, M. J. (1970b). *J. Colloid Interface Sci. 34,* 488.
Berlin, E., Kliman, P. G., and Pallansch, M. J. (1972). *Thermo-Chim. Acta 4,* 11.
Berlin, E., Anderson, B. A., and Pallansch, M. J. (1973a). *J. Dairy Sci. 56,* 685.
Berlin, E., Kliman, P. G., Anderson, B. A., and Pallansch, M. J. (1973b). *J. Dairy Sci. 56,* 984.
Briggs, D. R. (1931). *J. Phys. Chem. 35,* 2914.
Briggs, D. R. (1932). *J. Phys. Chem. 36,* 367.
Brunauer, S., Emmett, P. H., and Teller, E. (1938). *J. Am. Chem. Soc. 60,* 309.
Brunauer, S., Deming, L. S., Deming, W. E., and Teller, E. (1940). *J. Am. Chem. Soc. 62,* 1723.
Deitz, V. R., and Bitner, J. L. (1973). *Carbon 11,* 393.
Dietz, V. R., and Turner, N. H. (1971). *J. Phys. Chem. 75,* 2718.
Dewan, R. K., Chudgar, A., Bloomfield, V. A., and Morr, C. V. (1974). *J. Dairy Sci. 57,* 394.
Frenkel, J. (1946). "Kinetic Theory of Liquids." Oxford Univ. Press, London.
Halsey, G. D. (1948). *J. Chem. Phys. 16,* 931.
Hill, T. L. (1952). *Advan. Catalysis 4,* 211.
Kuntz, I. D., and Brassfield, T. S. (1971). *Arch. Biochem. Biophys. 142,* 660.
Kuntz, I. D., and Kauzmann, W. (1974). *Advan. Protein Chem. 28,* 239.
McDonough, F. E., Hargrove, R. E., Mattingly, W. A., Posati, L. P., and Alford, J. A. (1974). *J. Dairy Sci. 57,* 1438.
Mrevlishvili, G. M., and Privalov, P. L. (1969). *In* "Water in Biological Systems" (L. D. Kayushin, ed.), p. 63. Plenum Press, New York.
Pierce, C. (1960). *J. Phys. Chem. 64,* 1184.
Rao, K. S., and Das, B. (1968). *J. Phys. Chem. 72,* 1223.
Ruegg, M., and Blanc, B. (1976). *J. Dairy Sci. 59,* 1019.
Ruegg, M., Luscher, M., and Blanc, B. (1974). *J. Dairy Sci. 57,* 387.

San Jose, C., Asp, N., Burvall, A., and Dahlqvist, A. (1977). *J. Dairy Sci. 60*, 1539.

Sedlacek, Z. (1972). *Collec. Czech. Chem. Commun. 37*, 1765.

Tarodo De La Fuente, B., and Alais, C. (1975). *J. Dairy Sci. 58*, 293.

Thompson, M. P., Boswell, R. T., Martin, V., Jenness, R., and Kiddy, C. A. (1969a). *J. Dairy Sci. 52*, 796.

Thompson, M. P., Farrell, H. M. Jr., and Greenberg, R. (1969b). *Comp. Biochem. Physiol. 28*, 471.

Thompson, M. P., Gordon, W. G., Boswell, R. T., and Farrell, H. M. Jr. (1969c). *J. Dairy Sci. 52*, 1166.

PROTEASE ACTION ON PROTEINS AT LOW WATER CONCENTRATION

Soichi Arai
Michiko Yamashita
Masao Fujimaki

I. INTRODUCTION

Enzymes classified as proteases have an ability to catalyze
hydrolysis of proteins. Historically, many studies with proteases
have dealt with how they effectively catalyze digestion of dietary
proteins, thus playing an important role in the physiology and
nutrition of living things (Baldwin, 1952). Since natural substrates for proteases are proteins that in many instances constitute a major part of food, it is no wonder that food scientists
were and are generally concerned with these enzymes and their applications. Proteases are now recognized as some of the most
common food-related enzymes (Whitaker, 1974). Techniques for
their production and utilization were developed long ago (Reed,
1966, 1975).

Proteases are ubiquitous, occurring endogenously in almost
every kind of food. The enzymes are active in most fresh foods
and they often affect postharvest quality of foods. Occasionally,
palatable flavors may arise as a result of the action of proteases
in foods. This is especially true in proteinaceous foods such as
meat. Exogenous proteases are sometimes used to improve eating
quality of foods and thus increase their utility for human consumption. This approach has been developing in many countries of
the world, and now constitutes an important aspect of food industry
(Hata and Doi, 1971).

Almost all of the current industrial processes involving proteases, either endogenous or exogenous, seem to be directed toward utilization of their ability to catalyze hydrolysis of peptide bonds. Frequently, the protein substrate in these reactions is present at an extremely low concentration. In other words, water is present there at a predominantly high concentration, even at a level of nearly 55.5 M.

During the past 20 years applications of proteases have expanded greatly. Simultaneously, numerous basic studies on proteases have been conducted, disclosing that these enzymes have a potential ability to catalyze several unusual reactions besides protein hydrolysis. These include ester and amide hydrolysis (Glazer and Smith, 1971), transesterification (Glazer, 1966; Lake and Lowe, 1966; Sluyterman and Wijdenes, 1972), transamidation (Mycek and Fruton, 1957; Durell and Fruton, 1954), transpeptidation (Glazer and Smith, 1971), and even condensation (Bender and Kemph, 1957; Kezdy *et al.*, 1964). The plastein reaction, which is recognized to be a type of protease-induced clotting of peptides, should also be included in this list. The above unusual reactions, several of which have had practical use in peptide synthesis, are all characterized by their association with the unit process called "aminolysis." It may be quite possible that the aminolysis process competes with the usual hydrolysis process and is therefore influenced by the water concentration of the reaction system. Such an influence of water on protease action, though not yet completely defined, is regarded as important based on results of several experiments involving unusual protease-catalyzed reactions, including peptide synthesis.

II. PEPTIDE SYNTHESIS WITH PROTEASES

The usual hydrolysis process catalyzed by serine proteases (such as trypsin and α-chymotrypsin) and sulfhydryl proteases (such as papain and bromelain) can be written minimally by the following kinetic scheme:

$$E + S \underset{}{\overset{K_S}{\rightleftharpoons}} ES \xrightarrow{k_2} ES' \xrightarrow{k_3} E + P_2 \qquad (1)$$

$$+$$
$$P_1$$

where ES is an enzyme-substrate complex often called the Michaelis complex, K_S the dissociation constant, ES' the acyl-enzyme, P_1 the first product released with the formation (acylation) of ES', P_2 the second product formed as a result of ES' decomposition (deacylation), and k_2 and k_3 rate constants for acylation and deacylation, respectively. The hydrolysis reaction completes when the acyl-enzyme intermediate is hydrolytically deacylated, the water concentration being expected to be an important factor contributing to the deacylation process.

An unusual reaction may be possible when the system contains a nucleophilic amine that competes with water for the acyl-enzyme (Brubacher and Bender, 1966; Kortt and Liu, 1973). In such a case Eq. (1) must be rewritten as

$$E + S \underset{}{\overset{K_S}{\rightleftharpoons}} ES \underset{}{\overset{k_2}{\rightleftharpoons}} ES' \underset{\underset{\searrow k_4[N]}{}}{\overset{\nearrow k_3[W]}{}} \begin{matrix} E + P_2 \\[2mm] E + P_3 \end{matrix} \qquad (2)$$

$$+$$
$$P_1$$

where N is the nucleophile, P_3 the third product resulting from the aminolytic deacylation of the acyl-enzyme, and k_4 its reaction rate constant. There is a possibility that an added nucleophile binds to the acyl-enzyme and forms a ternary enzyme-substrate-nucleophile complex, since the molar reactivity of many nucleophiles as compared to water is much greater than would be expected from nonenzymatic reactions (Werber and Greenzaid, 1973). Discussion of this matter is beyond the scope of the present chapter.

The process involving such nucleophiles can be thought of as ester aminolysis in a chemical sense, because acyl-enzyme either from serine proteases or from sulfhydryl proteases are generally recognized as acyl-serine esters or acyl-cysteine thioesters by nature (Smellie, 1970). This type of amonolysis occurs when proteases are used to synthesize peptides. If an N-masked amino acid ester is used as substrate, an alcohol is first produced as P_1 [see Eq. (2)] by the esterase activity of the protease used. The resulting acyl-enzyme, i.e., N-masked aminoacyl-protease, can be attacked by a C-masked amino acid (such as an amino acid ester) added as nucleophile. Though a normal hydrolysis product (P_2) is inevitably formed since the reaction must, in actual situations be carried out in the presence of water, it is possible that an aminolysis product (P_3) occurs at the same time. Tauber (1952) may be the first who succeeded in synthesizing dipeptides in this way using the well-defined protease, α-chymotrypsin. As nucleophiles, amino acid amides could be used as well. Morihara and Oka (1977) have recently reported that a number of dipeptides are produced in various yields depending on the specificities of substrates as well as on the specificities of nucleophiles toward α-chymotrypsin.

The condensation reaction also is usable for peptide synthesis. N-masked amino acids, which are still able to acylate proteases with liberation of water as P_1 can be used as substrates. A Japanese group has used several kinds of N-carbobenzoxy-*L*-amino acids as substrates and *L*-amino acid diphenylmethyl esters as nucleophiles, and succeeded in obtaining coupling products in generally satisfactory yields (Isowa *et al.*, 1977). Papain is regarded as one of the most effective enzymes for catalyzing con-

densation reactions of this kind. It is particularly interesting
that Isowa (1978) successfully synthesized Val[5] angiotensin by
fragment condensation with the aid of papain. A similar experiment was conducted earlier by Saltman *et al.* (1977). An efficient
reaction of this kind requires at least two conditions. One is
that both the substrate, Boc-Val-Tyr(Bzl)·OH, and the nucleophile,
H·Val-His(Bzl)-Pro-Phe-OEt, be present at very high concentrations. Second, it is necessary that the reaction medium consist
of a 50:50 mixture of water and methanol. The latter condition
may be necessary because the half-decreased water concentration
promotes efficient aminolysis of the presumed acyl-enzyme, Boc-
Val-Tyr(Bzl)-Papain, by the nucleophile, with the result that the
condensation product is formed in a greater yield.

III. PLASTEIN REACTION

    More than 70 years have elapsed since the term "plastein" was
coined (Sawjalow, 1907). Interest in the plastein reaction
probably is as old as the history of protein biosynthesis research.
During a few decades prior to 1960, extensive studies were conducted to elucidate the relationship between the plastein reaction
and protein synthesis in biological systems (Haurowitz and Horowitz, 1955; Virtanen *et al.*, 1949; Wasteneys and Borsook, 1930).
Hwever, since the mechanisms were established for peptide chain
elongation *in vivo*, plastein studies have been directed toward
subjects of enzymological interest (Determann and Koehler, 1965).
    Although the mechanism of the plastein reaction is compli-
cated and not completely defined, data accumulated during the
past three decades indicate that both condensation and transpeptidation reactions are involved (Horowitz and Haurowitz, 1959;
Determann, 1965; Yamashita *et al.*, 1970a, 1974; Tanimoto *et al.*,
1972). These reactions change a substrate (protein hydrolyzate
or oligopeptide mixture) into a substance of greater molecular

weight called plastein. This substance is generally characterized by its low solubility in water and sometimes by its clotting nature.

Interestingly, pH optima for plastein formation by most proteases lies in the narrow range of pH 4-7 (Yamashita *et al.*, 1971a). Pepsin, for example, is an acid protease that promotes protein hydrolysis at low pH. Even at pH 1 this enzyme is able to degrade proteins. However, no appreciable amount of plastein is formed at such an acidic pH even though all other conditions are favorable. The optimum pH for pepsin-induced plastein formation is known to be about 4.5. In contrast, α-chymotrypsin, a typical alkaline protease, has maximum activity for protein degradation near pH 8, whereas this same enzyme promotes plastein formation most effectively at pH 5-6 (Yamashita *et al.*, 1970b).

Substrate concentration is another important factor influencing plastein formation. The highest possible concentration may be required to produce plastein in a very high yield. Tsai *et al.* (1972) have shown that peptic hydrolyzates of soy proteins should be used at concentrations of 20-40% for best results. In addition, there is a critical concentration. When this substrate is incubated at a concentration between 7 and 10% with an adequate amount of α-chymotrypsin, no reaction occurred as measured by trichloroacetic acid solubility/insolubility (Tanimoto *et al.*, 1975). Lower substrate concentrations favored the usual degradation reaction rather than plastein synthesis.

Enzymologically, it is probable that during the plastein reaction a peptidyl-enzyme intermediate is formed, with liberation of water in the condensation reaction and liberation of a smaller peptide fragment in the transpeptidation reaction. In both cases the peptidyl-enzyme is then attacked by another peptide acting as a nucleophile amine, with formation of a new larger peptide as an aminolysis product. If water rather than the nucleophile peptide attacks the peptidyl-enzyme, then a normal hydrolysis product is formed. Whether usual hydrolysis or unusual resynthesis

predominates depends, among other things, on the nucleophile concentration.

While the plastein reaction has practical use in obtaining
purified products from crude protein materials, this reaction is
even more useful for covalently incorporating some desired amino
acids (Fig. 1). Our group has successfully incorporated
L-methionine into soy protein hydrolyzate by means of the plastein
reaction with papain (Yamashita *et al.*, 1971b). A 10:1 mixture of

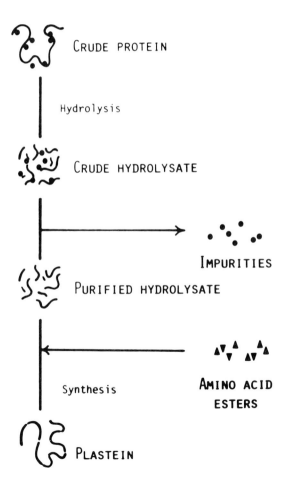

Fig. 1. A combined process of enzymatic protein hydrolysis
and resynthesis for producing a plastein with an improved quality.

this hydrolyzate and L-methionine ethyl ester hydrochloride was
incubated in the presence of papain under conditions favoring
plastein formation. The methionine content of the resulting plastein was found to be 7.22%, nearly seven times the original
methionine content of the soy protein. End group analyses demonstrated that 84.9% (molar basis) of the C-terminals of the plastein molecules were occupied with methionine, whereas only 14.4%
of the N-terminals contained methionine (Yamashita *et al.*, 1972).
This result indicates that the methionine ethyl ester used in the
plastein reaction acted as a nucleophile, probably attacking a
peptidyl-papain intermediate and becoming incorporated in the C-
terminal position of the deacylation product.

An important use of the plastein reaction would be to incorporate several essential as well as nonessential amino acids in
various proteins. Examples so far offered are summarized in Table
I. Any other L-amino acid could be likewise incorporated to a
similar extent. Aso *et al.* (1977) carried out a plastein experiment using an ovalbumin hydrolysate with a molecular weight
ranging from 500 to 4500. A 30% solution of this peptide mixture
in an aqueous medium, pH 6.0, was formulated with each of a
series of amino acid esters, each present at 20 mM. Each sample
was subjected to the plastein reaction using papain in the presence of 10 mM L-cysteine at 37°C. During the reaction, the
L-amino acid esters were incorporated with different initial velocities depending on their amino acid side chain structures. The
velocities tended to increase with increasing hydrophobicities
of the side chains, except for β-branched-chain amino acids (Fig.
2). However, even moderately hydrophilic and β-branched-chain
amino acids could be effectively incorporated folowing esterifica-
tion with longer chain alcohols such as n-butanol and n-hexanol.

TABLE I. Improvement of Food Proteins by Enzymatic Protein Hydrolysis and Resynthesis with Incorporation of Amino Acids

Food protein	Enzyme for hydrolysis	Enzyme for resynthesis	Incorporated amino acids	Improved property	Reference
Soy protein	Pepsin	Papain	Methionine	Protein efficiency ratio	d,e
Soy protein	Pepsin	Papain	Glutamic acid	Water solubility	f
Zein	Pepsin	Papain	Lysine Threonine Tryptophan	Essential amino acid pattern	g
Fish protein	Pepsin Pronase	Papain	Tyrosine Tryptophan	Suitability for dietetic use	h
Algal protein[a]	Pepsin	Papain	Lysine Methionine Tryptophan	Essential amino acid pattern	i,j
Bacterial protein[b]	Pepsin	Papain	Lysine Methionine Tryptophan	Essential amino acid pattern	i,j
Leaf protein[c]	Pepsin	Papain	Lysine Methionine Tryptophan	Essential amino acid pattern	j

[a]From Spirulina maxima.
[b]From Rhodopseudomonas capsulatus.
[c]From Triforium repens L.
[d]Yamashita et al. (1971a).
[e]Arai et al. (1974).
[f]Yamashita et al. (1975).
[g]Aso et al. (1974).
[h]Yamashita et al. (1976).
[i]Arai et al. (1976).
[j]Fujimaki et al. (1977).

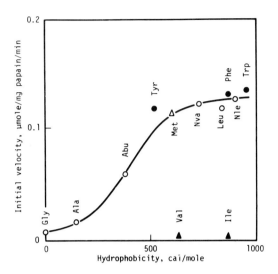

Fig. 2. Relationship between hydrophobicities of amino acid side chains and initial velocities of incorporation of amino acid ethyl esters (Aso et al., 1977). Abu, α-aminobutyric acid; Nva, norvaline; Nle, norleucine.

IV. AMINOLYSIS VERSUS HYDROLYSIS

Incorporation of amino acids via the plastein reaction can be predicted by using a simple model process. In most cases our group has been using ethyl hippurate, i.e., *N*-benzoylglycine ethyl ester, as a simple substrate instead of a protein hydrolysate (Aso *et al.*, 1978). Analysis of the products formed from this reaction indicated that the following reaction is probable:

Bz-Gly-Papain + AA-OR $\longrightarrow$ Bz-Gly-AA-OR + ethanol + papain

(3)

where AA-OR refers to the amino acid ester. Consequently, it is concluded that the amino acid ester acts as a nucleophile, as in the plastein reaction, thereby reacting with the acyl-enzyme (Bz-Gly-papain) to form the aminolysis product Bz-Gly-AA-OR. A clear specificity existed in this case; a more hydrophobic amino acid side chain was more effective except in the case of the β-branched-

chain amino acid L-valine. It was found, however, that even a
sterically unfavorable amino acid such as L-valine would become
reactive when esterified with the hydrophobic alcohol, n-butanol.
Similarly, n-butyl esters of alanine, α-amino acid, and phenylala-
nine were more reactive than their ethyl esters. It should be
noted that the reactivity of these amino acid esters as nucleophiles is well related to their affinity for papain, and can be
explained by the magnitude of K_N, the Michaelis constant in the
nucleophilic reaction (Fig. 3). There was excellent correlation
between K_N values (abscissa) and the initial velocities of incorporation into an ovalbumin hydrolyzate (ordinate).

According to Brubacher and Bender (1966), it is probable that
the acyl-enzyme Bz-Gly-Papain, present in a system containing an
added nucleophile, can undergo first-order deacylation by either
water (W) or the nucleophile (N). The observed deacylation rate

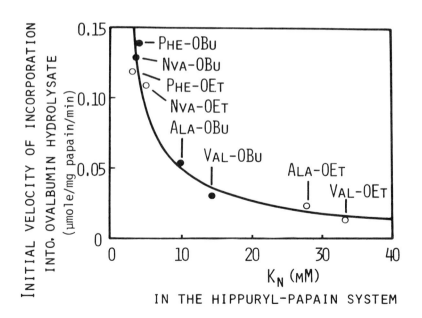

Fig. 3. Relationship between K_N values of amino acid esters
in the model system and their initial velocities of incorporation
into ovalbumin by papain (Aso et al., 1978). Et, ethyl: Bu,
butyl.

constant is given by

$$k_{obs} = k_3[W] + k_4[N] \tag{4}$$

Accordingly, k_{obs} should be enhanced by increasing [N]. To confirm the effect of [N] we conducted an experiment using Bz-Gly-OEt as substrate and *L*-methionine *n*-butyl ester as nucleophile (N). A 10% acetone solution in 0.1 *M* carbonate, pH 9, was prepared. The substrate, at a concentration of 50 m*M*, was dissolved in the buffer. To 1 ml of the solution were added 0.1 mg of papain (twice crystallized) and 1 mg of *L*-cysteine as its activator. After the mixture was incubated at 25°C for 5 min, a given amount of the nucleophile, already dissolved in a very small volume of the above buffer, was added and the incubation was continued for another 5 min. The incubation mixture was quantitatively analyzed for the hydrolysis product *N*-benzoylglycine and the aminolysis product *N*-benzoylglycyl-*L*-methionine *n*-butyl ester. The result showed that the hydrolysis reaction rate decreased significantly with increasing [N], whereas the aminolysis reaction increased (Table II), suggesting that the nucleophile competed with water for the acyl-papain.

Subsequently, we decreased the concentration of water by replacing part of it with glycerol. The glycerol was used at various concentrations in 0.1 *M* carbonate (pH 9). Each of the aqueous glycerol solutions containing *N*-benzoylglycine ethyl ester (50 m*M*), papain (0.1 mg/ml) and *L*-cysteine (1 mg/ml) was incubated at 25°C for 5 min. *L*-Methionine ethyl ester was added at a concentration of 5 m*M*, and the incubation was continued for another 5 min. Analysis of the products indicated that hydrolysis of the substrate was greatly affected by glycerol concentration. The observed rate of hydrolysis using a glycerol concentration of 50% was approximately one-fifth that observed in the absence of glycerol (Table III). The rate of the aminolysis reaction, on the other hand, was affected to only a moderate degree. By calculating relative velocities of aminolysis and hydrolysis, it is clear that a higher

TABLE II. *Papain-Catalyzed Reactions for Hydrolysis of N-Benzoylglycine Ethyl Ester Followed by Aminolysis by L-Methionine n-Butyl Ester: Compariosn of Their Initial Velocities as a Function of the Concentration of This Nucleophile Amine*[a]

| Nucleophile concentration (mM) | Initial velocity of formation (μmole/mg papain/min) | | A/H |
	Hydrolysis product (H)	Aminolysis product (A)	
0	5.62	–	
1	5.49	0.11	0.020
5	5.38	0.38	0.071
10	4.85	0.47	0.097
25	3.95	0.50	0.127
50	2.42	–	

[a]*Conditions: medium, 10% acetone in 0.1 M carbonate (pH 9); enzyme concentration, 0.1 mg/ml; substrate concentration, 50 mM; incubation temperature, 25°C; incubation time, 5 min.*

glycerol concentration favors incorporation of L-methionine.

In this context, it is interesting to note the work of Tanizawa and Bender (1974), because they have disclosed the effect of an aprotic solvent on acylation and deacylation. According to these authors, a higher solvent concentration in the reaction system has an important influence on k_3 without appreciably effecting k_2 at a solvent concentration up to 95% (Table IV. This result might suggest that the deacylation process is regulated by water concentration. It would be interesting to determine how such an extremely low water concentration favors aminolysis over hydrolysis when the system contains an added nucleophile amine.

V. PROTEASE-CATALYZED AMINOLYSIS OF PROTEINS

As discussed before, the plastein reaction becomes more valuable when applied as a tool for incorporating wanted amino acids. However, practical application may be economically limited,

TABLE III. Papain-Catalyzed Reactions for Hydrolysis of N-benzoylglycine Ethyl Ester Followed by Aminolysis by L-Methionine Ethyl Ester: Comparison of Their Initial Velocities as a Function of Glycerol Concentration[a]

Glycerol concentration (%)	Water concentration (%)	Initial velocity of formation (μmole/mg papain/min)		A/H
		Hydrolysis product (H)	Aminolysis product (A)	
0	90	5.26	0.34	0.065
10	81	5.55	0.27	0.049
20	72	4.39	0.25	0.057
30	63	1.90	0.22	0.116
40	54	1.61	0.21	0.130
50	45	1.02		

[a]Conditions: medium, a mixture of glycerol and 0.1 M carbonate (pH 9) containing 10% acetone; enzyme concentration, 0.1 mg/ml; substrate concentration, 50 mM; nucleophile concentration, 5 mM; incubation temperature, 25°C; incubation time, 5 min.

TABLE IV. The Effects of Dioxane Concentration on k_2 in the Porous Glass, α-Chymotrypsin-Catalyzed Hydrolysis of N-Benzoyl-L-Tyrosine p-Nitroanilide and on k_3 in That of N-Acetyl-L-Tryptophan p-Nitrophenyl Ester[a]

Water concentration (%)	k_2 (sec^{-1})	$k_3' \times 10$ (sec^{-1})
100	0.14	26.5
95	0.13	26.2
90	0.24	26.2
85	0.33	22.0
80	0.28	20.2
70	0.28	10.2
60	0.28	3.4
50	0.28	2.7
40	0.28	4.8
30	0.40	3.1
20	0.40	0.08
10	0.40	0.12
5	0.4	-
1	-	-

[a]Adapted from Tanizawa and Bender (1974).

since the entire process is composed of two different steps of enzymatic protein degradation and resynthesis (Fig. 1). The first step is generally carried out at a very low protein concentration and the second at a very high concentration of protein hydrolyzate in the presence of an added amino acid ester. If a proper set of reaction conditions is employed, it may be possible to unify both processes and thus incorporate amino acids in proteins in a single step. Though the product cannot be called plastein, the one-step process would be more economical and convenient than the conventional plastein reaction.

We have been successfully developing this new process for incorporating L-methionine directly into soy protein. Data accumulated to date indicate that amino acid incorporation is influenced, among other things, by substrate concentration and pH. The most suitable conditions are as follows: concentration of soy

protein (N 6.25 = 92.0%) in reaction medium, 20%; initial pH of
the medium, 9-10; ratio of papain (twice crystallized) to soy protein, 1:100 (w/w), concentration of L-cysteine, 1 mM; incubation
time at 37°C, 8-24 hr.

Figure 4 depicts the one-step process in which we have used
racemic DL-methionine ethyl ester for an economic reason. The pH
of the reaction system was adjusted using sodium carbonate and bi-
carbonate. After incubation under the stated conditions, the entire reaction mixture was treated with 80% ethanol to obtain a
refined product as a precipitate. Amino acid analysis showed that
the methionine content of the refined product was 8.21%, almost
seven times that of the original soy protein (Table V). It was
confirmed that the refined product carried no significant amount
of free amino acids. A gas chromatographic investigation of amino
acid isomers (Parr and Howard, 1971) indicated that no detectable
amount of D-methionine was present in the product. A degradation
experiment with leucine aminopeptidase (Hill and Smith, 1957)
showed that the rate of liberation of L-methionine from the N-

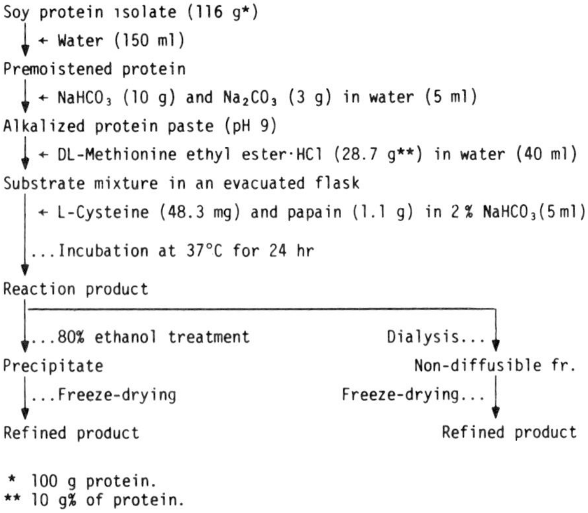

Soy protein isolate (116 g*)
$\downarrow$ ← Water (150 ml)
Premoistened protein
$\downarrow$ ← NaHCO$_3$ (10 g) and Na$_2$CO$_3$ (3 g) in water (5 ml)
Alkalized protein paste (pH 9)
$\downarrow$ ← DL-Methionine ethyl ester·HCl (28.7 g**) in water (40 ml)
Substrate mixture in an evacuated flask
$|$ ← L-Cysteine (48.3 mg) and papain (1.1 g) in 2 % NaHCO$_3$(5 ml)
$\downarrow$ ...Incubation at 37°C for 24 hr
Reaction product
$\downarrow$ ...80% ethanol treatment   Dialysis...$\downarrow$
Precipitate   Non-diffusible fr.
$\downarrow$ ...Freeze-drying   Freeze-drying...$\downarrow$
Refined product   Refined product

* 100 g protein.
** 10 g% of protein.

Fig. 4. One-step process for producting a refined product.

TABLE V. *Amino Acid Composition of the Original Soy Protein Concentrate and the L-Methionine-Incorporated Product Refined by the 80% Ethanol Treatment*

| Amino acid | Composition (wt.%) | |
	Soy protein concentrate	Methionine-incorporated product
Lysine	5.86	5.17
Histidine	2.65	3.12
Arginine	8.44	7.01
Aspartic acid	12.22	11.59
Threonine	4.30	3.21
Serine	5.27	4.47
Glutamic acid	20.00	17.77
Proline	6.20	4.66
Glycine	4.42	3.45
Alanine	4.41	3.73
Valine	5.37	4.47
Methionine	1.30	8.21
Isoleucine	5.21	5.28
Leucine	7.82	8.50
Tyrosine	4.09	3.56
Phenylalanine	5.51	5.43

terminals of the product was extremely slow compared to rates of liberation of other amino acids. On the other hand, when the product was degraded with carboxypeptidase Y (Hayashi *et al.*, 1973), L-methionine was liberated more rapidly. These results suggest that the L-methionine residues are located primarily at or near the C-terminals. Consequently, it is suspected that L-methionine ethyl ester acts as a nucleophilic amine in the one-step process. Probably in this way the soy protein can undergo aminolysis by L-methionine ethyl ester, with its incorporation in the C-terminals of the resulting peptide fragments.

As shown in Fig. 5, a clear specificity exists for hydrophobic amino acids, which is almost exactly similar to that observed for the amino acid incorporation during the plastein reaction by papain (see Fig. 2).

If an oversimplified visualization is allowed, the one-step process may be drawn as in Fig. 6. The protein undergoes amino-

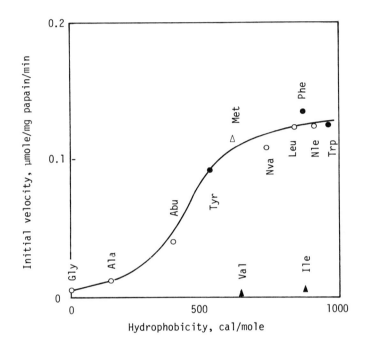

Fig. 5. *Relationship between hydrophobicities of amino acid side chains and initial velocities of incorporation of amino acid ethyl esters in the one-step process with papain. For the symbols see Fig. 2.*

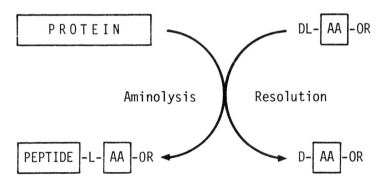

Fig. 6. *Incorporation of L-amino acid into protein by papain-catalyzed aminolysis with concurring resolution of the DL isomers.*

lysis to a smaller polypeptide fragment attaching just the L-amino acid at the C-terminal, and the D-amino acid remains behind.

As a matter of course, protein hydrolysis can take place at the same time, since the one-step process has to be carried out in the presence of a moderate concentration of water. Both the degree of protein degradation as measure by trichloroacetic acid solubility/insolubility and the rate of methionine incorporation were found to depend on the protein concentration in the medium. A lower protein concentration caused a higher degree of degradation as well as a lower rate of incorporation, and vice versa (Table VI). Interestingly enough, however, it was observed that even at a low protein concentration the rate of methionine incorporation was not affected so significantly as the rate of protein degradation when a part of the water in the system had been replaced by glycerol (Table VII). This again may indicate that a low water concentration favors aminolysis over hydrolysis.

TABLE VI. Effect of Protein Concentration on Proteolysis and Methionine Incorporation by Papain[a]

Protein concentration (%)	Approximate water concentration (%)	Degree of proteolysis (%)	L-Methionine incorporation (%)
1	99	60.0	50.5
20	78	46.1	80.4
30	67	37.7	81.0
40	56	26.4	54.6
50	45	19.4	52.3

[a]Conditions: medium, 1 M carbonate (pH 9); protein (soy globulin) concentration, described above; enzyme (papain)-protein ratio, 1:100 (wt/wt); L-methionine ethyl ester hydrochloride-protein ratio, 1:10 (wt/wt); incubation time, 24 hr; incubation temperature, 37°C.

TABLE VII. *Effect of Glycerol Concentration on Protein Degradation and L-Methionine Incorporation by Papain*[a]

Glycerol concentration (%)	Initial velocity of protein degradation (mg protein/mg papain/min)	Initial velocity of methionine incorporation (μmole/mg papain/min)
0	0.102	0.0098
25	0.073	0.0074
50	0.002	0.0073

[a]*Conditions: medium, a mixture of glycerol and 0.1 M carbonate (pH 9); protein (soy globulin) concentration, 1%; enzyme concentration, 0.1 mg/ml; concentration of L-methionine ethyl ester, 5 mM; incubation temperature, 25°C; incubation time, 1-3 hr.*

VI. CONCLUSION

Besides the protein hydrolysis reaction, several types of unusual reactions are catalyzed by proteases. These include the plastein reaction, which is characterized by its requirement for an extremely high substrate (protein) concentration. In biological systems, such as the human small intestine, a very high level of partially digested proteins can exist for several hours after eating. A similar environment can occur in some fermentation foods when a_w is very low and endogeneous proteases remain active. Detailed studies on proteases acting in such biological and food systems will provide information never before seen in the field of nutrition and food sciences.

In particular, the protease-catalyzed aminolysis reaction can have practical use in synthesizing useful peptides as well as in altering the amino acid composition of food proteins. It would be interesting to determine in greater detail how the aminolysis reaction rate varies with water concentration. The investigation would provide information on how to employ proper environmental conditions for this kind of unusual reaction. Furthermore, it will eventually be feasible to treat food proteins with immobilized proteases which are stable in an organic solvent system having an extremely low a_w, and this should facilitate the reaction. Such an unconventional treatment might disclose some unused specificity of proteases.

References

Arai, S., Aso, K., Yamashita, M., and Fujimaki, M. (1974). *Cereal Chem. 51,* 143.
Arai, S., Yamashita, M., and Fujimaki, M. (1976). *J. Nutr. Sci. Vitaminol. 22,* 447.
Aso, K., Yamashita, M., Arai, S., and Fujimaki, M. (1974). *Agri. Biol. Chem. 38,* 679.
Aso, K., Yamashita, M., Arai, S., Suzuki, J., and Fujimaki, M. (1977). *J. Agri. Food Chem. 25,* 1138.
Aso, K., Tanimoto, S., Yamashita, M., Arai, S., and Fujimaki, M. (1978). In preparation.
Baldwin, E. (1952). "Dynamic Aspects of Biochemistry," 2nd ed., pp. 81-94. Cambridge Univ. Press, England.
Bender, M. L., and Kemph, K. (1957). *J. Am. Chem. Soc. 79,* 116.
Brubacher, L. J., and Bender, M. L. (1966). *J. Am. Chem. Soc. 88,* 5871.
Determann, H. (1965). *Justus Liebig's Ann. Chem. 690,* 182.
Determann, H., and Koehler, R. (1965). *Justus Liebig's Ann. Chem. 690,* 197.
Durell, J., and Fruton, J. S. (1954). *J. Biol. Chem. 207,* 487.
Fujimaki, M., Arai, S., and Yamashita, M. (1977). "Food Proteins: Improvement through Chemical and Enzymatic Modification" (Advan. Chem. Ser. 160) (R. E. Feeney and J. R. Whitaker, eds.). Am. Chem. Soc., Washington, D.C.
Glazer, A. N. (1966). *J. Biol. Chem. 241,* 3811.
Glazer, A. N. and Smith, E. L. (1971). In "The Enzymes," Vol. III (P. D. Boyer, ed.), pp. 501/546. Academic Press, New York.
Hata, T., and Doi, E. (1971). *Proc. Intl. Symp. Conversion Manufacture of Foodstuffs by Microorganisms, pp.* 13-16. Saikon Publ. Co., Japan.
Haurowitz, F., and Horowitz, J. (1955). *J. Am. Chem. Soc. 77,* 3138.
Hayashi, R., Moore, S., and Stein, W. H. (1973). *J. Biol. Chem. 248,* 2296.
Hill, R. L., and Smith, E. L. (1957). *J. Biol. Chem. 228,* 557.
Horowitz, J., and Haurowitz, F. (1959). *Biochim. Biophys. Acta 33,* 231.
Isowa, Y. (1978). *Yuki Gosei Kagaku 36,* 195.
Isowa, Y., Ohmori, M., Ichikawa, T., Kurita, H., Sato, M., and Mori, K. (1977). *Bull. Chem. Soc. Japan 50,* 2762.
Kezdy, E. J., Clement, G. E., and Bender, M. L. (1964). *J. Am. Chem. Soc. 86,* 3690.

Kortt, A. A., and Liu, T.-Y. (1973). *Biochemistry 12,* 328.

Lake, A. W., and Lowe, G. (1966). *Biochem. J. 101,* 402.

Morihara, K., and Oka, T. (1977). *Biochem. J. 163,* 531.

Mycek, M. L., and Fruton, J. S. (1957). *J. Biol. Chem. 226,* 165.

Parr, W., and Howard, P. (1971). *Chromatographia 4,* 162.

Reed, G. (1966). "Enzymes in Food Processing." Academic Press, New York.

Reed, G. (1975). "Enzymes in Food Processing," 2nd ed. Academic Press, New York.

Saltman, R., Vlach, D., and Luisi, P. L. (1977). *Biopolymers 16,* 631.

Sawjalow, W. W. (1907). *Z. Physiol. Chem. 54,* 119.

Sluyterman, L. A. Æ., and Wijdenes, J. (1972). *Biochim. Biophys. Acta 289,* 194.

Smellie, R. M. S. (1970). "Chemical Reactivity and Biological Role of Functional Groups in Enzymes." Academic Press, New York.

Tanimoto, S., Yamashita, M., Arai, S., and Fujimaki, M. (1972). *Agri. Biol. Chem. 36,* 1595.

Tanimoto, S., Arai, S., Yamashita, M., and Fujimaki, M. (1975). *Abstracts of Papers,* p. 152, Annual Meeting of the Agricultural Chemical Society of Japan, Sapporo, Japan.

Tanizawa, K., and Bender, M. L. (1974). *J. Biol. Chem. 249,* 2130.

Tauber, H. (1952). *J. Am. Chem. Soc. 74,* 847.

Tsai, S.-J., Yamashita, M., Arai, S., and Fujimaki, M. (1972). *Agri. Biol. Chem. 36,* 1045.

Virtanen, A. I., Kerkkonen, H. K., Laaksonen, T., and Hakala, M. (1949). *Acta Chem. Scand. 3,* 520.

Wasteneys, H., and Borsook, H. (1930). *Physiol. Rev. 10,* 111.

Werber, M. M., and Greenzaid, P. (1973). *Biochim. Biophys. Acta 293,* 208.

Whitaker, J. R. (1974). "Food Related Enzymes" (Adv. Chem. Ser. 136). Am. Chem. Soc., Washington, D.C.

Yamashita, M., Arai, S., Matsuyama, J., Gonda, M., Kato, H., and Fujimaki, M. (1970a). *Agri. Biol. Chem. 34,* 1484.

Yamashita, M., Arai, S., Matsuyama, J., Kato, H., and Fujimaki, M. (1970b). *Agri. Biol. Chem. 34,* 1492.

Yamashita, M., Arai, S., Tsai, S.-J., and Fujimaki, M. (1971a). *J. Agri. Food Chem. 19,* 1151.

Yamashita, M., Tsai, S.-J., Arai, S., Kato, H., and Fujimaki, M. (1971b). *Agri. Biol. Chem. 35,* 86.

Yamashita, M., Arai, S., Aso, K., and Fujimaki, M. (1972). *Agri. Biol. Chem. 36,* 1353.

Yamashita, M., Arai, S., Tanimoto, S., and Fujimaki, M. (1974). *Biochim. Biophys. Acta 358,* 105.

Yamashita, M., Arai, S., Kokubo, S., Aso, K., and Fujimaki, M. (1975). *J. Agri. Food Chem. 23,* 27.

Yamashita, M., Arai, S., and Fujimaki, M. (1976). *J. Food Sci. 41,* 1029.

AUTOXIDATION-INITIATED

REACTIONS IN FOODS

M. Karel

S. Yong

I. INTRODUCTION

Lipid oxidation in foods is associated almost exclusively
with unsaturated fatty acids and is often autocatalytic with
oxidation products themselves catalyzing the reaction so that
the rate increases with time. The reaction may also be catalyzed
by metals or the enzyme lipoxygenase. Both the catalyzed and
autocatalytic reactions are initiated by decomposition of
hydroperoxides to free radicals. In pure systems monomolecular
decomposition occurs up to at least .1-2% oxidation (molar basis)
with the rate and time course proportional to the square root of
the extent of oxidation. During this period all oxygen absorbed
forms hydroperoxides. Above 1% oxidation bimolecular decomposition of the hydroperoxides becomes controlling, the rate being
proportional to the peroxide concentration.

The concept that bimolecular decomposition of hydroperoxides
is the predominant mode of initiation at relatively high hydroperoxide concentrations has been challenged by Hiatt and McCarrick
(1975), who presented evidence that 3-hydroperoxy-2,3-dimethyl-1-

511

butene decomposition in solvents involves addition of a peroxy
radical to a hydroperoxide and proceeds as follows:

$$RO_2^{\cdot} + RO_2H \longrightarrow RO_2R + \cdot OH \qquad (1)$$

$$RO_2R \longrightarrow 2RO\cdot \qquad (2)$$

This reaction has a concentration dependence similar to the
bimolecular decomposition, but the mechanism is different. It
is not known whether this mechanism is relevant in peroxidation
of food lipids.

A major factor controlling lipid oxidation in dehydrated
systems is water. Several hypotheses have been advanced to
explain the retarding effect water exerts on lipid oxidation.
In purified systems, water interferes with the normal bimolecular
decomposition of hydroperoxides by hydrogen-bonding with amphi-
polar hydroperoxides formed at the lipid-water interface (Karel
et al., 1967). In the presence of trace metals added to model
systems of lyophilized emulsions, humidification at moderate
levels retards oxidation because of hydration of metal ions
(Labuza et al., 1966; Karel et al., 1967). The reduction in
rate depends on the type and hydration state of the metal salt
added, as well as water content. The monomolecular rate period
is primarily affected by this mechanism, although bimolecular
rates are also decreased. As water contents are increased above
monolayer coverage in foods and in model systems, resistance to
diffusion of solutes decreases and solubilization becomes significant. In systems containing chelating agents and antioxidants, high water contents allow solubilization of chelating
agents, which sequester metals, as well as solubilization of
antioxidants. This effect lowers the rate of oxidation. However, at high water activities (a_w = 0.5-0.6) water may

accelerate oxidation by solubilizing catalysts or by inducing swelling of macromolecules, such as proteins, to expose additional catalytic sites (Heidelbaugh and Karel, 1970; Chou *et al.*, 1973). At still higher water activities (a_w = 0.75-0.85) dilution of catalysts may again retard the oxidation rate.

The presence of free radicals in lipid peroxidation is well established on the basis of reaction kinetics and the type of products formed, but the detection of free radicals in lipids is difficult because of their extremely short half-life. Schaich (1974) failed to detect free radicals in either peroxidized methyl linoleate or in the same lipid dispersed on cellulose, apparently because of high mobility and consequent short half-life and low steady-state concentration of the free radicals in the lipid phase. Attempts to detect free radicals in peroxidizing lipids at room temperature were unsuccessful in other studies reported in the literature. No ESR signals were exhibited by methyl linoleate exposed to γ-radiation or oxidized in bulk or on a cellulose support. Since it is known that lipid radicals are present under each of these conditions, the implications of these results are that either lipid radical concentrations are not high enough for detection or that linewidths of the signals are so broad that detection is not possible. It is known that short-lived as well as persistent radicals exist in both low moisture and aqueous food systems. The presence of water results in quenching of the radicals that are more stable in low-moisture systems. Decay of radicals upon addition of water was observed by Rockland (1969) in uv-irradiated gelatin, by Evans and Windle (1972) in various wheat products, by Guex (1975) in irradiated freeze-dried materials, and by many other authors. In our studies on free radicals produced in lysozyme exposed to peroxides of methyl linoleate, we observed that ESR signal intensity decreased with increasing water activity in the range 0-0.7.

Water may be expected to influence free-radical interactions between proteins and oxidizing lipid by influencing the concentrations of initiating radicals present, the degree of contact and mobility of reactants, and the relative importance of radical transfer vs. recombination reactions.

II.   WATER AND FOOD STRUCTURE

We have seen that water has several effects that may affect the course of autoxidation in foods. A little-explored aspect is the effect of water on structure of foods and on the ability of oxygen to reach oxidation-susceptible food components. This aspect has been investigated thoroughly in recent years by Professor Flink and his co-workers at MIT (To, 1978; Gejl-Hansen, 1977). They studied the influence of physical structure of freeze-dried emulsified systems on the oxidation behavior of the lipid component. The partitioning of lipid into encapsulated and surface fractions was evaluated microscopically and chemically in emulsion systems of varying composition and concentrations prepared under selected processing conditions and drying behavior. In these studies the influence of structure formation and lipid partitioning on subsequent oxidation behavior was demonstrated. The influence of environmental stresses on structural changes was evaluated, as well as the influence of loss of structure on lipid oxidation. Methods were developed for microscopic characterization of structure of freeze-dried emulsions and for evaluation of the properties of the dried emulsions. Emulsions were prepared by homogenizing a lipid, water, nonvolatile solute (matrix former), and an emulsifier. Solutes included maltodextrins, maltose, microcrystalline cellulose, and gelatin, while lipids included linoleic acid and triolein. Optical and electron-microscopic techniques (scanning

electron microscope and electron microprobe) were used to determine the location of the lipid phase in the freeze-dried material.
Methods were developed for quantitative determination of lipid
on the matrix surface and of encapsulated lipid *within* the matrix
without alteration of the matrix structure. Methods were also
developed to determine if the encapsulated lipid is accessible to
gases and vapors, including osmic acid, organic vapors, and
oxygen.

The distribution of the lipid in freeze-dried emulsions was
found to depend on the characteristics of the matrix-forming
solute. Proteins are very effective in encapsulating emulsion
droplets. The soluble carbohydrates examined (maltose and
matodextrin) also gave significant encapsulation, but in all
cases there was some nonencapsulated surface oil present. Insoluble carbohydrates (ungelatinized starch granules and microcrystalline cellulose) gave dried powders with all the lipid
present in the surface. Distribution of the surface lipid
differed depending on which solute was being used as the
matrix-forming element. Evaluations based on optical and
electron microscopic observations indicate that the lipid on
the maltodextrin surface is not present as a uniform film but
rather as pools located at surface irregularities such as
depressions, ridges, or grooves. A similar distribution was
found for maltose-based emulsions, though the sensitivity of
the maltose to "collapse" during freeze-drying results in a
complex behavior.

Foods prepared by rapid dehydration or by dehydration from
the frozen state (freeze-drying) often have a structure that
is metastable. In the absence of the plasticizing effect of
added moisture or of very high temperatures, this structure
persists. Addition of water and/or increase of temperature,
however, cause the phenomenon of collapse. Collapse may be
defined as the appearance of adequate mobility of the system

components to produce rearrangements of internal structure,
which may include crystallization. When collapse is avoided,
the surface lipid in maltose-based systems is distributed in a
manner similar to that for maltodextrin. When collapse occurs
the encapsulated lipid flows from the matrix interior to the
surfaces and is deposited in locations adjacent to points of
disruption of the matrix. Matrix thinning because of flow of
the solute that accompanies collapse also results in encapsulated droplets becoming deposited as surface fat.

Insoluble carbohydrates were observed to have only surface
lipid and no encapsulation. This lipid is more uniformly distributed than with the soluble carbohydrates but still did not
give completely uniform layers on all matrix surfaces even at
phase volumes of 20%. For example, with Avicel it was noted that
although there was some lipid present on each microcrystal,
this lipid was sometimes concentrated at one end of the microcrystal while the other end would be free of lipid. Behavior
of starch granules was somewhat different with the distribution
varying from one granule *cluster* to the next. Freezing rate
and oil/solute ratio influenced the degree of encapuslation by
maltose and maltodextrin especially at low phase volumes.

The "availability" of surface and encapsulated lipids to
vapors under nonoxidative conditions was studied by measuring
the sorption of vapors of osmic acid and of *n*-propanol. Freeze-
dried emulsions, with or without surface oil, were exposed to
saturated *n*-propanol vapors and the quantitative uptake measured
gravimetrically; or the emulsions were exposed to osmic acid
vapor, the uptake of which was measured qualitatively by degree
of darkening. Microscopic views showed that osmic acid vapors
strongly stain the surface lipid. Encapsulated lipid is unavailable as a site for sorption of either osmic acid or propanol.

When exposed to air, surface lipid is readily oxidized,
whereas encapsulated lipid is well protected. Breakdown of

TABLE I. Oxygen Absorbed after 90 hr by Hexane-Washed Freeze-
Dried Emulsions Exposed to Different Relative Humidities[a]

	Oxygen absorbed (µl/g)	
R.H. (%)	Linoleic acid (17.4%) entrapped in maltodextrin	Linoleic acid (20%) entrapped in maltose
0	0	0
43	0	190
75	0	200
93	100	280

[a]After Gejl-Hansen (1977).

the protective matrix by addition of water, however, makes the
encapsulated lipid labile to oxidation (Table I). These studies
show that comparisons of oxidation behavior for dehydrated systems of differing composition and/or concentration require consideration of the influence of physical structure, especially
the extent of lipid encapsulation. It was demonstrated that
surface lipid on both amorphous and crystalline matrices was not
uniformly distributed but rather present as "puddles" associated
with distinct morphological features of the matrix surface. It
is important that accelerated storage studies in which moisture
content and/or temperature are increased be carefully monitored
to ensure that structure transformations resulting in release of
encapsulated lipid do not occur.

III. PEROXIDIZING LIPIDS

     Reactions between peroxidizing aspect of oxidation in low-
moisture systems. As shown in Fig. 1, peroxidation of lipids
may lead to reactions with proteins and (1) lipid hydroperoxides,
(2) free radicals, (3) peroxide breakdown products such as
aldehydes, and (4) oxygenated polymers. We need to consider,

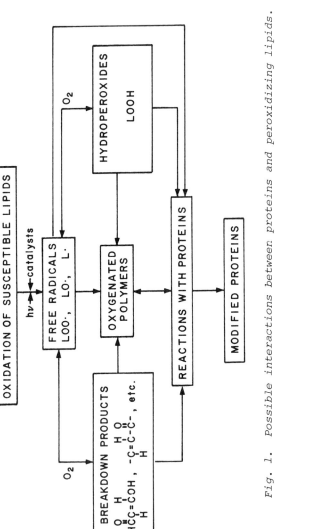

Fig. 1. Possible interactions between proteins and peroxidizing lipids.

therefore, effects of water on formation and stability of lipid oxidation intermediates, products, and polymers; the mobilization of reactive species and their substrates by water; and the effects of water on the secondary reactions in proteins.

In recent years increasing attention has been focused on hydroperoxides and transient free radicals from peroxidizing lipids as a major protein-damaging species. There are similarities between ionizing radiation-induced protein damage and that resulting from reactions of proteins with peroxidizing lipids. In radiation-induced protein reactions, high water activities and the presence of sulfur in the protein tend to promote cross-linking, whereas proteins with no sulfur (e.g., collagen, gelatin) tend to undergo peptids chain scission in the dry state (Friedberg, 1969; Bailey *et al.*, 1964).

We studied the reactions between peroxidizing methyl linoleate and gelatine or lysozyme in low-moisture model systems and observed a similar pattern of reactions (Zirlin and Karel, 1969; Kanner and Karel, 1976). In the gelatin-methyl linoleate system incubation in the dry state led to scission manifested by decrease in the viscosity of gelatin solutions, increase in the solubility of gelatin in ethanol-rich solvent mixtures, and increase in free-amide groups. With increasing water content of the gelatin-methyl linoleate samples, less peptide chain scission took place, and there was loss of solubility. When lysozyme was reacted with peroxidizing methyl linoleate in a similar low-moisture model system, free-radical concentrations in lysozyme (as determined by electron spin resonance) decreased whereas the degree of cross-linking (Table II), protein insolubilization, and loss of enzyme activity increased with increasing water activity.

Tappel's group in California also found similarity between effects of peroxidizing lipids and those of ionizing radiations on proteins. They reported that in aqueous solution, enzymes and other proteins (e.g., cytochrome *c*, hemoglobin, catalse, α-chymo-

TABLE II. Cross-Linking of Lysozyme after Incubation with Methyl Linoleate at 37°C for Seven Days[a]

Water activity	Formation of polymers as determined by SDS polyacrylamide gel electrophoresis
0	Monomer
0.11	Monomer, dimer
0.43	Monomer, dimer, trimer, etc.
0.75	Monomer, dimer, trimer, etc.

[a]Kramer and Karel (1976).

trypsin, ovalbumin, ribonuclease) undergo major reactions of polymerization, polypeptide chain scission, and chemical changes in individual amino acids when subjected to lipid peroxidation (Roubal and Tappel, 1966a, b). Lipid-centered radicals were estimated to be 1/10 as damaging as free radicals formed by ionizing radiations (Chio and Tappel, 1969). It is interesting to note that protein changes, which are caused by either peroxidizing lipids or ionizing radiations, are much less pronounced in systems with low water activity and in frozen systems, which in a sense are also "dehydrated" because the bulk of water is immobilized in ice crystals (I. A. Tabu, J. W. Halliday, and M. D. Sevilla, personal communication, 1978; S. H. Yong, and M. Karel, unpublished data, 1978; Yong, 1978; Kanner and Karel, 1976).

Amino acid residues that are readily destroyed by ionzing radiations or peroxidizing lipids include methionine, histidine, cysteine, tryptophan, and lysine (Roubal and Tappel, 1966b; Ambe et al., 1961). These amino acids are also among the most labile when proteins, peptides, or amino acid derivatives are photooxidized in aqueous solutions (Foote, 1976). In many cases, similar amino acid degradation products are observed whether the degradation is effected by peroxidizing lipids, ionizing radiations, or photooxidation (Table III).

When we recently studied the reactions of peroxidizing methyl linoleate with histidine, tryptophan, and their derivatives in

TABLE III. Comparison of the Products Formed upon the Exposure of Labile Protein Functional Groups to Peroxidizing Lipids, Ionizing Radiations, and Photooxidation

Labile side chains	Peroxidizing lipids	Ionizing radiations (aerobic)	Photooxidation
Histidyl	Asparagine, aspartic acid, ammonia (Yong and Karel, 1978a; Yong, 1978)	Asparagine, aspartic acid, ammonia. glutamic acid (Liebster and Kopoldova, 1964)	Asparagine, aspartic acid, urea, ammonia (Tomita et al., 1969; Johns and Jaskewycz, 1965)
Tryptophan	Kynurenine, other products (Yong and Karel, unpublished book 1978)	Kynurenine, N-formyl-kynurenine, 3-hydroxy-kynurenine, 3-hydroxy-anthranilic acid (Peter and Rajewsky, 1963; Jayson et al., 1954)	Kynurenine, N-formyl-kynurenine, 3-hydroxy-kynurenine (Benassi et al., 1967; Nakagawa et al., 1977)
Cysteinyl	Cystine, cysteic acid, alanine, H_2S, cysteine-lipid hydroperoxide adducts, cystine disulfoxide (Roubal and Tappel, 1966b; Lewis and Wills, 1962; Gardner et al., 1977)	Cystine, cysteic acid, alanine, H_2S, cysteine sulfinic acid, sulfur, sulfate ion, cystine disulfoxide (Markakis and Tappel, 1960; Grant et al., 1961; Rotherham et al., 1952)	Cystine, cysteic acid, H_2S (Gomyo and Fujimaki, 1970; Schocken, 1952)
Methionyl	Methionine sulfoxide (Njaa et al., 1968; Karel and Tannenbaum 1966)	Methionine sulfoxide, methionine sulfone, homo-cysteine, homocysteic acid, methyl mercaptan, amino-butyric acid (Kopoldova et al., 1958; Kumta et al., 1957)	Methionine sulfoxide, methionine sulfone, homo-cysteic acid (Sysak et al., 1977; Ray and Koshland, 1962)

low-moisture model systems, we observed much greater degradation of the amino acids and their derivatives at water activity of 0.75 and either 37° or 51°C as compared to the dry state (Yong and Karel, 1978a and unpublished data, 1978; Yong, 1978).

We have also been able to detect protein-centered free radicals formed during the reaction with peroxidizing lipids. Using electron spin resonance (ESR) we have succeeded in the detection and characterization of protein-centered free radicals in lysozyme and other proteins reacted in low-moisture model systems (Schaich and Karel, 1975, 1976; Karel et al., 1975). Based on the g value and signal geometry, the observed signals that persisted after solvent extraction of lipids are tentatively assigned to α-carbon radicals in the polypeptide backbone of the protein. The concentration of free radicals ranged from 10^{15} to 10^{17} per milligram of protein or approximately 0.02 to 2 per molecule of lysozyme. Concentrations were higher in samples incubated at water activities of 0 and 0.07 as compared to those incubated at higher water activities (0.40 and 0.75). Poly-acrylamide gel electrophoresis of reacted lysozyme showed that extensive cross-linking took place only at higher water activities (Schaich and Karel, 1975; Kanner and Karel, 1976). ESR signals were also observed when lyophilized lysozyme or other proteins were exposed to ionizing radiation in air, to t-butylperoxy radicals, or to heating to 160°C. Taub et al. (private communication, 1978) have recently reported the detection of similar α-carbon free radicals located in the peptide backbone when frozen protein samples were exposed to ionizing radiations. Upon thawing these radicals disappeared rapidly. We can conclude that free radicals being transferred from peroxidizing lipids to proteins are, at least in part, responsible for changes occurring in proteins that are incubated with peroxidizing lipids and that water plays an important role in these free radical reactions.

Free-radical transfer from lipid to protein may occur via complex formation. The fate of the resulting protein free radicals will be determined to some extent by water activity of the lipid-protein systems. Lack of peptide chain mobility in the dry state because of constraints on the flexing and torsional motions prevents cross-linking of protein radicals and favors rearrangements leading to peptide chain scission. On the other hand, constraints on the mobility of protein radicals would decrease with increasing water activity with the result that cross-linking predominates over oxidative scission at higher water activities. Aqueous radicals were shown to produce dimerization or ribonuclease (Mee and Adelstein, 1974). El-Zeany et al. (1975) found that lipid-centered free radicals accelerate oxidative polymerization of egg albumin. Thus, water may "quench" protein-free radicals by promoting radical-radical recombinations leading to cross-linking. However, we also have some experimental evidence for another mechanism by which water may contribute to the disappearance of protein radicals, namely, by proton donation (Schaich and Karel, 1975; Schaich, 1974).

To account for the observation that the destruction of labile amino acids (as residues in proteins, amino acid derivatives, or free amino acids) by peroxidizing lipids is greatly enhanced at higher water activities (e.g., 0.75) as compared to that in the dry state, we can hypothesize that water increases the *effective reaction surface* between labile amino acids and lipid-originated free radicals by (1) increasing the mobility of lipid hydroperoxides and lipid-centered radicals to the lipid-protein or lipid-amino acid interface so that their chances of reacting with proteins or amino acids instead of other lipid molecules are enhanced and (2) making the labile amino acids more accessible to attack by lipid hydroperoxides or lipid-centered radicals by increased mobility of peptide chains and increased solubility of proteins or amino acids at the interface.

IV. PRESENT AND FUTURE STUDIES

We are currently studying (1) the nature of the reactions
leading to the production of protein-centered free radicals,
(2) the important lipid species (whether primary lipid hydroperoxides and lipid-centered free radicals or those from their
decomposition products) responsible for the radical transfer,
(3) effects of environmental factors, especially water activity
and the presence of anti- or prooxidants on the transfer of free
radicals. Free radical signals could only be from samples
containing histidine, lysine, tryptophan, cysteine, and arginine
when different amino acids were incubated with methyl linoleate
(Schaich and Karel, 1976). We would like to know whether these
amino acids, which are also among the most labile upon exposure
to peroxidizing lipids or ionizing radiations, serve as the
specific sites for the transfer of free radicals from peroxidizing
lipids to proteins via hydrogen abstraction, addition, or other
free radical reactions.

Of the reactions between proteins and lipid oxidation products,
those involving malonaldehyde have been the subject of most
studies. Figure 2 summarizes the reactions that are known to
occur between malonaldehyde and different amino acid residues
in proteins. Among these reactions, 1,4 addition of the nucleophilic ε-nitrogen of lysine to the enolic carbon atom of the
α,β-unsaturated carbonyl system of malonaldehyde is the most
important, and this reaction leads to rapid loss of lysyl residues,
protein cross-linkings, loss of enzyme activities, and possibly
the formation of lipofuscin age pigments. Malonaldehyde usually
results from the oxidation of polyunsaturated lipids with three
or more double bonds (e.g., linolenate and arachidonate), but it
has not been isolated from oxidized linoleic acid or its esters
(Cobb and Day, 1965a, b; Kwon *et al.*, 1966), making it unlikely
to cause polymerization and other damages in the previously discussed protein-linoleate model systems. At present practically

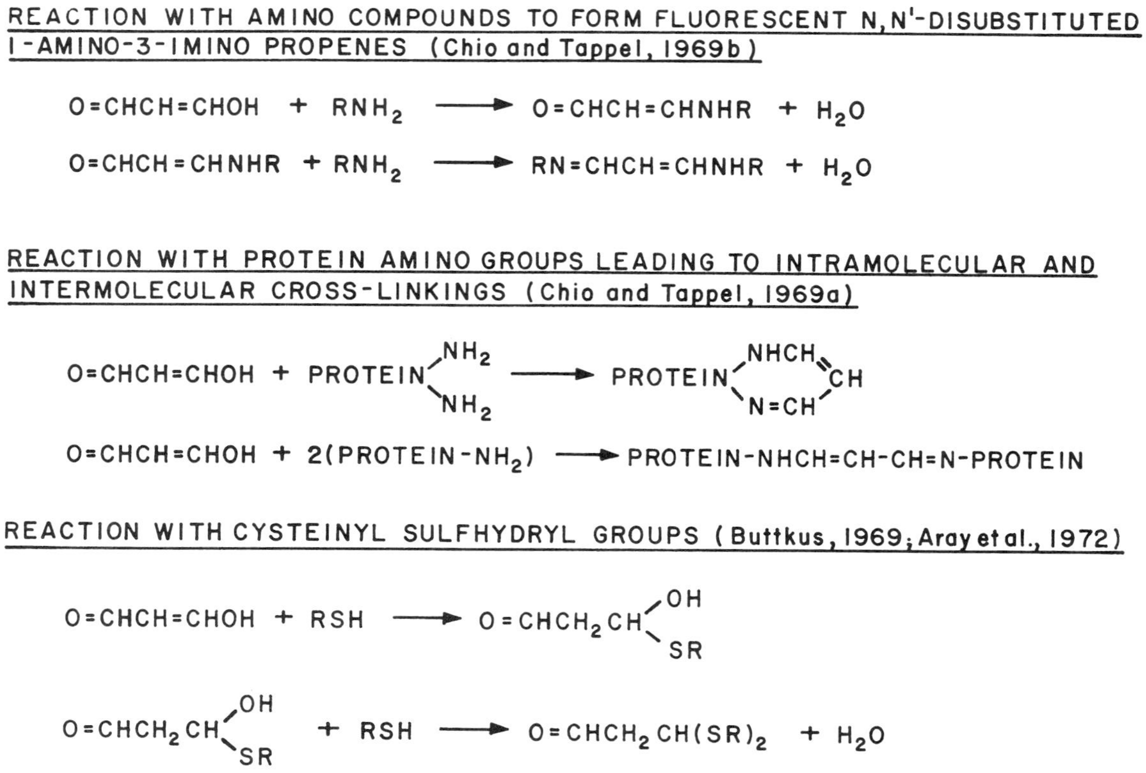

REACTION WITH AMINO COMPOUNDS TO FORM FLUORESCENT N,N'-DISUBSTITUTED
1-AMINO-3-IMINO PROPENES (Chio and Tappel, 1969b)

$$O=CHCH=CHOH + RNH_2 \longrightarrow O=CHCH=CHNHR + H_2O$$

$$O=CHCH=CHNHR + RNH_2 \longrightarrow RN=CHCH=CHNHR + H_2O$$

REACTION WITH PROTEIN AMINO GROUPS LEADING TO INTRAMOLECULAR AND
INTERMOLECULAR CROSS-LINKINGS (Chio and Tappel, 1969a)

$$O=CHCH=CHOH + PROTEIN \underset{NH_2}{\overset{NH_2}{\diagdown}} \longrightarrow PROTEIN \underset{N=CH}{\overset{NHCH}{\diagdown}} CH$$

$$O=CHCH=CHOH + 2(PROTEIN-NH_2) \longrightarrow PROTEIN-NHCH=CH-CH=N-PROTEIN$$

REACTION WITH CYSTEINYL SULFHYDRYL GROUPS (Buttkus, 1969; Aray et al., 1972)

$$O=CHCH=CHOH + RSH \longrightarrow O=CHCH_2CH\underset{SR}{\overset{OH}{\diagdown}}$$

$$O=CHCH_2CH\underset{SR}{\overset{OH}{\diagdown}} + RSH \longrightarrow O=CHCH_2CH(SR)_2 + H_2O$$

Fig. 2. Reactions between malonaldehyde and proteins or amino acids.

no data are available to assess the effects of water activity on the reactions involving malonaldehyde because all the experiments were performed in aqueous solutions. However, since all the reactions involving malonaldehyde are condensation-type reactions, effects of water activity are expected to be similar to those encountered in nonenzymatic browning, which also proceeds via condensation reactions.

Reactions of proteins with other lipid oxidation products have been studied sporadically. Cavins and Friedman (1967) reported that α,β-unsaturated carbonyls reacted strongly with free amino groups of lysine. Available experimental results also indicate that basic functional groups in proteins (e.g., amino, imidazole) catalyze the aldol condensation of carbonyls resulting from lipid oxidation into highly oxygenated brown pigments (Yong and Karel, 1978b; Pokorny et al., 1975; Montgomery and Day, 1965). These pigments contain very little nitrogen, and the rate of their formation increases with increasing water activity to a maximum point in the range of intermediate-moisture foods (e.g., 0.6–0.8). As humidity is further increased, the rate of browning decreases because of the "dilution" effects of water (Labuza et al., 1969). The rate of browning also increases with increasing pH of the system (Yong and Karel, 1978b; Venolia and Tappel, 1958).

Very little is known about the interactions of oxidized lipid *polymers* with proteins. Melanoidins, oxidized polymers that are obtained by the reaction of amino acids with carbonyls, have been observed to act as hydrophilic colloidal electrolytes, causing remarkable precipitation of proteins in aqueous solutions (Horikoshi and Gomyo, 1976) and retarding lipid peroxidation by chelating prooxidative metal ions (Kajimoto et al., 1975; Kawashima et al., 1977). Similar effects were observed with polymers that are formed during lipid peroxidation and have chemical properties resembling those of melanoidins (Pokorny et al., 1976; Montgomery and Day, 1965; Venolia and Tappel, 1958). Because of the lack of data concerning the effects of water activity on the interactions of oxidized lipid polymers with proteins, trace metals, or other components of foods, we can only conjecture that these interactions would be enhanced with increasing water activity because of the mobilization of these polymers as well as their potential substrates (e.g., proteins, trace metals) by water.

REFERENCES

Ambe, K. S., Kumata, U. S., and Tappel, A. L. (1961). *Radiat. Res. 15,* 709.
Bailey, A. J., Rhodes, D. N., and Cater, C. W. (1964). *Radiat. Res. 22,* 606.
Benassi, C. A., Scoffone, E., Galiazzo, G., and Iori, G. (1967). *Photochem. Photobiol. 6,* 857.
Cavins, J. F., and Friedman, M. (1967). *Biochemistry 6,* 3766.
Chio, K. S., and Tappel, A. L. (1969). *Biochemistry 8,* 2827.
Chou, H. E., Acott, K. M., and Labuza, T. P. (1973). *J. Food Sci. 38,* 316.
Cobb, W. Y., and Day, E. A. (1965a). *J. Am. Oil Chem. Soc. 42,* 1110.
Cobb, W. Y., and Day, E. A. (1965b). *J. Am. Oil Chem. Soc. 42,* 420.
El-Zeany, B. A., Pokorny, E., Smidkalova, E., and Davidek, J. (1975). *Nahrung 19,* 327.
Evans, J. J., and Windle, J. J. (1972). *Chem. Ind.,* 126.
Foote, C. S. (1976). *In* "Free Radicals in Biology" (W. A. Pryor, ed.), Vol. 2, p. 85. Academic Press, New York.
Friedberg, F. (1969). *Radiat. Res. 38,* 34.
Gardner, H. W., Kleiman, R., Weisleder, D., and Inglett, G. E., (1977). *Lipids 12,* 655.
Gejl-Hansen, F. (1977). Microstructure and stability of freeze-dried solute containing oil-in-water emulsions, Sc.D. Thesis, MIT Department of Nutrition and Food Science, Cambridge, Massachusetts.
Gomyo, T., and Fujimaki, J. (1970). *Agr. Biol. Chem. (Tokyo) 34,* 302.
Grant, D. W., Mason, S. N., and Link, M. A. (1961). *Nature 192,* 352.
Guex, M. M. (1975). *Proc. 6th Int. Course Freeze-drying Advanced Food Technol., Burgenstock-Lucerne, June 1973.* Academic Press, New York.
Heidelbaugh, N. D., and Karel, M. (1970). *J. Am. Oil Chem. Soc. 47,* 539.

Hiatt, R., and McCarrick, T. (1975). *J. Am. Chem. Soc. 97*, 5234.
Horikoshi, M., and Gomyo, T. (1976). *Agri. Biol. Chem. (Tokyo) 40*, 33.
Jayson, G. G., Scholes, G., and Weiss, J. (1954). *Biochem. J. 57*, 386.
Johns, R. B., and Jaskewycz, (1965). *Nature 206*, 1149.
Kajimoto, G., Yoshida, H., and Yamashoji, S. (1975). *Yukagaku, 24*, 15.
Kanner, J., and Karel, M. (1976). *J. Agri. Food Chem. 24*, 468.
Karel, M., and Tannenbaum, S. R. (1966). U. S. Army Natick Laboratory Contract No. DA-19-129-AMC-254-N.
Karel, M., Labuza, T. P., and Maloney, J. F. (1967). *Cryobiology 3*, 288.
Karel, M., Schaich, K., and Roy, R. B. (1975). *J. Agri. Food Chem. 23*, 159.
Kawashima, K., Itoh, H., and Chibata, I. (1977). *J. Agr. Food Chem. 25*, 202.
Kopoldava, J., Kolousek, J., Liebster, J., and Babicky, A. (1958). *Nature 182*, 1074.
Kumta, U. S., Gurnani, S., and Sahasrabudhe, M. B. (1957). *J. Sci. Ind. Res. (India) 16C*, 25.
Kwon, T. W., Mengel, D. B., and Olcott, H. S. (1966). *J. Food Sci. 31*, 552.
Labuza, T. P., Maloney, J. F., and Karel, M. (1966). *J. Food Sic. 31*, 885.
Labuza, T. P., Tannenbaum, S. R., and Karel, M. (1969). *Food Technol. 24(5)*, 35.
Lewis, S. E., and Wills, E. D. (1962). *Biochem. Pharmacol. 11*, 901.
Liebster, J., and Kopoldova, J. (1964). *In* "Advances in Radiation Biology" (L. G. Augenstein, R. Mason, and H. Quastler, eds.), Vol. 1, p. 157. Academic Press, New York.
Markakis, P., and Tappel, A. L. (1960). *J. Am. Chem. Soc. 82*, 1613.
Mee, L. K., and Adelstein, S. J. (1974). *Radiat. Res. 60*, 422.
Montgomery, M. W., and Day, E. A. (1965). *J. Food Sci. 30*, 828.
Nakagawa, M., Watanabe, H., Kodato, S., Okajima, H., Hino, T., Flippen, J. L., and Witkop, B. (1977). *Proc. Natl. Acad. Sci. (USA) 74*, 4730.
Njaa, L. R., Utne, F., and Braekkan, O. R. (1968). *Nature 218*, 571.
Peter, G., and Rajewsky, B. (1963). *Z. Naturforsch. 18b*, 110.
Pokorny, J., Kolakowska, A., El-Zeany, B. A., and Janicek, G. (1975). *Z. Lebensm. Unters.-Forsch. 157*, 323.
Pokorny, J., Tai, P. T., and Janicek, G. (1976). *Nahrung 20*, 149.
Ray, W. J., Jr., and Koshland, D. E., Jr. (1962). *J. Biol. Chem. 237*, 2493.
Rockland, L. B. (1969). *Food Technol. 23*, 1241.

Rotherham, M., Todd, N., and Whitcher, S. L. (1952). *Naturwissenschaften 39*, 450.

Roubal, W. T., and Tappel, A. L. (1966a). *Arch. Biochem. Biophys. 113*, 150.

Roubal, W. T., and Tappel, A. L. (1966b). *Arch. Biochem. Biophys. 113*, 5.

Schaich, K. (1974). Free radical formation in proteins exposed to peroxidizing lipids, Sc.D. Thesis, MIT, Cambridge, Massachusetts.

Schaich, K., and Karel, M. (1975). *J. Food Sci. 40*, 456.

Schaich, K., and Karel, M. (1976). *Lipids 11*, 393.

Schocken, K. (1952). *Science 116*, 544.

Sysak, P. K., Foote, C. S. and Chiang, T. Y. (1977). *Photochem. Photobiol. 26*, 19.

Tomita, M., Irie, M., and Ukita, T. (1969). *Biochemistry 8*, 5149.

To, E. C. H. (1978). Collapse, a structural transition in freeze-dried matrices, Sc.D. Thesis, MIT Department of Nutrition and Food Science, Cambridge, MA.

Venolia, A. W., and Tappel, A. L. (1958). *J. Am. Oil Chem. Soc. 35*, 135.

Yong, S. H. (1978). Reactions between peroxidizing methyl linoleate and histidine or histidyl residue analogues, Ph.D. Thesis, MIT, Cambridge, Massachusetts.

Yong, S. H., and Karel, M. (1978a). *J. Food Sci.* (in preparation).

Yong, S. H., and Karel, M. (1978b). *Lipids 13*, 1.

Yong, S. H., and Karel, M. (1978c). *J. Am. Oil Chem. Sco. 55*, 352.

Zirlin, A., and Karel, M. (1969). *J. Food Sci. 34*, 160.

INFLUENCE OF WATER ACTIVITY ON STABILITY
OF VITAMINS IN DEHYDRATED FOODS

James R. Kirk

I. INTRODUCTION

The desire to maintain the nutrient quality of foods during
processing and storage is an extension of our desire to preserve
foods. Changes in the quality of foods as a function of nutrient
destruction, development of off-flavors, color deterioration, and
changes in textural properties have become important in the determination of the shelf-life of foods.

The chemical, nutritional, and microbiological quality of dehydrated foods has been shown to be a function of storage temperature, light, oxygen, total moisture content, and the physiochemical state of water in the food system (Labuza, 1968, 1975a,b;
Lund, 1973; Singh *et al.*, 1976; Karel, 1975; Riemer and Karel,
1978; Kirk *et al.*, 1977; Dennison and Kirk, 1978).

The influence of total moisture content and water activity on
the stability of fat and water soluble vitamins is very complex
and appears to relate to the physiochemical state of the water in
the food system.

II. FAT-SOLUBLE VITAMINS

Little information is available concerning the destruction of the fat soluble vitamins over the wide range of moisture content, water activities, temperatures, oxygen partial pressures, and metal catalyst concentrations that exist in dehydrated food systems. From the limited information available, it appears that fat-soluble vitamins, particularly vitamins A and E, exhibit stability curves in low-moisture dehydrated systems characteristic of unsaturated fats (Labuza, 1975a,b). As shown by the data in Table I, the stability of retinyl acetate in a model food system described by Kirk *et al.* (1977) (Table II) decreased as a function of water activity and storage temperature. The deviate character of the rate constant reported for 45°C and 0.65 a_w is believed to be due to the analytical method for determining vitamin A in the study.

TABLE I. *Rate Constants and Half-Lives for Vitamin A Degradation as a Function of Water Activity and Storage Temperature in a Dehydrated Model Food System Stored in 303×404 Enameled Metal Cans*

Temp./a_w ($^\circ C$)	κ^a	σ^b	τ^c
30/0.24	−0.29	0.07	239
30/0.40	−0.65	0.02	106
30/0.65	−3.16	0.50	22
37/0.24	−0.70	0.05	100
37/0.40	−0.76	0.05	91
37/0.65	−4.59	0.82	15
45/0.24	−2.01	0.16	34
45/0.40	−5.94	1.09	12
45/0.65[d]	−2.32	0.65	30

[a]*First-order rate constant* × 10^{-2}, *days*$^{-1}$.
[b]*Standard deviation* × 10^{-2}.
[c]*Half-life, days.*
[d]*Correlation coefficient 0.9001.*

TABLE II. Composition of Model Food System

Component	%[d]
Protein[a]	10.2
Fat	1.0
Carbohydrate[b]	76.6
Reducing sugar[c]	5.1
Sucrose	5.1
Salt	2.0

[a]Soya protein, Promine E. Central Soya.
[b]Food grade powdered starch, A. E. Staley, Inc., and corn syrup solids 15 D.E., American Maize.
[c]Supplies by the corn syrup solids percent dry weight basis.
[d]Calculated on dry weight basis.

Temperature exhibited the greatest effect on the rate of retinyl acetate destruction. Calculation of the activation energy (E_a) from the Arrhenius equation yields values of 24.7, 28.6, and 10 kcal/mole for the a_w = 0.24, 0.40, and 0.65 samples, respectively. The E_a values for retinyl acetate degradation in the model food system at a_w = 0.24 and 0.40 were not significantly different. There was, however, a significant difference in E_a for the samples having a water activity of 0.65. This could represent a change in the degradation mechanism for retinyl acetate. The thermodynamic activation parameters were calculated to further investigate a possible change in degradation mechanism, using equations based on absolute reaction rate theory (Eyring, 1935):

$$k = \frac{k_b T}{h} \exp(\Delta S\ddagger) - \exp(\Delta H\ddagger/RT) \tag{1}$$

where k is the reaction rate constant at temperature T, k_b = 1.38×10^{-16} erg K^{-1} (Boltzmann constant), h = 6.63×10^{-27} erg sec (Planck's constant), R = 1.987 cal/°K-mole (gas constant), $\Delta H\ddagger = E_a - RT$ (for solutions). The free energy of activation was then calculated by

$$\Delta G\ddagger = \Delta H\ddagger - T \Delta S\ddagger \tag{2}$$

TABLE III. Activation Parameters for Retinyl Acetate Degradation in a Fortified Model System Stored in TDT Cans

a_w	E_a (kcal/mole)	$\Delta H\ddagger$ (kcal/mole)	$\Delta S\ddagger$ (eu)	$\Delta G\ddagger$ (kcal/mole)
0.24	25	24	-195	81
0.40	29	28	-168	78
0.65	10	9	- 80*	33*

As shown by the activation parameters in Table III, there is a compensation effect between $\Delta H\ddagger$ and $\Delta S\ddagger$ in the a_w = 0.24 and 0.40 samples similar to the effect observed for ascorbic acid degradation, which is discussed later. This compensation effect results in a constant $\Delta G\ddagger$ for the degradation of retinyl acetate in the two food systems that were not significantly different, indicating that retinyl acetate degrades by the same mechanism in samples with water activities of 0.24 and 0.40.

The calculation of activation parameters for the a_w = 0.65 model system indicates a change in the degradative mechanism for retinyl acetate associated with this water activity. This is only speculation, however, because the E_a value generated from the constant data for the 0.65 model system was based on a linear regression value because of the low rate constant associated with the 0.65 model system stored at 45°C.

The effect of metal catalyst, water activity, and temperature on the stability of retinyl acetate was studied in a model food system containing no antioxidants. The results are presented in Table IV. Kinetic data associated with the model system equilibrated to the lowest a_w indicate that all of the minerals tested had an effect on retinyl acetate stability. Iron exhibited the most significant effect on the stability of retinyl acetate. No significant difference was observed in the degradation half-lives of retinyl acetate when 10 or 25% of the US RDA of the mineral supplement was added to the model system. This was interpreted to occur because of the insolubility of the metal salts in the

TABLE IV. Rate Constants and Half-Lives for Vitamin A Degradation as a Function of Water Activity, Storage Temperature and Minearal Fortification in a Dehydrated Model Food System Stored in 303×404 Enameled Metal Cans[a]

Temp/a_w/mineral (°C)	κ	σ	τ
30/24/Control	-0.29	0.07	239
30/24/FeSO$_4$ 10	-0.84	0.12	83
30/24/FeSO$_4$ 25	-0.74	0.35	95
30/24/Zn O 10	-0.47	0.27	148
30/24/Zn O 25	-0.36	0.10	193
30/24/CaCO$_3$ 10	-0.42	0.15	165
30/24/CaCO$_3$ 25	-0.44	0.12	158
30/40/Control	-0.65	0.17	106
30/40/FeSO$_4$ 10	-0.71	0.17	98
30/40/FeSO$_4$ 25	--	--	--
30/40/ZnO 10	-0.72	0.10	96
30/40/ZnO 25	-0.77	0.08	90
30/40/CaCO$_3$ 10	-1.09	0.20	64
30/40/CaCO$_3$	-0.91	0.09	76
30/65/Control	-3.16	0.51	22
30/65/FeSO$_4$ 10	-2.18	0.43	32
30/65/FeSO$_4$ 25	-1.65	0.32	42
30/65/ZnO 10	-1.69	0.50	41
30/65/ZnO 25	-2.07	0.27	33
30/65/CaCO$_3$ 10	-1.08	0.18	64
30/65/CaCO$_3$ 25	-1.63	0.44	43

[a]See Table I for details.

soluble-solids fraction of the model system. A similar effect was also observed during degradation of ascorbic acid. This interpretation of the data is also supported by the absence of any significant difference in the half-life of retinyl acetate in an identical model system supplemented with zinc oxide, which has extremely poor solubility. Thus, these studies involving vitamin A and mineral fortification indicate that while the addition of minerals decreases the stability of retinyl acetate, the level of fortification does not appear to have a significant affect on the rate constant. This result is most likely due to solubility characteristics of the mineral supplements.

In both oat- and rice-based cereals the stability of retinyl
acetate decreased with an increase in the water activity at a
storage temperature of 45°C (Table V). At this storage condition,
temperature had an overriding influence on the reaction rate.
At the 37°C storage temperature, the stability of vitamin A was
lowest in samples that were equilibrated to a water activity of
0.40. These data are in agreement with the findings reported by
Labuza (1975a,b), concerning oxidative changes in foods as a function of water activity. He pointed out that as the water activity
is increased above the monomolecular layer level the viscosity of
the soluble solids decreases, and reactant and catalyst have more
mobility within the food system. The increase in half-life associated with the 0.65 a_w samples stored at 37°C likewise are characteristic of a food system where the effect of dilution of reactants and catalysts exceed the effects of a decrease in viscosity.
Quast and Karel (1972) reported that the rate of lipid oxidation
was less affected by a decrease in oxygen level in the package
atmosphere than by low water activity. This effect is most likely
the result of a greater concentration of oxygen being potentially
available to the reaction. This concept is discussed more compre-
hensively in Section IV.

Although information is available in the literature concerning
the degradation of vitamin A as retinyl acetate, retinyl palmitate,
and retinol, little if any information is available concerning the
factors that result in a loss of biological availability as a result of isomerization of the all-trans molecule to the cis form.
The most common of the cis isomers are the cis-9, cis-11, cis-13,
and various permutations of these isomers. Data available from
the work of Ames (1966) indicate a loss of biological potency of
the retinyl esters or the free alcohol as the all-trans form is
converted to the cis isomers. The kinetics of the conversion and
the factors affecting the rate of loss of biological potency are
not well known, with the exception of heat, which is known to increase the rate of cis isomer formation.

TABLE V. Rate Constants and Half-Lives for Vitamin A Degradation as a Function of Water Activity and Storage Temperature in Ready-to-Eat Oat Flake (OF) and Rice Base (RB) Cereals Stored in 303×404 Enameled Metal Cans[a]

Temp/a_w ($^\circ$C)	κ	σ	τ
OF 37/0.24	−0.53	0.09	131
37/0.40	−0.87	0.21	80
37/0.65	−0.62	0.18	112
RB 37/0.24	−0.51	0.23	136
37/0.40	−1.23	0.37	56
37/0.65	−0.97	0.27	65
OF 45/0.24	−0.69	0.09	100
45/0.40	−0.96	0.15	72
45/0.65	−1.77	0.18	39
RB 45/0.24	−1.17	0.27	59
45/0.40	−1.25	0.45	55
45/0.65	−2.22	0.43	31

[a]See Table I for details.

At the present time few data are available in the literature concerning the stability, or factors affecting the stability, of the tocopherols and tocopheryl esters in dehydrated food systems. The antioxidant potential exhibited by tocopherol does establish to some degree the factors that affect its stability. These, of course, include the factors that enhance lipid oxidation, namely, temperature, moisture content, a_w, metal catalyst, and oxygen. The stability associated with the tocopheryl esters, as a function of processing and storage, has not been well characterized for dehydrated foods. The affect of a_w on the rate of destruction of the tocopheryl esters has not been studied to any significant extent. Thus, the thermodynamics associated with vitamin A destruction have not been reported. This is probably due to problems associated with the rapid quantitation of tocopherol and tocopheryl ester losses in food systems until the advent of HPLC.

III. B COMPLEX VITAMINS

The effect of water activity and moisture content on the
stability of water-soluble vitamins is an area that received some
attention by researchers in the past. Farrer (1955) and Devivedi
and Arnold (1973) have reviewed the factors relating to the thermal destruction of thiamin. The effect of heat on the destruction
of thiamin can be described by the Arrhenius equation and is affected by pH, oxygen, trace minerals, and the form of thiamin.
Hollenbeck and Obermeyer (1952) reported that thiamin mononitrate
exhibited greater storage stability than thiamin chloride hydrochloride in enriched flour. Both thiamin salts were subjected to
increased rate of degradation with increasing moisture content.
Bookwalter et al. (1968) reported the stability of thiamin in a
corn-soy-milk (CSM) high-protein food supplement was a function of
both temperature and moisture content, as shown by the data in
Table VI. Although sorption isotherms were not reported for the
CSM system at the various temperatures studied ($-18°$, $25°$, $37°$,
and $49°C$), an equilibrium moisture content of 10% in the CSM system was reported to correspond to a water activity of 0.54 at $23°C$.

TABLE VI. Thiamin Content in Corn-Soy-Milk High-Protein Supplement as a Function of Moisture Content and Storage Temperature;
Product Stored in Glass Containers with Foil-Lined Closures[a]

Moisture	Storage conditions		Thiamin
(%)	Days	Temp ($°C$)	(ppm)
5	28	-18	4.8
10	28	-18	4.9
13.5	28	-18	4.8
5	28	49	5.2
10	28	49	4.1
13.5	28	49	1.6
5	182	37	5.5
10	182	37	4.2
13.5	182	37	1.9
5	365	25	-
10	365	25	5.2
13.5	365	25	4.9

[a]Bookwalter et al. (1968).

Studies in our laboratory to determine the kinetics of thiamin degradation were carried out with the model food system previously described in Table II and with commercially manufactured ready-to-eat breakfast cereals (Dennison, 1978). Sorption isotherms for the model system were determined at 10°, 20°, 30°, 37°, and 45°C according to the method of Palnitkar and Heldman (1971) and Bach (1974) in a closed system using a Cahn electrobalance and free water surfaces. Equilibration of the model system and ready-to-eat breakfast cereals to the proper a_w was accomplished by placing the samples in an equilibration chamber and forcing conditioned air from an Aminco-Aire unit through the closed system. Model system and cereal samples were immediately packaged in metal cans (TDT, 208×006 mm) and stored isothermally. Immediately after freeze-drying, unequilibrated-model-system and cereal samples were packaged in one-ounce commercial paperboard breakfast cereal boxes (3×7×10.3 cm) containing waxed liners (0.0009 cm thickness) and stored at 10, 40, and 85% RH in controlled-atmosphere cubicles at 30°C. The moisture transfer coefficients for the liner and box plus liner were equal (7.25×10^{-5} g H_2O cm/m^2/hr mmHg). As shown in Fig. 1 for an eight-month storage study, destruction of thiamin in the model system was less than 5% at storage temperatures ≤37°C and was independent of water activity at $a_w \leq 0.65$. Under these conditions rate constants describing thiamin losses, calculated according to first-order kinetics, were very small, with large standard deviations and correlation coefficients of <0.90. Treatment of data using other kinetic functions to describe the loss of thiamin did not result in a better fit. Thus, the loss of thiamin mononitrate under these storage conditions could be described either as zero or first order. As shown by the data in Table VII and Fig. 1, a significant increase in the rate of thiamin loss occurred in the model food system stored at 45°C when the water activity was at or above 0.24. Thiamin losses were greatest at water activities where nonenzymatic browning was visibly pronounced. These data support the findings of Van der Poel (1956) and Lhoest et al. (1958), who

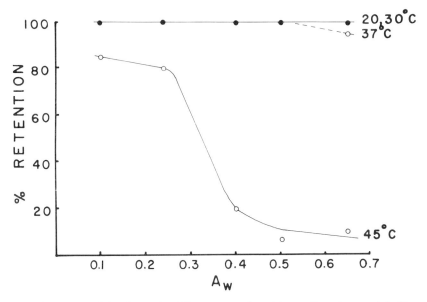

Fig. 1.   Thiamin retention as a function of water activity at selected storage temperatures.

TABLE VII. *Rate Constants and Half-Lives for Thiamin Degradation in a Model System Fortified with Thiamin or Thiamin, Ascorbic Acid and Vitamin A as a Function of Water Activity; Packaged in TDT cans and Stored at $45°C$*

a_w	B_1 model system			A, B_1, C model system		
	k^a	σ^b	$t_{\frac{1}{2}}{}^c$	k^a	σ	$t_{\frac{1}{2}}{}^c$
0.10	0.66	0.77	1050	0.14	0.71	4900
0.24	0.91	0.50	762	0.65	0.65	1066
0.40	6.75	1.20	103	6.48	1.75	107
0.50	11.01	2.08	63	9.27	1.73	75
0.65	8.67	1.69	80	9.48	1.84	73

[a]*First-order rate constant,* $\times 10^{-3}$ *days*$^{-1}$.
[b]*Standard deviation,* $\times 10^{-3}$.
[c]*Half-life, days.*

reported the reaction of thiamin in a Maillard-type browning reaction in dry and aqueous products during heating. This may be an important reaction in the loss of thiamin during processing and storage.

The stability data for thiamin mononitrate in the model system as a function of water activity at 45°C are presented in Table VII. Thiamin stability under these conditions was clearly dependent upon water activity. Above the monomolecular moisture content there was no statistically significant difference between the thiamin destruction rates for a model system containing only thiamin or a multivitamin-fortified model system containing vitamins A, B_1, and B_2. Data obtained for the rate of thiamin destruction in both the thiamin and multivitamin-fortified model food systems stored at 20°, 30°, and 37°C indicate no significant difference in the stability of thiamin in the two systems.

A thiamine-fortified model system and a model system fortified with thiamin, riboflavin, and vitamin A were stored in cardboard boxes without prior equilibration and these samples showed losses of less than 2% after eight months of storage at 10 and 40% relative humidity and 30°C (Table VIII). These data indicate that the oxygen content in the package atmosphere did not play a significant role in thiamin degradation. It is also apparent that the rate of diffusion of moisture through the package influenced the results. The lower water activity resulting from the reduced rate of mois-

TABLE VIII. Thiamin Loss (Percentage)[a] in a Model Food System after Eight Months Storage at Constant Relative Humidity; Samples were Packaged in Cardboard Boxes and Stored at 30°C

RH (%)	B_1 model system (%)	A, B_1, C model system (%)
10	<2	<2
40	<2	<2

[a] % loss = $(1 - e^{-kt}) \times 100$.

ture transport into the model system was interpreted as the cause
of the lower rate of thiamin loss in the model·system stored in
paperboard boxes at 40% relative humidity as compared to samples
equilibrated to a_w = 0.40 and packaged in metal containers.

The stability of riboflavin was also investigated in this
study. Riboflavin is considered more heat stable than thiamin but
highly sensitive to degradation by light. The stability of riboflavin in dry products is considered to be excellent in the absence
of light (Borenstein, 1971). Unfortunately, few data concerning
the kinetics and thermodynamics of riboflavin destruction in low
moisture dehydrated foods at specific conditions of moisture content and storage temperature are available.

Dennison *et al*. (1977) studied the stability of riboflavin in
a multivitamin-fortified model food system (previously described).
They equilibrated the samples at a_w = 0.1-0.65 and stored them in
metal cans at 10, 20, 30, and 37°C. The results are presented in
Fig. 2. The rate constants describing the loss of riboflavin in

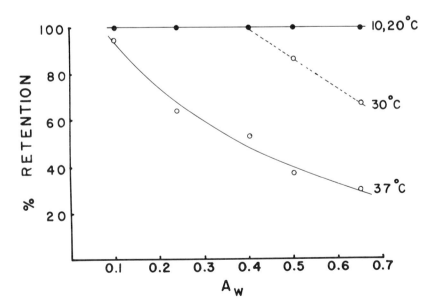

*Fig. 2. Riboflavin retention as a function of water activity
at selected storage temperatures.*

the model system packaged in TDT cans and stored at 10°, 20°, and 30°C and a_w = 0.10-0.65 were quite small ($k \leq 2 \times 10^{-3}$ days^{-1}) with large standard deviations and correlation coefficients of less than 0.90. A better fit of the riboflavin destruction data was obtained when other rate-order functions were used. Thus, for dehydrated foods of pH $\leq$ 7.0, riboflavin appears to have excellent storage stability. At a storage temperature of 37°C, riboflavin did exhibit decreasing stability with increasing water activity (Table IX).

Samples of the multivitamin-fortified model system were packaged in cardboard boxes and stored at 30°C in atmospheres of 10, 40, or 80% relative humidity. The results in Table X indicate that these half-lives for riboflavin destruction were significantly lower than those for the model system stored in metal TDT cans with limited oxygen but similar a_w and storage temperature. These data suggest the involvement of oxygen, either directly or indirectly, in the destruction of riboflavin. Increasing water activity had no detectable influence on the stability of riboflavin in the model system stored in cardboard boxes at 30°C.

Evaluation of riboflavin stability in commercially prepared cereal products stored for eight months at 10°, 20°, 30°, or 37°C in TDT cans at a_w = 0.24, 0.40, or 0.65 was similar to that observed for the model system ($k = 1 \times 10^{-3}$ days^{-1}) stored at 10° or 20°C. Cereal packaged in cardboard boxes showed a two- to three-

TABLE IX. *Influence of Water Activity on Rate Constants and Half-Lives at 37°C; Samples Packaged in TDT Cans*[a]

| | A, B$_2$, C model system | | |
a_w	k	σ	$t_{\frac{1}{2}}$
0.10	0.23	1.55	3013
0.24	1.88	1.68	369
0.40	2.63	2.11	264
0.50	4.11	1.97	169
0.65	5.03	2.25	138

[a]*See Table VII for details.*

TABLE X. Riboflavin Loss (Percentage)[a] in a Model Food System and in Commercial Cereals Packaged in Cardboard Boxes; Samples Were Stored for Eight Months at Constant Relative Humidity and 30°C

RH (%)	Multivitamin-fortified model system	Breakfast cereals fortified with vitamins A, B_2		
		Corn	Oat[b]	Oat[c]
10	65	21	<5	<5
40	64	21	<5	<5
85	64	36	<5	<5

[a] % loss = $(1-e^{-kt}) \times 100$.
[b] Oat-based cereal without folic acid.
[c] Oat-based cereal with added folic acid.

fold increase in riboflavin stability as compared to the boxed model system under similar conditions. The reason for this increased stability is not known; however, it may reflect the use of antioxidants in the commercially prepared cereal.

The stability and retention of biologically available B_6 vitamers in dehydrated foods has not been widely studied because of problems associated with accurate assay methods. Pyridoxine hydrochloride, the B_6 vitamer used for fortification purposes, has been found to be quite stable. Bunting (1965) reported that 90-100% of the pyridoxine added to corn meal and macaroni was retained following storage for one year at 38°C and 50% relative humidity. Cort et al. (1976) observed no loss of pyridoxine in corn meal that had been fortified with salts of iron, zinc, magnesium, and calcium, and then stored for six months at room temperature. Only 4% was lost during storage for 12 weeks at 45°C. Anderson et al. (1976) observed no loss of pyridoxine in iron fortified breakfast cereals during storage for three months at 40°C or 6 months at 22°C.

Although a systematic study of the stability of vitamin B_6 vitamers in dehydrated foods stored under various conditions has not been undertaken, Gregory and Kirk (1978) have reported the effects of high water activity and storage temperature on the stability and

bioavailability of vitamin B_6 vitamers in a model food system. A
model food system identical to that used in previously discussed
studies was used, with the exception that β-lactoglobulin and bovine serum albumin were used to replace the soya protein (promine
E). These proteins were selected on the basis of their solubility,
which facilitated their isolation from the model system. Model
systems containing pyridoxal phosphate monohydrate, pyridoxamine
dihydrochloride, pyridoxine hydrochloride, and a nonfortified control were used in the study. The model systems were equilibrated
to approximately a_w = 0.60 at 37°C. The equilibrated model systems
were then packaged in metal containers (TDT cans and 303 cans) to
prevent loss of moisture, and stored at 37°C. As would be expected
with this water activity and storage temperature, nonenzymatic
browning occurred readily. Plots of the time-dependent loss of
free amino groups (Fig. 3), and formation of melanoidins (Fig. 4)
associated with the soluble protein fraction, demonstrate the nonenzymatic browning was not a linear function during storage of the

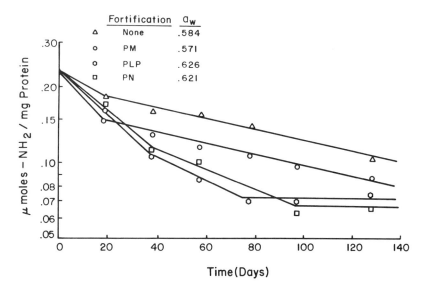

*Fig. 3. Loss of free amino groups in the soluble protein
fraction of dehydrated model systems during storage at 37°C in
TDT cans.*

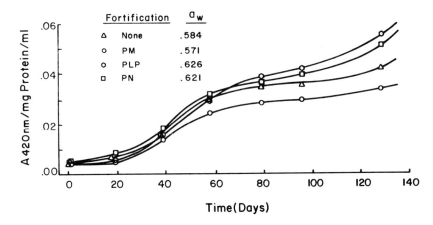

Fig. 4. Formation of melanoidins during storage of dehydrated model systems in TDT cans at 37°C.

model system in the TDT cans. Data presented in Table XI show that nonenzymatic browning occurred four to six times faster in the TDT cans than in the 303 cans, which had a greater oxygen content in the storage atmosphere.

The time-dependent loss of the B_6 vitamers in the model system stored in the TDT and 303 cans was monitored using an assay procedure involving high-performance liquid chromatography (HPLC) (Gregory and Kirk, 1978). The degradation of all B_6 vitamers could be described by the first-order equation. As shown by the

TABLE XI. Comparison of Browning Levels in Model Systems Stored in TDT and 303 Cans for 128 Days at 37°C.

Model system fortification	Free amino groups[a]		Browning pigments[b]	
	TDT	303	TDT	303
Nonfortified	0.108	0.111	0.0422	0.0082
PN	0.066	0.103	0.0523	0.0114
PM	0.088	0.117	0.0331	0.0071
PLP	0.075	0.099	0.0527	0.0095

[a]*Mean of triplicate determinations. Data, moles/mg protein.*
[b]*Mean of triplicate determinations. Data, A_{420} nm/mg protein/ml.*

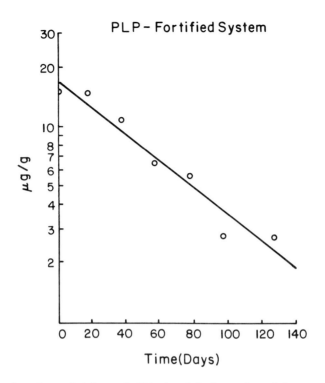

Fig. 5. Degradation of PLP in dehydrated model systems stored in TDT cans at 37°C.

data presented in Fig. 5 for pyridoxal phosphate, the rate of degradation was quite rapid, exhibiting a half-life of 44 days
(Table XII). Conversion of pyridoxal phosphate to pyridoxamine
phosphate during storage was not observed.

Pyridoxamine exhibited the greatest complexity in its degradation mechanism and kinetics for all of the forms monitored (Fig.
6). The disappearance of pyridoxamine followed first-order kinetics. The concurrent rapid formation of pyridoxal indicated the
occurrence of a transamination reaction. The pyridoxal curve,
showing a maximum pyridoxal concentration after approximately 35
days of storage (Fig. 6), indicated that the rate of pyridoxal
formation exceeded that of pyridoxal degradation. Therefore, at
a_w = 0.60 and 37°C, the rate-limiting step in the loss of total
vitamin B_6 in the pyridoxamine-fortified system was the degradation of pyridoxal.

TABLE XII. Kinetic Data for the Loss of the B_6 Vitamers from Dehydrated Model Systems Packaged in TDT Cans and Stored at 37°C

Model system fortification Reaction	k^a (days^{-1})	Correlation coefficient	$t_{\frac{1}{2}}^b$ (days)
Nonfortified			
Loss of inherent PL	0.0170±0.0027	−0.9876	41
Pyridoxine			
Loss of inherent PL	0.0151±0.0020	−0.9826	46
Loss of PN (>58 days)	0.0049±0.0005	−0.9889	141
Pyridoxal phosphate			
Loss of PL	0.0158±0.0019	−0.9667	44
Pyridoxamine			
Loss of PM	0.0200±0.0026	−0.9753	35
Loss of formed PL	0.0165±0.0040	−0.9667	42
Loss of total PM + PL	0.0165±0.0027	−0.9377	42

[a]First-order rate constant ± standard deviation.   [b]Half-life.

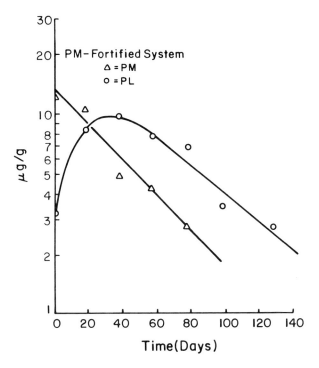

Fig. 6. Behavior of vitamin B_6 in a PM-fortified dehydrated model system stored in TDT cans at 37°C.

Pyridoxine hydrochloride was observed to be the most stable form of vitamin B_6 during storage at 37°C and a_w = 0.60. As shown in Fig. 7, no loss of pyridoxine was observed for the first 58 days of storage in TDT cans. Very slow degradation followed this storage period, and first-order kinetics appeared to apply. Conversion of pyridoxine to pyridoxal or pyridoxamine was not observed. Pyridoxal inherently present in the model system was observed to degrade at the same rate as the added pyridoxal phosphate.

Linear regression analysis of the data describing degradation of pyridoxine, pyridoxal, pyridoxal phosphate, and pyridoxamine confirmed that B_6 losses in the model system stored in TDT cans conformed to first-order kinetics with correlation coefficients greater than 0.93. Treatment of the data using other rate functions did not provide a better linear fit of the data.

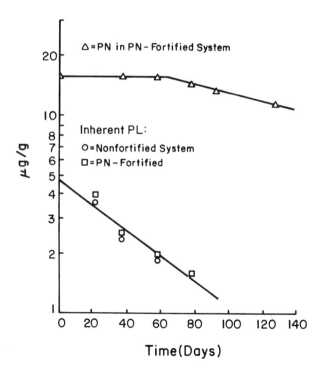

Fig. 7. Degradation of added PN and inherent PL in dehydrated model systems stored in TDT cans at 37°C.

Determination (HPLC) of vitamin B_6 in the model system, packaged in TDT or 303 cans and stored for 128 days at 37°C, indicated substantially greater stability of the B_6 vitamers in the larger cans.

Biological availability studies of B_6 vitamers in the model system stored in TDT or 303 metal cans for 128 days at 37°C indicated that the levels of B_6 vitamers were 100% biologically available (Table XIII).

The observed stability properties of pyridoxine are in agreement with several previous reports concerning pyridoxine stability in cereal products (Bunting, 1965; Anderson *et al.*, 1976; Cort *et al.*, 1976). The induction period observed prior to the slow loss of pyridoxine suggests that pyridoxine degradation is dependent on the occurrence of another reaction in the model system, possibly an activated complex. The rapid conversion of pyridoxamine to pyridoxal during storage was unexpected and indicates a rapid transamination reaction. Under the conditions of storage, the conversion of pyridoxamine to pyridoxal was unidirectional; thus, k_{-1} was much less than k_1 and could be neglected:

$$\text{pyridoxamine} \xrightarrow{k_1} \text{pyridoxal} \underset{k_{-1}}{\overset{k_3}{\rightleftharpoons}} \text{products}$$
$$\downarrow k_2$$
$$\text{products}$$

The observed rate constant for the net disappearance of pyridoxamine would represent the sum $k_1 + k_2$.

The increased stability of the B_6 vitamers in model systems stored in the larger cans (303) with large headspaces cannot be explained from these data. The retarded browning and reduced rate of vitamin B_6 loss in the larger cans provides evidence for a relationship between browning and vitamin B_6 degradation.

TABLE XIII. *Comparison of Fluorometric (Fluor.), Microbiological (Micro.), HPLC, and Rat Bioassays for Vitamin B_6; Model Systems Were Packaged in TDT and 303 Cans, Stored for 128 Days at 37°C, Then Samples Were Pooled[a,b]*

| Model system fortification | Rat bioassay response criterion (μg available B_6/g model system) | | | | Micro. | HPLC | Fluor. |
	Growth	Growth (g) per g feed consumed	AspAT Activity	AspAT PLP Stim.			
Nonfortified	-0.5	-0.7	-1.2	-0.3	0.5	0.5	13.5
PN	17.7	17.2	16.8	12.3	11.9	12.8	24.0
PM	4.5	3.7	9.1	10.2	6.3	7.5	16.3
PLP	4.2	3.2	1.8	1.6	2.9	4.2	13.4

[a]μg total vitamin B_6/g model system.
[b]Fluorometric values are significantly greater than all other estimates. No significant differences between microbiological and HPLC values. Microbiological and HPLC values were not significantly different from the bioassay estimates for each model system ($p < 0.05$).

IV. ASCORBIC ACID

Reports concerning the effects of moisture content, water activity, temperature, pH, oxygen content, and transition metals on
the oxidation of ascorbic acid in dehydrated food systems are
readily available in the literature. However, an understanding
of the way in which these factors affect the quality of dehydrated
foods is not completely understood and is subject to different interpretations by various researchers.

Karel and Nickerson (1964), Jensen (1967), Vojnovich and
Pfeifer (1970), Lee and Labuza (1975), Waletzko and Labuza (1976),
Kirk *et al.* (1977), Dennison and Kirk (1978), Dennison (1978),
and Riemer and Karel (1978) have studied the stability of reduced
ascorbic acid in various low and intermediate moisture foods and
model systems as a function of moisture content and water activity.
Results reported by Kirk *et al.* (1977) for the degradation of reduced and total ascorbic acid (Table XIV) show that the rate of
destruction increased as the total moisture content and water activity increased. The rates of loss of reduced and total ascorbic
acid is a function of water activity conformed to first-order
kinetics. The slow degradation of ascorbic acid in samples stored
at 10 C appeared to follow first-order kinetics; however, the extremely long half-lives associated with these conditions did not
permit studies to be carried out for even one half-life.

The temperature dependencies for reduced, dehyro, and total
ascorbic acid degradation have been studied and they conform to
the Arrhenius equation. Activation energies (E_a) calculated by
Lee and Labuza (1975), Kirk *et al.* (1977), and Dennison *et al.*
(1978) for the destruction of total and reduced ascorbic acid in
model food systems, equilibrated to water activities from 0.24 to
0.85, increased with increasing water activity from 12.5 to 20.0
kcal/mole and were not significantly different from one another.
An exception to this was recently reported by Riemer and Karel
(1978). In a study reported by Kirk *et al.* (1977), where a model

TABLE XIV. *Rate Constants and Half-Lives for TAA and RAA Loss as a Function of Water Activity and Storage Temperature; Dehydrated Model Food Systems Fortified with Ascorbic Acid and Packaged in TDT Cans*

Temp. (°C)	a_w	TAA			RAA		
		k^a	σ^b	τ^c	k^a	σ^b	τ^c
10	0.10	0.31	0.03	224	0.43	0.07	161
	0.24	0.37	0.04	187	0.45	0.05	154
	0.40	0.42	0.05	165	0.47	0.09	147
	0.50	0.49	0.03	141	0.58	0.05	119
	0.65	0.50	0.02	139	0.55	0.05	126
20	0.10	0.45	0.06	154	0.65	0.01	107
	0.24	0.95	0.12	73	1.34	0.21	52
	0.40	1.28	0.11	54	1.69	0.22	41
	0.50	1.12	0.08	62	1.30	0.13	53
	0.65	1.44	0.28	48	1.93	0.54	36
30	0.10	0.91	0.08	76	1.11	0.15	63
	0.24	1.78	0.23	39	2.30	0.31	30
	0.40	3.13	0.26	22	3.84	0.42	18
	0.50	3.99	0.34	17	4.63	0.50	15
	0.65	11.28	1.30	6	11.64	1.00	6
37	0.10	0.98	0.11	71	1.23	0.20	56
	0.24	5.01	0.36	14	4.44	0.16	16
	0.40	7.03	0.24	10	7.87	0.28	9
	0.50	9.24	0.59	8	8.90	0.57	8
	0.65	15.74	0.66	4	16.85	1.11	4

[a] *First-order rate constant, $\times 10^{-2}$ days^{-1} (KINFIT analysis).*
[b] *Standard deviation, $\times 10^{-2}$.*
[c] *Half-life, days.*

TABLE XV. *Activation Energies for TAA and RAA Loss; Dehydrated Model Food Systems were Fortified with Ascorbic Acid or Multivitamins and Packaged in TDT Cans.*

	Activation energy (kcal/mole)			
a_w	Ascorbic acid model system		Multivitamin model system	
	TAA	RAA	TAA	RAA
0.10	8.1	7.1	11.5	13.1
0.24	15.9	14.2	13.5	11.8
0.40	17.6	17.8	17.8	16.6
0.50	19.2	18.1	15.9	16.1
0.65	24.0	23.3	20.0	19.3

system was packed in metal cans with a limited moisture content, the E_a for total and reduced ascorbic acid were 8.1 and 7.1 kcal/mole, respectively (Table XV). These E_a values were significantly lower than those reported for samples with water activities above the monomolecular moisture content of the model system.

Studies have been carried out where the oxygen concentration of the storage atmosphere of the dehydrated food system was varied from 0.0 to 20.1% and the rate constants for loss of reduced, dehydro, and ascorbic acid were observed to vary significantly. Intermpretation of the effect of oxygen concentration on the rate of reduced ascorbic acid destruction differs with the product being studied (Miers *et al.*, 1958; Karel and Nickerson, 1964; Rhee, 1976; Kirk *et al.*, 1977; Dennison *et al.*, 1978; Riemer and Karel, 1978. In studies reported by Kirk *et al.* (1977) and Dennison *et al.* (1978), in a model food system with a pH of 6.8, the rate of destruction of ascorbic acid was dependent upon the presence of oxygen in the storage atmosphere (Table XVI). The presence of a large excess of oxygen in the two packages (303×404 metal cans and one-ounce cardboard cereal boxes) caused not only an increase in the degradation rate constants of total and reduced ascorbic acid but also an increase in the energy of activation for ascorbic acid destruction in the model system with $a_w = 0.10$.

The discontinuity in E_a values for the destruction of reduced ascorbic acid and total ascorbic acid in a limited oxygen environment, and the effect of increased soluble solids on the decreased solubility of oxygen in aqueous systems (Joslyn and Supplee, 1949), suggests that at low water activities an anaerobic mechanism for ascorbic acid oxidation may exist. Although an energy of activation has not been reported for the destruction of ascorbic acid in a dehydrated food system at pH 6.8, Lee *et al.* (1977) studied ascorbic acid destruction in a liquid food system at pH 4.06 and found that E_a is a function of pH according to

$$E_a = 1.840(\text{pH}) - 4.178 \tag{3}$$

TABLE XVI. Rate Constants and Half-Lives for RAA Loss as a Function of Water Activity and Storage Temperature in a Dehydrated Model Food System

Temp (°C)	a_w	303 can[e]			TDT can[a,f]		
		k[b]	σ[c]	$\tau_{\frac{1}{2}}$[d]	k[b]	σ[c]	$\frac{1}{2}$[d]
10	0.10	0.34	0.51	204	0.43	0.07	161
	0.24	–	–	–	0.45	0.05	154
	0.40	0.63	0.61	110	0.47	0.09	147
	0.65	0.81	1.21	86	0.55	0.05	126
20	0.10	0.84	1.27	83	0.65	0.01	107
	0.24	1.03	0.90	67	1.34	0.21	52
	0.40	1.47	0.97	47	1.69	0.22	41
	0.65	2.04	2.00	34	1.93	0.54	36
30	0.10	1.29	1.09	54	1.11	0.15	63
	0.24	–	–	–	2.30	0.31	30
	0.40	3.12	2.04	22	3.84	0.42	18
	0.65	5.11	4.98	14	5.29	5.10	13
37	0.10	1.86	2.18	37	1.23	0.20	56
	0.24	–	–	–	4.44	0.16	16
	0.40	7.59	9.1	9	7.87	0.28	9
	0.65	11.67	26.2	6	16.85	1.11	4

[a]*Data from Kirk et al. (1977).*
[b]*First-order rate constant, $\times 10^{-2}$ days^{-1}.*
[c]*Standard deviation, $\times 10^{-3}$.*
[d]*Half-life, days.*
[e]*Stored in excess oxygen environment.*
[f]*Stored in limited oxygen environment.*

Applying this equation to the model food system having a pH of 6.8, a predicted E_a of 8.5 kcal/mole of ascorbic acid is obtained, which correlates well with the observed E_a. These data could be interpreted to reflect a change in the pathway of ascorbic acid degradation or the destruction of ascorbic acid by more than one pathway.

In an effort to answer this question concerning ascorbic acid degradation, the activation parameters for total ascorbic acid and reduced ascorbic acid degradation in the model food system were calculated from Eq. (3) (Eyring, 1935). The free energy of activation of ascorbic acid degradation was then calculated from:

$$\Delta G\ddagger = \Delta H\ddagger - T \Delta S\ddagger \tag{4}$$

These values are shown in Table XVII for ascorbic acid destruction in TDT containers and in Table XVIII for ascorbic acid destruction in 303 containers. Graphically depicted, both $\Delta H\ddagger$ and $\Delta S\ddagger$ increased as a function of water activity (Fig. 8), while $\Delta G\ddagger$ remained constant. The effect of water activity on the destruction of ascorbic acid in the dehydrated model system was interpreted as a solvent effect. These results are supported by the relationship between rates of ascorbic acid degradation and water activity, calculated from the data of Lee and Labuza (1975) (Fig. 9), which also can be related to the ratio of dissolved oxygen to ascorbic acid as a function of water activity. Assuming the concentrations of soluble solids and ascorbic acid can only decrease with an increase in water activity and the concentration of oxygen would remain constant, the relationship shown in Fig. 10 is obtained.

Activation parameters calculated for the degradation of total ascorbic acid and reduced ascorbic acid in the dehydrated model system of Kirk *et al.* (1977) and Dennison *et al.* (1978) indicate

TABLE XVII. Activation Parameters for TAA and RAA Degradation in an Ascorbic Acid Fortified Model System Stored in TDT Cans

a_w	E_a (kcal/mole)	$\Delta H^{\ddagger}$ (kcal/mole)	$\Delta S^{\ddagger}$ (eu)	$\Delta G^{\ddagger}$ (kcal/mole)
		TAA		
0.10	8.1	7.5	-65.6	27.0
0.24	15.9	15.3	-37.6	26.5
0.40	17.6	17.0	-30.6	26.1
0.50	19.2	18.6	-25.6	26.2
0.65	19.2	18.6	-24.6	25.9
		RAA		
0.10	7.1	6.5	-68.6	26.9
0.24	14.2	13.6	-42.6	26.3
0.40	17.8	17.2	-29.6	26.0
0.50	18.1	17.5	-28.6	26.0
0.65	19.3	18.7	-24.6	26.0

TABLE XVIII. *Activation Parameters for TAA and RAA Degradation in an Ascorbic Acid Fortified Model System Stored in 303 Cans*

a_w	E_a (kcal/mole)	$\Delta H^{\ddagger}$ (kcal/mole)	$\Delta S^{\ddagger}$ (eu)	$\Delta G^{\ddagger}$ (kcal/mole)
		TAA		
0.10	10.7	10.1	-55.6	26.7
0.40	16.0	15.4	-35.6	26.0
0.65	18.3	17.7	-27.6	25.9
		RAA		
0.10	10.7	10.1	-55.6	26.7
0.40	15.6	15.0	-50.6	30.1
0.65	17.0	16.4	-31.6	25.8

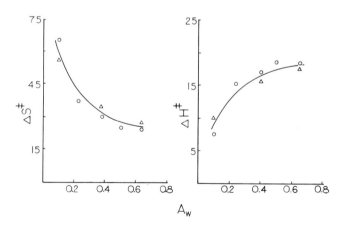

Fig. 8. *Entropy of activation (A) and enthalpy of activation (B) for TAA loss as influenced by water activity. A dehydrated model system stored in TDT cans (o) and 303 cans (Δ).*

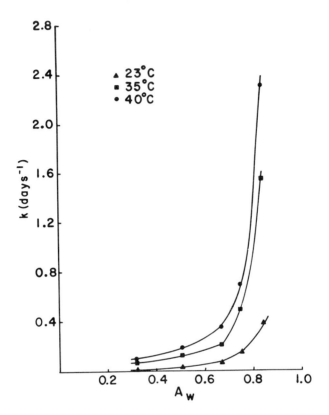

Fig. 9. Rate of ascorbic acid degradation versus water activity at constant temperature for an intermediate-moisture model food system (data from Lee and Labuza, 1975).

that degradation of ascorbic acid obeys the compensation law or isokinetic relationship

$$\Delta H^{\ddagger} = \beta \; \Delta S^{\ddagger} + \Delta H_0^{\ddagger} \tag{5}$$

$$\beta = {}^{\circ}K, \quad \Delta H_0^{\ddagger} - \text{intercept at } \Delta S^{\ddagger} = 0$$

reported by Leffer (1955). The isokinetic temperature β may be viewed as a constant, characteristic of a reaction series and dependent on the experimental temperature interval. As shown in Fig. 11, the isokinetic relationships for degradation of total ascorbic acid and reduced ascorbic acid in TDT and 303 metal cans

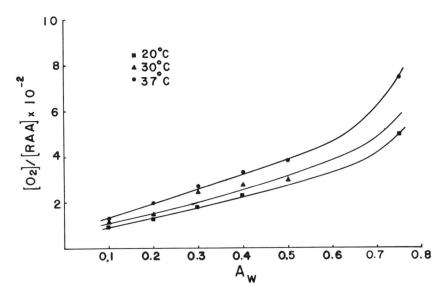

Fig. 10.   Ratio of O_2/AA as a function of water activity.

fall on the same line, yielding β values of 273 and 276°K for total
ascorbic acid and reduced ascorbic acid, respectively. The validity of this isokinetic relationship suggests that, although there is
a large change in $\Delta H^{\ddagger}$ (E_a) and $\Delta S^{\ddagger}$ as a function of water activity,
the constant $\Delta G^{\ddagger}$ indicates that the degradation reactions of reduced and total ascorbic acid in the model system follow the same
mechanism. In addition, the fact that the β values for total and
reduced ascorbic acid are less than the temperature range studied
implies that the degradation reaction is controlled by the entropy
of activation (Blackalder and Hinchelwood, 1958a,b).

Data presented for the involvement of oxygen and a_w in the degradation of ascorbic acid may suggest a possible rationale for the
dependence of $\Delta S^{\ddagger}$ and $\Delta H^{\ddagger}$ on a_w. Because $\Delta S^{\ddagger}$ is a measure of randomness of the activated complex, a decrease in $\Delta S^{\ddagger}$ with increasing
a_w may be due to increased solvation of the reactants and the activated complex, or a decrease in the effective charge of the transition compound. At a_w = 0.10, $\Delta S^{\ddagger}$ for total ascorbic acid destruction was -43 and -33 cal/deg (eu), respectively (Tables XVII and

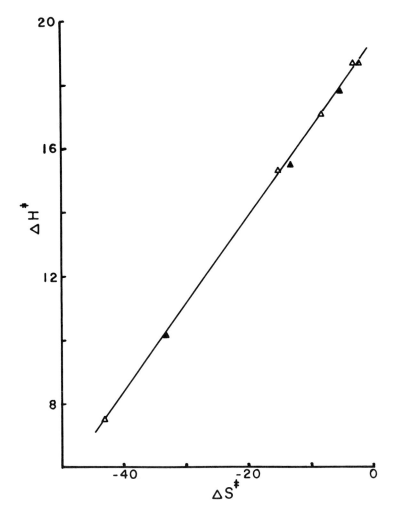

Fig. 11. Isokinetic relationship of TAA degradation in a de-hydrated model system stored in TDT cans (Δ) and 303 cans (▲).

XVII), in the TDT and 303 cans, indicating little randomness in the reaction and a tightly bound activated complex. As water activity increases in the model system through a_w = 0.65, the entropy of activation approaches zero, indicating little difference in the internal degrees of freedom of the activated state or the reactants. Similarly, the increase in $\Delta H^{\ddagger}$ as the water activity increases, most likely represents the increased solvation and greater energy requirement for formation of the transition state, and this, in turn, may be due to a decrease in effective charge. However, based on the data generated from this study, the degradation of ascorbic acid in the model system appears to be entropy controlled.

The absence of an oxygen effect on ascorbic acid degradation in high-acid dehydrated foods has been reported by Karel and Nickerson (1964) and Riemer and Karel (1978). They reported no significant variation in the rate of ascorbic acid destruction in dehydrated orange juice crystals and tomato juice powder when samples were stored in an oxygen-free environment or in air at lower water activity and 37°C (Fig. 12). The exact effect of oxygen on the stability of total and reduced ascorbic acid in dehydrated tomato juice may be somewhat different than shown in Fig. 12 because data for reduced ascorbic acid must be corrected to 100% to present this type of plot. As the data are presented, they are not randomly scattered and appear to fit a curved rather than a straight line. This could be interpreted to indicate a

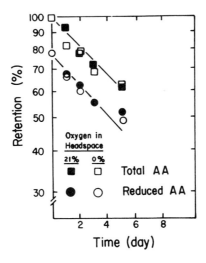

Fig. 12. Retention of total ascorbic acid (AA) and of reduced AA in dehydrated tomato juice stored at a water activity of 0.32 at 37°C (Riemer and Karel, 1978).

second rate function for loss of ascorbic acid in the dehydrated
tomato juice when the oxygen concentration drops to a level where
it becomes rate limiting for degradation of ascorbic acid. These
data could indicate that there may be two first-order rate constants that describe the destruction of ascorbic acid. The first
rate would represent the effect of residual oxygen in the system,
which would be consumed during the first few days of storage. The
data point at day 5 may represent the first data point where oxidation of ascorbic acid occurred under anaerobic conditions. A
similar pattern of ascorbic acid degradation was reported by Lee
et al. (1977) in tomato juice, which was heat treated to 129.5°C
for 9 sec, cooled to 93.3°C, filled in 8 ounce enameled metal
cans, sealed and cooled to 37.8°C, and then stored for 20 days.
Following this initial storage time, the rate at which reduced ascorbic acid was converted to dehydroascorbic acid decreased significantly with time, but was accurately described by first-order
kinetics.

Until recently no information was available concerning the effects of mineral fortification on the stability of ascorbic acid
in low-moisture dehydrated food systems. Khan and Martell (1967a,
b) and Blackburn and Jameson (1975) have reported extensive work
concerning metal-ion-catalyzed oxidation of ascorbic acid in aqueous solutions. Khan and Martell (1967a,b) postulated an ascorbate-
metal-oxygen complex as the intermediate for copper and iron catalysis of ascorbic acid. Recently, a metal-metal dinuclear ascorbate-
oxygen complex was suggested as the intermediate by Blackburn and
Jameson (1975).

The influence of mineral fortification on the stability of total and reduced ascorbic acid in a food system packaged in 303 cans
was studied at 30°C and a_w = 0.10, 0.40, and 0.65 by Dennison
(1978). As shown by the kinetic parameters (treated as first order presented in Tables XIX and XX for total and reduced ascorbic
acid, it is apparent that mineral fortification at either 10 or
25% of the RDA value exhibited no catalytic effect on degradation

TABLE XIX. *Rate Constants and Half-Lives for TAA Loss as a Function of Mineral Fortification and Water Activity; A Dehydrated Model Food System Stored at 30°C in 303 cans*

Mineral supplement	RDA (%)	$a_w = 0.10$		$a_w = 0.40$		$a_w = 0.65$	
		k^a	$t_{\frac{1}{2}}^b$	k^a	$t_{\frac{1}{2}}^b$	k^a	$t_{\frac{1}{2}}^b$
None	-	1.92	36	3.50	20	5.09	14
FeSO$_4$	10	1.31	53	2.20	31	11.22	6
	25	1.81	38	2.11	33	11.62	6
FeCl$_2$	10	1.33	52	2.57	27	10.48	7
	25	1.59	44	2.94	24	9.41	7
Fe	10	1.31	53	2.69	26	8.65	8
	25	1.53	45	2.25	31	17.28	4
ZnCl$_2$	10	1.40	50	2.38	29	17.25	4
	25	1.62	43	2.47	28	25.65	3
ZnSO$_4$	10	1.44	48	2.91	24	18.60	4
	25	1.75	40	2.30	30	17.53	4
ZnO	10	1.34	52	2.25	31	3.19	22
	25	1.63	43	2.50	28	3.91	18
CaCO$_3$	10	1.45	48	2.97	23	10.57	7
	25	1.46	47	2.78	24	17.39	4
CuCl$_2$		1.36	51	3.86	18	5.51	13
		1.38	50	4.50	15	11.67	6
CuSO$_4$		1.17	59	5.25	13	5.00	14
		1.52	46	6.37	11	9.00	8

[a]First-order rate constant, 10^{-2} days^{-1}.
[b]Half-life, days.

of total or reduced ascorbic in the model system, which was equi-librated to a water activity of $a_w = 0.10$. This was anticipated because a water activity of 0.10 is below the BET monomolecular moisture content and the mobility of the metal ions would be extremely limited. Similar data were also reported by Dennison (1978) at $a_w = 0.40$, with the exception of a slight catalytic effect by CuCl$_2$ and CuSO$_4$. This catalysis may result from a slightly greater mobility by the Cu(II) ion than the other mineral cations. The lack of any catalytic effect by calcium and zinc may relate to the theory of Khan and Martell (1967). They suggested that the rate-determining step of metal-catalyzed ascorbic acid oxidation involves a one-electron transfer from the ascorbate anion

TABLE XX. Rate Constants and Half-Lives for RAA Loss as a Function of Mineral Fortification and Water Activity; A Dehydrated Model Food System Stored at 30 C in 303 Cans

Mineral supplement	RDA (%)	$a_w = 0.10$		$a_w = 0.40$		$a_w = 0.65$	
		k[a]	$t_{\frac{1}{2}}$[b]	k[a]	$t_{\frac{1}{2}}$[b]	k[a]	$t_{\frac{1}{2}}$[b]
None	–	2.04	33	3.45	20	5.01	14
$FeSO_4$	10	1.74	40	2.47	28	11.48	6
	25	2.16	32	2.44	28	11.73	6
$FeCl_2$	10	1.70	41	2.97	23	10.07	7
	25	1.72	40	2.69	26	9.87	7
Fe	10	1.32	52	2.64	26	9.38	7
	25	1.65	42	2.20	31	17.51	4
$ZnCl_2$	10	1.96	35	2.55	27	17.06	4
	25	2.13	33	2.47	28	23.19	3
$ZnSO_4$	10	1.87	37	2.93	24	18.13	4
	25	1.77	39	3.02	23	18.65	4
ZnO	10	1.74	40	2.52	28	3.37	21
	25	1.63	43	2.46	28	3.90	17
$CaCO_3$	10	1.61	43	3.30	21	10.41	7
	25	1.81	38	3.11	22	17.72	4
$CuCl_2$		1.28	54	3.01	23	5.94	12
		1.75	40	5.93	12	12.01	5
$CuSO_4$		1.40	50	4.00	17	4.72	15
		1.80	39	5.92	12	9.18	8

[a]First-order rate constant, $\times 10^{-2}$ days^{-1}.
[b]Half-life, days.

to the metal ion, thus requiring a stable lower valence form of the metal ion for electron acceptance.

The stability of ascorbic acid in the copper- and iron-fortified model systems at low water activities was interpreted to result from the inability of the aqueous phase to solubilize and/or mobilize the metal ions.

At $a_w = 0.65$, which is the capillary region of the adsorption isotherm, a two- to threefold increase in the degradation rate for both total and reduced ascorbic acid is noted for all added minerals except zinc oxide, which did not exhibit any catalytic effect. With the exception of copper, it was not possible to discern an effect of mineral concentration on the rate of ascorbic acid degradation (Tables XIX and XX). These data may suggest that free mobility of the Cu(II) and Fe(III) ions may require the complete hydration of the metal ion in its octahedral configuration, and this may be possible only in the presence of free water in the capillary region. Below this region, the metal ions may have one or more charge species from the product matrix, effectively immobilizing or limiting migration of the ions.

The catalytic effects of soluble zinc and calcium salts were unexpected because of the lack of stable lower valence states (Khan and Martell, 1967). The absence of any effect by zinc oxide was attributed to its insoluble character. The rate enhancement for ascorbic acid degradation with $CaCO_3$, $ZnCl_2$, and $ZnSO_4$ may be explained by the effect of added electrolytes on the activity of dissolved oxygen (Long and McDevit, 1951). The activity coefficients of oxygen have been shown to increase linearly with electrolyte concentration and are dependent upon both the anion and cation present.

From the data presented, trace mineral addition to low-moisture dehydrated food should have a negligible effect. Maximum catalytic effect would be evidenced in food with water activities in the capillary region.

References

Ames, S. R. (1966). *J. Am. Oil Chem. Soc. 49,* 1071.

Anderson, R. H., Maxwell, D. L., Mulley, A. E., and Fritsch, C. W. (1976). *Food Technol. 30,* 110.

Bach, J. A. (1974). M.S. Thesis, Michigan State University, East Lansing, Michigan.

Blackadder, D. A., and Hinchelwood, C. (1958a). *Chem. Soc.* p. 2720.

Blackadder, D. A., and Hinchelwood, C. (1958b). *Chem. Soc.* p. 2728.

Blackburn, ., and Jameson, (1975). *J. Inorg. Nucl. Chem. 37,* 809.

Bookwalter, G. M., Moser, H. A., Pfeifer, V. F., and Griffin, E. L. (1968). *Food Technol. 22,* 1581.

Borenstein, B. (1971). *Crit. Rev. Food Technol. 2,* 171-186.

Bunting, W. R. (1965). *Cereal Chem. 42,* 569.

Cort, W. M., Borenstein, B., Harley, J. H., Oscada, M., and Scheiner, J. (1976). *Food Technol. 30,* 52.

Dennison, D. B., and Kirk, J. R. (1978). *J. Food Sci. 43,* 609.

Devivedi, B. K., and Arnold, R. G. (1973). *J. Agri. Food Chem. 21,* 54.

Dye, J. L., and Nicely, V. A. (1971). *J. Chem. Educ. 48,* 443.

Eyring, H. (1935). *J. Chem. Phys. 3,* 107.

Farrer, K. T. (1955). *Adv. Food Res.* (E. M. Mrak and G. F. Stewart, eds.), p. 311. Academic Press, New York.

Gregory, J. F., and Kirk, J. R. (1978). *J. Food Sci.,* in press.

Hollenbeck, C. M., and Obermeyer, H. E. (1952). *Cereal Chem. 29,* 82.

Jensen, A. (1967). *J. Sci. Food Agri. 20,* 622.

Joslyn, M. A., and Supplee, H. (1949). *Food Res. 14,* 216.

Karel, M. (1975). *In* "Water Relations of Foods" (R. B. Duckworth, ed.), pp. 639-658. Academic Press, New York.

Karel, M., and Nickerson, J. T. R. (1964). *Food Technol. 18,* 104.

Khan, M. M. T., and Martell, A. E. (1967a). *J. Am. Chem. Soc. 89,* 4176.

Khan, M. M. T., and Martell, A. E. (1967b). *J. Am. Chem. Soc. 89,* 7104.

Kirk, J. R., Dennison, D. B., Kokoczka, P., and Heldman, D. R. (1977). *J. Food Sci. 42,* 1274.

Labuza, T. P. (1968). *Food Technol. 22*(3), 15.

Labuza, T. P. (1975a). *In* "Water Relations of Foods" (R. B. Duckworth, ed.), pp. 155-172. Academic Press, New York.

Labuza, T. P. (1975b). *In* "Water Relations of Foods" (R. B. Duckworth, ed.), pp. 455/474. Academic Press, New York.

Lee, S. H., and Labuza, T. P. (1975). *J. Food Sci. 40,* 370.

Lee, Y. C., Kirk, J. R., Bedford, C. L., and Heldman, D. R. (1977). *J. Food Sci. 42,* 640.

Leffer, J. E. (1955). *J. Org. Chem. 31,* 533.

Lhoest, W. J., Busse, L. W., and Baumann, C. A. (1958). *J. Am. Pharm. Assoc. 49,* 254.

Long, F. A., and McDevit, W. F. (1952). *Chem. Rev. 57,* 119.

Lund, D. B. (1973). *Food Technol. 27(1),* 16.

Miers, J. C., Wong, F. F., Harris, J. G., and Dietrich, W. C. (1958). *Food Technol. 12,* 542.

Palnitkar, M. P., and Heldman, D. R. (1971). *J. Food Sci. 35,* 799.

Quast, D. G., and Karel, M. (1972). *J. Food Sci. 37,* 584.

Rhee, S. H. (1976). *Hanguk Yong Hakhoe Chi 9,* 318. [Abstracted in *Chem. Abstr. 88,* 135022W, May 8, 1978.]

Riemer, J., and Karel, M. (1978). *J. Food Processing Preservation 1,* 293.

Singh, R. P., Heldman, D. R., and Kirk, J. R. (1976). *J. Food Sci. 40,* 164.

Van der Poel, G. H. (1956). *Chem. Abstr. 54,* 3770A, 1957.

Vojnovich, C., and Pfeifer, V. F. (1970). *Cereal Sci. 19,* 317.

Waletzko, P., and Labuza, T. P. (1976). *J. Food Sci. 41,* 1338.

FORMATION AND DECOMPOSITION OF BROWNING

INTERMEDIATES AND VISIBLE SUGAR-AMINE BROWNING REACTIONS

K. Eichner
and
M. Ciner-Doruk

I. INTRODUCTION

Physical preservation methods like sterilization and drying
are associated with the application of heat. In these cases —
because of their high temperature coefficients — sugar-amine
browning reactions are dominant deteriorative reactions. These
so-called nonenzymic reactions (Maillard reactions) are caused
by reactions between reducing sugars and amino groups of amino
acids and proteins (Hodge, 1953; Reynolds, 1963, 1965), leading
to losses of nutritive value, brown discoloration, and the
formation of off-flavors. These reactions usually have maximum
reaction rates in the low-moisture range. At lower
moisture levels there is a slow diffusion of the reactants,
while at higher-moisture contents dilution effects and the inhibitory effect of water (being a reaction product) by laws of
mass action prevail (Eichner and Karel, 1972; Eichner, 1974).
Because of this maximum in reaction rate, the Maillard reactions
are promoted by lowering the water content during drying. A
series of investigations have been performed to elucidate the
combined influence of water content and temperature on the rate
of Maillard reactions. These studies provide a basis for cal-

culating the reaction's extent at the end of a drying process as
well for optimizing processing conditions. For example, Hendel
et al. (1955) established a temperature-moisture profile for
browning of white potato during drying (Fig. 1), from which they
were able to calculate the amount of browning at different drying
periods. The last drying period maximized browning, whereas the
interval of the browning maximum makes only a minor contribution
because of the small inherent time intervals (Fig. 2). As a

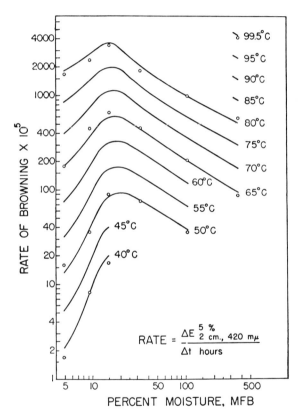

*Fig. 1. Effect of moisture content and of temperature on
rate of browning of white potato (Hendel et al., 1955).*

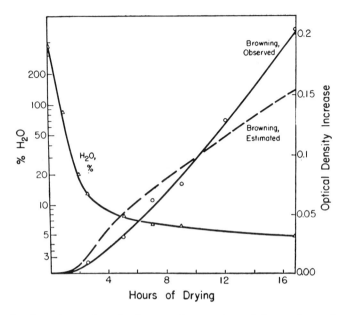

Fig. 2. Estimated and observed amounts of browning of white potato during cabinet drying. Air temperature: 75°C dry bulb, 45°C wet bulb (Hendel et al., 1955).

consequence, the temperature should be lowered during the last drying period and the drying potential.

Using a glucose-glycine browning model, Kluge and Heiss (1967) evaluated the permissible reaction time t_{zul} for a given permissible extent of browning for each temperature-water content combination that occurs during drying (Fig. 3). If we know the temperature-water content profiles during a given drying process we can add up the reciptrocals of the permissible browning times for each drying interval dt, in this way getting integral values quite analogous to the F values used for expressing the lethality effect of a heat sterilizing process. According to the described calculation method, the permissible browning limit is reached

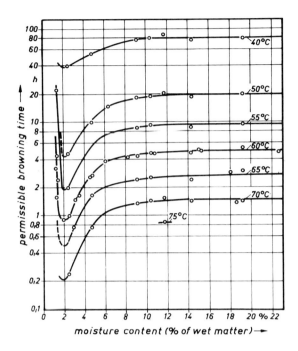

Fig. 3. *Influence of moisture content on permissible brown-ing time* (t_{zul}) *of a freeze-dried glucose-glycine-cellulose model system at different temperatures (Kluge and Heiss, 1967).*

when the hatched area in Fig. 4,   becomes 1.

$$F_Q \int_0^{t_r} \frac{1}{t_{zul}}\, dt$$

In this example, representing an air drying experiment, this value was reached after 137 min. Here drying could be stopped after about 90 min, where no more loss of moisture occurs and well before the permissible browning limit is reached.

From the experiments described, general conclusions can be drawn with respect to an improvement in the quality of dried products. These investigations are based on the measurement of browning rates under different temperature-water content con-ditions, corresponding to the last step of a multistage deteriora-

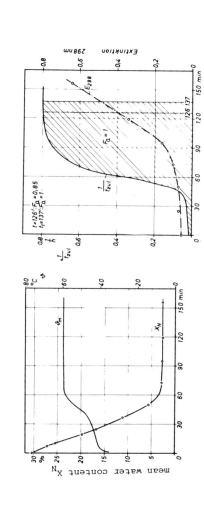

Fig. 4. Effect of drying conditions and drying time on the browning extent of a freeze-dried glucose-glycine-cellulose model system (the permissible extent of browning is corresponding to $F_Q = 1$) (Kluge and Heiss, 1967).

tive reaction accompanied by the formation of off-flavors. Therefore, an analytical method based on the evaluation of the end products of deteriorative reactions that lower the sensory quality are not satisfactory. We were looking for analytical methods that can detect the onset of sugar-amine browning reactions well before detrimental sensory changes occur.

II. FORMATION OF BROWNING INTERMEDIATES AS AN EARLY INDICATION OF MAILLARD REACTION IN PRODUCTS OF PLANT ORIGIN.

It is well known that the sugar-amine browning reaction in foods is initiated with the formation of reducing intermediates, which are formed by a condensation reaction between reducing sugars and the amino groups of amino acids and proteins, followed by an Amadori rearrangement (Fig. 5). This reaction converts aldosyl-amino acids formed by condensation between aldoses (e.g., glucose) and amino acids to ketose-amino acids (Amadori compounds, e.g., fructose-amino acids), which can be detected in foods because of their reducing properties (Abrams et al., 1955). During further progress of the reaction 3-deoxyhexosulose (III in Fig. 5), a very reactive dicarbonyl sugar derivative is formed, which on the one side undergoes further condensation reactions resulting in brown pigments (Hodge, 1953; Reynolds, 1963, 1965), and on the other side may react with free amino acids to produce volatile Strecker-degradation products (Hodge, 1967; Reynolds, 1970) exerting detrimental sensory effects. In this context the great variety of side reactions is not discussed because our investigations were restricted to the formation of Amadori products and browning. As shown in Fig. 5, the formation of Amadori compounds is an acid-base-catalyzed reaction and, as expected, is promoted in the presence of acid-base catalysts like proton-donating and proton-accepting buffers of weak organic and inorganic acids.

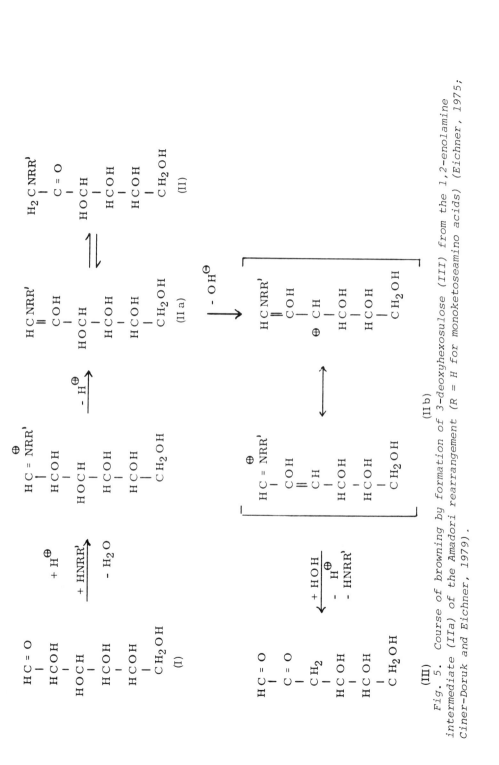

Fig. 5. Course of browning by formation of 3-deoxyhexosulose (III) from the 1,2-enolamine intermediate (IIa) of the Amadori rearrangement (R = H for monoketoseamino acids) (Eichner, 1975; Ciner-Doruk and Eichner, 1979).

It can also be seen that the Maillard reaction may proceed
directly via the 1,2-enaminol intermediate of the Amadori rearrangement or via its end product, the ketose-amino acid. As
reported by Eichner (1974) the direct reaction path is favored
at higher water contents. This has been verified in recent experiments with low-moisture, sugar-amino acid model systems,
which are outlined later in this chapter.

Furthermore, it has already been shown that Amadori products
formed in low-moisture, sugar-amino acid model systems can be
detected by amino acid analysis (Eichner, 1974). Figure 6 shows
the separation of different fructose-lysine derivatives from unreacted lysine. (As a general principle, Amadori compounds always appear before the corresponding amino acids.) The formation
of fructose-lysine derivatives by reaction of glucose residues
with the ε-amino groups of protein-bound lysine has been used
for an early detection of quality changes in proteins caused by
Maillard reactions (Reynolds, 1965; Erbersdobler and Zucker,
1966; Erbersdobler, 1970). In this case the proteins have to be
hydrolyzed, where about 50% of the sugar-bound lysine is split
back and about 50% is converted to the lysine derivatives furosine
and pyridosine (Heyns et al., 1968; Finot et al., 1968, 1969;
Sulser and Büche, 1969) (Fig. 7), which can be detected by amino
acid analysis. In this way no exact picture of the real reaction extent can be obtained.

Although Amadori compounds of free amino acids have been
found in animal and plant tissue (Abrams et al., 1955; Borsook
et al., 1955; Anet and Reynolds, 1956, 1957; Huang and Draudt,
1964), they were not used systematically for an early detection
of Maillard reactions in these substrates. Since all foods of
plant origin contain free amino acids, it seemed promising for
us to look for Amadori compounds in heated or stored vegetable
products and to investigate the conditions for their formation
and decomposition to brown pigments and volatile products causing
sensory changes.

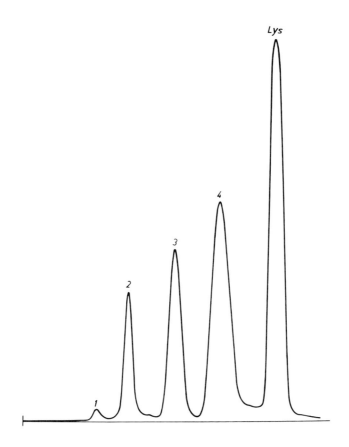

Fig. 6. Separation of fructoselysine derivatives formed in a heated low-moisture glucose-lysine-Avicel model system by amino acid analysis (Eichner, 1975; Ciner-Doruk and Eichner, 1979). Peak 2, α,ε-difructoselysine; peak 3, α-fructoselysine; peak 4, ε-fructoselysine; peak 1, unknown compounds.

As an example, we first chose tomato powder, a very sensitive product with regard to browning. By amino acid analysis of water extracts of spray-dried tomato powder, a series of peaks (Fig. 8, capital letters) appeared that cannot be associated with amino acids, whereas in tomato concentrate and freeze-dried tomato powder none or only minor quantities of these compounds could be found. For the purpose of identification of these unknown peaks, low-moisture model systems containing the individual reducing

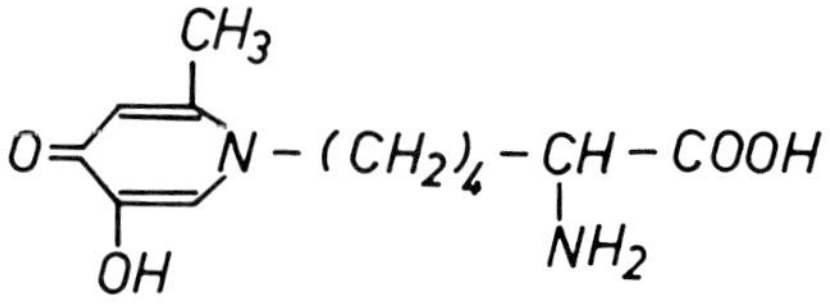

Fig. 7. *Furosine (I) and pyridosine (II) as formed by acid hydrolysis of heated milk protein.*

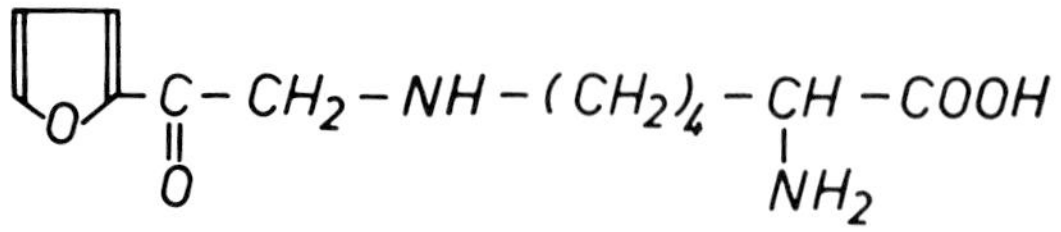

sugars (glucose, fructose, and galacturonic acid) and amino acids of tomato were heated and the resulting reaction products were analyzed and related to the unknown compounds formed in spray-dried tomato powder by thermal processing. It turned out that peak C comprised the Amadori products fructose-glutamic acid together with small amounts of fructose-asparagine, fructose-serine and fructose-threonine, whereas peak B contained the corresponding reaction products with galacturonic acid (tagaturon-amino acids; Heyns and Schultz, 1962). Peaks A_1 and A_2 correspond to reaction products of galacturonic acid and glucose with aspartic acid, peaks D and H to the respective reaction products with γ-aminobutyric acid. No browning intermediates could be found by reaction of fructose with amino acids.

It should be emphasized that for the analysis of these amino acid derivatives, simple water extraction could be applied. Thus no decomposition of Amadori compounds took place as in the case of protein derivatives, where acid hydrolysis has to be used.

The rate of amino acid loss and of formation of Amadori products is increased by increasing temperature and water content.

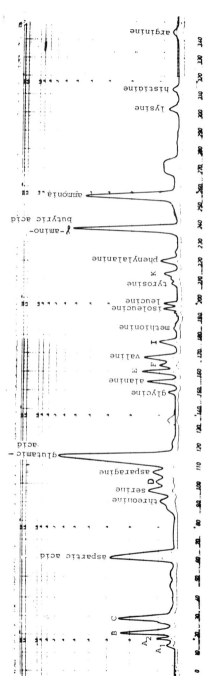

Fig. 8. Amino acid chromatogram of heat processed tomato powder. Peak: Reaction product from A_1, galacturonic acid + aspartic acid; A_2, glucose + aspartic acid; B, galacturonic acid + glutamic acid, asparagine, serine, threonine; C, glucose + glutamic acid, asparagine, serine, threonine; D, galacturonic acid + γ-aminobutyric acid; I, glucose + NH_3; K, glucose + lysine.

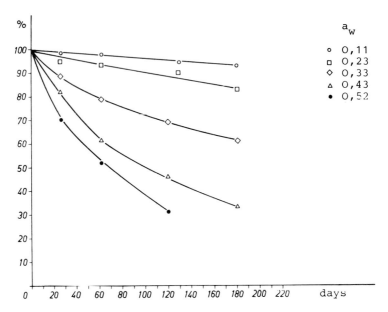

Fig. 9. Glutamic acid loss in freeze-dried tomato powder during storage at 23°C dependent on water activity.

Figures 9 and 10 show the decrease of glutamic acid in to-mato powder dependent on the equilibrium relative humidity at 23° and 40°C, adjusted by placing in vacuum exsiccators over different concentrated salt solutions (Rockland, 1960). Figures 11 and 12 show the corresponding increase of the peak C area in the amino acid chromatogram (Fig. 8) having a characteristic approximately specular to the amino acid loss. This relationship is less significant with the other Amadori products because of their lower stability (cf. Table I and II).

It is remarkable that there was no induction period in the formation of Amadori compounds in contrast to browning, as demonstrated in Figs. 13-15 for 23°, 40°, and 55°C, respectively. It can be seen that the browning characteristic dependent on equilibrium relative humidity is quite similar at the temperatures

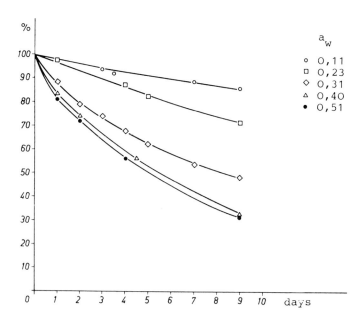

Fig. 10.   Glutamic acid loss in freeze-dried tomato powder
at elevated storage temperature (40°C) dependent on water
activity.

investigated, the induction period always being more pronounced
at the lower moisture contents.

   At the lowest relative humidities the browning rate was very
slow compared with the rate of formation of browning intermediates.
This may be attributed to the fact that, because of the high
diffusion resistance in the reaction medium, which lowers reactant mobility (Eichner and Karel, 1972; Eichner, 1974; Duckworth, 1962), a multistage reaction such as browning is inhibited
to a higher extent than a simple two-step reaction such as the
formation of Amadori compounds.   Therefore, at low moisture
contents they can accumulate over longer periods of time without
any noticeable browning.   This is shown in Fig. 16 for relative
humidities between 11 and 33% and temperatures of 23° and 55°C.

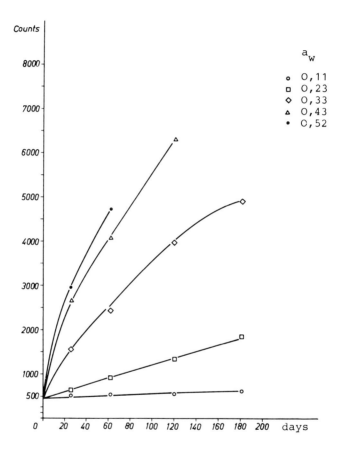

Fig. 11.   Increase of the area (counts) of peak C (Fig. 8)
during storage of freeze-dried tomato powder at 23°C dependent
on water activity.   (The initial concentration of glutamic acid
corresponds to a peak area of 27,000 counts; for comparison
purposes the counts of peak C must be multiplied by a ninhydrin-
color factor of 2.)

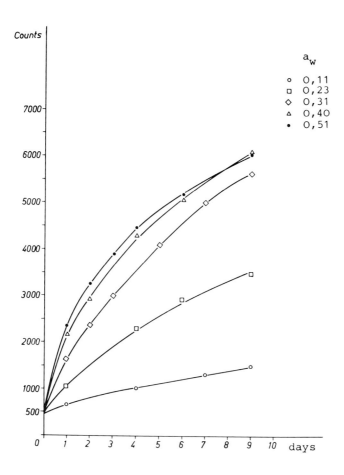

Fig. 12. Increase of the area (counts) of peak C (Fig. 8) during storage of freeze-dried tomato powder at 40°C dependent on water activity.

Table I[a]

Water activity	Decomposition of Amadori compounds (%)	Browning ext. 420 nm	Amino acid loss, conc. of Amadori compounds (%)	Browning ext. 420 nm
	Fructose-glutamic acid		Glucose + glutamic acid	
0.23	–	0.008	0.6	0.016
0.31	–	0.008	5.5	0.031
0.40	1	0.012	9.2	0.060
0.51	13	0.025		
	Tagaturon-glutamic acid		Galacturonic acid + glutamic acid	
0.23	–	0.007	4.1	0.018
0.31	8	0.009	13.7	0.074
0.40	12	0.025	31.9	0.552
0.51	26	0.066		

[a]The extinction values are related to 3.35×10^3 M concentrations of amino acids or ketoseamino acids (Amadori compounds), respectively.

TABLE II[a]

Water activity	Decomposition of Amadori compounds (%)	Browning ext. 420 nm	Amino acid loss, conc. of Amadori compounds (%)	Browning ext. 420 nm
	Fructose-γ-aminobutyric acid		*Glucose + γ-aminobutyric acid*	
0.23	–	0.006	1.9	0.045
0.31	7	0.034	16.2	0.172
0.40	19	0.160	33.9	0.232
0.51	20	0.178		
	Tagaturon-γ-aminobutyric acid		*Galacturonic acid + γ-aminobutyric acid*	
0.23	17	0.336	5.0	0.115
0.31	65	0.681	23.8	1.140
0.40	95	0.914	50.4	2.110
0.51	99	1.224		

[a]The extinction values are related to 3.35×10^3 M concentrations of amino acids of ketoseamino acids (Amadori compounds), respectively.

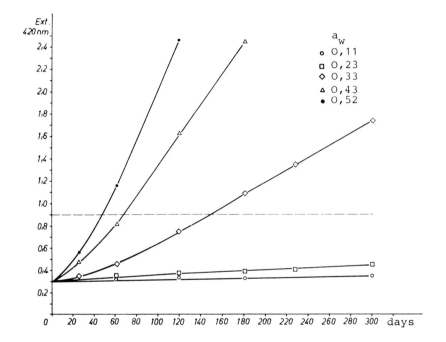

Fig. 13. Browning of freeze-dried tomato powder during
storage at 23°C and different water activities. Ext.420 = 0.9
corresponds to the limit of acceptability (For determination
of the extinction values at 420 nm about 1 g (0.957 g dry
matter) of tomato powder was extracted with 25 ml water).

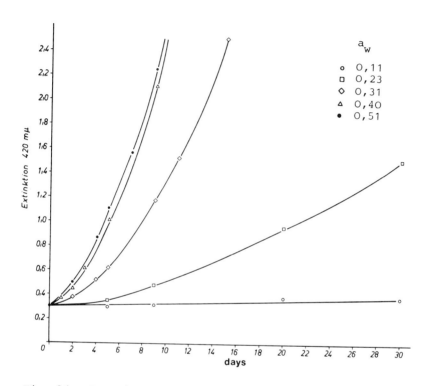

Fig. 14.   Browning of freeze-dried tomato powder during
storage at 40°C and different water activities.

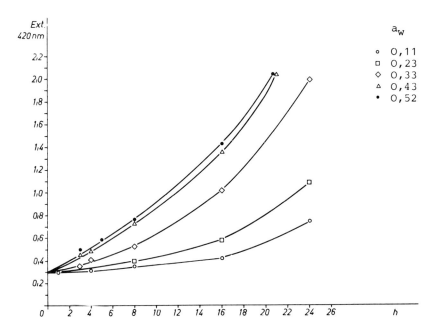

Fig. 15. Browning of freeze-dried tomato powder equili-brated to different water activities at 23°C during heating at 55°C.

In contrast to browning, Amadori compounds are formed without any induction period at all moisture contents. Therefore they can be used for the early indication of the onset of Maillard reactions. In this respect the reaction products contained in peak C (Fig. 8) are most suitable because of their higher stability compared with other Amadori compounds. Furthermore the temperature coefficients (Q_{10} values) for the formation of these browning intermediates are significantly higher than the Q_{10} values of browning, as shown in Table III.

Therefore, the Amadori compounds of peak C can be used as very sensitive indicators for the overall heat impact during processing.

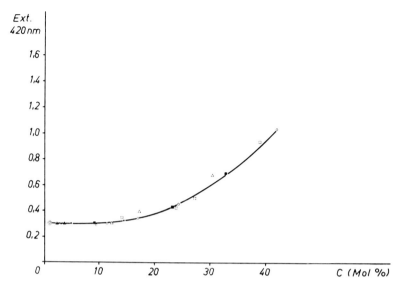

Fig. 16. Correlation between browning and the concentrations of browning intermediates (peak C in Fig. 8) formed at different temperatures and water activities. ▲ a_w at 23°C: ● , 0.11; △, 0.23; ▪ , 0.33. a_w at 55°C: ○ , 0.11; ○, 0.23; □ , 0.33.

TABLE III. Q_{10} Values for the Formation of Browning Intermediates (1) and Visible Browning (2) between 23 and 40°C

a_w	(1)	(2)
0.23	11.3	6.8
0.32	6.1	5.6
0.41	5.1	4.4
0.52	4.5	3.8

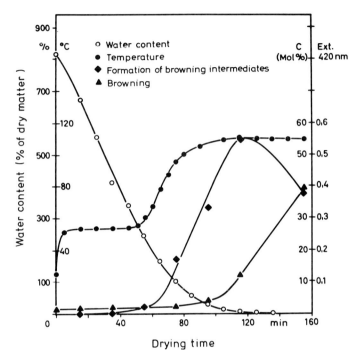

Fig. 17. Formation of browning intermediates and browning during air drying of carrot cubes (air temperature: 110°C).

Figure 17 shows an air-drying experiment with carrot cubes using an air temperature of 110°C (K. Eichner and W. Wolf, unpublished results). It can be seen that from a water content equivalent to about 70% (related to wet matter) the product temperature rises and at the same time the rate of formation of Amadori compounds of peak C (Fig. 8) increases and parallels the temperature increase until a maximum (reaction extent at the maximum: about 30% in relation to the corresponding amino acids) is reached, whereas browning does not become significant until this point, where decomposition of browning intermediates begins to predominate.

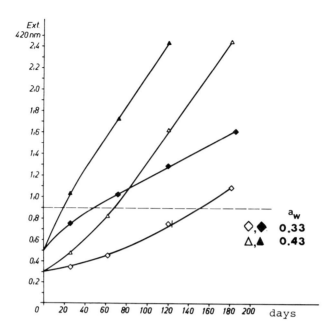

Fig. 18. *The influence of heat treatment on the shelf life of freeze-dried tomato powder stored at 23°C and different water activities (Ext.$_{420}$ = 0.9 corresponds to the limit of acceptability.)* ◆ , ▲ , *preheated at 40°C and a water activity of 0.11 (about 30% of glutamic acid having reacted to fructose-glutamic acid);* ◇ , △ , *unheated.*

From this experiment it becomes clear that optimization of drying processes based on measurements of browning in different drying periods is unsatisfactory because the Maillard reaction may have proceeded to a great extent during drying before any noticeable browning occurred. As a consequence, if part of the induction period of browning has been used up by the formation of browning intermediates, it has to be expected that the storage stability of the dried product will be decreased, depending on the concentration of these intermediate. As an example, Fig. 18 shows the browning rates of unheated and preheated, freeze-dried tomato powder, where about 30% of the amino acids have been reacted to Amadori compounds of peak C in Fig. 8, at different equilibrium relative humidities. It can be seen that with the preheated product there is no more induction period of browning in contrast to the unheated product; thus the shelf life is greatly reduced.

From these results it may be inferred that for achieving the highest possible quality and storage stability of dried products of plant origin, the formation of colorless browning intermediates should be studied, depending on temperature and product moisture content, and the results can then be applied for optimization of drying.

III. FORMATION AND DECOMPOSITION OF BROWNING INTERMEDIATES AND
      VISIBLE BROWNING IN MODEL SYSTEMS

A. *Formation of Browning Intermediates and Browning with Different Sugars and Amino Acids in Tomato Powder*

After having investigated the browning characteristics of tomato powder, we studied appropriate model systems made up of reducing sugars and amino acids, which mainly contribute to overall browning (K. Eichner and M. Ciner-Doruk, unpublished results). It turned out that galacturonic acid and γ-aminobutyric acid have the highest browning potential, while fructose and glutamic acid give only a minor contribution to browning, Glucose has a position between fructose and galacturonic acid. The model systems investigated were prepared by adjusting solutions containing reducing sugars, amino acids, and citric acid (molar ratio: 1.2:1:0.5) to pH 4.2 (corresponding to the pH value of tomato extracts), thoroughly mixing with Avicel and freeze-drying. For studying the browning behavior these model systems were stored at 40°C in vacuum desiccators over

appropriate salt solutions (Rockland, 1960) and analyzed after
varying periods of time.

Figure 19 shows the total reaction of glutamic acid with
glucose at two water activities (open signs), the corresponding
increase in fructose-glutamic acid (filled signs), and browning
(dashed lines). It can be seen that fructose-glutamic acid is
rather stable at a medium water activity of 0.4, where only little
browning was observed, whereas at this water activity tagaturon-
γ-aminobutyric acid (formed in the galacturonic acid-γ-aminobutyric
acid model system) proved to be very unstable and a large difference between the amount of total amino acid reacted and the

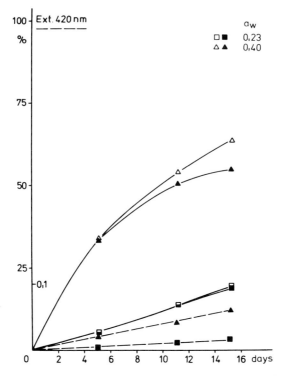

*Fig. 19. Total reaction of glutamic acid (□ , △), formation
of fructose-glutamic acid and browning (■ , ▲) of a glucose-
glutamic acid-citric acid-Avicel model system (1.2:1:0.5 mol,
3.5 g Avicel per mmol of amino acid) at different water activities.
For browning measurements 250 mg (dry matter) of the model system
were extracted with 10 ml water.*

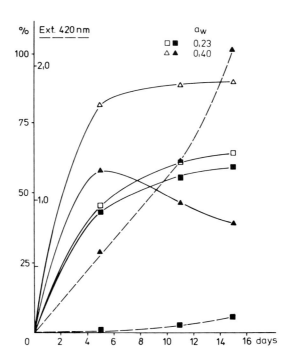

Fig. 20. Total reaction of γ-aminobutyric acid, formation of tagaturon-γ-aminobutyric acid, and browning of a galacturonic acid-γ-aminobutyric acid-citric acid-Avicel model system (1.2:1: 0.5 mol) at different water activities (symbols as in Fig. 19).

corresponding Amadori compound formed in accordance with a high browning rate was observed (Fig. 20). The combinations galacturonic acid-glutamic acid and glucose-γ-aminobutyric acid have a position between the two model systems described (Figs. 21 and 22). In general, at the lower water activity only small differences between the total reaction of amino acids and the formation of Amadori compounds together with a low browning rate was observed. At this water activity glutamic acid shows a lower reaction rate than γ-aminobutyric acid, which must be attributed to its lower solubility in water.

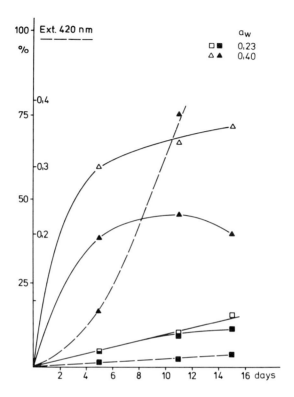

Fig. 21.   Total reaction of glutamic acid, formation of
tagaturon-glutamic acid and browning of a galacturonic acid-
glutamic acid-citric acid-Avicel model system (1.2:1:0.5 mol) at
different water activities (symbols as in Fig. 19).

In Fig. 23 the browning of glucose, fructose, and galacturonic
acid in combination with γ-aminobutyric acid at 40°C and a water
activity of 0.31 are compared, showing that fructose has the
lowest browning potential.

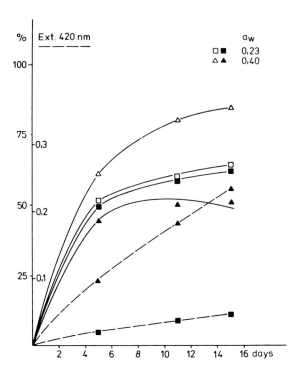

Fig. 22.   Total reaction of γ-aminobutyric acid, formation of fructose-γ-aminobutyric acid, and browning of a glucose-γ-amino-butyric acid-citric acid-Avicel model system (1.2:1:0.5 mol) at different water activities (symbols as in Fig. 19).

B.   *The Influence of Organic Acid Buffers and pH on the Formation of Browning Intermediates and Browning*

Since all foods of plant origin contain organic acids in buffered form, their influence and the effect of pH on the reaction rate were studied.   Acid-base catalysis is supposed to play a major role in the Maillard reaction.   The signification of this principle was investigated using a low-moisture, glucose-glycine-citric acid-Avicel model system as an example.   Figure 24 shows the total amino acid reaction at pH 4.2 and 7.0 in the absence and presence of citric acid (open signs) as well as the formation

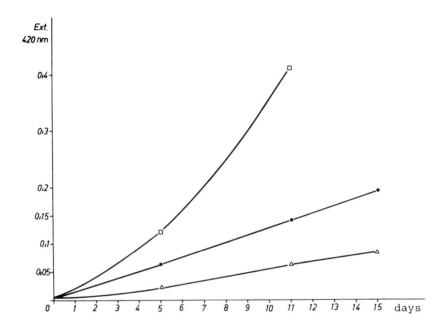

Fig. 23.   Browning of Avicel model systems containing γ-amino-
butyric acid and citric acid in combination with glucose (●),
fructose (Δ), and galacturonic acid (□) respectively at 40°C
and a water activity of 0.40 (molar ratios, see Figs. 19-22).

of fructose-glycine under the same conditions (filled signs).
Reaction conditions were 40°C, a_w = 0.40. It is remarkable that
an increase in pH caused only a relatively small increase in
reaction rate compared with the effect of citrate ions, apparently
because of the greatly accelerating effect of acid-base catalysis.
A comparison of Fig. 24 with Fig. 25 shows again that the browning
rate is correlated with the difference between the total amount
of amino acid reacted and fructose-amino acid formed.   This
difference apparently is due to decomposition of Amadori compound
causing browning.

      Using glutamic acid instead of glycine, the accelerating
effect of added organic acid ions was much less, because glutamic

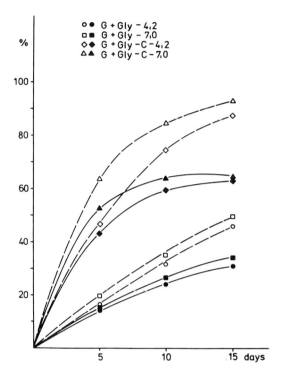

Fig. 24. Total reaction of glycine (dashed curves) and formation of fructose-glycine (solid curves) in glucose-glycine-citric acid-Avicel model systems (5:1:1 mol 1.95 g Avicel per mmol of glycine) in the absence and presence of citric acid (C) at pH 4.2 and 7.0 (40°C, a_w = 0.40).

acid *per se* obviously exerts an effect by means of its second acid group.

C.   *Decomposition of Amadori Compounds of Tomato Powder and*
     Browning

The browning rate of tomato powder apparently depends on the stability of the individual browning intermediates as indicated in Fig. 19 with Fig. 20. For further elucidation of these findings the most important Amadori compounds formed in tomato powder were prepared in order to study their stability

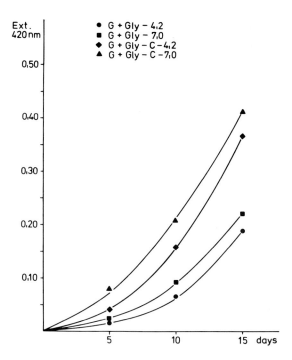

Fig. 25. Browning of glucose-glycine-citric acid-Avicel model systems (5:1:1 mol) in the absence and presence of citric acid (C) at pH 4.2 and 7.0 (40°C, a_w = 0.40). For browning measurements 430 or 370 mg (dry matter) of the model systems with or without citric acid were extracted with 10 ml water.

and browning potential dependent on water activity. In Tables I and II the degree of decomposition and resultant browning of the isolated browning intermediates after a reaction time of 5 days at 40°C are compared with browning of model systems containing the corresponding free sugars and amino acids. Browning of these model systems is set in relation to the difference between the amount of amino acids reacted and Amadori compounds being formed, which should be equivalent to the proportion of decomposed Amadori compounds (reaction time: 15 days at 40°C).

As shown in Tables I and II, in accordance with Figs. 19 and 20, fructose-glutamic acid has the highest and tagaturon-γ-amino-

butyric acid derivatives generally have a higher browning potential than the corresponding glutamic acid derivatives.

By comparing the tagaturon-amino acid models and the models
containing galacturonic acid and amino acids it becomes clear that
in the latter cases — especially at higher water activities —
the reaction pathway leading to brown pigments predominantly must
proceed the "direct way," i.e., the 1,2-enaminol intermediate of
the Amadori rearrangement (Fig. 5), because decomposition of the
corresponding Amadori compounds gave a comparatively small contribution to browning.

During decomposition of Amadori compounds and resulting
browning, part of the reacted amino acids is set free, which also
applies in direct browning (see Fig. 5). By this means they can
again react with reducing sugars whereby a higher molar ratio of
sugar molecules compared to amino acid molecules present may be
converted to brown pigments. The portion of reacted amino acids
not being converted to Amadori compounds (difference between reacted amino acids and identified Amadori compounds) may be incorporated in brown pigments and/or react to volatile products
by Strecker degradation (Hodge, 1967).

D. *The Influence of Organic Acids, Organic Acid Buffers, and*
pH on the Decomposition of Browning Intermediates and Browning

Figure 26 shows that Amadori compounds — for example, tagaturon-γ-aminobutyric acid — are stabilized by increasing amounts
of citric acid, whereby browning is greatly diminished. Since
the addition of free acids is associated with a drop in pH, the
effect of citric acid at constant pH had to be investigated.
Figure 27 and 28 show that, in the presence of citrate and with an
increase in pH, decomposition of Amadori compounds and browning
are promoted. Jointly with the fact that these dependences in

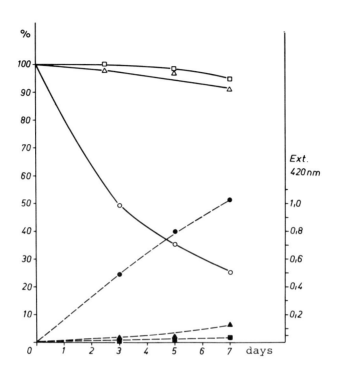

Fig. 26. Decomposition of tagaturon-γ-aminobutyric acid (open signs) and resultant browning (filled signs) (cf. Table III) in the absence and presence of citric acid at 40°C and a_w = 0.31. Decomposition, open symbols; browning (ext. 420 mg) solid symbols. Without citric acid (pH 4.95) O, ●; with citric acid (pH 4.4) △, ▲, (pH 4.25) □ , ■ .

principle also apply to the formation of browning intermediates (Fig. 24), it becomes clear that overall browning is enhanced in a similar way by these parameters (Fig. 25).

IV.   CONCLUSIONS

From the experiments described it may be inferred that the quality and storage stability of dried products of plant origin

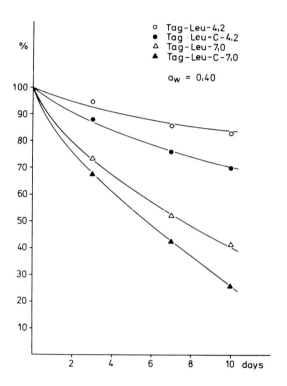

Fig. 27. Decomposition of tagaturon-leucine in the absence and presence of citric acid (C) (1:1 mol) at pH 4.2 and 7.0 (40°C, a_w = 0.40) in an Avicel model system (1.95 g Avicel per mmol of the Amadori compound).

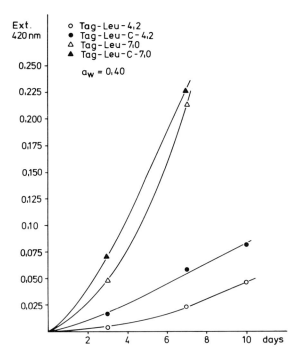

Fig. 28. Browning of tagaturon-leucine in the absence and presence of citric acid (C) (1:1 mol) at pH 4.2 and 7.0 (40°C, a_W = 0.40) in an Avicel model system (see Fig. 27). For browning measurements 300 mg (dry matter) of the model system were extracted with 10 ml water.

can be improved if the formation of Maillard reaction intermediate (Amadori) compounds are minimized during heat processing. After these intermediates are formed by thermal treatment, the induction period, leading to sensory changes, is decreased.

Furthermore, it was demonstrated that the overal reaction rate depends not only on environmental conditions, such as temperature and water content, but also on the chemical composition of the food.

On the basis of further investigations it may be possible to select the raw materials most suitable for quality retention during processing and storage.

REFERENCES

Abrams, A., Lowy, P. H., and Borsook, H. (1955). Preparation of 1-amino-1-deoxy-2-ketohexoses from aldohexoses and α-amino acids. *J. Am. Chem. Soc.* 77, 4794.

Anet, E. F. L. J., and Reynolds, T. M. (1956). Reactions between amino acids, organic acids and sugars in freeze-dried apricots. *Nature 177,* 1082.

Anet, E. F. L. J., and Reynolds, T. M. (1957). Chemistry of non-enzymic browning. I. Reactions between amino acids, organic acids, and sugars in freeze-dried apricots and peaches. *Austral. J. Chem. 10,* 182.

Borsook, H., Abrams, A., and Lowy, P. H. (1955). Fructose-amino acids in liver: stimuli of amino acid incorporation in vitro. *J. Biol. Chem. 215,* 111.

Ciner-Doruk, M., and Eichner, K., (1979). Bildung and stabilität von Amadori-verbindungen in wasserarmen lebensmitteln. *2 Lebsm. Unters. Forsch. 168,* 9.

Duckworth, R. B. (1962). Diffusion of solutes in dehydrated vegetables. *In* "Recent Advances in Food Science" (J. Hawthorn and J. M. Leitch, eds.), Vol. 2, p. 46. Butterworths, London.

Eichner, K. (1975). The influence of water content on nonenzymic browning reactions in dehydrated foods and model systems and the inhibition of fat oxidation by browning intermediates. *In* "Water Relations of Foods" (R. B. Duckworth, ed.), pp. 417-434. Academic Press, New York.

Eichner, K., and Karel, M. (1972). The influence of water content and water activity on the sugar-amino browning reaction in model systems under various conditions. *J. Agr. Food Chem. 20,* 218.

Erbersdobler, H. (1970). Zur Schädigung des Lysins bei der Herstellung und Lagerung von Trockenmilch. *Milchwissenschaften 25,* 280.

Erbersdobler, H., and Zucker, H. (1966). Untersuchungen zum Gehalt an Lysin und verfügbarem Lysin in Trockenmagermilch, *Milchwisenschaften 21,* 564.

Finot, P. A., Bricout, J., Viani, R., and Mauron, J. (1968). Identification of a new lysine derivative obtained upon acid hydrolysis of heated milk. *Experientia 24,* 1097.

Finot, P. A., Viani, R., Bricout, J., and Mauron, J. (1969).
    Detection and identification of pyridosine, a second lysine
    derivative obtained upon acid hydrolysis of heated milk.
    Experientia 25, 134.
Hendel, C. E., Silveira, V. G., and Harrington, W. O. (1955).
    Rates of nonenzymatic browning of white potato during de-
    hydration. *Food Technol. 9*, 433.
Heyns, K., and Schulz, W. (1962). Die Umsetzung von D-Glucuron-
    säure und D-Galacturonsäure mit Aminosäuren zu 1-*N*-Aminosäure-
    1-desoxy-fructuronsäuren und 1-*N*-Aminosäure-1-desoxy-tagaturon-
    säuren. *Chem. Ber. 95*, 709.
Heyns, K., Heukeshoven, J., and Brose, K.-H. (1968). Zwischen-
    produkte von Bräunungsreaktionen. Der Abbau von Fructose-
    Amino-säuren zu *N*-(2-Furoylmethyl-)aminosäuren. *Angew. Chem.
    80*, 627.
Hodge, J. E. (1953). Chemistry of browning reactions in model
    systems. *J. Agr. Food Chem. 1*, 928.
Hodge, J. E. (1967). Origin of flavor in foods. Nonenzymatic
    browning reactions. *Proc. Symp. Foods*, "The Chemistry and
    Physiology of Flavors" (W. H. Schultz, E. A. Day, and L. M.
    Libbey, eds.), pp. 465-491. AVI Publ. Westport, Connecticut.
Huang, I.-Y., and Draudt, H. N. (1964). Effect of moisture on
    the accumulation of carbonyl-amine browning intermediates in
    freeze-dried peaches during storage. *Food Technol. 18*, 124.
Kluge, G., and Heiss, R. (1967). Untersuchungen zur besseren
    Beherrschung der Qualität von getrockneten Lebensmitteln unter
    besonderer Berücksichtigung der Gefriertrocknung. *Verfahrens-
    technik 1*, 251.
McDonald, F. J. (1966). Available lysin content of dried milk.
    Nature (London) 209, 1134.
Reynolds, T. M. (1963). Chemistry of nonenzymic browning. I. The
    reaction between aldoses and amines. *Advan. Food Res. 12*, 1.
Reynolds, T. M. (1965). Chemistry of nonenzymic browning. II.
    Advan. Food Res. 14, 168.
Reynolds, T. M. (1970). Flavours from nonenzymic browning re-
    actions. *Food Technol. Australia 22*, 610.
Rockland, L. B. (1960). Saturated salt solutions for static
    control of relative humidity between 5° and 40°C. *Anal.
    Chem. 32*, 1375.
Sulser, H. (1973). Die Bedeutung des Fructoselysins und seiner
    Abbauprodukte Furosin und Pyridosin für die Qualitätsbeurtei-
    lung von Lebensmitteln. *Lebensm. Wiss. Technol. 6*, 66.
Sulser, H., and Büche, W. (1969). Abbauprodukte von Fructosely-
    sin in pflanzlichen Trockenlebensmitteln und in einem Modell-
    gemisch nach thermischer Behandlung. *Lebensm. Wiss. Technol.
    2*, 105.

THE NONENZYMATIC BROWNING REACTION
AS AFFECTED BY WATER IN FOODS

Theodore P. Labuza
Miriam Saltmarch

I. INTRODUCTION

The primary nonenzymatic browning reaction that can occur during storage of dehydrated or semimoist foods is the Maillard reaction. The reaction results from reducing compounds, primarily sugars, reacting with proteins or free amine groups and results in changes in both the chemical and physiological properties of the protein. This, in turn, affects the nutritional value of the food as well as the color and texture since it gets darker and becomes tougher.

Both temperature and water content have specific effects on the browning reaction as demonstrated in this review. The water content of a food is especially important to the food industry because it controls the rate and type of deteriorative reaction. The influence of water content of dehydrated and semimoist food products during both processing and storage may result in a change in color, flavor, and more importantly, the nutritional quality of the protein.

The influence of temperature and moisture content on the browning reaction during processing and storage has been extensively studied over the last ten years. However, many of these studies have not utilized a kinetic approach when evaluating the deleterious effects of the browning reaction on various food systems. By

analyzing the affect of water content or water activity (a_w) on
the kinetic parameters of the browning reaction as demonstrated
in various studies, a better understanding of the role of water
in the Maillard reaction should result. An understanding of this
role would be useful in controlling and predicting the shelf life
of susceptible products.

II. GENERAL EFFECTS OF WATER ON BROWNING

In order to understand the affect of water on the nonenzymatic browning reaction, it is first necessary to define the state
of water in foods.

Water contributes to the textural characteristics of a food
through its physical state. In addition, its interaction with
food components can be directly related to the chemical reactions
that take place (Labuza, 1970; 1975; Rockland, 1969).

The term by which this interaction is quantified is called the
water activity of the food. This is a measure of the relative vapor pressure above the food as defined in

$$a_w = \frac{p}{p_0} = \frac{\%RH}{100} \tag{1}$$

where a_w is the water activity, p the vapor pressure above food
at temperature T, p_0 the vapor pressure of pure water at T, and
%RH the equilibrium relative humidity at which food neither gains
nor loses water. Water activity or relative humidity is related
to the moisture content of the food through the sorption isotherm
shown in Fig. 1.

Methods to determine isotherms and measure water activity have
been reviewed elsewhere (Labuza, 1968, 1975; Gal, 1975). The lower part of the curve in Fig. 1 is the moisture region of dehydrated
foods (a_w = 0 to 0.5) and the upper part (a_w = 0.6 to 0.9) applies to
semimoist food products. Most natural tissue foods such as meats,

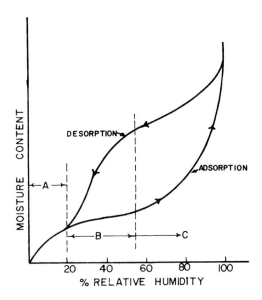

Fig. 1.   Typical water sorption isotherm showing hysteresis.

fish, vegetables, and fruits have $a_w \approx 1.0$ and moisture contents
greater than 40% water on a wet basis.   A knowledge of the region
of the moisture sorption isotherm where a particular food product
belongs permits conclusions to be made about the state of the
water in the food product and the possible reactions that might
take place.

     To understand the effect of a_w on nonenzymatic browning, this
state or degree of boundness of water must be examined.   At low
a_w water is tightly bound to surface polar sites by chemisorption
and is generally unavailable for reaction and solution.   The upper
limit of this region is called the BET monolayer value, which occurs at about a_w = 0.2-0.3 for most foods.   This value is the most
stable moisture content for most dehydrated foods (Salwin, 1959).

     At or above the monolayer, water is held to a varying degree
in multilayers, in capillaries, and possibly entrapped in various
structural components.   Dissolved solutes also reduce the freedom

of movement of water due to colligative properties as defined by
Raoult's law and the Flory-Huggins theory (Lewicki *et al.*, 1978).
These factors all account for a reduction in the relative vapor
pressure of water as solids content increases, but they do not
completely inhibit the ability of water to act as a solvent, reaction medium and as a reactant itself (Labuza, 1975).

As a result, many deteriorative reactions increase exponentially in rate as a_w increases above the monolayer. However, the rate
may level off at high a_w for some reactions or even decrease again.
Three general patterns of the effect of a_w upon reaction rates are
shown in Fig. 2. In general, nonenzymatic browning follows pattern I, in which a maximum is present.

With respect to browning, water can retard the rate of the initial glycosylamine reaction in which water is a product. This results in product inhibition. Eichner and Karel (1972) found this
to be the case in studies of browning reaction between glucose and
glycine in glycerol/water mixtures. Other reactions in the sequence may also be inhibited since three moles of water are produced per mole of carbohydrate used. On the other hand, water may
enhance deamination reactions in the browning reaction sequence as
observed by Reynolds (1963) for the production of furfural or hy-
droxymethylfurfural.

A second important factor that could decrease reaction rates,
is the dilution of reactive components with increasing water content (Eichner, 1975; Labuza, 1970). Since the overall rate of a
reaction can be represented by

$$dB/d\theta = k(R)^n$$

where $dB/d\theta$ is the rate in moles/liter/hr, k the rate constant,
R the concentration of reacting species, and n the order, if the
water content increases without any change in the amount of dissolved reacting species, then the value of R decreases since it is
the number of moles per liter of water. At high a_w, a small change
in a_w leads to a large increase in moisture content (Fig. 2) and

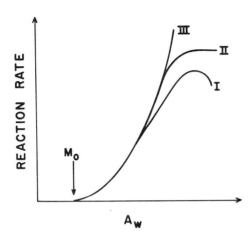

Fig. 2.   Effect of a_w on reaction rate, showing (I) a rate
increase to a maxima then a decrease, (II) an increase to a maxima then leveling off, (II) a continuous rate increase.

thus a large dilution effect.   Of course at lower a_w the increased
water content can dissolve new reacting species.

The decrease of the viscosity of the aqueous phase with increasing moisture content can result in faster mobility of the
reactive species.   This will increase the value of k in the above
equation and lead to a greater browning rate.   The first two factors eventually overcompensate for the decreased viscosity at high
a_w levels and thus the overall rate of browning decreases as is
generally observed.

Specifically, as shown in Fig. 3, studies on many dry foods
humidified over a wide range of water activity have shown a maxima for the nonenzymatic browning reaction in the intermediate-
moisture food (IMF) water activity range (Lea and Hannan, 1949;
Loncin et al., 1968; Labuza, 1972; Sharp, 1957; Sharp and Rolfe,
1958; Hendel et al., 1955; Wolfrom et al., 1974; Potthast et al.,
1976).

The region where the maxima occurs is usually near a_w = 0.65-
0.70.   However, it is important to remember that the overall effect of water content can be modified by the presence of various

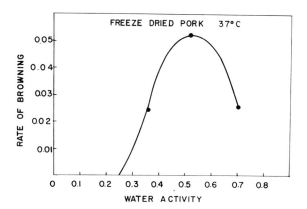

Fig. 3. Typical browning curve showing maximum for pork bites (Labuza, 1970).

substances like liquid humectants such as glycerol. Eichner and Karel (1972) and Warmbier (1976) found that both liquid and solid model systems containing glycerol had nonenzymatic browning rate maximas in the a_w range 0.41-0.55 as shown in Fig. 4. They con-

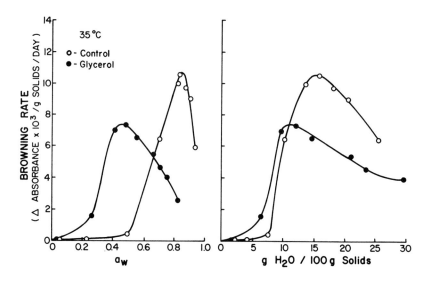

Fig. 4. Effect of glycerol on browning rate in a casein-glucose model system (Warmbier et al., 1976a).

cluded that glycerol can influence the rate of browning at lower
a_w values by acting as an aqueous solvent and thereby allowing
reactant mobility at much lower moisture values than would be expected for water alone. However, as a_w increases, the water acts
to decrease the browning rate by the mass action effect noted
above. The overall effect of glycerol or other liquid humectants
on the maximum for nonenzymatic browning is to shift it to a lower
a_w. Obanu *et al.* (1977), on the other hand, observing browning in
glycerol-amino acid mixtures stored at 65°C concluded that glycerol
itself might participate in the browning reaction. No specific
mechanism was proposed however, and this would not explain the a_w
shift.

III. GENERAL SCHEME FOR NONENZYMATIC BROWNING REACTIONS

A. *Types of Reactions*

There are three major pathways by which nonenzymatic browning
can occur: high temperature carmelization, ascorbic acid oxidation, and the Maillard reaction.

Carmelization, the browning reaction of sugars heated above
their melting point in the absence of proteins or amino acids, can
be both beneficial or detrimental to the quality of a food product.
Carmelization can be prevented by the avoidance of high-temperature
processing and storage temperatures. It is enhanced in alkaline or
acid conditions and is used to make commercial caramel colorings
and flavors.

Ascorbic acid (vitamin C) oxidation, a second kind of browning
reaction, is catalyzed by low pH and elevated temperatures. The
decomposition products resulting from the oxidation of ascorbic
acid results in a brown discoloration, as well as a decreased nutritional value.

The Maillard reaction, of special interest here because of its

effect on product acceptability and protein quality, is the third
type of nonenzymatic browning reaction. The Maillard reaction is
especially important to the food processor because it is the major
cause of browning developed during the processing and storage of
dehydrated and semimoist protein-containing foods.

In general, the accumulation of brown pigments is the most obvious indication that Maillard browning has occurred in a food containing both carbohydrates and protein. It is used as an indicator
of excessive thermal processing in the milk industry as shown by
Choi et al. (1949) and Patton (1955). The dry cereal, animal feed,
and pet food industries also experience difficulties with undesirable browning in their products.

B. Mechanism of the Maillard Reaction

The Maillard reaction, first reported in 1912 by the French
chemist Maillard (1912), occurs in the three phases briefly described in Fig. 5 as elucidated by Hodge (1953).

The first step in this scheme involves a condensation reaction
between the free amino group and the carboxyl group of a reducing
sugar resulting in a product known as a Schiff's base, as well as
a molecule of water. Subsequent cyclization and isomerization under acidic conditions (Amadori rearrangement) results in a 1-amino-
1-deoxy-2-ketose derivative. This component and its precursors are
colorless (Eichner, 1975) and this first phase is believed to be
reversible (Hurrel and Carpenter, 1974).

In the initial reaction both mono- and disaccharide reducing
sugars can react with either free amino groups or amino groups on
proteins. The reactivity of these reducing sugars varies with al-
dopentoses being more reactive than aldohexoses, which are in
turn more reactive than disaccharides (Spark, 1969). A nonreducing
disaccharide such as sucrose, if hydrolyzed by acid or an enzyme
during storage into a reducing moiety can also lead to browning
(Karel and Labuza, 1968). Amino groups must be free to react with
the sugars, with primary amines reacting most rapidly.

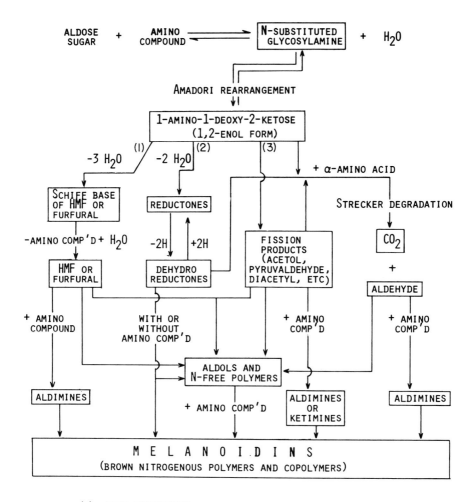

(1)   ACID CONDITIONS
(2)   BASIC CONDITIONS
(3)   HIGH TEMPERATURE

Fig. 5.   Maillard Browning pathway (from Hodge, 1953).

The second or intermediate reaction phase involves the removal
of the amino groups from the reducing sugar complex with subsequent
dehydration and cyclization, fragmentation, or amine condensation.
Depending upon environmental conditions such as pH and temperature
three general pathways exist during this phase. Under acidic conditions, hydroxymethyl furfural or furfural is produced; under mild

to basic conditions, reductones and dehydroreductones result, which
can undergo Strecker degradation; and at high temperatures fragmentation products from the Amadori product are produced.

Polymerization of the products from the second phase yields
brown melanoidin pigments in the third and final phase of the
Maillard reaction. These reactions can also lead to a toughening
of the protein as was found by Labuza (1973) for an intermediate-
moisture food at a_w = 0.75 undergoing Maillard browning.

C. Kinetics of Maillard Browning

Various kinetic models have been utilized to define and predict the specific effects of water content and a_w upon the Maillard reaction. The assumption implicit in these studies is that
by understanding the influence of water on kinetic parameters, a
more detailed picture of water effects will result.

The initial browning reaction producing brown pigment precursors is generally viewed as an overall second-order mechanism,
where

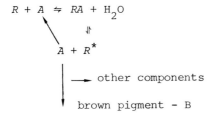

where R is the reducing sugar concentration, A the reactive amine
group concentration, RA the intermediates, R^* the reactive reducing
compound intermediates, and B the brown pigments. The rate of
brown pigment formation is related to the formation of R^* by

$$\frac{dB}{d\theta} = \frac{dR^*}{d\theta} = k_B (R)^a (A)^b \tag{2}$$

where k_B is the overall rate constant, which depends on a_w and is

inversely proportional to phase viscosity, and a, b are specific
orders for each reactant usually considered to be equal to 1.

As a_w increases above the BET value, R and A will initially
increase as more reactants are dissolved into the aqueous phase
from the crystalline state (Sloan and Labuza, 1975). However,
once the aqueous phase is saturated, concentrations of R and A
will decrease continuously with increasing water content. This
decrease can be very dramatic in the IMF range for most foods,
where a change of a_w from 0.7 to 0.9 results in a doubling to
tripling of the water content. This decrease in reactant concentration can lead to a four- to ninefold decrease in reaction rate
if no change in viscosity occurs. However, there generally is a
decrease in viscosity as water content increases that permits
greater reactant mobility (Lee and Labuza, 1975). Since the reaction rate constant is inversely proportional to viscosity, decreasing the phase viscosity acts to increase the value of k in
the reaction scheme shown above. The exact effect has not been
elucidated as a function of solids concentration or type of dissolved solids. It would depend also on whether the water/solute
interaction resulted in a pseudoplastic or Newtonian fluid. In
general, there is a decrease in reaction rate at high a_w. Thus,
this indicates that the concentration and possible local product
inhibition effects by water are greater than any increase in
reaction rate brought about by a decrease in viscosity.

D. Kinetic Models for Maillard Browning

A number of kinetic approaches have been proposed to account
for the influence of temperature and water content on the Maillard
reaction during processing and storage of food products. Haugaard
et al. (1951), after studying the browning rate of an aqueous mixture of D-glucose and glycine at reflux temperatures, concluded
that the brown pigment produced was proportional to the square of
the amino acid concentration A, the concentration of the reducing

sugar R, and the square of the reaction time as shown by

$$B = K(A)^2(R)\theta^2 \tag{3}$$

where k is the rate constant. Differentiating this equation gives

$$dB/d\theta = KA^2R\theta \tag{4}$$

demonstrating that the browning rate increases with time. This model mechanism implies that there are two moles of amine reacting rather than the one mole of amine, which is generally accepted by other researchers. The temperatures at which this study was conducted were very high (refluxing), so that the application to foods held at normal storage conditions is limited. In addition, most storage studies at low temperatures (20-50°C) shows that the browning rate is constant and thus is not proportional to time.

Song *et al.* (1966) and Song and Chichester (1966, 1967) developed kinetic schemes for the Maillard reaction of glucose-glycine. The overall browning rate expression was

$$dB/d\theta = k_2\left[k_1(G)/k_2^{1}(I)\right]^{\frac{1}{2}} + k_3(g_0-B)(I-B) \tag{5}$$

where B is the brown melanoidin pigments, G the glucose, I the intermediates, θ time, and k_i are constants. They found that browning was linear over time after an initial induction period, i.e., a constant browning rate. The complexity of Eq. (5) was designed to account for the nonlinearity of the induction period, but limits its utility for practical use in complex food systems.

Mizrahi *et al.* (1970a,b) studied the browning reaction of unsulfited dehydrated cabbage as a function of moisture content and temperature. It was found that at constant conditions the reaction followed a zero-order scheme. In order to be useful over a broad moisture range, two equations relating browning rate to moisture content were derived through the use of curve-fitting techniques:

$$dB/d\theta = K_1 \left[1 + \sin(-\pi/2 + m\pi/m_x) \right]^n \qquad (6)$$

$$dB/d\theta = K_2 \left(\frac{A+M}{B+M} \right)^s \qquad (7)$$

where browning is in Klett units/day, and K_1, K_2, A, B, n, and s are constants fit by computer analysis, m_x the moisture content at which browning is at a maximum, and m the moisture content on a dry basis (g H_2O/100 g solids) at the water activity of the study.

Equation (6) gave a minimum statistical variance in predicting browning rate at 30 and 37°C, whereas Eq. (7) was the best fit for samples stored at 45 and 52°C. The equations accurately predicted the browning rate of unsulfited dried cabbage and greatly shortened the time needed to predict browning at other moisture contents from a minimum of data. Since the constants in the equations are limited to freeze-dried cabbage, studies similar to their (several water activities and several temperatures, usually a 3×3 experiment) would have to be done for other foods. Unfortunately, such data are usually not found in the literature.

The study of Jokinen et al. (1976) developed a browning rate equation with respect to lysine loss as a function of system composition and temperature. As will be noted later, lysine loss in such systems is primarily due to nonenzymatic browning. From the work of Warmbier et al. (1976a), Warren and Labuza (1977), and Eichner and Karel (1972), it appears that much of this loss precedes the formation of brown pigments and can be considered a measurement of the impact of the Maillard reaction on protein quality.

Jokinen et al. (1976), utilizing samples composed of soybean protein, glucose, sucrose, potato starch, microcrystalline cellulose and water, derived the following equation:

$$L/L_0 = 0.581 + 0.047(\text{pH}) - 0.093(G) - 0.59(T) - 0.0068\theta$$

$$+ 0.035(a_w)^2 + 0.025(a_w)(S) + 0.0331(S)(S_T) \qquad (8)$$

where L/L_0 is the fraction of available lysine remaining as
measured by the FDNB reaction, G the glucose content, T the temperature, θ the time, S the sucrose content, and S_T the starch
content.

This study was the first to use kinetics to predict lysine
loss as a function of system composition and temperature. Jokinen
et al. found that during extrusion of the model system, the maxima
for lysine loss occurred in the a_w range of 0.65-0.7, which parallels that found for browning in storage studies as noted earlier.
Unfortunately this equation is limited, when predicting losses at
normal storage temperatures, because it applies to extrusion processing where high temperatures (80°-130°C) are used. As noted
earlier, the pathways for browning can change at high temperature.
In addition, the equation was developed by curve fitting and thus
does not help to understand the basic mechanism. However, it
does show a useful approach for product development.

Mizrahi and Karel (1977) evaluated a kinetic model for predicting the extent of Maillard browning under continuously changing
moisture content (i.e., moisture gain through a package). Actual
data for browning of dehydrated cabbage was taken from the study
conducted by Mizrahi et al. (1970a,b) discussed above. They concluded that their model was adequate as a predictor of browning in
a package although inferior to the traditional constant-condition
models such as that used by Mizrahi et al. (1970a,b).

Each of the kinetic approaches reviewed here has limitations
to the prediction of browning during storage of foods because they
are based either upon high-temperature conditions encountered in
specific extrusion processes or because they describe conditions
unique to a particular product. It would be useful on a practical
level to have a simple kinetic model which would describe the effects of water upon browning.

As found by Mizrahi et al. (1970a) for cabbage, Warmbier et
al. (1976a) for a model casein/glucose system and Waletzko and Labuza (1976) for a semimoist food, the simplest model for prediction

of browning is that of a zero-order reaction, where

$$dB/d\theta = k_B = \text{rate of browning} \tag{9}$$

No induction period is assumed (which is usually a fairly good
assumption) and the dependence of the rate of pigment formation
on reactants is accounted for in k_B. Again, this means that k_B
is specific to a specific food system. As Warmbier *et al.* (1976b)
showed, k_B increased as the amount of reducing sugar was increased
up to a point. The absolute magnitude of k_B can, however, be compared between products since it is a direct measure of amount of
pigment produced per unit of time.

E. The Temperature Dependence of the Specific Effects of Water on Browning Reaction

Labuza *et al.* (1970) and Loncin *et al.* (1968) found that for
the Maillard browning reaction, the progress of brown pigment development could be considered to be a zero-order reaction when
reactant concentrations were not limiting for the rate of formation of brown pigment. Assuming, then, a zero reaction rate for
browning development or a linear increase of browning with time
(i.e., assuming no real induction period) the following equation,
described before, can be utilized:

$$\int_{B_0}^{B} dB = \int_{0}^{\theta} k_B \, d\theta \tag{10}$$

Integrating both sides of Eq. (10) between the limits of B_0 at
time zero and B at time θ gives

$$B - B_0 = k_B \theta \tag{11}$$

or

$$B = B_0 + k_B \theta \tag{12}$$

where B is the brown pigment color concentration developed, and B_0 the original brown pigment concentration. A plot of B vs. t should give a straight line where the slope is equal to k_B and the y intercept is equal to B_0. Studies made at various temperatures and a_w as done by Mizrahi et al. (1970a) can then be used to evaluate k as a function of temperature. This dependence is described through the use of the Arrhenius equation, where

$$k = k_0 e^{-E_a/RT} \tag{13}$$

where k is the rate constant, k_0 the absolute rate constant, E_a the activation energy, R the gas constant (1.986 cal/mole $^\circ$K), and T the absolute temperature ($^\circ$K). If the Arrhenius equation is plotted as ln k vs. t/T ($^\circ$K) the slope is equal to $-E_a/R$.

Activation energies (E_a) can be determined for browning reactions in various food systems as a function of water content from which the temperature dependence of the browning reaction at different water contents can be elucidated.

Another way that temperature effects on the browning reaction at various water contents can be measured is through the use of the term Q_{10}; defined as the increase in rate for every 10°C increase in temperature and related to activation energy (E_a) by

$$\log Q_{10} = \frac{2.189E_a}{(T)(T+10)} \tag{14}$$

The Q_{10} value may also be obtained from a shelf life plot as shown in Fig. 6, where the time to some unacceptable browning value is plotted against temperature. Assuming a zero-order reaction, the time to some endpoint is inversely proportional to the reaction rate, and thus

$$Q_{10} = \theta_{s(T^\circ C)}/\theta_s(T+10^\circ C) \tag{15}$$

where $\theta_{s(T^\circ C)}$ is the shelf life at a given temperature and $\theta_{s(T+10^\circ C)}$ the shelf life at 10°C above that temperature. The activation energy E_a can be obtained by solving Eq. (14) for E_a.

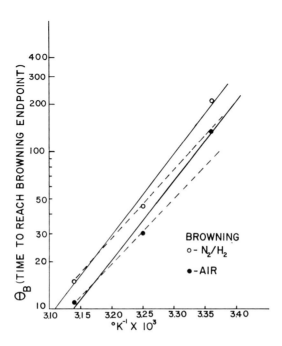

Fig. 6. Typical shelf life plot of time to reach the point of undesirable browning (Waletzko and Labuza, 1976) in an IMF system at a_w = 0.84. Dotted line shows deviation in prediction to lower temperature from higher temperature.

This approach is useful where minimal shelf life data are available for a food.

Table I indicates E_a and Q_{10} values at various a_w or water content levels in different food products for the browning reaction. These values have been obtained both from studies that have utilized a kinetic approach and from those which did not use this approach but where it was possible to utilize the simplified kinetic approach outlined above. In most cases studies as a function of both a_w and temperature were not done.

The activation energy levels in Table I generally fall in the range 20-40 kcal/mole, except for a few values. This range is consistent with the range of nonenzymatic chemical reaction activation energies indicated by Bluestein and Labuza (1975).

The Q_{10} values range from 2 to 8 for the browning reaction.

TABLE I. Kinetic Parameters of Maillard Browning

Food System	(g H$_2$O/g solids)	a_w	Temperature range (°C)	E_a (kcal/mole)	Q_{10}	Study	Comments
Cabbage (freeze dried)	0.18	0.62	37	28	(30–40°C) 4.42	Mizrahi et al. (1970a)	Determined colorimetrically
	0.117	0.51		29	4.66		
	0.089	0.43		32	5.47		
	0.056	0.32		34	6.08		
	0.032	0.20		38	7.52		
	0.014	0.01		40	9.36		
Dried potatoes	3.70	1.00	40–80	26	(50–60°C) 3.38	Hendel et al. (1955)	Colorimetric
	1.10	0.98		25	3.22		
	0.33	0.90		25	3.22		
	0.15	0.70		28	3.71		
	0.09	0.53		32	4.47		
	0.05	0.15		37	5.65		
IMF model system air N$_2$/H$_2$	0.3	0.84	25–45	30	(30–40°C) 4.92	Waletzko and Labuza (1976)	Spectrophotometric
	0.3	0.84		26	3.98		
Casein/glucose/ glycerol model system	0.15	0.41	45	33	(40–50°C) 5.17	Warmbier et al. (1976a)	Spectrophotometric humectant added
Glucose-glycine system	Aqueous	1.0	57–99	16.1 [1] 22.1 [2]	2.0 2.7	Song (1966)	Two E_a: (1) for induction period and (2) for steady state

Material	Water condition					Reference	Remarks
Dry casein		0.0 0.11 0.32	70–110	47	5.3	Flink (1974)	Heated casein at constant temperature in freeze-drier
Processed cheese	~35% Water	1.0	5–40	24	4.7	Thomas et al. (1977)	Color measurement (based on endpoint after 6 months)
Goat's milk	~88% (Wet)	1.0	93–121	27	$(100–110°C)$ 2.59	Burton (1963)	–
Apple juice	88% (Wet)	1.0	37–130	27	$(40–50°C)$ 3.84	Hermann and Andrae (1963)	–
Lemon crystals (freeze-dried)		0.0 0.6 0.11 0.22	4–35	18.37 24.67 26.67 26.75	2.74 3.87 4.32 4.34	Kopelman et al. (1977)	Air storage conditions
		0.0 0.06 0.11 0.22		29.91 31.75 32.16 27.59	5.16 5.71 5.84 4.55		N_2 storage conditions
Figs		0.6–0.8	21	19.8	$(20–30°C)$ 3.10	Copley and Van Arsdel (1964)	
Golden raisins		0.6–0.8	21	27.5	4.80		
Regular raisins		0.6–0.8	21	23.9	3.90		

Labuza (1972) indicated a range of 2-15 for the browning
reaction Q_{10} values; the high values, however, occurring only
in frozen foods. In general, a range of 2-6 is what would be
expected for semimoist and dry foods. Since Q_{10} is a function
of T^{-2}, one would expect an increased value in the frozen-food
range. Thus according to Eq. (14) a reaction with an
E_a = 30 kcal/mole has Q_{10} = 8.3 in the frozen-food range,
Q_{10} = 5.5 in the 30 -40°C range, and Q_{10} = 2.9 in the high-
temperature (100°-110°C) processing range. Therefore, E_a is
a more reliable index of temperature sensitivity since it
covers the whole temperature range. One should view Q_{10} in
the literature with caution because of this. In addition,
some literature mistakenly report Q_{10} as a 10°F ratio without
even stating such.

In Table I there appears to be a trend toward lower E_a and
Q_{10} values with higher water activity or water content levels for
the same food system. This trend is shown most dramatically in
the case of freeze-dried cabbage (Mizrahi *et al.*, 1970a) and dried
potatoes (Hendel *et al.*, 1955). The Kopelman *et al.* (1977) work
with lemon crystals does not clearly show this trend probably due
to the fact that the a_w range examined was so narrow (0-0.22).
The effect of water then seems to be to decrease the temperature
sensitivity of the reaction with increasing levels of water
present. The mechanism by which this occurs has not been elucidated. However, since from kinetic theory the E_a is an overall
measure of the limiting E_a for each step in the reaction scheme,
a lower E_a suggests that as water content increases, those previously limiting steps can now proceed easier. This may be either
due to better diffusion to and from the reaction site, increased
proton or electron mobility at the site, or some other factor.

It is interesting to note that although humectants have an
effect on the a_w maximum for the browning reaction, the glycerol
in the casein-glucose-glycerol model system studied by Warmbier
et al. (1976a) did not appear to alter the temperature dependence

of the browning reaction. In Table I the E_a value of 33 kcal at a_w = 0.41 is not significantly different than the E_a values found in cabbage and dried potetoes at a similar a_w.

Several experimental factors other than water content could contribute to variations in E_a values. For example, the variations could be the variations due to the differences in the methodology used to determine the extent of browning. Some values in the table (e.g., cheese and raisins) were based on an endpoint analysis from shelf life data. A different endpoint might affect the value of k_B if the reaction is not exactly zero order, especially as has been found at higher temperature.

Obviously the variations in activation energy also result from differences in composition and physical structure of the complex food systems. A protein with a structure that allows easier access to the lysine might show a different E_a depending on its degree of folding as a function of temperature. In addition, it is known that the reducing sugars react at different rates. However, no data are available as to their different effects on E_a. One would expect differences to be shown in this parameter and thus depending on the sugars present in the food the E_a will change.

Work with dehydrated potatoes shows that at normal (<35°C) temperature lipid oxidation is the main route of deterioration. At high temperature, however, because of the greater Q_{10} for browning (5-7 vs. 2 for oxidation) the development of off colors leads to deterioration. This is especially true in drying. The higher the reducing sugar content (Copley and Van Arsdel, 1964) of the potato, the greater the rate of browning during storage. Sulfite can increase the shelf life by two to five times. The lower the reducing sugar level the greater the increase in shelf life. A content of less than 1% reducing sugars is best for preparing dehydrated potato.

These factors would also help to explain the differences in E_a values at similar a_w levels of about 0.2 between lemon crystals (E_a = 26.8 kcal/mole) (Kopelman et al., 1977), dried cabbage

(E_a = 38 kcal/mole) (Mizrahi *et al.*, 1970a) and dried potatoes
(E_a = 37 kcal/mole) (Hendel *et al.*, 1955).

Another effect on activation energy could be due to the presence or absence of oxygen, which could change the reaction pathway. Labuza (1973) found that oxidative rancidity led to formation of carbonyl reducing compounds capable of producing browning in an IMF chicken/glycerol system. Addition of antioxidants such as BHA or EDTA inhibited both browning and oxidation. To further corroborate this effect the results (Fig. 7) of Waletzko and Labuza (1976) showed greater browning rate in an IMF system (a_w = 0.84) exposed to air vs. stored in an oxygen-free atmosphere. As seen in Table I the air system had a higher E_a, suggesting a different reaction pathway. Since vitamin C was added to this product, it is possible that the difference was due to the big difference in its rate of destruction. Kopelman *et al.* (1977), on the other hand, found the situation reversed for lemon crystals, with higher E_a values in the N_2 atmosphere. Since the lemon crystals also contain vitamin C some other factor must be considered.

Song *et al.* (1966), when kinetically analyzing the browning reaction in a glucose-glycine model system determined the tempera-

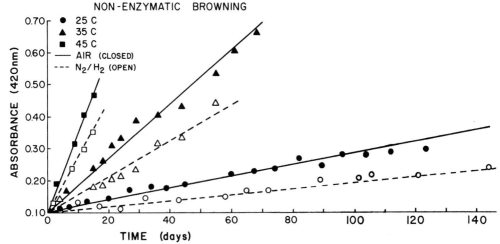

Fig. 7.   *Browning in an IMF system at a_w = 0.84 as a function of O_2 level. The residual O_2 is <0.001% in the N_2/H_2 system (Waletzko and Labuza, 1976).*

ture dependence of both the induction and linear increase periods. Their results show that during the induction period before the development of colored pigments, E_a is lower (E_a = 16.1 kcal/mole) than during the steady-state coloration period (E_a = 22.1 kcal/mole). This is as might be expected if the reaction building up to reducing compound intermediates is not the limiting step in the overall reaction. Warmbier *et al*. (1976a) found a similar situation for a casein/glucose model system stored at 25°-45°C. The E_a for glucose loss was about 25 kcal/mole, whereas for browning it was about 33 kcal/mole. This indicates that the controlling reactions are in the latter phase of the overall reaction leading to pigment formation. It also suggests that at low-temperature storage, the steps leading to initial reactions of the reducing compounds with lysine may occur without any appearance of brown pigment. As was found by Warmbier *et al*. (1976b) over 50% of the lysine was lost in 20 days in a model system stored at 25°C, whereas color formation did not become visually detectable until after 80 days.

F. *Specific Effects of Water on the Browning Rate Maximum*

Many studies of browning have analyzed the specific a_w at which the maximum rate of browning occurs. Lea and Hannan (1949), Labuza (1970), Labuza *et al*. (1972), Loncin *et al*. (1968), and Karel and Labuza (1969) have all reported an a_w between 0.65 and 0.70 for the browning rate maximum.

Sharp (1957) studied the deterioration of dehydrated pork as a result of the Maillard reaction. The pork when stored at various RHs at both 35° and 50°C showed a maximum browning at 57% RH vs. 70, 38, and 16% RH. Although this is a lower a_w than reported above, since the pork was not stored between a_w of 0.57 and 0.65 the maximum could be in that range.

Eichner (1975) indicated that in a glucose-lysine-avicel model system held at 40°C, the browning rate increased rapidly from

a_w = 0.23 to 0.4, with a slightly greater increase from a_w = 0.4
to 0.75, where it reached a maximum.   At a_w = 0.82 there was a
slight decrease in reaction rate.   This fits the above data.

In contrast to these findings, Han et al. (1973) reported a
maximum browning rate at a_w = 0.93 in dried anchovy held at 20°C
for two months at a_w ranging from 0.11 to 0.93.   In addition, Han
found the minimum for browning at a_w = 0.32-0.45, while the monolayer value for the anchovy was a_w = 0.21.   This is also in contrast to the findings of Labuza (1973), who found the maxima for
browning to be at the monolayer in dried milk.

Feillet et al. (1976) reported that browning in pasta stored
at 30, 60, and 90°C was more rapid at 100% RH than at 50% RH with
the browning at 90°C and 100% RH being almost double that at 90°C
and 50% RH.   This would seem to indicate that 100% RH would be
the water content maxima for browning in pasta, but since Feillet
did not measure browning rates at 65-70% RH it is possible that
this maxima did occur in this zone and was not measured.   In addition, he may have been measuring an enzymatic reaction, since
the enzymes are not destroyed in processing.   Enzymatic reactions
most likely would not show the IMF range maximum.   They might be
denatured at the higher storage temperatures where the reaction
was at a maximum so that these results might not show the intermediate maximum.

Wolfrom (1974) found that the browning rate decreased with
water content increases from 65 to 95% H_2O on a wet basis for a
dehydrated orange juice model system held at 65°C.   The sharpest
drop in rate occurred in a 65-80% moisture range.   Unfortunately,
no a_w were reported but one would expect these to be at a_w ≈ 1.

Potthast et al. (1976) found that in prerigor freeze-dried
beef held at 80°C, nonenzymatic browning reactions of the Maillard
type reach a maxima near 60% RH, which is similar to the previous
results in meat.

The variations in a_w or water content rate maxima for browning
can be attributed to the physical and/or chemical nature of the

food itself as was discussed with respect to E_a. Karel and Labuza (1969) found a difference in the a_w browning rate maxima between dehydrated pork sausage bits (a_w = 0.62), dry pea soup mix (a_w = 0.69), and dry corn chowder mix (a_w = 0.79) stored at 37°C.

The fact that many workers have reported a maxima for browning in the a_w = 0.60-0.80 range makes this reaction a significant one with respect to intermediate moisture foods (IMF) that are in the a_w = 0.60-0.85 range (Labuza, 1975). In many IMF, sugar-type humectants that are reducing compounds (glucose, fructose, corn syrup solids, etc.) are used to control the a_w level. Thus browning can be a serious problem. If liquid humectants such as glycerol or propylene glycol are used, however, they lower the a_w maximum for browning from around 0.7 to a value around 0.4 to 0.5. Eichner and Karel (1972), Warmbier *et al.* (1976a) (Fig. 8), and Warren and Labuza (1977) found this to be the case in IMF model systems held at normal storage temperatures. Thus these humectants should be the ones of choice if high quality is to be maintained. The problem

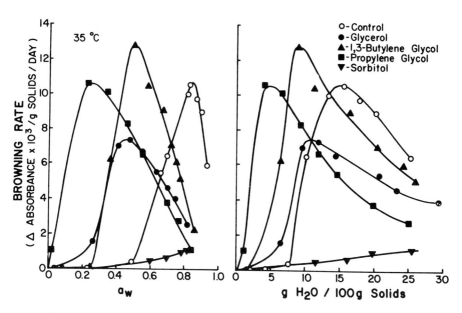

Fig. 8. Effect of humectants on rate of browning as a function of humectants (Warmbier et al., 1976a).

is their taste and cost.

Warren and Labuza (1977) and Kopelman et al. (1977) have also stated that the induction period, defined as the time to visual detectable browning, was also inversely proportional to the a_w. Since this merely reflects the rate as a function of browning this would be expected since, as noted before, zero-order kinetics usually prevail.

IV. EFFECTS OF WATER ON PROTEIN QUALITY LOSS

A. Kinetics of the Reaction

Studies that have analyzed the loss of protein quality with respect to browning are usually based on the loss of the essential amino acid lysine. Goldblith and Tannenbaum (1966) reviewed the literature prior to 1965, which pointed out that lysine loss in foods is primarily due to the Maillard reaction and can be used as an index of the extent of browning. Lysine is usually lost more rapidly than other essential amino acids in the first stages of the Maillard reaction because of its free epsilon amino group. Warren and Labuza (1977), Warmbier et al. (1976a,b), and Eichner and Karel (1972) demonstrated a substantial loss in available lysine before the visual development of brown pigments (Fig. 9).

In order to understand the specific effects of water on the kinetic parameters of lysine loss, a simplified kinetic model similar to that for the browning reaction can be utilized. In contrast to the browning reaction, which is generally considered a zero-order reaction, lysine loss has been generally considered to be a first-order reaction at least for up to 75% loss. This means that the rate of loss of lysine is directly dependent on the concentration of the remaining lysine. Although Dvorshak and Hegedus (1974) reported a fourth-order decrease in lysine with time in milk powder systems when using curve-fitting techniques, most other research has not supported this idea.

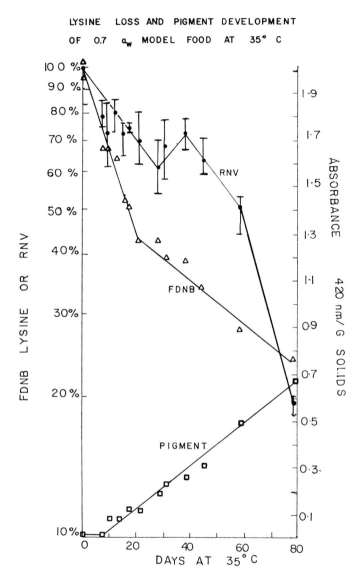

LYSINE   LOSS AND PIGMENT DEVELOPMENT
OF   0.7 a_w MODEL FOOD   AT   35° C

*Fig. 9. Kinetics of browning reaction showing the large de-
crease in lysine before significant pigment production. Color
change becomes visible at an OD value of 0.2/g solids (Warren and
Labuza, 1977).*

If a first-order reaction is assumed, then the loss of lysine
can be described by

$$dA/d\theta = -k_L (A) \tag{16}$$

where A is the concentration of lysine, k_L the rate constant, and
θ the time. If $A = A_0$ at time zero, this equation can be integrated to obtain

$$\ln(A/A_0) = -k_L \theta \tag{17}$$

where A is the concentration lysine at time measured and A_0 the
original concentration of lysine. From this equation, a plot of
the logarithm of the percentage of lysine remaining vs. time can
be constructed, which will yield a straight line. The slope of
this straight line is equal to k, the rate constant. With studies
at different temperatures, the activation energy can be obtained
as for browning from Eq. (13).

The activation energies and Q_{10} values from several studies
are listed in Table II. The activation energies for lysine loss
range from 10 to 38 kcal/mole, while Q_{10} values range from as low
as 1.5 to 4.7. It is not possible to establish a trend of decreasing activation energy values with increasing water content
for lysine loss as was the case for browning. None of the research
for which it has been possible to obtain kinetic parameters has
evaluated this phenomenon with the exception of Jokinen et al.
(1976), who did not find a significant difference between activation energy values at different a_w levels in a soy model system
held at 80°C. Thus only the average activation energy value was
presented for the entire water activity range 0.33-0.93. Evaluation of the data of Ben Gara and Zimmerman (1972) indicates a
higher activation energy (19.4 kcal/mole) at a_w = 0.6 in nonfat
dried milk than at a_w = 0.4, where a value of 14.6 kcal/mole was
found. At a_w = 0.2, the E_a was only 7 kcal/mole. These results
would seem to suggest that the reaction becomes more temperature
sensitive with increasing water content, which is opposite to the
effect of a_w on the overall browning reaction. However, since
very few data points were presented in the paper the validity of
this conclusion is in doubt.

TABLE II. Kinetic Parameters of Protein Quality Loss

Food system	(g H_2O/g solids, %)	a_w	Temperature range (°C)	(Kcal/mole)	Q_{10}	Study	Comments
Soybean meal	1		100–120	30	2.73	Taira et al. (1966)	FDNB
Dry cod muscle	14		105–115	23.6	2.21	Miller (1956)	Avail. lysine for chick growth
			95–115	37.5	3.75		FDNB-lysine
Herring press cake	4.9–14		85–145	25	(100–110°C) 2.4	Carpenter et al. (1962)	FDNB-lysine
Bovine serum albumin	14.4 (Wet) 14.1 (Wet)		85–145 85–145	24 31	2.3 3.0		
Nonfat dried milk		0.4 0.6	20–40	14.6 19.4	2.18 2.80	Ben Gara and Zimmerman (1972)	FDNB-lysine
Model systems 1% glucose	6–18		80–130	35.1	(100–110°C) 3.45	Thompson et al. (1976)	Soy/glucose/ MCC/potato starch

Table 2 con't.

						Reference	Method
4% glucose	6-18			33.9	3.30		FDNB-lysine (assumed zero order kinetics)
Casein	70 RH	0.70	0-70	29.0	(30-40°C) 4.66	Lea and Hannan (1949)	Biological activity
Model soy system		0.33-0.93	80-30	28.5	3.06	Jokinen et al. (1976)	FDNB-lysine
Fortified rice meal	15-20 (Wet)		115-184	12.5	(120-130°C) 1.5	Tsao et al. (1978)	2 + 4% lysine fortified TNBS-determination
Pasta		0.5	40-90	13.3	2.12	Fabriani and Frantoni et al. (1972)	FDNB

Although the results in Table II do not conclusively indicate any particular trend toward a change in temperature sensitivity with water content, they do indicate that water content has a smaller affect on the temperature dependence of the lysine loss reaction as compared to the browning reaction. This is probably due to the fact that the reaction mechanism for lysine loss is different than that for browning. Lysine participates both in the initial step as well as in the final steps where pigment is formed. Depending on how available lysine is measured, very different results could be obtained.

The results of Tsao *et al.* (1978) indicate an activation energy value of only 12.5 kcal/mole, which is somewhat lower than the values obtained from the results of other studies. This could be due to the physical structure of the components of the food system studied or to a change in reaction mechanism due to the high temperatures (115°-184°C) at which lysine loss was measured. In addition, they found no effect of moisture on the reaction and pooled all the results as did Jokinen *et al.* (1976). Since the residence time in the extruder was so short this may not have been a good assumption.

Another factor that could cause variations in temperature dependence is the methodology employed to measure lysine loss. For example, the activation energy values obtained from the work of Miller (1956) for dry cod muscle at 14% H_2O (wet basis) shows an activation energy of 23.6 kcal/mole for chemically determined available lysine (FDNB method) vs. a value of 37.5 kcal/mole for available lysine as measured by chick growth.

An unusual phenomenon was noted by Thompson *et al.* (1976) when measuring lysine loss in a soy model system in an extrusionlike process. Lysine was initially lost rapidly with the data following a semilog plot as a pseudo-first-order reaction. After a maximum lysine loss of slightly more than 50% was reached however, the amount of available lysine actually appeared to increase for a short period of time. The lysine level then appeared to level out

and no further losses were observed. This they term a no-loss
period.

Although Thompson et al. (1976) and Wolf et al. (1977a,b)
were the first workers to observe a recovery or period of increase
for lysine content, other workers have observed a gradual leveling
out of lysine loss after the initial 50% loss. Warren and Labuza
(1977) noted a leveling out of lysine loss after the initial 50%
loss even though the brown pigments continued to increase linearly
with time. The results of Lea and Hannan (1950) indicated that, at
70% RH and 37°C at various glucose levels (1-8 moles/mole lysine)
FDNB lysine losses leveled out after the initial 50% loss. Warmbier et al. (1976b) concluded that after a minimum amount of pigment precursors is formed, the browning rate proceeds irrespective
of the remaining available lysine content. In addition, above a
ratio of 3 moles glucose/mole lysine, the rate of browning did not
increase. Warmbier et al. (1976b) also noted that lysine does not
always obey first-order kinetics. This could occur because lysine
after being initially made unavailable is later released from gly-
cosylanine after it undergoes the Amadori rearrangement reaction.
Lysine may then later combine with unreacted sugar moieties and/or
reactable pigment precursors. The fact that lysine is again released temporarily could cause the amount measured as lost to level
off.

With respect to the a_w maximum for lysine loss Loncin et al.
(1968), Lee and Hannan (1950), Eichner and Karel (1972), Jokinen
et al. (1976), and Hendel et al. (1955) have found that the maxima
in an a_w range 0.65-0.75. Loncin et al. (1968), when examining
nonfat dry milk powders held at 40°C, found that the lysine loss
rate as a function of a_w paralleled that of the browning reaction.
Similarly, Potthast et al. (1977) found that the activity of glycolytic enzymes decreased during the storage of freeze-dried muscle
in an a_w range 0.25-0.65. The fastest rate of loss occurred at
a_w = 0.65 with over a 50% loss in 30 days, while it took 120 days
at a_w = 0.4. No loss in activity took place at a_w = 0.25. This

pattern was ascribed to glycolytic metabolites reacting with the amino acids of the enzymes causing them to lose their activity.

Other workers have obtained a_w rate maximas ranging from 0.3 to 0.7. It is important to remember that as with the browning reaction, the rate maxima can be influenced by the physical and/or chemical nature of the food system being measured, as well as the method of measurement.

Rolls and Porter (1973) recommended that nonfat dry milk be kept at moisture levels below 5% (wet basis) in order to minimize lysine loss. Measurements of a_w for whey powders in our laboratory indicate that this moisture would correspond to the monolayer a_w value of 0.2-0.3, where the reaction rate would be at a minimum. Recently Turner *et al.* (1978) have shown that the mode of reaction in milk is predominantly between lactose and the κ-casein protein fraction. However, this was done in the wet state and may not be true at low a_w.

In foods that contain a humectant, such as glycerol as used by Warmbier *et al.* (1976a) in a model system stored at 35° and 45°C, the a_w rate maxima for lysine loss shifts below a_w = 0.5. This was measured by the FDNB reaction. This is consistent with the results of Warren and Labuza (1976), who noted that lysine loss rates were greatest at a_w = 0.50 in a model system containing propylene glycol stored at 35°C and measured with a microbiological test.

Carpenter *et al.* (1962) found that in freeze-dried herring press cakes heated to temperatures ranging from 85° to 145°C for 27 hr the cakes with moistures ranging from 0 to 11.8% had similar losses of lysine. Although there was a greater loss at the 4-12% moisture content range, the high temperature and long time probably masked any moisture effect.

Miller (1956) however, reported that there was a gradual decrease in net protein utilization (NPU) as a_w increased in codfish meals throughout the whole water content range when heated at 105°C for 24 hr. He found no maximum at 65-70% RH as previously reported by Lea and Hannan (1949) at lower temperature. This difference in water content effect could be due to a change in reaction mechanism brought about by the elevated temperature conditions utilized in the study.

The effect of water content on lysine loss in a model food system (a_w = 0.33-0.94) consisting of soybean protein, glucose, microcrystalline cellulose, and water during processing at elevated temperatures (80°-130°C) was examined by Wolf *et al.* (1977a,b). Their data indicate that a maxima for lysine loss did not occur at water activities near 0.65. They concluded that this could be due to the rapid shift in water activity caused by high processing temperatures but offered no theoretical proof for this. In addition, they concluded that the effects of temperature and glucose content would be masking the effects of water activity. Most likely as noted previously, the high temperature is the probable cause.

In a fortified-rice meal heated at temperatures from 115° to 185°C and at moisture contents of 15-20% (wet basis), Tsao *et al.* (1978) also evaluated the effect of water content. They concluded that water content did not have a significant effect on the rate constants for available lysine degradation. In spite of this conclusion, it is possible as in the other studies, that the effects of specific water content levels could have been masked by the short processing time, high temperature, and some effect of pressure under extrusion conditions. Finally, processing at high temperatures such as those employed in this study could alter the lysine loss reaction mechanism resulting in a different reaction controlling the loss of lysine.

B. *Practical Implications of Protein Quality Loss*

Thus far, browning and lysine loss reactions have been considered from an analytical standpoint. However, the practical implications of protein quality loss in food products during processing and storage cannot and should not be ignored.

The loss of protein quality increases dramatically at elevated temperature conditions such as those employed in some food processes but the amount lost depends on the food system. Bender (1972) when discussing protein quality concluded that, in general, high-temperature processes had the most effect on protein nutritional quality.

Burvall *et al.* (1977) reported that in lactose-hydrolyzed milk, UHT (ultra-high-temperature) sterilization and evaporation at 25°C did not significantly lower protein nutritional quality. However, spray drying decreased lysine content by 49% as measured by NPU in rats and by 45% as measured by BV in rats. However, in unhydrolyzed milk the lysine loss as determined by NPU measurements with rats was only 14% and was only 9% when measured as BV in rats. This difference is specifically due to the presence of the reducing sugars as formed during lactose hydrolysis. The lactose-hydrolyzed milk had more reducing sugars in the form of glucose and galactose available for reaction with lysine. Recently (*Nutr. Rev. 36*:133, 1978) a report on a study of school children in Surinam was published. In this study children were given lactose-hydrolyzed skim milk powder and compared to two other groups: one given skim milk powder and one group given the powder supplemented with glucose and galactose on the day of administration of the milk. After one year it was found that the students (6-12 years old) receiving the lactose-hydrolyzed milk powder showed the poorest growth rate and weight gain. Although this could not be confirmed in animal studies, the FDNB lysine was lower in the hydrolyzed milk. It was suggested that the Maillard reaction as influenced by the high humidity and temperature was the cause. This is the first human study showing detrimental effects of browning on growth.

Holsinger *et al.* (1973) when evaluating losses of chemically available lysine in whey products dehydrated by different processes, concluded that excessive or prolonged exposure to heat is chiefly responsible for lysine loss. They found that spray-drying was preferable to roller-drying and that prolonged holding times at warm temperatures prior to drying should be avoided. Rolls and Porter (1973) found that in milk, chemically available lysine was reduced by only 3-10% for spray-drying and by 5-35% by roller-drying. In rats, they found that the NPU for spray-dried milk products was 100%, while for roller-dried milk products it was between 64 and 86%.

The effect of Instantizing on the amino acid content of milk was examined by Posati *et al.* (1974). No significant losses were found in dry milk powder that had been instantized at 212°F and 90% RH for 1 to 2 min. Hackler *et al.* (1965) concluded that in soymilk heated at 93° and 121°C for different time periods the loss of protein quality was dependent upon both time and temperature of heat treatment.

Studies like these indicate that as time and temperature and moisture content are increased up to some maximum, the rate of protein quality loss increases. The practical implication of this is that the time to a significant decrease in protein quality is shortened. If a first-order reaction is assumed, then a plot of the percentage of protein quality (or percentage of lysine) vs. time based upon Eq. (17) should result in a straight-line relationship on semilog paper. The slope of this line, the rate constant k, can then be used to calculate the time to a significant, i.e., 50% decrease in quality. The rate constant then is

$$k_L + 0.693/\theta_{\frac{1}{2}} \tag{18}$$

where $\theta_{\frac{1}{2}}$ is equal to the time for 50% loss of protein quality loss or lysine.

An example of this kind of plot is shown for lysine loss in nonfat dry milk as determined from the results of Ben Gara and Zimmerman (1972) in Fig. 10 and as derived from Kramer (*Food Technol.*, Jan., 1974) in Fig. 11.

Table III lists k values and half-lives ($\theta_{\frac{1}{2}}$) for a number of

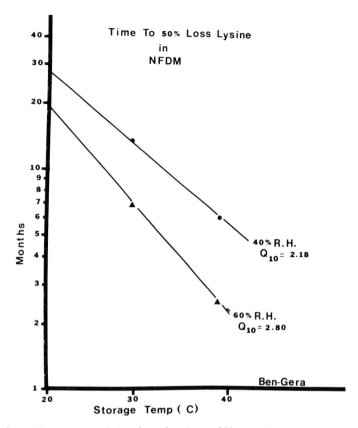

Fig. 10. Loss of lysine in dry milk powder as a function of % RH and temperature (Ben-Gara and Zimmerman, 1972).

studies of foods. The half-lives obviously show the same kinetic pattern as previously described. For example, the water content at which herring press cakes held at 130°C (Carpenter et al., 1962) have the shortest half-life ($\sim$30 hr) is in the 10-14% moisture (wet basis) range, which would have a_w = 0.6-0.8. An IMF casein-glucose model system studied by Warren and Labuza (1977) had the shortest half-life at 12 days for chemically determined protein quality at a_w = 0.50. A value of 22 days was found for the biological determination at the same a_w. This was done at 35°C with glycerol in the system, which also reduced the browning maximum to a_w = 0.5. It should be noted that no data were col-

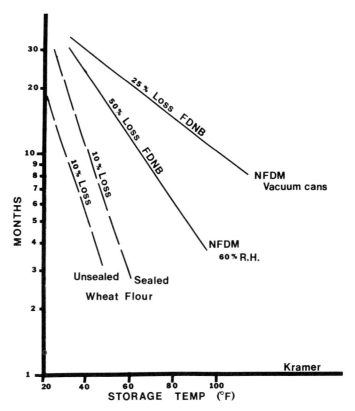

Fig. 11. Loss of protein nutritional value in wheat flour and milk as a function of temperature and % RH (Kramer, Food Technol., Jan. 1974).

lected between a_w = 0.5 and 0.75. Thus the maximum could be in that range.

Miller (1956) found that as moisture content increased from zero to 17% (wet basis) at 105°C in cod fish meal, the half-life shortened dramatically from 140 to 28 hr. It is interesting that most of the decrease in half-life occurred between the 0 and 9% (wet basis) moisture content levels with a decrease in half-life from 140 to 33 hr.

A half-life of 48 hr for 10.6% H_2O (wet basis) vacuum-dried fish fillets held at 105°C (Carpenter *et al.*, 1957) is somewhat longer than that found by Miller under similar conditions for cod.

TABLE III. Time to 50% Loss Protein Quality

Food System	Water Content	a_w	Temperature (°C)	Time to 50% Loss ($\theta_{\frac{1}{2}}$)	k	Study	Comments
Whey powder, acid, cottage cheese	1.6% (Wet)		90	177 Min (2.95 hr)	(min^{-1}) 0.004	Holsinger et al. (1973)	Accelerated storage test, lysine—TNBS—
			109	42 Min	0.017		Posati modifi-
			125	0.6 Min	0.116		cation
	8.18% (Wet)		90	114 Min (1.9 hr)	0.006		
			109	30 Min	0.023		
			125	6 Min	0.116		
Casein	18.34 g H$_2$O/g solids	0.70	35	22 Days	(day^{-1}) 0.032	Schnickels et al. (1976)	FDNB lysine de- termination used
Whey	18.53	0.78	35	25 Days	0.028		
Soy (TVP)	20.00	0.73	35	7 Days	0.099		
Egg albumin	19.76	0.63	35	>60 Days	0.011		
Wheat gluten	19.05	0.66	35	40 Days	0.017		
Fish conc.	18.25	0.68	35	19 Days	0.036		
Herring press cake (de- fatted)	∿0% (Wet)		130	144.8 Hr	(hr^{-1}) 0.005	Carpenter et al (1962)	% H$_2$O calculated on fat free basis
	4.9%		130	34.5 Hr	0.020		
	10.6%		130	30.0 Hr	0.023		Carpenter's FDNB lysine
	14.4%		130	30.0 Hr	0.023		
	29.6%		130	37.0 Hr	0.019		
	55.8%		130	48.0 Hr	0.014		

TABLE III (Continued)

Sample	aw	Temperature (°C)	Time	(mo^{-1})	Reference	Notes
Pasta		45	36 Mo	0.0192	Fabriani and Frantoni (1972)	FDNB
		60	15.1 Mo	0.046		
		70	6.7 Mo	0.104		
		80	4.7 Mo	0.147		
Casein-glucose model system	0.3	35	14 Days (FDNB chemical lysine)	0.050	Warren and Labuza (1977)	Used propylene glycol as humectant
			23 Days (Tetrahymena biological value)	0.030		
	0.5		12 Days (FDNB)	0.058		
			22 Days (Tetrahymena)	0.032		
	0.7		20 Days (FDNB)	0.035		
			61 Days (Tetrahymena)	0.011		
Whole herring meal	6.2% (Wet)	25	78 Mo	0.009	Lea et al. (1958)	FDNB
	11.1% (Wet)	25	67 Mo	0.010		
Cod fish meal	0% (Wet)	105	140 Hr	0.005	Miller (1956)	NPU determination
	9% (Wet)		33 Hr	0.021		
	17% (Wet)		28 Hr	0.025		

Food System	Water Content	a_w	Temperature (°C)	Time to 50% Loss ($\theta_{\frac{1}{2}}$)	k	Study	Comments
Vacuum dried fish fillets	10.6% (Wet)		105	48 Hr	0.014	Carpenter et al. (1957)	FDNB
	1.7-2.3% (Wet) (0% RH)	0	100	144 Hr	0.005		
	1.7-2.3% (Wet) (95% RH)	0.95	105	41 Hr	0.02		
Nonfat dry milk	40% RH	0.4	40	6 Mo	0.116	Ben Gara and Zimmerman (1972)	FDNB
	60% RH	0.6		2.5 Mo	0.278		
	40% RH	0.4	30	13 Mo	0.053		
	60% RH	0.6		7 Mo	0.099		
	40% RH	0.4	20	27.5 Mo	0.025		
	60% RH	0.6		19.5 Mo	0.036		

This difference could be due to differences in the amount of natural reducing sugars present or to variations between the methodologies used in these studies to measure protein quality loss. The half-lives of the products examined by Lea *et al*. (1958), Holsinger *et al*. (1973), Miller (1956), Ben Gara and Zimmerman (1972), and Carpenter *et al*. (1957) all decrease with increasing moisture content. None of these studies evaluated a_w that were above the maxima for protein quality loss. The half-lives of the products studied by Carpenter *et al*. (1962) and Warren and Labuza (1974) increase above a certain water content or a_w level as is found for browning. Lea *et al*. (1958) reported that at 25°C herring meals (11.1% H_2O wet basis) lost 4% of chemically available lysine after two and one-half months and 9% after 12 months in air storage. In nitrogen atmosphere storage at 25°C, less than 2% lysine was lost in 12 months. This may be similar to the effect on lipid oxidation as reported earlier.

Figure 10 shows the time to reach 50% lysine loss for nonfat dry milk held at 40 and 60% RH (Ben Gara and Zimmerman, 1972). From a plot such as this it should be possible to make predictions for a product stored at any temperature and relative humidity within the range of the shelf-life plot using the techniques described earlier.

Work is now in progress in our laboratory to develop kinetic models that will permit accurate predictions of protein quality loss under non-steady-state temperature and humidity conditions such as those found in food distribution systems. Through the development of data such as in Figs. 10-12 under steady-state conditions, studies are being made on various whey protein products and on pasta. Pasta is distributed in the U.S. through normal ambient temperature channels and is usually in a fairly moisture permeable package. Fabriani and Frantoni (1972) have shown that considerable loss can occur at high temperatures to which the product may be subjected, as seen in Fig. 12. By a combination of the methods for predicting moisture gain, a knowledge of the a_w and

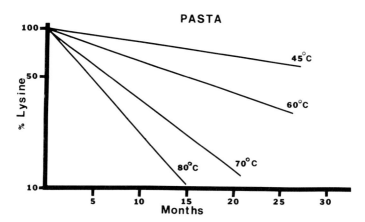

Fig. 12. Affect on temperature on lysine loss in pasta (Fabriani and Frantoni, 1972).

temperature effect on reaction kinetics, and a knowledge of distribution conditions, useful predictions of the extent of browning and protein quality loss could be made.

Acknowledgments

This study was supported in part by the University of Minnesota Agricultural Experiment Station Projects No. 18-72 and 18-78.

References

Bender, A. E. (1972). *J. Food Technol.* 7, 239.
Ben-Gara, I., and Zimmerman, G. (1972). *J. Food Sci. Technol.* 9, 113.
Bluestein, P., and Labuza, T. P. (1975). *In* "Nutritional Evaluation of Food Processing" (E. Karmas and R. Harris, eds.), 2nd ed. AVI, Westport, Connecticut.
Burvall, N. G., Asp, N. G., Dahlquist, A., and Oste, R. (1977). *J. Dairy Res.* 44, 549/553.
Carpenter, K. J., Ellinger, G. M., Minro, M. I., and Rolfe, E. J. (1957). *Br. J. Nutr.* 11, 162-173.
Carpenter, K. J., Morgan, C., Lea, C., and Parr, L. (1962). *Br. J. Nutr.* 16, 451.

Choi, R. P., Koncus, A., O;Malley, C., and Fairbanks, B. (1949).
    J. Dairy Sci. 32, 580.
Cook, B. B., Fraenkel-Conrat, J., Singer, B., and Morgan, A. F.
    (1951). J. Nutr. 44, 217.
Copley, M. J., and Van Arsdel, W. B. (1964). "Food Dehydration,"
    Vol. II, p. 540.
Dvorshak, E., and Hegedus, M. (1974). Acta Aliment. Acad. Sci.
    Hung. 3, 337.
Eichner, K. (1975). In "Water Relations in Foods" (R. Duckworth,
    ed.). Academic Press, New York.
Eichner, K., and Karel, M. (1972). J. Agri. Food Chem. 20, 218.
Fabriani, G., and Frantoni, A. (1972). Biblioteca Nutritia Dieta
    17, 196.
Feillet, P., Jeanjean, M. F., Kobrehel, K., and Laignelet, B.
    (1976). Tecnica Molitoria 27(3), 81-86.
Flink, J. (1974). J. Food Sci. 39, 1244.
Gal, S. (1975). In "Water Relations in Foods" (R. Duckworth, ed.).
    Academic Press, New York.
Goldblith, S. A., and Tannenbaum, S. R. (1966). Proc. 7th Int.
    Congr. Nutr., Vols. 1-5.
Hackler, L. R., Van Buren, J. P., Steinkraus, K. H., El Rawi, I.,
    and Hand, D. B. (1965). J. Food Sci. 30(4), 723-728.
Han, S. B., Lee, J. H., and Lee, K. H. (1973). Korean Fisheries
    Soc. Bull. 6, 37-43.
Haugaard, G., Tumerman, L., and Silvestri, H. (1951). J. Am.
    Chem. Soc. 73, 4594.
Hendel, C. E., Silveira, V., and Harrington, W. O. (1955). Food
    Technol. 2, 433.
Herrman, J., and Andrea, W. (1963). Papierchromatogr. Nashweis.
    Nahrung. 7, 243.
Hodge, J. E. (1953). J. Agri. Food Chem. 1, 928.
Holsinger, V. H., Posati, L. P., and Pallansch, M. J. (1970).
    J. Dairy Sci. 53, 1638.
Holsinger, V. H., Posati, L. P., Devilgiss, E. D., and Pallansch,
    M. J. (1973). J. Dairy Sci. 56(12).
Hurrell, R. F., and Carpenter, K. J. (1974). Br. J. Nutr. 32,
    589.
Jokinen, J. E., Reineccius, G., and Thompson, D. R. (1976). J.
    Food Sci. 41, 816.
Karel, M., and Labuza, T. P. (1968). J. Agr. Food Chem. 16, 717.
Karel, M., and Labuza, T. P. (1969). Contract Report, U.S. Air-
    force, Brooks Air Force Base, Texas.
Kopelman, I. J., Meydau, S., and Weinberg, S. (1977). J. Food
    Sci. 42, 403-410.
Labuza, T. P. (1968). Food Technol. 22, 263.
Labuza, T. P. (1970). Proc. 3rd Int. Congr. Inst. Food Technol.,
    p. 618.
Labuza, T. P. (1972). CRC Crit. Rev. Food Techn., pp. 217-240.
Labuza, T . P. (1973). NASA Contract Phase I, Lyndon Johnson Space
    Center, Houston, Texas.

Labuza, T. P. (1975). *In* "Water Relations in Foods" (R. Duckworth, ed.). Academic Press, New York.

Labuza, T. P., Mizrahi, S., and Karel, M. (1972). *Trans. Am. Soc. Agri. Eng. 15*, 150.

Labuza, T. P., Tannenbaum, S. R., and Karel, M. (1970). *Food Technol. 24*, 35-42.

Lea, C. H., and Hannan, R. S. (1949). *Biochim. Biophys. Acta 3*, 313.

Lea, C. H., and Hannan, R. S. (1950). *Biochim. Biophys. Acta 4*, 518.

Lea, C. H., Parr, L., and Carpenter, K. J. (1958). *Br. J. Nutr. 12*, 297.

Lea, C. H., Parr, L. J., and Carpenter, K. J. (1960). *Br. J. Nutr. 14*, 91.

Lee, S., and Labuza, T. P. (1975). *J. Food Sci. 40*, 370.

Lewicki, P., Busk, G., and Labuza, T. P. (1978). *J. Colloid Interface Sci. 64*, 501.

Loncin, M., Bimbenet, J. J., and Lenges, J. (1968). *J. Food Technol. 3*, 131.

Maillard, L. C. (1912). *Compt. Rend 154*, 66.

Miller, P. S. (1956). *J. Sci. Food Agri. 7*, 337.

Mizrahi, S., and Karel, M. (1977). *J. Food Sci. 42*, 958-963.

Mizrahi, S., Labuza, T. P., and Karel, M. (1970a). *J. Food Sci. 35*, 799.

Mizrahi, S., Labuza, T. P., and Karel, M. (1970b). *J. Food Sci. 35*, 804.

Obanu, Z. A., Ledward, D. A., and Lawrie, R. A. (1977). *Meat Sci. 1*, 177-183.

Patton, S. (1955). *J. Dairy Sci. 38*, 457.

Posati, L. P., Holsinger, V. H., DeVilbiss, E. D., and Pallansch, M. J. (1974). *J. Dairy Sci. 57*, 258-260.

Potthast, K., Hamm, R., and Acker, L. (1976). *Z. Lebensm. Unters. Forsch. 162*, 139-148.

Potthast, K., Hamm, R., and Acker, L. (1977). *Z. Lebensm. Unters. Forsch. 165*, 18.

Reynolds, T. M. (1963). *Advan. Food Res. 14*, 168.

Rockland, L. B. (1969). *Food Technol. 23*, 1241.

Rolls, B. A., and Porter, J. W. G. (1973). *Proc. Nutr. Soc. 32*, 9-15.

Salwin, H. (1959). *Food Technol. 13*, 594.

Schnickels, R. A., Warmbier, H. C., and Labuza, T. P. (1976). *J. Agri. Food Chem. 24*, 901-903.

Sharp, J. G. (1957). *J. Sci. Food Agri. 8*, 21.

Sharp, J. G., and Rolfe, E. J. (1958). *In* "Fundamental Aspects of Dehydration of Foodstuffs," p. 197. Soc. Chem. Ind., London.

Sloan, A. E., and Labuza, T. P. (1975). *Food Prod. Devel. 9*(9), 75.

Song, P. S., and Chichester, C. O. (1966). *J. Food Sci. 31*, 906.

Song, P. S., and Chichester, C. O. (1967). *J. Food Sci. 32*(98), 107.

Song, P. S., Chichester, C. O., and Stadtman, F. H. (1966). *J. Food Sci. 31*, 914.

Spark, A. A. (1969). *J. Sci. Food Agri. 20*, 308.

Taira, H., Taira, H., and Sukukai, Y. (1966). *Japan J. Nutr. Food 18*, 359.

Thomas, M., Turner, A. D., Abad, G., and Towner, J. M. (1977). *Milchwissenschaft 32*, 12-15.

Thompson, D. R., Wolf, J. C., and Reineccius, G. A. (1976). *Trans. ASAE 19*(5), 989-992.

Tsao, T. F., Frey, A. L., and Harper, J. M. (1978). *J. Food Sci. 43*, 1106-1108.

Turner, L. G., Swaisgood, H., and Hansen, A. (1978). *J. Dairy Sci. 61*, 384.

Waletzko, P. T., and Labuza, T. P. (1976). *J. Food Sci. 41*, 1338.

Warmbier, H. C. (1975). Ph.D. Thesis. Univ. of Minnesota, St. Paul.

Warmbier, H. C., Schnickles, R. A., and Labuza, T. P. (1976a). *J. Food Sci. 41*, 528.

Warmbier, H. C., Schnickles, R. A., and Labuza, T. P. (1976b). *J. Food Sci. 41*, 981.

Warren, R., and Labuza, T. P. (1977). *J. Food Sci. 42*, 429.

Wolf, J. C., Thompson, D. R., and Reineccius, G. A. (1977a). *J. Food Sci. 42*, 1540.

Wolf, J. C., Thompson, D. R., and Reineccius, G. A. (1977b). *J. Food Processing Preserv. 1*, 271.

Wolfrom, M. L., Kahimura, N., and Horton, D. (1974). *J. Agri. Food Chem. 22*, 796-800.

ACCELERATION OF CHEMICAL REACTIONS DUE TO FREEZING

K. Porsdal Poulsen
Flemming Lindeløv

I. INTRODUCTION

At temperatures between the freezing point of aqueous systems
and the temperature at which ice crystallization stops, water activity is directly related to temperature. This is true when the
definition of water activity is the ratio of the water vapor
pressure of the system to that of pure water under identical conditions (primarily identical temperature).

The first formation of ice occurs at the freezing point of a
system, then when the temperature is lowered more ice separates
leaving a more concentrated unfrozen solution. The water activity
at any given temperature in the above-mentioned temperature interval can then be calculated from a phase diagram for water, or the
values can be found in tables. A precondition is that a certain
amount of ice is in equilibrium with a certain concentration of
unfrozen solution and consequently both exert the same water
vapor pressure. For cellular systems (tissues) some time is necessary before equilibrium is achieved.

Vapor pressure values for water and ice may be obtained from
handbooks, and a_w values calculated between $-1°$ and $-50°C$ (See
Table I.). Extrapolation from the values shown in Table I yields
$a_w = 0,1$ at about $-150°C$.

Several investigations have been carried out to determine the
water activity at subfreezing temperatures for complex food sys-

TABLE I. a_w of Water and Aqueous Systems below Freezing Point

Temp. (°C)	Vapor pressure (mm Hg) Pure water	Ice	$a_w = P_{ice}/P_{water}$
0	4.579	4.579	1.00
-5	3.163	3.013	0.953
-10	2.149	1.950	0.907
-15	1.436	1.241	0.864
-20	0.943	0.776	0.823
-25	(0.607[a]	0.476	(0.784)
-30	(0.383)	0.286	(0.75)
-40	(0.142)	0.097	(0.68)
-50	(0.048)	0.030	(0.62)

[a]Values in parentheses are not accurately known.

tems and other biological materials. Dyer et al. (1966) found that the equilibrium vapor pressure of beef is approximately 20% lower than the vapor pressure of pure ice at the same temperature. Similar results have been reported by Hill and Sunderland (1967), who examined the equilibrium vapor pressure of chopped sirloin, lamb, veal, pork, chicken, and beef fat in a temperature range of -4° to -26°C and found that they exhibited between 13 and 20% lower vapor pressure than that of pure ice. However Storey and Stainsby (1970) measured the equilibrium water vapor pressure of frozen cod and finding values equal to that of ice and obtained different results. Later data by Fennama and Berny (1974) using beef, potatoes, apples, 15% sucrose solutions, and pure ice showed that their equilibrium vapor pressures did not differ significantly from the equilibrium vapor pressure of ice. The same results were obtained by Rödel and Krispein (1977) and are shown in Table II. They show rather different water activities at 25°C but all become equal to that of pure water at -25°C.

The conclusion of these various experiments is that water activity is temperature related only at subfreezing temperatures and that the values given in Table I can be used in explaining the T-TT data in this chapter. Scheeberger et al. (1977) reach

TABLE II. *Water Activities of Various Meat Products at Freezing Temperatures Depending on Their aw Measured at +25°C* [a]

Temp. (°C)	Fresh meat	aw of product examples				Cured dried meat
		Sausages, German type				
		Brühwurst	Kochwurst	Rohwurst		
+25	0.993	0.980	0.970	0.880		0.820
+5						
±0	essentially unchanged					
−1		essentially unchanged				
−2	0.981		essentially unchanged			
−3	0.971	0.971				
−4	0.972	0.962	0.962			
−5	0.953	0.953	0.953	essentially unchanged		essentially unchanged
−10	0.907	0.907	0.907			
−15	0.864	0.864	0.864	0.864		
−20	0.823	0.823	0.823	0.823		
−25	0.784	0.784	0.784	0.784		0.784

[a] From Rödel and Krispein (1977).

the same conclusion concerning temperature and water activity but add that biological, chemical, and physical phenomena cannot be described solely by a_w because dissolved solutes also influence diffusion, growth of microorganisms, and enzymatic reactions.

"Frozen" foods might be found in the temperature range between their freezing point and $-273°C$. In the upper part of the range there is the possibility for the growth of microorganisms. Leistner (1977) states that many organisms in foods are inhibited solely by temperature, but the growth of some is probably not limited by temperature but by a_w.

Figure 1 represents the well-known curve developed by Heiss and Eichner (1971), where the intensity of sensory changes and

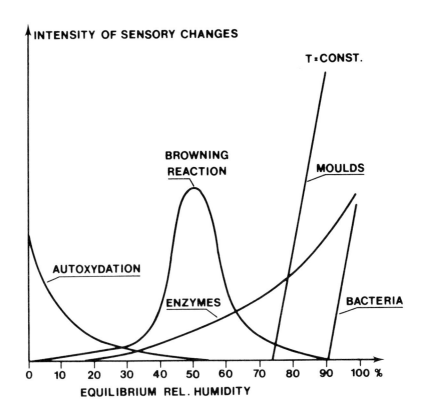

Fig. 1. Sensory changes and degradation in foods as a function of water activity (Heiss and Eichner, 1971).

water activity are related. Figure 2 is a partly theoretical
counterpart, where the temperature is converted to water activity
and the influence of the temperature on the reaction rate is included.

At very low water activities the analogy might help to explain some of our findings, but at higher temperatures it will be
more realistic to look directly at the concentration, which takes
place in the liquid portion of the "frozen" product. Pincock and
Kiovsky (1966) describe the relationship between reactions in
liquids and in frozen solutions of bimolecular reactions. According to them, a second-order reaction (A + B → P) in the liquid
phase of a frozen solution is given by

$$\frac{d[A_S]}{dt} = -k_2[A_A][B_S]\frac{V_S}{V_f} = -k_2[A_A][B_S]\frac{C_f}{C_S} \tag{1}$$

where k_2 is the second-order rate constant, which depends only on

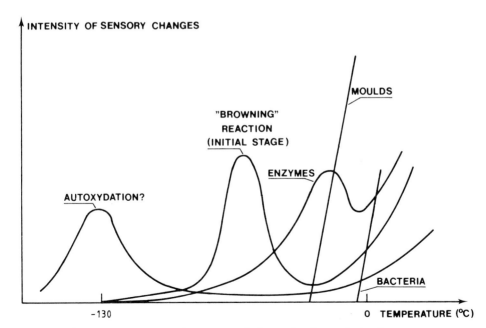

*Fig. 2. Sensory changes and degradation of foods as a func-
tion of temperature (partly theoretical).*

temperature, A_S and B_S are the molar concentrations of the reactants A and B in the thawed solution, V_S and V_f are the volumes of the thawed solution and the volume of liquid regions in frozen solutions, respectively; and C_S and C_f are the total concentration of solutes, with similar indices.

From the second part of Eq. (1) it can be concluded that the product $k_2 C_f$ has a maximum at a temperature below the freezing point of the solution (Fig. 3). The ratio C_f/C_S is never less than one and consequently the "frozen" rate is never less than the "unfrozen" rate at the same temperature.

In the systematic considerations given by Fennema et al. (1973) it is taken into account that $C_S = A_S + B_S + P_S + I_S$, where P_S and I_S are concentrations of the resulting products and the concentration of inert or inhibiting components. The final influence may be:

1. Normal stability: A temperature decrease results in a slower reaction rate (and better stability when foods are stored).

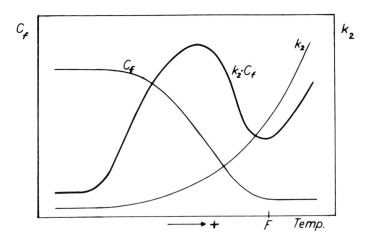

Fig. 3. Relation of $k_2 C_f$ to temperature. (k_2 is the rate constant of a second-order reaction and C_f the concentration of solutes in the liquid portion of the product; F is the freezing point.

2. Neutral stability: The temperature has no influence on the reaction rate.

3. Reversed stability: A temperature decrease results in an increased reaction rate.

The mathematical solution shows that $k_2 C_f$ has a maximum at a temperature normally lower than the intersection of the exponential k_2 line and the C_f line. The use of mathematics is, however, of little relevance because the organoleptic limitation for foods is caused by chain reactions of a high order.

In addition to the simple concentration effect described by Pimcock and Kiovsky (1966) the solution can be influenced by crystallization of components that control pH and strongly influence factors such as ionic strength. Larsen (1974) uses the terms primary and secondary concentration effects. The primary (concentration) effect is the influence on the chemical reaction, which can be solely ascribed to the crystallization of the solvent. The secondary (crystallization) effect is the influence on the chemical reaction, which can be ascribed to crystallization of the solvent together with one or several of the solutes. Another complication to the elementary considerations given by Pincock and Kiovsky (1966) is that some components (mainly organic) inhibit crystallization and lead to solutions that do not crystallize below the eutectic temperature. This phenomenon is known as glass transition or vitrification, and is caused by the high viscosity that is a result of reducing the temperature of the solution.

Oxidation of lipids is strongly influenced by the potential activity of metallic catalysts. According to Labuza *et al.* (1974) the effect of water on fat as well as a swelling system can be explained on the basis of mobility and solubility of catalysts. At high water activity the aqueous phase becomes less viscous and the mobility of catalysts is thus increased. At lower water activities water inactivates the catalysts, and hydrogen bond peroxides show an antioxidative effect. While the in-

fluence of the temperature on the reaction rate constant is known
to be exponential, the temperature-concentration relationship is
more complex and varies from a linear to a higher order relation
(Fig. 4).

The possible influence of concentration and/or water activity
during freezing has been examined by experiments with frozen solutions and frozen foods.

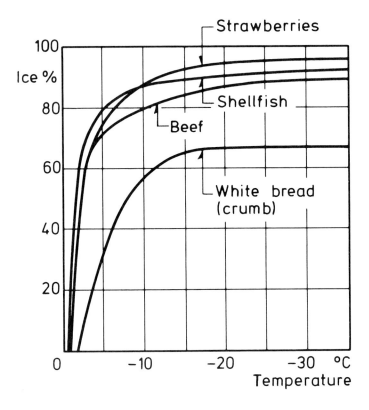

Fig. 4. Amount of water that has changed into ice as a function of temperature for various products.

II. REACTIONS IN A FROZEN SOLUTION

Malonaldehyde is one of the decomposition products of auto-
oxidized, polyunsaturated lipid materials (Loury, 1972; Kuusi *et
al.*, 1975) and occurs in foods and biological preparations only
after lipid oxidation. Because of its high reactivity it is an
interesting compound from the viewpoint of food chemistry. Of
special importance are the reaction of malonaldehyde with functional groups of proteins (Davidkova *et al.*, 1973), and the fact
that malonaldehyde is the main component in the measurement of
the TBA value (2-thiobarbituric acid) that is used to evaluate
the degree of oxidation of lipids (Marcuse and Johansson, 1973).
It has been shown by Buttkus (1967) that malonaldehyde reacts
with myosin from trout. The rate of reaction with α-amino groups
of myosin was greater at -20°C than at 0°C and almost equal to
that of 20°C. In another paper Buttkus (1970) concluded that the
denaturation of myosin increased in a frozen solution. The rate
of formation of the insoluble, high-molecular-weight protein ag-
gregates increased as the temperature decreased below the freezing point and reached a maximum near the eutectic point of the
solution. At -20° and -30°C, which are below the eutectic point
and where only water bound to the protein remains unfrozen, the
rate of aggregation and consequent insolubilization decreased
again and approached the rate observed at 0°C.

A. Experimental

Myosin was prepared from young rabbit according to the procedure of Kessler *et al.* (1952) and Buttkus (1966). Before use,
the myosin solutions were recentrifuged for 1 hr at 37,000 rpm
(at 0°C) in a Sorval Ultracentrifuge, Model RC 2-B.
Malonaldehyde was prepared from 1,1,3,3-tetraethoxypropane
(Fluka) by acid-catalyzed hydrolysis of the acetal. After two
distillations of the 1,1,3,3-tetraethoxypropane a colorless

liquid with a boiling point at 211.5°C was obtained. To a glass-stoppered tube was added 1.0 ml of the distilled tetraethoxypro-pane, 0.10 ml, 0.1 N HCl, and 10.0 ml water. The tube was held at 50°C for 1 hr. After the incubation period 0.50 ml, 0.1 M NaOH was added and the volume then made up to 100.0 ml with a pH 6.8 buffer containing 0.45 M KCl and 0.026 M phosphate. Mixtures of malonaldehyde and myosin were prepared to determine the amount of malonaldehyde reacting with myosin at 0°C, since it was found that this temperature causes the least changes of the myosin molecule: 1.0 ml of 1.0% myosin solution, in phosphate buffers of adjusted pH and ionic strength, or in phosphate buffer for a standard of the same pH and ionic strength, was reacted with 0.05 ml, 8.34×10^{-4} M malonaldehyde solution in KCl-phosphate buffer. Reaction mixtures were filled into glass tubes (5 ml, 15 mm diam) and stored at 45°, 22°, 0°, −5°, −12°, −18°, −24°, −30°, and −40°C. At each temperature, samples from the full range of pH (5.2, 7.0, 9.2; made from 0.067 M monopotassium phosphate and 0.067 M disodium phosphate) and ionic strengths (0.3 and 0.6) were examined.

After freezing, the glass tubes were stored at the nine tem-peratures mentioned above. Initially a sample was then removed for every storage temperature at hourly intervals for the first 6 hr and thereafter every 8–12 hr for determination of the amount of malonaldehyde bound to the protein.

The tubes were placed at room temperature half an hour before the analysis was to be conducted (thawing time 20 min). To the reaction mixture 1.0 ml 5% TCA was added to precipitate the myosin and the myosin-malonaldehyde complex. After 5 min in an ice bath the precipitate was compacted by centrifugation for 10 min at 3000 rpm. Then 1.0 ml of the supernatant was pipetted into a test tube and 0.50 ml of 2-thiobarbituric acid solution (0.50 g of 2-thiobarbituric acid, 48.0 ml of water, and 1.65 ml of 2.0 N NaOH) together with 0.20 ml of citrate buffer (14.75 g $Na_3C_6H_5O_7 \cdot 2\ H_2O$ and 12.5 ml concentrated HCl made up to 100 ml with water) were added. All the glass tubes were transferred to a boiling water bath at the same time, heated for 60 min at 100°C, and then cooled in tap water. To the pink solution 4 ml of water was added, and the optical density read at 534 nm against the buffer solution.

B. Results and Discussion

It was demonstrated that the reactions did not follow a simple rate law (i.e., not a first-, second- or third-order reaction). It was therefore decided to use the "initial rate" method (Frost and Pearson, 1961), which is commonly used to determine the reaction rate constant for reactions of unknown order. The initial rate of the reaction, which is known to be proportional to the reaction rate constant, was estimated as the initial slope of the kinetic curve (i.e., $\Delta x/\Delta t$ at $t \sim 0$). This slope was determined by fitting (using the method of least squares) a third-order polynomial to the experimental points.

In Fig. 5 it can be seen that the rate of reaction at pH 7 decreased as the temperature was reduced from 45°C to approximately the freezing point. When the mixture freezes, however, the reaction rate increases once again to the level found at 22°C and the maximum reaction rate for the system is found at about -24°C. Further decrease in the temperature to -40°C caused a reduction in the reaction rate. The temperature profiles obtained for the other pH values (5.2 and 9.2) were similar (Table III).

Over the full temperature range studied, there was no statistical difference (95% level) in the reaction rate for different ionic strengths. However, between -5°C (the freezing point for the solution with $I = 0.3$) and -10°C (the freezing point for the solution with $I = 0.6$) a difference was observed in the low-ionic-strength solution which had a partially reaction rate due to the freeze concentration effect.

III. TT EXAMINATIONS OF CURED AND UNCURED PORK BELLIES

Today more and more frozen foods are sold as ready-to-eat dishes with different ingredients, salt, and spices in the mixture. Such foods must be assumed to be complex in their behavior during frozen storage at different temperatures. Normally the literature does not mention the possibility of decreasing temperatures leading to reduced stability; however, in the annual reports for the years 1931 and 1932 of the Food Investigation Board, Low Temperature Research Station, Callow (1931; 1932) stated that fat in bacon kept at between -5° and -20°C for 10 weeks was more rancid after defrosting than that kept at temperatures ranging from -3° to 15°C for 10 weeks. The reason for the difference, according to Callow, could be a minimum uptake of oxygen by the fat at the higher temperatures because of microbiological activity. After large-scale experiments, Callow observed that fat from mild-cured bacon frozen at -30°C and stored at -10°C become rancid after only 8 weeks. Finally Callow concluded, "Thus, the transport of mild cured, green bacon in the frozen condition from Australia and New Zealand cannot be carried out by ordinary commercial methods."

Gibbons et al. (1951, 1953) observed in Canadian Wiltshire bacon that decreasing the storage temperature to -40°C accelerated the development of oxidative rancidity. He concluded that bacon kept better near its freezing point than at -40°C.

Experiments carried out at the Danish Meat Research Institute (1955), the Danish Meat Products Laboratory (1968), and the Swedish Institute for Food Preservation Research (private communication) with salted meat showed that in the temperature range of -6° to -23°C temperature has no influence, or that the lowest temperature gave the shortest shelf life.

Recent results from our laboratory (Poulsen and Lindeløv, 1975; Lindeløv and Poulsen, 1975) have shown that smoked bacon has an abnormal temperature profile between -5° and -60°C. From

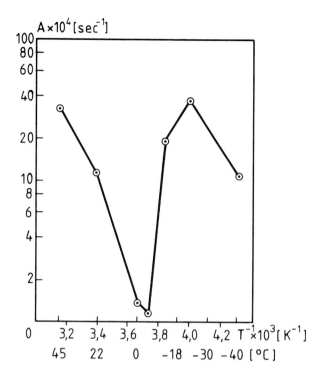

Fig. 5. Rate of reaction as a function of the reciprocal of temperature for the reaction between myosin and malonaldehyde (pH 7.0, ionic strength 0.6).

TABLE III. Reaction Rate Number (a × 10⁴ sec⁻¹) for the Reaction of Myosin with Malonaldehyde[a]

	Temp. (°C)							
pH	45	22	0	−12	−18	−24	−30	−40
5.2	27.2	10.5	2.1	42.4	70.0	45.5	3.5	1.5
9.2	28.8	10.8	2.3	52.0	67.0	69.2	18.6	2.6

[a] pH_{init} = 5.2/9.2, ionic strength = 0.6 (KCl).

taste panel scores it was concluded at the 99.95% confidence level that -30° to -40°C results in a shorter shelf life than any other temperature between -5° and -60°C. Another paper also reported a faster oxidation of bacon by lowering the temperature; in fact, maximum degradation occurred below -50°C and a very short shelf life at -25°C (about 10 days) was observed (Nilsson and Hällsäs, 1973).

The reactions in food given as examples of abnormal temperature influence above were all in salted meat. However similar reactions in milk (McWeeny, 1968), butter (Poulsen *et al.*, 1976), fish (Lea *et al.*, 1958), and drugs (Larsen, 1973) have been reported.

A. *Experimental*

The experimental materials used in this study were four different products prepared from 42 pig gilts, all from the same farm: (I) fresh pork bellies; (II) unsmoked bacon (7.75 g salt[*] 100 g H_2O); (III) smoked pork bellies; (IV) smoked bacon (6.92 g salt[*] 100 g H_2O).

All samples were sliced; about 150 slices were cut from each back, forward of the last rib. The slices were vacuum-packed consecutively in numbered pouches (laminate of polyethylene and polyester with Saran coating). After freezing by placing the pouches (each containing three slices) in a room at -24°C for 24 hr in still air, the products were stored at the following temperatures: -12°, -18°, -24°, -30°, -40°, and -60°C.

Organoleptic tests, including taste, odor and saltiness, were carried out by a standardized procedure on a soft-fried product once a month by an expert sensory panel (eight members). Consecutive pouches were placed at 5°C for 18 hr before examination. The products, which had been heated in an electric oven at 220°C for

[*]*98,70% NaCl + 0.85% KNO_3 + 0.45% $NaNO_2$.*

10 min in individual capsules, were assessed using a hedonic scale [ranging from -5 (bad, dislike extremely) through zero to +5 (ideal, like extremely)]. The numerical values were subjected to statistical analysis (balanced incomplete block design). The order of serving the products was randomized and no more than four products were served at the same time to a single panel member.

B. Results and Discussion

The initial chemical, physical and microbiological data for the products are listed in Table IV.

Table IV. Values for Chemical, Physical, and Microbiological Tests for Experimental Pork Belly Products

Protein (%)	12.4 ± 1.1					
Fat (%)	52 ± 11					
Water (%)	33 ± 3					
Ash (%)	0.65 *(product I)* ± 0.09 3.31 *(product IV)*					
Iodine number	61 ± 1					
pH	6.5 *(product I)* ± 0.1 5.6 *(product III)*					
Bacterial number (initial)[a]	1.1×10^{2} *(product IV)* 1.3×10^{4} *(product II)*					
Fatty acids (initial) (%)	*14:0*	*16:0*	*16:1*	*16:2*	*18:0*	*18:1*
	1.3	*27.9*	*2.9*	*0.4*	*14.7*	*41.7*
(%)	*18:2*	*20:0*	*20:1*	*20:2*	*20:4*	*20:0*
	7.0	*0.5*	*0.8*	*0.3*	*0.2*	*0.4*

[a]*Standard plate count (incubation temperature 30°C).*

C. Sensory Evaluation

Based on the statistical analyses it was possible to plot taste panel scores for the different products vs. storage time. Straight lines were obtained with different slopes (Q) indicating a loss of quality per time unit (dQ/dt).

Good correlation was found between odor and taste. For the cured products (products II and IV) a rise in saltiness took place a few weeks before the reduction in taste quality, possibly because of protein denaturation. The aftertaste of the pork products ranged from metallic, oily, and chemical to old, fishy, and rancid. Because taste was the main factor deciding the quality of the products it was possible to calculate the high-quality life (HQL)[*] and practical storage life (PSL)[†] from the taste scores for the different products. Figures 6-9 give the HQL and PSL for products I, II, and III and product IV, respectively. The data are plotted semilogarithmically.

The results support the suggestion of dividing foodstuffs into the three main groups and, on the basis of our observations product I (fresh pork bellies) and product II (unsmoked bacon) can be classified into group 1 (normal stability) and group 3 (reversed stability), respectively. The shortest HQL for product II is at -60°C (70 days) and the longest at -5°C (150 days), while product I demonstrated an increase in HQL for decreasing temperature (100 days at -5°C and 210 days at -60°C). All products, however, can be classified into all three groups, depending on the temperature interval examined; but generally the cured products have the shortest HQL at temperatures between -30 and -40°C.

[*]High-quality life (HQL): the time required to detect a "just noticeable difference" in quality between a sample of frozen food and the same freshly packed food, e.g., a one-point reduction on the +5 to -5 scale.
[†]Practical storage life (PSL): the time required for various deteriorative changes in a freshly packed product to accumulate to such an extent that the market quality is somewhat lower than excellent but still marketable. This degradation might be represented by a score of -2 in a scale from +5 to -5.

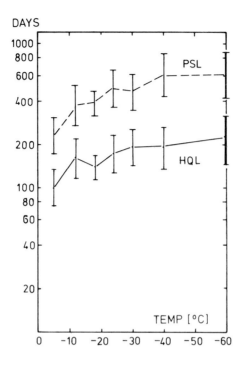

Fig. 6. High-quality life (HQL) and practical-storage-life (PSL) for vacuum-packed fresh pork bellies as a function of temperature.

IV. T-TT EXAMINATION OF VIENNA SAUSAGES

In order to examine the storage bahavior of more complex foods in frozen condition a number of investigations have been undertaken. The temperatures used were the same as mentioned in Section III for most of the experiments, and the same panel took part. However, in some experiments other trained panels were used in order to confirm the findings. In the vienna sausage storage test, a panel from the Danish Meat Products Laboratory was used. The sausages were produced mainly from pork, fat, water, and veal. The composition is given in Table V. The sausages were packed with and without vacuum.

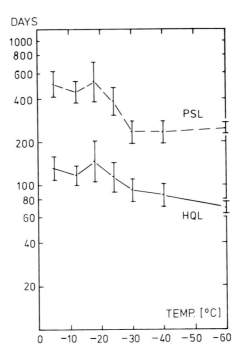

Fig. 7. *HWL and PSL for vacuum-packed unsmoked bacon*
(7.75 g salt/100 g H_2O) as a function of temperature.

Results

HQL and PSL quality estimates from the sensory evaluation of
the sausages packaged in 70 µl polyethylene pouches are shown in
Fig. 10. The two panels have independently observed the same
temperature/stability relationship. Vienna sausages showed increasing stability in the range $-5°$ to $-24°C$ and decreasing stability between $-24°$ and $-60°C$. Vacuum packaging improved the
shelf life considerably, but did not change the shape of the
curves.

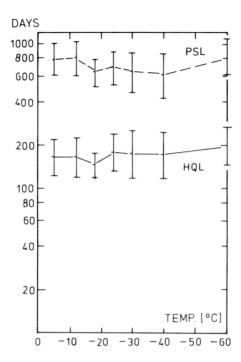

Fig. 8. HQL and PSL for smoked pork bellies as a function of temperature.

V. APPLE AROMA

When apple juice is to be stored for long periods the aroma components are often stripped off and kept separate. The aroma fraction varies in composition but a common commercial grade is a 200-fold concentrate that was used in the experiments to be described. The aroma fraction is constituted of approximately 95% water, approximately 5% ethyl alcohol, and a number of volatile organic flavor compounds of which esters and aldehydes like hexanal and trans-2-hexanal are essential. During storage at temperatures below 0°C it must be assumed that the organic components are concentrated in the liquid portion of the aroma concentrate.

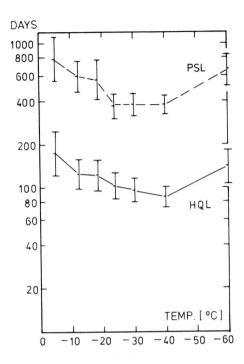

Fig. 9. HQL and PSL for smoked bacon (6.92 g salt/100 g H₂O) as a function of temperature.

TABLE V. Composition of Experimental Vienna Sausages

Fat (%)	35.87 ± 1.1
Water (%)	54.82 ± 0.8
Protein (%)	12.05 ± 0.4
Ash (%)	7.27 ± 0.1
Salt (g/100 g H₂O)	3.72 ± 0.2

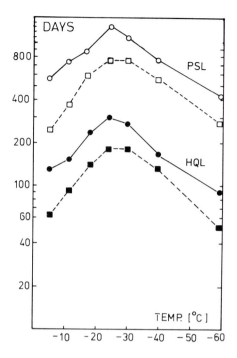

Fig. 10. HWL and PSL for vienna sausages packaged without vacuum as a function of temperature. Findings of the two panels are indicated with squares and circles, respectively.

A. Experimental

A number of 10 ml samples of apple aroma concentrate were filled into glass ampules. After flushing the ampules with nitrogen to exclude oxygen, the ampules were closed by melting the necks. The samples were then placed at the following temperatures: $37°$, $5°$, $-5°$, $-12°$, $-18°$, $-24°$, $-30°$, $-40°$, and $-60°C$. Time-temperature-tolerance data were obtained by organoleptic testing and gas chromatographic analyses. Redilution for organoleptic evaluation was done by adding 0.4% aroma concentrate, 0.5% citric acid, and 9% sugar to distilled water. The judges were asked to rank samples for overall impression. No attempt was made to estimate HQL or PSL, but only to find the relative influence of the temperature. The samples proved to be microbially stable, so that organoleptic degradation alone could be related solely to the chemical reactions taking place between components present in the concentrate. Organoleptic examinations were carried out after 21 days, 3 months, 18 months, and 30 months of storage. The judges were trained in apple juice testing and agreed well in their findings.

B. Results

The statistical evaluation of the sensory data was done by use of tables worked out by Kramer et al. (1973). Evaluations after 21 days of storage samples held at 5°, -12°, and -30°C were compared. Five out of 10 judges found a difference between the samples. The result of those five was that there was 99% chance that storing at -30°C was the worst condition and 95% chance of -12°C being the best.

Examinations after 3 months of storage indicated a preference for sample held at -24°C vs. -18°, -30°, and -40°C. Those at -40°C got the lowest score. (Significance was at the 90% level.)

Figure 11 shows data for the ranking test after 18 months of storage using 10 judges. When the total temperature range from 37 to -60°C is considered, a general trend of improved shelf life with decreasing temperature is found. However, deviation from a simple relationship is obvious. Statistical analysis of the data shows at a 99% significance level that -24 and -60°C were the best storage temperatures and at a 95% significance level for storage at 37°C to be the worst.

Examination after 30 months of storage where -18°, -24°, and -40°C were compared, shows, at a 95% significance level, that -24°C was a best storage temperature. Seven trained judges took part in this examination.

Another examination was carried out after 30 months of storage where -18°, -24°, and -30°C were compared. Ten judges took part, and the result was again that at a 95% significance level the

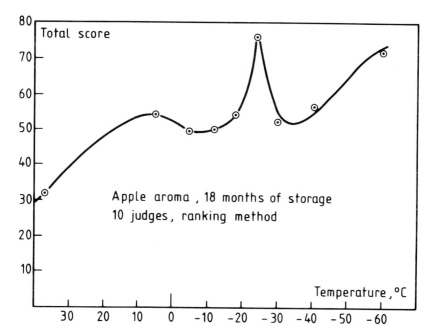

Fig. 11. Apple aroma scores after 18 months of storage at various temperatures.

sample stored at -24°C was the best.

Gas chromatographic analyses, using the head-space technique and tenax adsorbent on rediluted samples as well as direct injections of the concentrate, did not show any significant difference among samples.

VI. ACCELERATION OF CHEMICAL REACTIONS DUE TO FREEZING

The influence of temperature on the rate of chemical processes was successfully interpreted by van't Hoff in 1884, and by Arrhenius in 1889 (Hulton, 1955). As a rough guide, the rate of reaction doubles for every 10°C rise in temperature. However, satisfactory estimates of reaction rates (and thus shelf life) at subfreezing temperatures generally cannot be obtained simply by

extrapolating data obtained at above-freezing temperatures. As
the temperature is lowered and freezing ensues, reaction rates
may follow, and decrease more or less than the change predicted
by extrapolating Arrhenius plots of above-freezing data. What is
occurring in the frozen state to produce faster reactions and subsequently shorter shelf life in certain systems? Well-substantiated
explanations are available for some of the observed effects
(Fennema et al., 1973) and some hypotheses have been advanced.
These explanations are widely diverse and illustrate the wide
range of circumstances under which abnormal temperature effects
can be encountered. Some of them are presented below:

Concentration of Non-aqueous Constituents

Freezing of an aqueous solution or tissue system results in
the formation of essentially pure ice crystals. Since nearly all
of the non-aqueous constituents are left behind in the unfrozen
water and fat phase, one effect is simple, that of concentration.
Any reaction with higher than first-order kinetics will be accelerated when the solvent crystallizes out and leaves the solutes
at higher concentrations in the remaining liquid phase. For reactions that accelerate as a result of the "concentration effect," a
large acceleration is likely to occur if the sample is initially
dilute rather than concentrated and if inert solutes are absent.
Further, it is possible that the kinetic order will change during
freezing.

Regardless of the type of aqueous system, concentration during
freezing causes the unfrozen portion to undergo marked changes in
such physical properties as ionic strength (faster protein denaturation), viscosity, a_w, surface and interfacial tension, and oxidation-reduction potential. It is important to note that oxygen
is almost totally rejected from ice crystals as they are formed
(Scholander et al., 1957).

Different salts and salt concentrations vary in their influence on the rancidity process in meat products (Ellis and

Gaddis, 1970) but it appears that an increase in the concentration of salt ("concentration effect") produces a decreased stability of fat (Gaddis, 1952). It is well established that frozen cured products become rancid much more rapidly than frozen uncured sides (Lindaløv and Poulsen, 1975; Wiesmann and Ziemba, 1946). How important the observed effect of salt concentrations is on oxidation of cured products remains to be determined, but it is suggested that nitrite concentration is one of the important variables. It is even suggested that salt also can have an antioxidative effect (Marcuse and Johansson, 1973).

pH Shift

Because of the concentration effect pH can undergo changes during freezing. A decrease in pH to more favorable values for degradation processes could be another explanation of the observed phenomena. Experiments by Labuza et al. (1972) indicate that lowering the pH actually accelerates the oxidation process in lard.

Enzyme Reactions

Some enzyme-catalyzed reactions accelerate as freezing takes place. This behavior is far more common in cellular systems than in noncellular ones. Freezing damage to cellular systems often results in a likelihood of profound changes in the rates of enzyme-catalyzed reactions, maybe because of a release of calcium ions that accelerates the activity of certain enzymes.

Formation of Antioxidants

An increase in reaction rate, following a period of storage at lower temperatures, due to a shift in the "equilibrium" point of a chemical system and a consequent increase in the concentration of a product having an antioxidative effect, is possible. Marcuse (1962) concluded that many amino acids are antioxidants and also it has been proved that the products formed between some amino acids and aldehydes are antioxidants (Kawashima et al., 1977). This fact is very interesting because the reaction between

amino acids and aldehydes increases in rate when the temperature
decreases from $0°$ to $-24°C$ (anti-Arrhenius system) (Lindeløv, 1978)
and a maximum of a new product (antioxidant) is formed at the lower
temperatures. The opposite is also possible in that an antioxidant can be destroyed, e.g., vitamin E in lard, after which the degradation process accelerate.

Ice Surface

The reactant molecules might be frozen into a geometrical configuration highly favorable for their interaction and find themselves, as a result of temperature and structural barriers, unable
to assume a more random state before reacting (Grant, 1966). The
surface of ice crystals might behave catalytically in somewhat the
same manner as other inert surfaces (Grant, 1967).

Proton Mobility

Most of the reactions studied depend on the transfer of posi-
tively charged hydrogen ions or protons, and the proposal has been
made that proton mobility in ice may be much greater than in water
(McWeeny, 1968).

The Dielectric Constant

The dielectric constant of ice, being more favorable than that
of water, promotes nucleophile associations (Fennema et al., 1973).

Other factors may also play a role, and there is reason to be-
lieve that a single explanation will not be adequate for all the
cases so far studied.

VII. CONCLUSIONS

When complex foods are frozen, deviations from the simple assumption that any lowering of temperature improves shelf life
often appear. A number of explanations can be found, among them
a change in water activity. In a review in Food Technology,

Rockland (1969) stated that "rancidity may develop more rapidly in frozen products after removal from low temperature storage than in freshly prepared material held at the same elevated temperature. An explanation for this phenomenon may be that the product undergoes LI-II to LI-I (lowering of a_w) transition after being placed in frozen storage."

For the results described in this chapter, vienna sausage may show acceleration of oxidation due to lowering of water activity. Other products that show neutral or reversed stability can better be described by concentration effects as long as temperatures lower than $-60°C$ are not included.

References

Buttkus, H. (1966). *J. Fish Res. Bd. Canada 23,* 563.
Buttkus, H. (1967). *J. Food Sci. 32,* 432.
Buttkus, H. (1970). *J. Food Sci. 35,* 558.
Callow, E. H. (1931). *Ann. Rep. Food Invest. Bd. (Great Britain).*
Callow, E. H. (1932). *Ann. Rep. Food Invest. Bd. (Great Britain).*
Davidkova, E., Svadlenka, J., and Rosmus, J. (1973). *Z. Lebensm. Unters.-Forsch. 153,* 13.
Dyer, D. F., Carpenter, D. K., and Sunderland, J. E. (1966). *J. Food Sci. 31,* 196.
Ellis, R., and Gaddis, A. M. (1970). *J. Food Sci. 35,* 52.
Fennema, O., and Berny, L. A. (1974). *Proc. 4th Int. Congr. Food Sci. Technol., Madrid, 1974.* Work Documents Topic 2, p. 12.
Fennema, O. R., Powrie, W. D., and Marth, E. H. (eds.) (1973). "Low Temperature Preservation of Foods and Living Matter." Marcel Dekker, New York.
Frost, A. A., and Pearson, R. G. (1961). "Kinetics and Mechanism," p. 45.
Gaddis, A. M. (1952). *Food Technol. 6,* 294.
Gibbons, N. E. (1953). "Advances in Food Research," Vol. IV. Academic Press, New York.
Gibbons, N. E., Rose, D., and Hopkins, J. W. (1951). *Can. J. Technol. 29,* 458.
Grant, N. H. (1966). *Discovery 27,* 27.
Grant, N. H. (1967). *Arch. Biochem. Biophys. 118,* 292.
Heiss, R., and Eichner, E. (1971). *Food Manufacture 46,* 37.
Hill, J. E., and Sunderland, J. E. (1967). *Food Technol. 21,* 1274.
Hultin, E. (1955). *Acta Chem. Scand. 9,* 1700.
Kawashima, K., Itoh, H., and Chibata, I. (1977). *J. Agri. Food Chem. 25,* 202.

Kessler, V., and Spicer, S. S. (1952). *Biochim. Biophys. Acta 8,* 474.

Kramer, A., Cooper, G. K. D., and Papavasiliou, A. (1974). *Chem. Senses Flavor 1,* 121.

Kuusi, T., Nikkila, O. E., and Savolainen, K. (1975). *Z. Lebensm. Unters.-Forsch. 159,* 285.

Labuza, T. P., and Chon, H. E. (1974). *J. Food Sci. 39,* 112.

Labuza, T. P., McNally, L., Gallagher, D., Hawkins, J., and Hurtado, F. (1972). *J. Food Sci. 37,* 154.

Larsen, S. S. (1973). *Arch. Pharm. Chem. Sci. Ed. 1,* 41.

Larsen, S. S. (1974). "Stabilisering af laegemidler ved nedfrysning." Eget forlag, Copenhagen.

Leistner (1977). *Proc. Symp.: How Ready Are Ready-to-Serve Foods? Karlsruhe, 1977,* p. 260. S. Karger, Basel.

Lea, C. H., Parr, L. J., and Carpenter, K. J. (1958). *Br. J. Nutr. 12,* 297.

Lindeløv, F., and Poulsen, K. P. (1975). Paper C2.52 presented at *XIV Int. Congr. Refrigeration, Moscow, 1975.*

Lindeløv, F. (1978). *Int. J. Refrig. 1,* 92.

Loury, M. (1972). *Lipids 7,* 671.

McWeeny, D. J. (1968). *J. Food Technol. 3,* 15.

Marcuse, R. (1962). *J. Am. Oil Chem. Soc. 39,* 97.

Marcuse, R., and Johansson, L. (1973). *J. Am. Oil Chem. Soc. 50,* 387.

Nilsson, R., and Hällsås, H. (1973). *Livsmedelsteknik 1,* 26.

Pincock, R. E., and Kiovsky, T. E. (1966). *J. Am. Chem. Soc. 88,* 4455.

Poulsen, K. P., and Lindeløv, F. (1975). *Scand. Ref. 4,* 165.

Poulsen, K. P., Danmark, H., and Mortensen, B. K. (1976). Paper presented at *IIR Congr. Melbourne, 1976.*

Rockland, L. B. (1969). *Food Technol. 23,* 11.

Rödel, W., and Krispein, K. (1977). *Fleischwirtschaft 57,* 1863.

Scholander, P. F., van Dam, L., Kanwisher, J. W., Hammel, H. T., and Gordon, M. S. (1957). *J. Cell. Comp. Physiol. 49,* 5.

Schneeberger, R., Voilley, A., and Weisser, H. (1977). Paper 73, presented at *IIR Congr. Karlsruhe 1977.*

Storey, R. M., and Stainsby, G. (1970). *J. Food Technol. 5,* 157.

Wiesmann, C. K., and Ziemba, J. V. (1946). *Food Ind. 18,* 95, 241.

THE FORMS AND ENERGY OF MOISTURE BINDING
IN FOODS AS A BASIS FOR CHOOSING RATIONAL METHODS
FOR PROCESSING AND STORAGE

A. S. Ginzburg

INTRODUCTION--CLASSIFICATION OF THE BINDING FORMS
AND MOISTURE STATE IN THE MATERIAL

Raw, half-finished, and cooked foods are naturally moist, colloidal, structurally porous bodies characterized by different forms of water binding. The classification of the forms and types of moisture binding adopted in the USSR was established by academician P. A. Rebinder. This classification was further developed by E. D. Kazakov and supplemented with regard to the biological properties of foods as given in Table I.

As seen, the form of water binding determines both the character of the moist foods and the properties of its water.

II. MOISTURE BINDING ENERGY

The binding energy is calculated by the thermodynamic formula as a decrease of free energy ΔF at a constant temperature or work L (joules/mole) required to remove 1 mole of water from the dry skeleton of the material (without changing its composition)

$$-\Delta F = L = RT \ln \frac{P_s}{P_u} = -RT \ln \phi \qquad (1)$$

TABLE I. *Classification of Water Bonding and State of Water I and II*

Character and forms of bonding	Ionic bond	Molecular bond (hydration water)	Adsorption bond (water bonding in hydrate shells)		Osmotic bond	Structure bond
1. Formation conditions of bonding forms with moisture	Chemical reaction (hydration)	Crystallization from solution (formation of crystalline hydrate); formation of crystalline hydrates (hydrotactoeds) of the unpolar side chains of protein molecules	Dissolution in water; adsorption limit of water by molecules and ions; formation of solvate shells		Selective diffusion through semipermeable shells (cell diagrams)	Gel formation
2. Cause of bonding	Privary valency electrostatic forces of interaction)	Secondary valencies	Molecular power field		Osmotic pressure, biological forces of interaction	Water entrapped during forming formation of gel structure
			All molecules	Molecules and micelles of the inner and outer surface	Molecules of outer surfaces	

	Bodies forming ion-dispersive or molecule-dispersive water solutions	Hydrophilic bodies	Hydrophobic bodies	Plant cells with concentrated solution where water penetrates from less concentrated solution	Gel producing bodies (1% solid phase, 99% water)
3. Examples of bodies forming bonds	$CuSO_4 + 5\,H_2O$ $Ca^{2+}{=}{=}O^{2-} +$ $\overset{H^+}{\underset{OH^-}{\longrightarrow}}$ Ca^{2+} with OH^-, OH^-, OH^-	$R_1{-}CO{-}NH{-}R_2 + H_2O \longrightarrow$ $\longrightarrow R_1{-}COOH{-}R_2{=}NH_2$	$-Co - NH -$ $-CH-$ $R \cdot nH_2O$		
4. Bonding intensity (energy)	Very powerful bond	Powerful bond	Bonding of medium intensity — Irreversible · Hard to reverse · Reversible	Weak reversible bond-holding	Mechanical holding of water, the monomolecular bound adsorptively
5. Biological state	Anabiosis, the state of reversible cessation of life	Lifeless state — Mesabiosis, intermediate state between anabiosis and vitality		Hypobiosis, slow or limited life	
6. Character of cell structure functioning	Nonfunctioning of cell structure; extremely slow and suspended enzymatic reactions	One-sided catalytic action of enzymoactive components of cell structure; termination of synthesis processes		Weak functioning of cell structure	

Table 1 con't.

	Lifeless state					
	Dissimilation (processes of substance disintegration)			Reduced exchange intensity (hypometabolism)		Combination of assimilation and destructive processes
	Lack of reactivity			Lower reactivity		
	Annealing or burning	Water evaporation	Desorption	Desadsorptions	Moisture removal, formation of a more concentrated solution outside the cell	Evaporation, pressing out of water by upsetting the structure
7. Character of exchange	Minor destructive processes which are not actual dissimilation					
8. Characteristics of reactivity and breaking the bond between the dry substance and water	Chemical interaction; water is not removed by heating to 120–150°C; water can be removed only by annealing or burning					
9. Body and water change for a given bonding form	A new body is formed; water as such disappears and becomes a composite part of a new substance	Solvent shells are formed; spontaneously dispersive bodies disintegrate into ions or molecules forming actual solutions; in living organisms water acquire a stable structure with a small coefficient of self-diffusion	The body changes its properties; it is said to be plasticized; water reduces solidity, acts as plasticizer	The surface layer of material changes its properties due to formation of surface layer of water (monomolecular layer is most tightly bound); the bound part of moisture loses the properties of free water	Body swelling	The body changes its properties; quasi-solid body is formed; there occurs immobilization of water in the structure; water does not change its properties

Table 1 con't.

Classification of Water Bonding and State of Water; III

III. Physicomechanical bonding (water holding indefinite correlations)

Character and forms of bonding	Bonding in microcapillaries	Bonding in macrocapillaries	Moistening bond
Types of bonding			
1. Formation conditions of bonding form with moisture	Moisture absorption from moist air or by direct contact	Water absorption by direct contact through open capillaries and absorption from moist air by closed capillaries	Penetration of water by direct contact with the body's surface
2. Cause of bonding	Capillary pressure caused by the curvature of liquid surface	Same cause	Surface tension with the wetting angle of 90°
3. Examples of bodies forming	Body with capillaries $r < 10^{-5}$ cm	Body with capillaries $r > 10^{-5}$ cm	Nonporous hydrophilic bodies
4. Bonding intensity (energy)	Mechanical bonding of water (the layer near the walls is adsorptively bound)		Mechanical holding of water (the layer near the body's surface is adsorptively bound)

Table 1 con't.

	Life state
	Bio-active life
5. Biological state	Full reactivity
6. Character of cell structure functioning	Full functioning of cell structure
7. Character of exchange	Dissimilation (combination of substance exchange processes), full intensity of exchange
8. Characteristic of reactivity and conditions of breaking the bond between the dry substance and water	Pressure is higher than capillary pressure; Mechanical methods; evaporation
9. Body and water change for a given bonding form	Major portion of held water is free and preserves its properties; the adsorptively bound layer of water changes its properties; the body preserves its main properties; only some bonds are changed due to formation of a layer of adsorptively bound water

where R is the gas constant, P_s the saturation vapor pressure of free water, and P_u the partial vapor pressure of equilibrium water over material with moisture content U. It is evident that

$\phi = P_u/P_s$.

The tighter the bond between water and the material the smaller is the value of P_u. On the contrary, P_u for free water reaches value P_s, ϕ becomes equal to 1, and the binding energy $L = 0$.

The data concerning the value of L can be obtained from the analysis of the sorption/desorption isotherms. Figure 1 shows the nomogram for calculation of the energy in some food materials and products studied by the author.

According to the Gibbs-Helmholtz equation,

$$-\Delta F = -\Delta E + T \ \Delta S \tag{2}$$

where ΔE is the decrease of internal energy (or enthalpy) in isochoric-isobaric processes due to water bonding during molecular

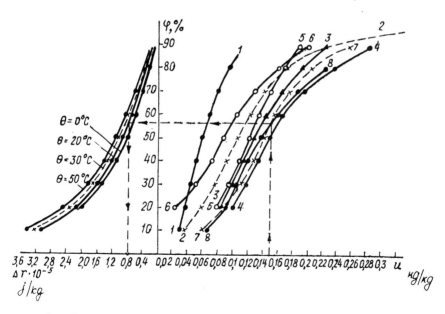

Fig. 1. Nomogram for calculation of the energy of moisture bonding in various foods. (1) sunflower seeds, $\theta = 20°C$; (2) wheat grain, $\theta = 50°C$; (3) maize, $\theta = 30°C$; (4) rye grain, $\theta = 0°C$; (5) shelled rice, $\theta = 20°C$; (6) flour, $\theta = 24°C$; (7) maize starch, $\theta = 20°C$; (8) macaroni, $\theta = 20°C$.

interaction, and ΔS the increase of bound water entropy in isothermal processes.

The entropy bond was usually considered weak. However, research
done by A. A. Buinov (1977), the author of this study, and V. I.
Syroedov shows that complex interaction between moisture and proteins is characteristic of biopolymers including, for example,
fish hydrolyzates. Specifically, swelling, "glassing," and dis-
solving processes occur in which the entropy bond is of importance.

The water solutions of fish food hydrolyzates (FFH) representing
a mixture of different peptides and free amino acids belong to the
group of difficult-to-dry products.

The isotherm sorption data obtained from the vacuum sorption
unit were processed for the thermodynamic analysis of the interaction process between FFH and water in the form of a modified
Bradley-Wilkinson equation

$$P_u/P_s = U_p \exp \sum_i C_i U_p^{i-1} \tag{3}$$

where U_p is the equilibrium moisture content of a food.

It has been found that the dependence of virial coefficients
on temperature is of a linear character and can be described by

$$C_i = B_i - Q_i/RT \tag{4}$$

The combined solution of Eqs. (3) and (4) produces ratio (5),
which links the values P_u and P_s with temperature:

$$P_u/P_s = U_p \exp\left(\sum_i B_i U_p^{i-1}\right) \exp\left(-\sum_i Q_i U_p^{i-1}/RT\right) \tag{5}$$

Virial coefficients C_i, B_i, Q_i with $i = 5$ were computer-calcu-
lated on the basis of experimental data.

The force of sorption/desorption is known to be the difference
of chemical potentials μ of free and bound water corresponding to
the differential change of the Helmholtz free energy. Provided

that steam behaves in accordance with the laws of ideal gases by
the Kelvin equation, considering expression (5) we have

$$\frac{\partial \; \Delta F}{\partial \; U_p} = \Delta\mu = RT \; \ln \; P_u/P_s = RT\left(\ln \; U_p + \sum_i B_i U_p^{i-1}\right) - \sum_i Q_i U_p^{i-1} \quad (6)$$

In differentiating (6) by T we get the expression for the differential change of the sorption entropy:

$$\frac{\partial \; \Delta S}{\partial \; U_p} = -\frac{\partial}{\partial T}\left(\frac{\partial \; \Delta F}{\partial \; U_p}\right) = -R\left(\ln \; U_p + \sum_i B_i U_p^{i-1}\right) \quad (7)$$

The second member of Eq. (6) is the differential change of
the internal energy (thermal effect) of sorption:

$$\frac{\partial \; \Delta E}{\partial \; U_p} = -\sum_i Q_i U_p^{i-1} \quad (8)$$

As a result of integration, considering Eqs. (7) and (8) we
get the above-mentioned Gibbs-Helmholtz equation (2).

Equation (5) is suitable for the approximation of isotherms
for the purpose of a thermodynamic analysis.

The analysis of constituents (7) and (8) of dependence (6)
shows (Fig. 2) that values $\partial \; \Delta F/\partial \; U_p$ and $\partial \; \Delta E/\partial \; U_p$ (curves 1-3)
decrease with an increase of the moisture content.

This is caused by the energetics of the sorption centers, the
molecular structure of the dry skeleton, and also (with
$U_p = 0.15-0.6$) by the mobility of individual links and FFH macromolecule as a whole, which take energetically advantageous conformations in sorption.

The value of the differential change of entropy member
$T \; \partial \; \Delta S/\partial \; U_p$ sharply decreases (curve 4) along with the increase of
the moisture content to 0.08, which is caused by the slowing of
the mobility of the water molecules in filling the monolayer. FFH
is a sorbent of small pores in which oriented sorption layers are
formed. In this case $T \; \partial \; \Delta S/\partial \; U_p < 0$ and decreases along with an

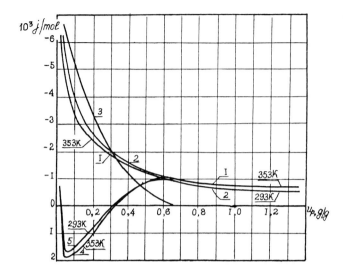

Fig. 2. Dependence of free ∂ ΔF/∂U (1,2), bound -T ∂ ΔS/∂U (4,5), and internal ∂ ΔE/∂U (3) energy of fish food hydrolyzates.

increase of U_p. Further on, due to swelling and subsequent dis-solving, $T \, \partial \, \Delta S / \partial \, U_p$ changes the sign that is characteristic of glassing biopolymers having flexible molecular chains.

Thus, in drying the FFH solutions from 1.4 to 0.08 g/g sorp-tion work $L = - \, \partial \, \Delta F / \partial \, U_p$ is used to break the entropy bond and to bring about conformation changes that are of irreversible character and compensate for the thermal interaction of desorption molecules of water from the surface of FFH.

In accordance with ideas based on electron microscopic research studies, biopolymer macromolecules form globules or packs of com-pact macromolecule arrangements which impedes diffusion of the moisture out of foods during desorption.

In order to decrease the hydrodynamic resistance of the dry skeleton and improve the drying, it is necessary to orient the flexible macromolecules of FFH. This can be achieved due to ad-sorption of macromolecules on the phase boundary during drying in the foamy state.

III. THERMODYNAMIC CHARACTERISTICS OF MOIST MATERIALS

Based on sorption isotherms, the following thermodynamic characteristics of moist foods can be calculated: specific mass (moisture) capacity and thermogradient coefficient.

A distinction is made between the average and local values of the specific mass capacity. The average specific moisture capacity $\overline{C}_m$ denotes changing of the average moisture content of body ΔU corresponding to the changing of the mass (moisture) transfer potential as per a unit of potential, i.e.,

$$\overline{C}_m = \Delta U / \Delta \theta_m \tag{9}$$

In the hygroscopic region ($U < U_h$, where U_h is the hygroscopic moisture content of a body corresponding to $\phi = 1$) the chemical potential μ is the potential of mass transfer.

In the region of the moist state of the body ($U > U_h$) μ is equal to the chemical potential of free water, which is a constant value since the pressure of vapor over the macrocapillary meniscus does not practically depend on the radius.

In this region the chemical potential cannot be the potential of the moisture transfer. Therefore, Lykov (1968) introduces the notion of a unified transfer potential θ_m for any moisture content of a body.

The transfer potential is determined experimentally by measuring the moisture content of investigated and standard (filter paper) bodies in contact, which are in a state of thermodynamic equilibrium. Similar to the heat transfer potential (temperature θ, the transfer potential θ_m of 100 mass exchange degrees corresponds to the maximum specific moisture content of a standard body $(U_{st\ max})$

$$\theta_{m(exp)} = \frac{U_{st}}{(C_{m\ st})_\theta} = \frac{U_{st} \times 100}{(U_{st\ max})_\theta} \ (^\circ M) \tag{10}$$

where U_{st} is the moisture content of a standard, and $C_{m\ st}$ the specific moisture capacity of a standard, at $U_{st} = U_{st\ max}$, $\theta_m(\exp) = 100°M$.

The data of the equilibrium moisture content of a body based on sorption isotherms $U_p = f_1(\phi)$ and data $\theta_m(\exp) = f_2(\phi)$ allow us to construct the function $\theta_m(\exp) = f_3(\phi)$ and then by graphic differentiation of this curve at constant temperature θ to find the function $C_m(\exp) = f(U_p)$. Table II shows the thermodynamic characteristics of some grain crops given by Krasnikov (1964).

Similar data were obtained by I. M. Savina and the author of this study for pasta dough (see Table III).

One can judge the extent to which the method of food processing influences its technological properties by the value of the average specific isothermal mass capacity (moisture capacity) $\overline{C}_m$. Such data are given by Nikitina (1968) in Table IV.

Table IV shows that the preliminary hygrothermal processing of foods (bananas, beef) lowers the specific moisture capacity of foods (i.e., their moisture-accumulating ability). This evidently is of particular importance for the subsequent drying and, mainly, storage of dried foods.

The drying of some foods in vacuum (cabbage, cod) causes an increase in C_m, but for carrots, for example, the dependence is the opposite. In any case there is a possibility for choosing the most rational method of preliminary processing and drying of any food by value C_m because the method of technological processing influences the value of U_p, upon which C_m depends. Besides, it is probably possible to choose the most suitable raw material by the optimum value of C_m of the final product, as the raw material naturally influences product C_m. Choosing the optimum value naturally requires relevant experiments and calculations. Some description of this dependence is given by Nikitina in Table V.

It is characteristic that C_m increases in the process of storage.

TABLE II. Thermodynamic Characteristics of Mass Transfer in Grain Crops

θ (°C)	U (kg/kg)	θ_m (°M)	$\overline{C}_m \times 10^2$ (kg/kg °M)	$\dfrac{\partial \theta_m}{\partial \theta} U$ (°M/°C)
		Rye		
0	9.6	16.5	0.59	
20	8.7	16.5	0.52	
30	8.4	17.0	0.50	
0-30	5.0	−	−	0.10
0	14.8	36.0	0.26	
20	14.6	37.5	0.28	
30	14.2	38.5	0.26	
0-30	10.0	−	−	0.15
0	22.5	64.5	0.28	
20	21.2	64.5	0.24	
30	20.4	64.5	0.24	
0-30	15.0	−	−	0.10
25	7.6	12.5	0.60	
25	14.4	35.5	0.30	
25	23.7	57.5	0.42	
25	36.3	100	0.30	
		Wheat		
0	8.8	15.5	0.56	
20	7.8	15.5	0.50	
30	7.4	15.5	0.47	
0-30	5.0	−	−	0.05
0	11.2	24.0	0.28	
20	10.7	24.0	0.35	
30	10.6	24.5	0.33	
0-30	10.0	−	−	0.12
0	21.2	65.0	0.24	
20	20.0	65.0	0.22	
30	19.4	64.5	0.21	
0-30	15.0	−	−	0.06
		Hard Wheat		
25	7.4	13.0	0.55	
25	13	33.5	0.28	
25	36.4	100	0.36	

Raw Rice

0	8.2	15.5	0.53	
20	7.5	15.5	0.48	
30	7.1	15.5	0.53	
0-30	5.0	–	–	0.07
0	11.0	23.5	0.34	
20	10.3	23.5	0.34	
30	10.0	23.5	0.34	
0-30	10.0	–	–	0.1
0	16.6	47.5	0.23	
20	15.2	47.5	0.22	
30	14.7	47.5	0.20	
0	19.2	64.5	0.15	
20	17.6	64.5	0.13	
30	17.1	64.5	0.14	
0-30	17.0	–	–	0.45

Rice

25	7.0	12.5	0.54
25	10.2	20.0	0.40
25	18.2	46.5	0.30
25	30.5	100.0	0.22

White Rice

0	8.7	15.5	0.55	
20	8.0	15.5	0.52	
30	7.6	15.5	0.48	
0-30	5.0	–	–	0.1
0	11.5	23.5	0.34	
20	10.9	23.5	0.34	
30	10.6	23.5	0.34	
0-30	10.0	–	–	0.14
0	16.8	47.5	0.22	
20	16.0	47.5	0.21	
30	15.3	47.5	0.20	
0-30	15.0	–	–	0.20

Buckwheat

25	9.4	18.0	0.52
25	21.3	55.0	0.32
25	32.0	100.0	0.22

		Maize		
0	9.4	15.5	0.60	
20	8.7	16.5	0.53	
25	9.6	16.0	0.60	
30	8.2	16.5	0.50	
50	5.0	-	-	0.10
0	17.6	47.5	0.25	
20	16.9	47.5	0.29	
30	15.8	47.5	0.25	
50	11.0	32.5	0.22	
0-50	11.0	-	-	0.25
0	20.1	64.5	0.15	
20	19.2	64.5	0.14	
30	18.3	64.5	0.14	
0-50	18.0	-	-	0.25
		Barley		
25	7.4	12.5	0.6	
25	15.0	37.5	0.30	
25	29.0	64.5	0.52	
25	36.5	100.0	0.22	
		Oat		
0	6.8	13.0	0.25	
20	5.7	12.5	0.45	
25	6.0	12.5	0.44	
30	5.0	12.5	0.40	
0-30	5.0	-	-	0.10
0	17.9	47.5	0.33	
	10.8	47.5	0.31	-
25	18.2	47.5	0.33	
30	16.2	47.5	0.32	
0-30	10.0	-	-	0.15
0	20.7	64.5	0.17	
20	19.9	64.5	0.18	
25	31.5	-	0.24	
30	19.0	64.5	0.16	
0-30	19.0	-	-	0.25

TABLE III. *Thermodynamic Characteristics of Pasta Dough and Hard Wheat Coarse-Granular First-Grade Flour*

Indices	T (°K)			
	293	305	321	333
	$W_p^C = 4\%$ and $\left(\dfrac{\partial \theta_{m(exp)}}{\partial T}\right)_U = 0.05°M/K$			
$C_{m\ exp} \times 10^2$ (kg/kg °M)	0.53	0.47	0.46	0.44
$\theta_{m\ exp}$ (°M)	7.5	8.5	8.6	9.0
	$W_p^C = 8\%$ and $\left(\dfrac{\partial \theta_{m(exp)}}{\partial T}\right)_U = 0.075°M/K$			
$C_{m\ exp} \times 10^2$	0.53	0.47	0.46	0.44
$\theta_{m\ exp}$	15.1	17.0	17.2	18.0
	$W_p^C = 12\%$ and $\left(\dfrac{\partial \theta_{m(exp)}}{\partial T}\right)_U = 0.175°M/K$			
$C_{m\ exp} \times 10^2$	0.53	0.47	0.46	0.44
$\theta_{m\ exp}$	22.7	25.5	26.5	27.2
	$W_p^C = 15\%$ and $\left(\dfrac{\partial \theta_{m(exp)}}{\partial T}\right)_U = 0.8°M/K$			
$C_{m\ exp} \times 10^2$	0.44	0.35	0.32	0.26
$\theta_{m\ exp}$	34.5	43.0	46.0	58
	$W_p^C = 17\%$ and $\left(\dfrac{\partial \theta_{m(exp)}}{\partial T}\right)_U = 0.82°M/K$			
$C_{m\ exp} \times 10^2$	0.44	0.35	0.32	0.26
$\theta_{m\ exp}$	38.5	49.0	53.0	66
	$W_p^C = 20\%$ and $\left(\dfrac{\partial \theta_{m(exp)}}{\partial T}\right)_U = 0.85°M/K$			
$C_{m\ exp} \times 10^2$	0.355	0.33	0.30	0.26
$\theta_{m\ exp}$	56	61	67	77

TABLE IV. *Dependence of Moisture Capacity of Foods on Processing Method*[a]

Food	Method of preliminary processing	Method of drying	Average moisture capacity in hygroscopic region $\overline{C}_{m\mu} \times 10^{-7}$
Bananas	Processed by steam	Vacuum-drying	0.108
	No processing		0.665
Beef	Boiled	Vacuum-drying	0.057
	No processing		0.480
Cabbage	Treated by boiling water	Air-drying	1.56
	Boiled	Vacuum-drying	3.08
Blood		Spray-drying	0.349
		Drum-drying	0.722
Carrots	Treated by boiling water	Vacuum-drying	0.576
		Air-drying	1.210
Cod	-	Air-drying	0.272
		Vacuum-drying	0.355

[a]*Given here are values of actual specific mass capacity expressed by chemical potential:*

$$C_{m\mu} = \left(\frac{\partial U}{\partial \mu}\right)_{\theta} \quad in \quad \frac{kg \ Kmole}{kg \ J}$$

TABLE V. *Dependence of Moisture Capacity on the Properties of Raw Materials*

Food	Raw material	$\overline{C}_{m\mu} \times 10^{-7}$ (kg kmole/kg J)
Vermicelli	Durum granular	0.290
	Durum forina	0.400
	Durum sovyetskaya	0.544
Pasta	Top-grade flour	0.443
	First-grade flour	0.483
Starch	Potato	0.885
	Maize	0.370
Strawberries	After vacuum-drying	0.77
	After 12 months of storage	1.77

The thermogradient coefficient δp (kg/kg K) shows a ratio between the drop in the moisture content ΔU of a body and the temperature drop $\Delta\theta$ in a stationary condition (when moisture flow $q_m = 0$):

$$\delta p = (\Delta U / \Delta\theta)_{q_m = 0}$$

Thus δp shows a drop in the moisture content of a body at $\Delta\theta = 1$.

It is not difficult to establish a relationship between δp and the average specific mass capacity $\overline{C}_m$:

$$\delta p = \overline{C}_m \left(\frac{\Delta\theta m}{\Delta\theta} \right)_{q_m = 0} \quad \text{or} \quad \delta p = C_{m\mu} \left(\frac{\partial\Delta\mu}{\partial T} \right)_{q_m = 0} \tag{12}$$

where $\Delta\theta_m / \Delta\theta$ is the temperature coefficient of moisture transfer potential (see Tables II and III).

Since the sorption isotherms of many foods, being colloidal capillary porous materials, are of typically smooth shaped. character, it is difficult to mark critical points on them that denote a transfer from one phase of moisture bonding to another. In this respect it is of significant interest to analyze how thermodynamic characteristics depend on the moisture content of a food, specifically, to analyze graphs $C_{m\mu} = f(U_p)$ and $\delta p = f_1(U)$.

The data of $C_{m\mu}$ and δp calculated on the basis of the value of the chemical potential of mass transfer μ for plums and cherries are given in Fig. 3 (research by V. F. Karazhia supervised by the author of this study and B. V. Zozulevich).

It is clearly seen that at $U_p = 0.7$ kg/kg there is a salient point indicating a boundary between the adsorptively bound moisture and the moisture bound by the capillary forces and osmosis.

In this connection it is expedient to create two zones in the working chamber of the drier. The drying conditions will be different there in accordance with the changing properties of fruits in the course of processing.

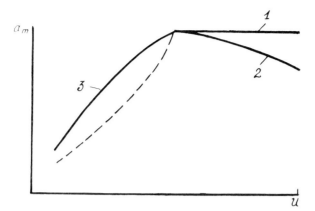

Fig. 4. The schematic dependence of the moisture diffusion coefficient on moisture content of a food.

Section 1 is indicative of the molar transfer of the capillary moisture in the form of liquid; section 2 is indicative of the removal of the osmotically adsorbed moisture due to diffusion. Along the lowering of the moisture content of a body (sections 3 and 3') the molar transfer of the capillary moisture mainly in the form of vapor is starting to play a more important role due to the deeper evaporation surface (the velocity of moisture supply inside the material is less than that of the evaporation of moisture from the surface). Due to the formation of a dry layer there is a greater resistance to moving moisture from the material. Such a phenomenon occurs, for instance, in bread drying. The adsorptively bound moisture is also removed in the form of vapor by the diffusion forces. Curve 3 is indicative of the summary effect of transfer.

Different types of moisture transfer occur simultaneously. The experimental graphs, shown here as an example, reflect the summary effect when one or another type of transfer prevails at various stages of the drying process.

The dependence of the moisture diffusion coefficient on the temperature of the material was studied by Miniovich and other researchers. It is expressed by the formula

$$a_m = a_{mo}\left(\frac{T}{273 + t_o}\right)^n \qquad (15)$$

(n reaches 10-14).

The dependence of the moisture diffusion coefficient of white malt on temperature and humidity W^C was investigated by Popov and is shown in Fig. 5. It is characteristic that despite the lowering of the moisture content, the value of a_m sharply increases due to the temperature rise. Figure 5 also depicts the changing of the external moisture exchange coefficient α_{mu} depending on the humidity and temperature.

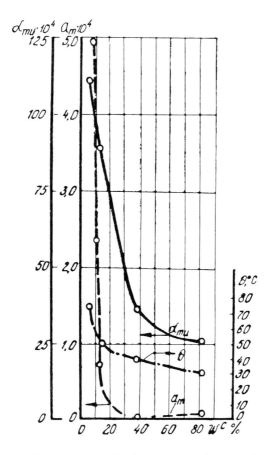

Fig. 5. The dependence of the mass exchange characteristics of white malt on temperature and humidity.

V. THE CHOICE OF RATIONAL METHODS OF FOOD PROCESSING

Special attention should be paid to the sharp dependence of the moisture diffusion coefficient on temperature of preheating (before drying) of a number of moisture inertial materials, i.e., materials with a low moisture diffusion coefficient (including the majority of foods) can be of great practical importance for quicker drying.

Thus, calculations made by Lykov (1968) show that the moisture diffusion coefficient of bread is approximately $a_m = 0.45 \times 10^{-5} m^2$/hour at air temperature

$$t_c = 40°C \text{ and } a_m = 12 \times 10^{-5} m^2/\text{hour at } t_c = 120°C.$$

The author of this study and V. P. Dubrovskii have determined the moisture diffusion coefficient for wheat grain with moisture content $U = 0.28\text{-}0.43$ kg of moisture/kg of dry substance at temperatures of 20, 40, and 55°C.

The research was conducted on the basis of the method by Lykov of substance nonstationary flow under isothermal conditions. This method is based on solving thermal conductivity problems for two semibounded bodies under a terminal condition of the fourth type. The investigations have shown that the dependence among the moisture diffusion coefficient of wheat grain, its moisture content, and its temperature can be described by

(a) with $\bar{U} = 0.283\text{-}0.360$ kg/kg,

$$a_m = \frac{1}{A\bar{u}^{-2} + B\bar{u} + c}\left(\frac{T}{293}\right)^K \times 10^{-9} \quad (m^2/c) \tag{16}$$

(b) with $\bar{U} = 0.360\text{-}0.430$ kg/kg,

$$a_m = \left(0.147 + 0.055\bar{U}\right)\left(\frac{T_{grain}}{293}\right)^K \times 10^{-9} \quad (m^2/c) \tag{17}$$

where A, B, C, and K are coefficients depending on the moisture content of the material, and T_{grain} is the grain temperature (K).

With $\overline{U}$ = 0.283-0.340 kg/kg, A = 558, B = -382, C = 67.8.

With $\overline{U}$ = 0.324-0.360 kg/kg, A = 3620, B = -2340, C = 380.4.

The considerable value of index K, which changes from 8 to 18, deserves attention. Such a high value of K indicates that it is expedient to heat grain prior to drying.

The criterion of Lykov, Ly = a_m/a can be taken as a characteristic criterion determining a relationship between the moist and thermal inertial properties of the material.

For grain Ly $\approx 10^{-3}$, which means that it is quickly heated to the extreme permissible temperature but lets out moisture slowly. For such materials alternating drying conditions are used, i.e., they are subject to intermediate cooling after heating and the drying cycle is then repeated.

The recirculation isothermal method of grain drying with preliminary heating and drying at the optimum temperature is considered the most efficient in the USSR.

The author has noted the important role of the initial condition of the material and the initial impulse of the external influence, which determines the reaction of the moist material and development of respective internal fields influencing the intensity of processing.

Such a phenomenon is specifically characteristic of drying, being a typically irreversible process tending to equilibrium. In accordance with the universal Le Chatelier-Braun principle, the stronger the external influence on the drying object in the initial stage, the more intensive are the internal processes trying to bring the system back to the equilibrium state.

Recent research by Militser, Strauss, and Brink (1976) shows that the theory of "initial impulse" is applicable to many mass exchange processes (extraction, adsorption, desorption).

Because of considerable thermolability, it is necessary to prepare the material prior to the initial impulse. Therefore, various technological methods of preliminary preparation of the material acquire importance, such as dispersion, vibration processing, preheating, foaming, the use of surface active substances, the combination of drying with other technological operations (inside-the-material transfer of water-soluble minerals, enzymes, etc.; oxidation, recovery, acceleration of relaxation of the internal stresses, etc.), which is especially important for food and biomaterials.

The problem of controlling the mechanism of moisture transfer inside the material in the process of drying with directed localization of the zone of phase changes also deserves much attention. Thus, for many materials it is desirable to secure moisture transfer mainly in the form of liquid (i.e., the evaporation zone should be located close to the surface of the material). For example, in drying grain under these conditions, minerals dissolved in water are transferred to the germ and the germinating power of seeds improves in the process of drying. In other cases (during sublimation drying, for example) moisture moves inside the material in the form of vapor, which helps to preserve valuable flavor compounds, enzymes, etc.

VI. INFLUENCE OF WATER ON PRODUCT PROPERTIES

Thus, the condition of moisture, its molecular structure, and its bonding with the dry skeleton of the body are the decisive factors that influence the mechanism of the moisture transfer, process kinetics, and energy consumption in processing moist products. There is a need to further develop the theory of the forms and energy of moisture bond on the basis of research data obtained from the analysis of the physical properties of water having an electronic molecular structure, as well as studies in the field of molecular interaction between water and the surface of the adsorbent-the solid skeleton of the material.

Besides changing the properties of the bound moisture it is

necessary to take into account the changing nature of the solid
skeleton proper, which influences the character and energy of bonding.

An interesting description of water is given by the well-known
Hungarian researcher Dierdi "Water is a strange substance.... It
possesses properties that change the molecular structure in a ca-
pricious way...." Studying the structure of water "we find ourselves in a fantastically charming world ... water is not only the
mother it is also the matrix of life. Probably, biology has not
yet succeeded in understanding the most important functions just
because it concentrated its attention on substance in the form of
particles separating it from other matrices-water and the electrical field."

Important research has been carried out in the USSR in analyzing water properties during the past years and studies have been
conducted in the field of interaction between the molecules of
water and the surface of the adsorbent-the solid skeleton of the
body (B. V. Deryagin, M. M. Dubinin, M. P. Volarovich, A. V.
Kiselev, G. N. Zatsepina, P. V. Churaev, N. I. Gamayukov, P.
Broyer, D. P. Poshkus, etc.). The studies by Duckworth (1975) and
other researchers presented at the First International Symposium
on Water Properties, Glasgow, are also of great interest.[*]

It is important to take into account that during the thermal
treatment of materials there is an increase of the number of functional groups, the active centers of adsorption.

The investigation of the above interaction, which is electromagnetic, provides the approach for the study of the microproces-
ses developing in the phase boundary. It is important to note
that frequently these processes can be characterized by macrono-
tions (coordinate, impulse, energy), which establish a relationship
between the static and dynamic regularities. This will allow us to

[*] *The papers of this symposium are now being translated into
Russian by the Food Industry publishing house, Moscow.*

study and analyze the properties of microscopic bodies proceeding
from the microscopic model of the substance structure. Such an
approach will lead to a more detailed study of various properties
of materials as drying objects (this will make it possible to
directionally change their properties) and to a deeper understanding of the molecular mechanism of the complex phenomena (physical,
physicochemical, biological) developing in the materials during
their processing and storage.

Theoretical and experimental research is underway for this
purpose based on application of more precise methods and instruments such as infrared spectroscopy, radioactive indicators, paramagnetic and nuclear magnetic resonance, X-ray structure analysis,
gammascopy, and other similar methods of modern physics. The data
obtained from the studies of the moisture state at temperatures
below zero (down to $-30^{\circ}C$), and specifically, the data relating
to the determination of the quantity of nonfreezable water are of
interest.

The research conducted by the Moscow Technological Institute
of Food Industry (E. I. Ryzhova, V. I. Kovnatskii, V. I. Syroedov,
V. P. Seregin, and the author) covers the influence of coffee extract concentration on the cryoscopic temperature and determines
how the quantity of nonfreezing water depends on the extract concentration and the temperature of the cooling medium.

In the process of extract freezing, the structure of the drying body is formed and its thermophysical and physicochemical
characteristics are changed. These changes may vary depending on
the concentration of the extracts, the cooling rate, and accordingly the freezing time. Concentration overcooling and the emergence of the concentration gradient in liquid phase (before the
crystallization front), which is characteristic of nonequilibrium
processes, are of great importance. The experimentally discovered
effect of some "critical" concentration above which the average
freezing temperature begins to increase is apparently connected
with this phenomenon.

In investigating the freezing of the water solution of common salts the situation is similar. Salt crystals settle out of the solution, as if there occurs "self-separation" of phases. The remaining liquid is of lower concentration and freezes at a higher temperature.

Significant studies of the water structure and properties were conducted by Zatsepina (1974) on the biological functions of water by Kazakov of the Moscow Technological Institute of Food Industry, of the structure of the water solutions of non-electrolytes by Shahparonov and his associates of the Moscow State University, of the adsorbed water properties by Gamayunov and others of the Kalinin Polytechnical Institute.

It is necessary to note the discrepancies between the data related to the density of the adsorbed water. In some investigations it has been reported as higher, in others as lower than 1. Such discrepancies are mainly caused by inaccurate methods of measuring density using liquid density meters, when relatively large liquid molecules could not penetrate all pores of the small-pore material under investigation. Therefore, the density of this material turned out to be lower, and accordingly the density of the adsorbed substance was higher. Most accurate data on the density of adsorbed water were obtained by Volarovich, Gamayunov, Vasilyeva, and Vasilyev (1975) using a helium density meter. Helium is known to be an inert gas that does not interact with metal and is not sorbed at room temperature. Helium atoms being comparatively small (diam ≈ 2.15 Å) can penetrate a considerable portion of pores of such biopolymers as starch, gelatin and cellulose. However, the penetration of helium atoms into pores depends on the structure of the material, which is determined by the conformation changes of the sorbent-sorbent system in the processes of water sorption-desorption.

Figure 6 shows the dependence of the measured density ρ_{eff} (the authors call it "effective" density) on the moisture content of biopolymers. As seen, $\rho_{eff} \approx 1.0$ with $W^c = 10\%$ when water mole-

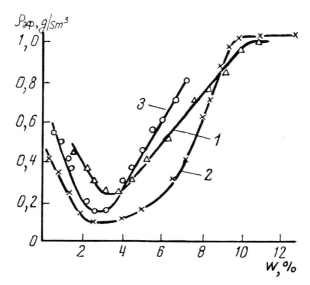

Fig. 6. The dependence of the effective density of sorbed water on the moisture content of starch (1), gelatin (2), and cellulose (3).

cules in the region where monosorption is exceeded move the macro-molecules (and their packs) of sorbent kerns apart and cause their swelling. This disturbs the orderly arrangement of the mate-rial microstructure and helps helium atoms to penetrate micropores. In this case the "effective" density of adsorbed water is probably close to the true value.

In the process of moisture desorption due to the conformation changes causing a "contraction" of polymer particles, the polymer structure becomes less permeable by helium atoms (w^C = 3-4%) and ρ_{eff} decreases. Upon removal of the remaining water the structure again becomes less orderly arranged due to the decrease of the number of bonds between the functional groups and water molecules, which helps the penetration of helium atoms and ρ_{eff} to increase.

It is characteristic that the investigations of the properties of sorbed water using the nuclear magnetic resonance-spin echo method conducted by the authors have shown that with w^C = 3-4% the molecular mobility of sorbed water for such polymers really

becomes minimal, which is caused by the maximum "contraction" of
the system. In this process there occurs a peculiar "dissolution"
of the low-molecular substance-water in the high-molecular biopolymer, and water molecule packs do not form a continuous phase
of sorbed substance. Thus, the notion "density" of water makes
no physical sense. At a moisture content of about 10%, micropores
accessible to water molecules are filled up, forming a phase whose
density is measured by the helium density meter. Thus, a conclusion can be made that the real density of water is changed little
in adsorption bonding.

VII. ELECTROMAGNETIC FIELD EFFECT ON MOIST MATERIALS

In studying how different methods of power supply influence
the processes of the thermal treatment of food products, we have
noted that the best effect is achieved in the resonance state,
i.e., with the frequency of the electromagnetic field close to
the frequency of oscillations of water molecules proper, i.e., in
the infrared and high-frequency range of the spectrum.

In the past years, a number of researchers have published information about the frequency of oscillations of water molecules.

Thus, academician I. N. Plaksin and others cite data by Hiben,
who established, as a result of studying the spectrum of dispersion of light by water, that the wave number of oscillations of
water molecules in the equilibrium state is equal to 150 cm^{-1}.
Hence, the frequency of oscillations of water molecules proper
will make

$$\nu_{water} = 150 \times C \ \text{sec}^{-1}$$

where C is the velocity of light, i.e.,

$$\nu_{water} = 4.5 \times 10^{12} \ \text{sec}^{-1}.$$

The authors arrived at the conclusion that the intensity of extinction of light by water is affected by the electrical field, which is added to the existing effects of influence of various external fields caused by solar radiation, radio waves, and other electromagnetic fields. The above influence causes structural changes due to the violation of the frequency of oscillations of water molecules proper, which is close to the frequency of the electrical field.

Duckworth (1971) using nuclear magnetic resonance-spin echo method established that the time of molecular rotation of "free" water is 10^{-12} sec. Haberditzl (1972) notes that it has been experimentally established that oscillations with frequency 10^{14} sec^{-1} correspond to an energy of approximately 1000 K per degree of freedom.

Zatsepina (1974), in analyzing the processes of molecular dispersion of the electromagnetic field energy by water, has found that if uniformity is characteristic of water in the range of visible light frequencies, heterogeneity is characteristic of it in the range of frequencies $\nu_{water} \approx 10^{13}$ sec^{-1}, which is caused by the distant dispersion interaction due to the great anisotropic polarizability of water molecules in the liquid water.

In the research conducted at the Moscow Technological Institute of Food Industry by the author, V. V. Krasnikov, N. G. Selyukov, V. I. Syroedov, and S. G. Ilyasov, on application of infrared radiation, it was established that the best effect is achieved with λ_{max} = 1 mkm, i.e., with $\nu = 3 \times 10^{14}$ sec^{-1} for a number of processes (bread baking, drying of some food products).

Modern ovens for baking foods using an infrared power supply were developed on the basis of these studies jointly with the Kiev Technological Institute of food industry (A. A. Mihelev, M. N. Sigal, A. V. Volodarkii, and others). The scientifically based selection of shortwave radiation generators and of their operating conditions will considerably reduce the process time, improve the quality of products, and the technical and economic indices of oven operation. The use of electromagnetic field energy in various units for drying, frying, sterilization, and other kinds of thermal processing of moist foods produced good results.

The above examples testify to the fact that further theoretical and experimental research of water properties and their changes caused by external factors, besides being scientifically important, are also of practical value for solving the important task of raising the efficiency of the thermal processing of foods and improving their quality.

References

Buinov, A. A. (1977). *In* "New Developments in Drying Technology of Different Foods and Materials." Moscow Technological Institute of Food Industry, Moscow.
Duckworth, R. B. (1971). *J- Food Technol. 6.*
Duckworth, R. B., ed. (1975). "Water Relations in Foods." *Proc. of Int. Symp. Water Relations in Foods, Glasgow, September 1974.* Academic Press, New York.
Ginzburg, A. S. (1969). "Application of Infrared Radiation in Food Processing," p. 412. Leonard Hill, London.
Ginzburg, A. S. (1973). *Food Ind.,* 528.
Ginzburg, A. S. (1976). *Food Ind.,* 248.
Ginzburg, A. S., and Dubrovskii, V. P. (1964). *Int. Chem. Eng. 4,* 2.
Ginzburg, A. S., and Kazakov, E. D. (1976). *Works All-Union Res. Inst. Grain Its Products 83.*
Ginzburg, A. S., and Lebedev, P. D. (1971). *In* "Progress in Heat and Mass Transfer," Vol. 4. Pergamon Press, Oxford and New York.
Ginzburg, A. S., Syroedov, V. I., Ryzhova, E. I., Kovnatskii, V. I., and Barabanov, M. I. (1975). *Proc. XIV Int. Congr. Cold,* Section 1.20, Moscow.
Haberditzl, W. (1972). "Bausteine der Materie und Chemische Bindung." Veb Deutscher Verlag der Wissenschaften, Berlin.
Krasnikov, V. V. (1964). *Bull. USSR Higher Educ. Inst. Food Technol. 3.*
Lykov, A. V. (1968). "Drying Theory," p. 471. Energy, Moscow.
Lykov, A. V. (1978). "Heat and Mass Exchange," p. 479. Energy, Moscow.
Militzer, K. E., Strauss, R., and Brink, E. (1976). *Wiss. Z. Tech. Univ. Dresden 25,* H.4.

Nikitina, L. M. (1968). "Thermodynamic Parameters and Coefficients of Mass Transfer in Moist Materials." Energy, Moscow.

Rebinder, P. A. (1958). *All-Union Sci. Tech. Conf. Drying*, Profizdat, Moscow.

Sent Dierdi, A. (1960). :Biogenetics." Physicomathematical Sciences, Moscow.

Rebinder, P. A. (1978). "Selected Works, Surface Phenomena in Disperse Systems. Colloidal Chemistry," p. 368. Science, Moscow, p. 368.

Volarovich, M. P., Gamayunov, N. I., Vasilyeva, L. Y., and Vasilyev, Y. M. (1975). *Colloid. J. 37*, 2.

Zatsepina, G. N. (1974). "Water Properties and Structure," p. 167. Moscow Univ., Moscow.

WATER ACTIVITY AT

SUBFREEZING TEMPERATURES

Owen Fennema

I. INTRODUCTION

Considerable research has been done on the water activity of foods and biological materials at above-freezing temperature, and important relationships have been established between this property and food stability (Acker, 1969; Rockland, 1969; Labuza, *et al.*, 1970). In unfrozen biological systems and foods, it is generally recognized that food stability is more closely related to water activity than to total moisture content. Conditions for optimum moisture content can be predicted with reasonable accuracy from plots of water activity vs. total moisture content, such plots being known as "moisture sorption isotherms" (Acker, 1969; Rockland, 1969; Caurie, 1971; Labuza, 1971).

Comparatively little effort has been devoted to the subjects of vapor pressure and water activity of complex foods and biological matter at subfreezing temperatures. This is somewhat surprising considering the usefulness of the water activity (a_w) data at above-freezing temperatures. Several points are of interest with regard to a_w values at subfreezing temperatures:

(1) What value does one choose for the reference state when defining a_w?

(2) What sorts of plots of a_w are most useful?

(3) Does the temperature dependence of a_w values change as a result of freezing?

(4) What meaning can be attached to a_w values derived from samples at subfreezing temperatures?

These questions are considered in this chapter.

II. DEFINITION OF WATER ACTIVITY AT SUBFREEZING TEMPERATURES

Water activity has been defined as the ratio of the fugacity of water in a sample to the fugacity of pure water at the same temperature (Ross, 1975). In practice, fugacity is approximated by vapor pressure, giving the following definition that applies at above-freezing temperatures and normal atmospheric pressure:

$$a_w = p_w/p_w^o$$

where P_w is the vapor pressure of water generated by the sample and p_w^o the vapor pressure of pure water at the same temperature.

At subfreezing temperatures, reanalysis of the definition of a_w is necessary, since a problem arises as to whether the vapor pressure of ice or the vapor pressure supercooled water is most appropriate for the denominator term p_w^o. The vapor pressure of supercooled water turns out to be the proper choice since (1) values of a_w at subfreezing temperatures can then, and only then, be accurately compared to a_w values at above-freezing temperatures, and (2) choice of the vapor pressure of ice as p_w^o would result, for samples that contain ice, in a meaningless situation whereby a_w would be unity at all subfreezing temperatures. The reason point (2) occurs is explained below. Acceptance of the above arguments results in the following definition of a_w as applied to samples at subfreezing temperatures:

$$a_{\mathrm{w}} = p_{\mathrm{w}}/p_{\mathrm{sw}}^{\mathrm{o}} \qquad\qquad (2)$$

where $p_{\mathrm{sw}}^{\mathrm{o}}$ is the vapor pressure of pure supercooled water.

Some controversy exists as to the value of the numerator term p_{w} in samples at subfreezing temperatures. Most investigators have argued that the vapor pressure of a frozen sample is equal to the vapor pressure of pure ice at the same temperature (Storey and Stainsby, 1970; Fennema and Berny, 1974). However, some investigators have reported deviations from this concept. Most noteworthy in this regard is the work of Sunderland and co-workers (Dyer et al., 1966; Hill and Sunderland, 1967). These investigators measured the vapor pressures of veal, lamb, beef, poulty, pork, fluid from beef, and ice over the temperature range -26° to $-1^{\mathrm{o}}\mathrm{C}$. The test procedure involved placing a frozen sample in a glass flask, reducing the pressure to about 0.3 mm Hg, establishing equilibrium at the temperature desired, recording the pressure, and correcting this pressure for the residual air in the flask (Dyer and Sunderland, 1964). Surprisingly, they found vapor pressure values of the various animal tissues, and the fluid expressed from beef tissue, ranged from 13 to 20% lower than the vapor pressures of pure ice at the same temperatures. The reduced vapor pressures of the animal products were attributed to the formation of solid solutions.

More recently, Storey and Stainsby (1970) reported vapor pressure values for cod, beef, and pure ice at subfreezing temperatures ranging from -23° to $-4^{\mathrm{o}}\mathrm{C}$. Measurements were made according to the previously described method of Dyer et al. (1966) with only minor modifications. They found that vapor pressures of code, beef, and ice all agreed well with published values for ice at the same temperatures. Thus, the results of Storey and Stainsby are in marked disagreement with those of Sunderland's group. Storey and Stainsby were unable to account for the cause

of the disagreement, thereby leaving this important matter
unresolved.

Fennema and Berny (1974), using a differential pressure
gauge, studied the vapor pressures of beef (pre- and postrigor),
potatoes, apples, a 15% sucrose solution, and pure ice at
temperatures of -5°, -10°, -15°, and -20^OC. The results, shown
in Table I, indicate that equilibrium vapor pressures of the
various samples studied did not differ significantly (95% level
of confidence) from the vapor pressure of ice, at any of the
temperatures studied.

These results agree well with those of Storey and Stainsby
(1970) and disagree with those of Sunderland's group. Since
the results by Fennema and Berny (1974) were obtained with a
more accurate and completely different type of apparatus than
that used by Storey and Stainsby (1970) and Sunderland's group,
it seems proper to conclude, as did Storey and Stainsby, that
the results of Sunderland's group are in error.

Why Sunderland's data for tissues and tissue fluids are
erroneously low, is not easy to ascertain, since many of the
experimental details were not published. Nevertheless one can
speculate about possible weaknesses in the procedures of Sunderland's group. They mentioned tests to assure conditions of true
equilibrium; however, failure to obtain true equilibrium must
be regarded as a possibility. Furthermore, the validity of the
method that Sunderland's group used to calculate residual air
pressure in the sample flask has been questioned by Bralsford
(1968), and this appears to be a possible source of error even
though Storey and Stainsby presumably used the same technique
and obtained results different from those of Sunderland's group.
Whatever the faulty procedure was, it must have resulted in
erroneous values for food samples and true values for ice, since
their values for ice agree well with literature values.

If one accepts the position that the equilibrium vapor
pressure of pure ice and foods are identical at any given sub-

TABLE I. Vapor Pressures of Ice and Various Foods[a]

Temp.[b] (°C)	Products[c]	Gauge readings		Vapor pressure Mean[e] (mm Hg)	Equilibration time (range in min)	Significance of differences between ice and foods (95% level of confidence)[g]
		Range	Mean[d]			
−5.00	Ice	3.40–3.42	3.41	3.01	35–40	N.S.
	Beef postrigor	3.42–3.43	3.42	3.02	100–110	N.S.
	Beef prerigor	3.41–3.43	3.42	3.02	100–110	N.S.
	Apples	3.38–3.42	3.40	3.00	80–85	N.S.
	Potatoes	3.36–3.40	3.39	2.99	80–85	N.S.
	15% Sucrose	3.38–3.43	3.40	3.00	63–68	N.S.
−10.00	Ice	2.16–2.19	2.18	1.95	35–40	N.S.
	Beef Postrigor	2.17–2.20	2.18	1.95	53–58	N.S.
	Apples	2.16–2.20	2.18	1.95	52–55	N.S.
	Potatoes	2.16–2.21	2.19	1.96	55–60	N.S.
	15% Sucrose	2.17–2.20	2.18	1.95	35–40	N.S.
−15.00	Ice	1.38–1.42	1.40	1.24	12–16	N.S.
	Beef postrigor	1.39–1.41	1.40	1.24	30–35	N.S.
	Beef postrigor[f]	1.37–1.38	1.38	1.22	30–35	N.S.
	Beef prerigor	1.38–1.39	1.39	1.23	25–27	N.S.
	Beef prerigor[f]	1.38–1.39	1.39	1.23	30–35	N.S.
	Apples	1.39–1.40	1.39	1.23	27–35	N.S.
	Potatoes	1.39–1.42	1.41	1.25	30–35	N.S.
	15% Sucrose	1.40–1.43	1.42	1.26	20–24	N.S.
−20.00	Ice	0.83–0.86	0.85	0.78	8–12	N.S.
	Beef postrigor	0.84–0.85	0.85	0.78	20–25	N.S.
	Apples	0.83–0.87	0.85	0.78	20–25	N.S.
	Potatoes	0.85–0.86	0.85	0.78	20–25	N.S.
	15% Sucrose	−.85–0.87	0.86	0.79	15–18	N.S.

Table 1 con't.

[a] From Fennema and Berry (1974); courtesy of the International Union of Food Science and Technology.

[b] Precision, ±0.03°C; accuracy, ±0.03°C.

[c] Frozen by immersion in liquid nitrogen, unless otherwise stated. During freezing samples weighed approximately 0.15 g, except for beef samples, which weighed approximately 30 g. Frozen beef samples were then cut into cylinders weighing about 0.15 g each.

[d] Mean of three replicates, unless otherwise stated.

[e] Taken from calibration curve.

[f] Mean of two replicates; samples slowly frozen (6 hr elapsed during passage from point of initial freezing to -5°C).

[g] Coefficient of variablity 0.9%.

freezing temperature down to at least -20°C, then it becomes
readily apparent why all subfreezing values of a_w would be unity
if the vapor pressure of pure ice were to be used as the denominator of the ration p_w/p_w^o.

A convenient situation develops when a_w values at subfreezing
values are defined as stated in Eq. (2). Since the vapor pressure
of ice is accurately known to very low temperatures and the vapor
pressure of supercooled water is accurately known to -15°C, and
can be extrapolated with reasonably accuracy to somewhat lower
temperatures, a_w values at subfreezing temperatures can be calculated rathern than measured. Calculated values for water activity over the range -1° to -40°C are listed in column 6 of
Table II.

Another important difference should be noted between a_w
values at above- and below-freezing temperatures. In the absence
of an ice phase, a_w is a function of sample composition and
temperature (constant pressure) with the former factor pre-
dominating. In the presence of an ice phase, a_w becomes independent of sample composition and depends solely on temperature.
This difference is important and is discussed below.

III.   PLOTTING OF a_w VALUES OBTAINED AT SUBFREEZING TEMPERATURES

a_w values pertaining to samples at subfreezing temperatures
can be plotted in at least two useful ways, and both are
illustrated. Plots of a_w vs. moisture content are shown in
Figs. 1 and 2, and although these plots do not appear very
similar to conventioanl moisture sorption isotherms at above-
freezing temperatures, some similarities do exist.

In Fig. 1, the family of four curves situated at the center
and lower-left portion of the graph represent samples of bovine
muscle that contained no ice, and these curves are therefore
analogous to moisture sorption isotherms at above-freezing

TABLE II. Relationship between Subfreezing Temperature and Water Activity in Samples Containing Ice

Temperature			Vapor pressure (mm Hg)		$\dfrac{p_{ice}}{p_{sw}}$	$\log \dfrac{p_{ice}}{p_{sw}}$
°C	°K	1000/T	supercooled water[a]	ice[b] or biological matter containing ice		
-1	272.15	3.674	4.258	4.217	0.9904	-0.00420
-2	271.15	3.688	3.956	3.880	0.9808	-0.00842
-3	270.15	3.702	3.673	3.568	0.9714	-0.01260
-4	269.15	3.715	3.410	3.280	0.9619	-0.01688
-5	268.15	3.729	3.163	3.013	0.9526	-0.02110
-6	267.15	3.743	2.931	2.765	0.9434	-0.02532
-7	266.15	3.757	2.715	2.537	0.9344	-0.02945
-8	265.15	3.771	2.514	2.326	0.9252	-0.03376
-9	264.15	3.786	2.326	2.131	0.9162	-0.03803
-10	263.15	3.800	2.149	1.950	0.9074	-0.04220
-11	262.15	3.815	1.987	1.785	0.8983	-0.04656
-12	261.15	3.829	1.834	1.632	0.8899	-0.05068
-13	260.15	3.844	1.691	1.490	0.8811	-0.05496
-14	259.15	3.859	1.560	1.361	0.8724	-0.05927
-15	258.15	3.874	1.436	1.241	0.8642	-0.06338
-20	253.15	3.950	0.9406	0.776	0.8250	-0.08354
-25	248.15	4.030	0.6053	0.476	0.7864	-0.10436
-30	243.15	4.113	0.3816	0.2859	0.7492	-0.12539
-35	238.15	4.199	0.2354	0.1675	0.7116	-0.14779
-40	233.15	4.289	0.1418	0.0966	0.6812	-0.16670

[a] Values for -15°C and warmer are observed data (Weast, 1969). Values for -20°C and colder are calculated data (Mason, 1957).

[b] From Weast (1969) except value for -35°C, from Mason (1957).

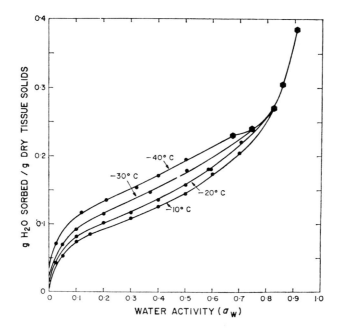

Fig. 1. Desorption isotherms and freezing point curve for
bovine muscle at subzero temperatures. From MacKenzie (1975),
with permission.

temperatures, except the S shape is less pronounced. Each of
these curves joins the "freezing-point curve" for bovine muscle,
which is situated in the upper right-hand corner of the graph.
The freezing-point curve represents samples that do contain ice
and for which a_w depends only on temperature. For samples containing ice, the temperature-a_w relationship is obviously not
effectively displayed in Fig. 1 since water content (grams
H_2O/grams dry solids) should not be a variable. However, the
appropriate information is present if properly interpreted. Thus,
the a_w at -40°C, for a sample containing ice, is determined by
locating the point of intersection of the -40°C isotherm and the
freezing point curve, and simply reading off the associated a_w
on the abscissa. In a similar manner, the a_w at -30°, -20°, and

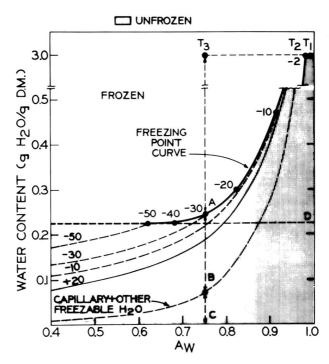

Fig. 2. Water activity of lean beef (75% water) at low temperatures. Dashed lines are hypothetical but realistic. Data from Riedel (1961), with permission.

$-10^{\circ}C$ can be determined, but it is evident that the accuracy with which these values can be determined decreases with increasing temperature.

From Fig. 1 is also apparent that each point on the freezing point curve is associated with a specific water content. This relationship may appear contrary to the previous statement that, "in the presence of an ice phase, a_w becomes independent of sample composition and depends solely on temperature." There is, however, no contradiction, just a problem of interpretation. Any given water content graphically associated with the freezing point curve should be regarded as the *minimum* water content that will yield a temperature-a_w relationship that it is independent

of water content. Thus, water contents in excess of the values
associated with the freezing point curves simply provide a
measure of the ice content of the sample.

   This point is more understandable in Fig. 2, which is a plot
similar to Fig. 1. Data in Fig. 2 apply to lean beef and were
originally reported in a different form by Riedel (1961). in
Fig. 2, freezing involves movement in a horizontal direction
from right to left (constant moisture). Since the beef sample
in Fig. 2 contained 75% moisture (3 g H_2O/g dry matter) the
starting conditions at room temperature are represented by point
T_1. Removal of sensible heat would cause a slight decline in
a_w and freezing would commence at about -2°C (point T_2 and
$0.98a_w$). Further cooling would result in additional ice formation and movement to the left along line T_1-T_3. During the
course of freezing, the freezing point of the unfrozen phase, and
the amounts of ice and unfrozen water can be derived from the
figure. This is done by drawing a vertical line from the appropriate point on line T_1-T_3 to the abscissa. For example, at
subfreezing temperature T_3, the freezing point of the unfrozen
phase is -30°C, the amount of ice is represented by the length
of line segment T_3-A and the amount of unfrozen water is represented by segment A-C. The unfrozen water will in turn
consist of firmly bound unfreezable water (segment A-B) and
capillary and other water that would freeze if the temperature
were lowered further (B-C). It is evident from Fig. 2 that
maximum ice formation occurs at about -40°C. If, prior to cooling,
meat is dried to a moisture content of less than 0.225 g H_2O/g
D.M. (point D) then no ice will form regardless of temperature.

   Conditions during freeze drying also can be determined from
Fig. 2. Consider, for example, that a product of normal water
content is placed in a freeze-dryer and dried at a constant
temperature at -30°C. Conditions would first change from T_1 to
T_3 as previously described for freezing (except that a slight
amount of moisture would be lost during this period, causing

T_3 as previously described for freezing (except that a slight amount of moisture would be lost during this period, causing T_3 to assume a slightly lower position than shown), then gradually change from T_3 to A as the frozen phase is sublimed. At the conclusion of sublimation, drying would continue (conversion of liquid water and sorbed water to water vapor) and the conditions would move from point A downward to the left along the dashed -30°C isotherm, where a_w is a function of both water content (sample composition) and temperature. This same sequence of events, during freeze-drying, can be followed on Fig. 1, except the ice phase is now shown.

Although the data in Figs. 1 and 2 are useful, a still more useful plot, for samples that contain ice, is that of a_w vs. temperature. For these samples a_w has been found to vary with temperature in accord with the relationship

$$a_w = S \exp(\Delta H_a/RT)$$

where s is a constant, ΔH_a the latent heat of fusion for ice, R the gas constant, and T absolute temperature. Thus a plot of $\log a_w$ vs. $\frac{1}{T}$ (Fig. 3) should be linear with a slope of $\Delta H_a/R = 2.303$. This is true over the range of subfreezing temperatures, or over the range of above-freezing temperatures, but not over both ranges combined. From Fig. 3, the a_w at any temperature over the range -1° to -15°C can be determined and the result is known to apply accurately to muscle tissues, and is believed to apply with equal accuracy to biological materials and foods in general (MacKenzie, 1975; Riedel, 1961). Thus, the a_w of a relatively high-moisture sample with an initial freezing point of -2°C (such as muscle) would decrease from an a_w of about 0.99 at 20°C to 0.98 at -2°C, and then to 0.86 at -15°C. Other products with different initial freezing points would intersect the -1° to -15°C line as a family of lines that are essentially parallel to the dashed line already shown.

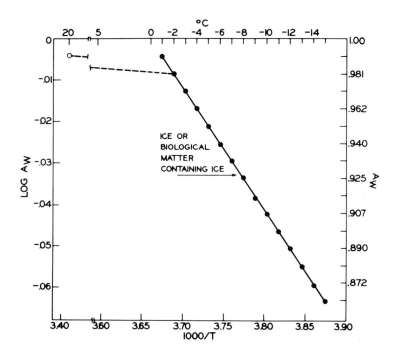

Fig. 3. Relationship between water activity and temperature for samples above and below freezing. From Fennema (1978).

IV.   TEMPERATURE DEPENDENCE OF a_w VALUES

The temperature dependence of a_w, griefly alluded to earlier, is clearly evident in Fig. 3. At above-freezing temperatures, a_w changes about 0.00002 to 0.002 units per degree centrigrade, depending on the temperature of the product (Ross, 1975; Van den berg, 1975; Rödel and Krispien, 1977). In the presence of an ice phase, a_w changes about 0.008 units per degree (Fig. 3 and Table II), or an amount of 4-400 times greater than that exhibited by samples at above-freezing temperatures.

V.   MEANING OF a_w VALUES AT SUBFREEZING TEMPERATURES

It is noteworthy that a product with an a_w of 0.86 at -15°C
(Fig. 3) would generally exhibit relatively slow rates of reactions
and no growth of microorganisms, whereas the same a_w at 20°C
would result in rapid rates of chemical reactions and moderate
growth of some microorganisms. This clearly indicates that
interpretation of a given a_w value in terms of chemical reactions
and microbial growth cannot be accurately accomplished unless
the temperature is known.

One point is of overriding importance when interpreting a_w
values at above- and below-freezing temperatures. As was stated
earlier, a_w values at above-freezing temperatures depend on both
temperature and sample composition, whereas in the presence of
an ice phase (at least over the range extending from the initial
freezing point of the sample to about -20°C) a_w values are
dependent on temperature but are *independent* of the kind or
amount of solute(s) present. This fact has an important bearing
on the usefulness of a_w values as indicators of the rates of
various chemical, physical, and physiological events at subfreezing temperatures. Consider first the consequences of bringing
an aqueous product to solid-liquid equilibrium at a given subfreezing temperature, say -10°C. Assuming the ice phase is pure
and solid-liquid equilibrium has been attained, the product,
regardless of its original composition, will have a fixed a_w
(0.91) and a fixed mole fraction of total solutes (expressed in
terms of osmotically active particles, i.e., ions, molecules,
dimers, etc.) existing in the unfrozen phase. Thus, subfreezing
temperature, a_w, and mole fraction of total solutes are related
in a fixed and known manner, regardless of the product. This
fact, while not particularly new or unusual, has an important
consequence, namely, that the kind (size, shape, chemical nature)
and ratios of osmotically active solute particles in the sample
can be changed without altering a_w at a given subfreezing

temperature. This, in turn, has additional ramifications some of which were stressed by Schneeberger et al. (1978).

Since the kinds and ratios of solute particles can be changed without altering a_w at a given subfreezing temperature, this means that: (1) diffusional properties (related to size and shape of the solute particles) of frozen samples can be altered dramatically without changing a_w at a given subfreezing temperature, (2) chemicals that activate or inhibit important chemical reactions (e.g., enzyme activators or inhibitors, cryoprotective agents, antimicrobial compounds, metal catalysts for oxidative reactions, chemicals that alter pH or oxidation-reduction potential) can be added or removed without changing a_w at a given subfreezing temperature, and (3) cellular systems can be partially or fully disrupted without affecting a_w at a given subfreezing temperature.

The relationships among molecular weight (molecular size), magnitude of the diffusion coefficient and a_w are indicated in Table III and Fig. 4. It is clear from these data that molecular weight has a profound influence on the diffusion coefficient and has no influence on a_w. Thus, at subfreezing temperatures, the a_w value will be of absolutely no help in predicting rates of physical and chemical events that are influenced by diffusion. [Diffusion-controlled events generally exhibit low activation energies and velocities that are dependent on viscosity (Gutfreund, 1965)]. Some enzyme-catalyzed reactions can be rate limited by diffusion and this is also true of recrystallization of ice and growth of microorganisms (Schneeberger et al., 1978).

Similarly, a_w values would not be expected to be useful indicators of the rates of oxidative reactions (oxidation of ascorbic acid, lipids, β-carotene, oxymyoglobins) at subfreezing temperatures, since these reactions are especially sensitive to the amounts and kinds of catalysts present (metals or enzymes). The fact that oxidative reactions do not behave in accord with predictions based on a_w or temperature has been amply documented

TABLE III. Diffusion Coefficients for Solutes of Various Molecular Weights

Substance	Molecular weight	D_{20-25° $cm^2\ sec^{-1} \times 10^{7a}$	Reference
Diglycine	132.12	79	b
Penta-erthritol	136.15	76.1	c
Mannitol	182.18	68.2	c
Triglycine	189.17	67	b
Sucrose	342.3	52.3	c
AL-1 protease (3.4.4.24)	33,000	14	d
Ribonuclease	13,400	11	b
Trypsin (3.4.4.4.)	23,800	9.40	d
Carboxypeptidase A (3.4.2.1)	34,300	8.82	d
Bromelain (3.4.4.24)	33,000	7.77	d
β-Lactoglobulin	18,300	7.6	b
Bovine serum albumin	69,000	6.7	b
Leucine Aminopeptidase (3.4.1.1)	326,000	3.75	d
Bovine fibrinogen	390,000	2.0	b

[a] Extrapolated to zero solute concentration in water.

[b] Gutfreund (1965)

[c] Weast (1969)

[d] Altman and Ditmar (1974)

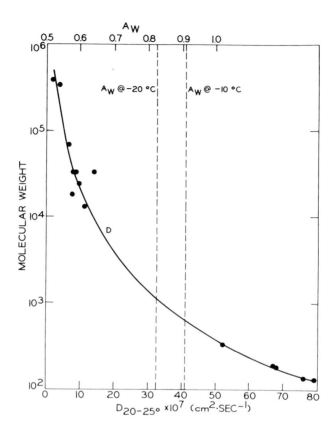

Fig. 4. Relationships among water activity, diffusion co-efficients, and molecular weight.

(Fennema, *et al.*, 1974; Fennema, 1975b). At subfreezing temperatures, it is also true that many other catalyzed (metals, enzymes,
acid) reactions, as well as reactions resulting in protein insolubilization, occur at rates that deviate substantially from
expectations based on a_w or temperature (Fennema, 1975a, b).

Furthermore, it is well known (Fennema, 1975a) that many
enzyme-catalyzed reactions are influenced by the presence or
absence of a cellular structure, particularly during freezing.
During freezing of cellular systems, many enzyme-catalyzed re-

actions tend to accelarate, whereas during freezing of noncellular systems, many enzyme-catalyzed reactions tend to decrease in
rate more than would be expected based on a_w and temperature.
Since the presence or absence of cellular structure will not
influence a_w at a given subfreezing temperature, a_w will not be
of use in predicting rates of reactions that are influenced by
cellular structure.

It should now be clear that the meaning and usefulness of
a_w values differ substantially depending on whether a sample is
above or below its freezing point, and that a_w values at subfreezing temperatures are far less useful indicators of chemical,
physical, and physiological activities than are a_w values at
above-freezing temperatures.

VI. SUMMARY

1. At subfreezing temperatures, water activity a_w is most
appropriately defined as $a_w = p_w/p_{sw}^O$, where p_w and p_{sw}^O are, respectively, the vapor pressures of water generated by the sample
and the vapor pressure of pure supercooled water at the same
temperature.

2. a_w values at subfreezing temperatures can be plotted in
two useful ways: (a) a_w vs. moisture content, with a freezing
point curve superimposed, and (b) a_w vs. $1/T$.

3. a_w values at above-freezing temperatures depend on temperature and sample composition, whereas a_w values at subfreezing
temperatures depend on temperature only (assuming a pure ice
phase and establishment of solid-liquid equilibrium).

4. a_w values are slightly temperature dependent (0.00002-0.002
units/C^O) at above-freezing temperatures and highly temperature-
dependent (0.008 units/C^O) at subfreezing temperatures.

5. Since a_w values at subfreezing temperatures are not influenced by the kind or ratio of solutes present, any event that

is influenced by the kind of solute (e.g., diffusion-controlled processes, catalyzed reactions, and reactions that are influenced by the absence or presence of cryoprotective agents, antimicrobial agents, and chemicals that alter pH and oxidation-reduction potential) present cannot accurately be forecast based on the a_w value. Thus a_w values at subfreezing temperatures are far less valuable indicators of physical, chemical, and physiological events than are a_w values at above-freezing temperatures.

ACKNOWLEDGMENTS

This contribution is from the College of Agricultural and Life Sciences, University of Wisconsin-Madison, Madison, Wisconsin.

REFERENCES

Acker, L. W. (1969). *Food Technol. 23*, 1257.
Altman, P. L., and Dittmer, D. S. (1974). "Biology Data Book," 2nd ed., Vol. III, p. 1476. Fed. Soc. Exp. Biol., Bethesda, Maryland.
Bralsford, R. (1968). *Food Technol. 22*, 130.
Caurie, M. (1971). *J. Food Technol. 6*, 85.
Dyer, D. F., and Sunderland, J. E. (1964). *Vacuum 14*, 396.
Dyer, D. F., Carpenter, D. K., and Sunderland, J. E. (1966). *J. Food Sci. 31*, 196.
Fennema, O. (1975a). *In* "Water Relations of Foods" (R. B. Duckworth, ed.), pp. 397-413. Academic Press, London.
Fennema, O. (1975b). *In* "Water Relations of Foods" (R. B. Duckworth, ed.), pp. 537-556. Academic Press, London.
Fennema, O. (1978). *In* "Dried Biological Systems" (J. H. Crowe and J. S. Clegg, eds.). Academic Press, New York.
Fennema, O., and Berny, L. A. (1974). *Proc. 4th Int. Congr. Food Sci. Technol. 2*, 27-35.
Fennema, O., Powrie, W. D., and Marth, E. H. (1974). "Low-Temperature Preservation of Foods and Living Matter." Marcel Dekker, New York.
Gutfreund, H. (1965). "An Introduction to the Study of Enzymes." Wiley, New York.
Hill, J. E., and Sunderland, J. E. (1967). *Food Technol. 21*, 1276.

Labuza, T. P. (1971). *Proc. 3rd Int. Congr. Food Sci. Technol.*
    3, 618.
Labuza, T. P., Tannenbaum, S. R., and Karel, M. (1970). *Food*
    Technol. 24, 543.
MacKenzie, A. P. (1975). *In* "Water Relations in Foods" (R. B.
    Duckworth, ed.), pp. 477-503. Academic Press, London
Mason, B. J. (1957). "The Physics of Clouds." p. 445. Oxford
    (Clarendon), London.
Riedel, L. (1961). *Kaltetecknik 13(3)*, 122.
Rockland, L. B. (1969). *Food Technol. 23*, 1241.
Rödel, W., and Krispien, K. (1977). *Fleischwirtschaft 10*, 1863.
Ross, K. D. (1975). *Food Technol. 29(3)*, 26-34.
Schneeberger, R., Voilley, A., and Weisser, H. (1978). *In*
    "Freezing, Frozen Storage and Freese-Drying." International
    Institute of Refrigeration, Paris.
Storey, R. M., and Stainsby, G. (1970). *J. Food Technol. 5*, 157.
Van den Berg, I. C. (1975). Course on Intermediate Moisture
    Foods Cycle, Seminaire E. 5. CPCIA Europe.
Weast, R. C., ed. (1969). "Handbook of Chemistry and Physics,"
    50th ed. Chemical Rubber Co., Cleveland, Ohio.

TRADITIONAL TECHNIQUES IN JAPAN FOR
FOOD PRESERVATION BY FREEZING, THAWING, AND DRYING

Tokuji Watanabe

I. INTRODUCTION

In Japan some processed foods are dried by freezing. High-
moisture foods, mainly composed of starch, protein, and other
high-molecular-weight compounds, are sometimes difficult to dry
because of case-hardening. Traditionally it was discovered that
freezing these foods outdoors in winter changed the state of
water, and made it easy to separate the water from the foods
when they were thawed. The water drips down with the soluble
components of the foods. Each day as the food gradually loses
water with repeated freezing and thawing, they become easy to
dehydrate by pressing and thus are dried by forced air. Since
they differ from freeze-dried foods and in order to avoid confusion with the so-called freeze-dried foods, I shall, hereafter,
call them "frozen-and-dried foods" as proposed by Matsuhashi
(1974).

These foods are still being processed by natural freezing,
but because of the amount of labor, limited production, and instability of quality in natural freezing, refrigeration is becoming
more common.

Five traditional frozen-and-dried foods, kori-tofu (dried tofu),
harusame (dried starch noodle), agar, kori-mochi (dried mochi), and
kori-konnyaku (dried glucomannan gel) will be discussed here. They
are, with the exception of agar, usually cooked with seasonings be-

733

fore eating, and it is difficult to recover the original state of
these frozen-and-dried foods by hydration. Agar is dissolved in
boiling water, and when cooled becomes so-called agar gel. Noodle-
like agar gel, called tokoroten in Japanese, is a favorite traditional dish. The annual production of kori-tofu is about 15,000
tons, harusame about 32,000 tons, and agar about 3000 tons. Production quantities of kori-mochi and kori-konnyaku are very low
and are only used as local foods.

This chapter will briefly describe the processes applicable
to these foods and food materials and discuss the chemical changes
observed in some of them.

II. KORI-TOFU

Kori-tofu is a tofu dried by freezing, the process of which
is shown in Fig. 1. Tofu is a curd prepared by coagulating soybean milk with calcium chloride and molding in the box to remove
excess whey. Tofu, which has almost 85% moisture and is cut to a
definite size, is frozen in forced chilled air by refrigerating
at -10°C, which takes about 1-2 hr. After freezing it is kept at
-1° to -3°C to promote a change in the texture. The frozen storage or aging requires 2-3 weeks. After aging the tofu is thawed
under running water or a shower. The thawed tofu becomes like a

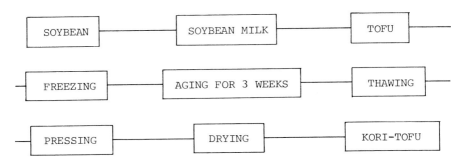

Fig. 1. Flowchart of kori-tofu production.

sponge and can be pressed to remove the water. Pressed tofu is dried first by forced hot air and then in open air at room temperature. In the case of natural freezing, time of freezing is longer and freezing and thawing are repeated during the aging period. Protein solubility in dilute alkaline decreases gradually during this period (Watanabe, unpublished, 1966), which certainly relates to the formation of spongelike texture.

When soybean protein solution, instead of the curd, is frozen and aged, protein becomes partly insoluble, forming precipitates. It seems to be similar to the sponge formation of the curd in the preparation of kori-tofu. It was found that when soybean protein solution was frozen and thawed, the protein seemed to denature partly without insolubilization. Acrylamide gel electrophoresis of frozen-and-thawed protein solution showed disappearance of the original 7 and 11 S components and appearance of new components having lower mobilities. By treatment of the protein solution with mercaptoethanol solution the new components disappeared and 7 and 11 S components reappeared. On the other hand by treating the protein solution with N-ethylmaleimide (NEM) the new components were not observed. These results suggest that the changes of soybean protein during frozen storage are based on the formation of intermolecular S-S bonding (Hashizume et al., 1971). The same might be also true in the case of sponge formation of soybean curd in kori-tofu preparation. According to the determination of water mobility using NMR, tightly bound water that is unfrozen at $-20°C$ increases for 10 days and then decreases with age (Nagashima, unpublished, 1978). On the other hand, weakly bound water that is unfrozen at $-1.4°C$ decreases gradually during aging. It seems to correspond to the sponge formation during this period. The increase of tightly bound water might be brought about by increased intermolecular bonding of the protein molecule, and its decrease, by blocking the hydration site of the protein molecule, resulting from their closer association.

III. HARUSAME

Harusame is a dried starch noodle. Originally it was made in China from the starch of mung beans. In Japan, however, potato and sweet potato are commonly used (Yamamura, 1969). As shown in Fig. 2, to make harusame a 3/4% suspension of starch is heated to gelatinize. This paste is mixed with raw starch by adding water to make a 45% solid slurry. It is then transferred to a perforated pan and extruded through the holes. The fine strings formed are successively placed in boiling water. The gelatinized noodles are then suspended on fine sticks and slowly frozen by refrigeration; it takes about 12 hr to reach a final temperature of about -10°C. After this treatment the noodles on the sticks are thawed with water. Each noodle can then be easily separated without sticking and thus effectively dried, which results in a porous texture. Compared with dried starch noodle without freezing, its cooking time is much shorter, it has a characteristic chewiness, and it is able to hold seasonings. Loss of starch during cooling is negligible. Those characteristics become apparent from the retrogradation during freezing (Yamamura, 1966).

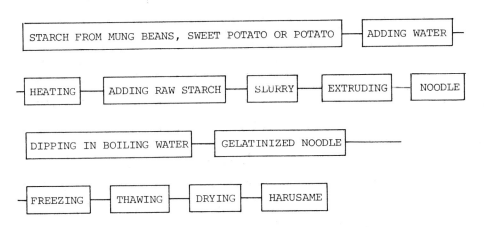

Fig. 2. Flowchart of harusame production.

IV. AGAR

Agar is a porous and light solid, used as the material for agar gel. It is made from *Gelidium, Gracilaria,* and other species of red algae. Agar, on a dry basis, contains over 90% carbohydrate. It is derived from a family of polysaccharides with the same backbone structure. It consists of an alternate (1–3%-linked D-galactose and (1-4)-linked anhydro-L-galactose residue, respectively. The three extremes of structure are neutral agarose, pyruvated agarose, and sulfated galactan.

Agar is dissolved in boiling water and after cooking and cooling forms a gel at such low concentrations as 0.1%. The gelling temperature of 1.5% agar solution is around 35°C, whereas its melting temperature is over 83°C. Although agar is used as a culture medium for microorganisms and other nonfood materials, it has been traditionally used as an ingredient in Japanese confectionery. It continues to find new food uses because of its gelling properties and its heat reversibility.

For agar production, as shown in Fig. 3, it is necessary to extract the carbohydrates from algae by cooking with dilute acid. The solution obtained by filtration forms gel by cooling. The gel is then frozen by refrigeration or by placing outdoors in winter.

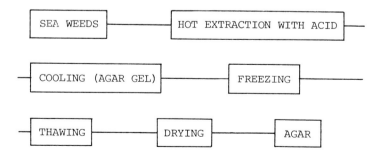

Fig. 3. Flowchart of agar production. Note: Natural freezing and thawing are repeated.

Natural freezing requires 2-3 days. After repeated freezing and
thawing the physical structure of the gel changes completely and
water drains. Impurities are removed with the water. Since it
still contains water it is dehydrated by pressing, if necessary,
and dried by hot air or by placing outdoors. The product has a
characteristic honeycomblike structure. In natural freezing the
whole procedure from the extraction to the drying requires about
2 weeks. There are several types of agar products: string, bar-
style, and powdered type. In the case of the former two, each
gel is cut into string or bar style and then frozen outdoors, resulting in the honeycomb structure. Powdered agar is generally
produced in part from the frozen-and-dried material using refrigeration and in part from directly dried material.

The products are soluble in boiling water and become gels by
cooling. Hardness of the gel had been considered to be lower than
that of the original unfrozen gel, based on the same solid content.
But a recent study shows that this is not correct. It has also
been shown that gel strength of agar by quick freezing is greater
than that by slow freezing (Matsuhashi, 1974).

The extraction rate from algae is 30% on a solid basis and
the yield of dry product from 100% solid gel is 80%. Thus the
yield of agar from the original material is 24%. In a laboratory
analysis the yield is expressed as the product of gel strength
and weight of dried agar obtained from 100 weight gel.

KORI-MOCHI

Kori-mochi is a frozen-and-dried product of mochi. Mochi is
a very popular Japanese cake base, made by kneading steamed glutinous rice. Fresh mochi is eaten with shoyu, sugar, or other
seasonings, but it becomes hard, unacceptable, and indigestible
with time, because of retrogradation of starch. For consumption
it is necessary to bake or cook it again. Kori-mochi is mochi

that is frozen and dried. Fresh mochi is not easily dried, but when it is frozen and thawed outdoors it loses water and finally dries when exposed to the air. Figure 4 shows a flowchart of the production. Kori-mochi is now a local food which is naturally frozen and thawed, although the use of refrigeration is currently under development by Nagano Prefectural Laboratory of Food Technology.

When fresh mochi is frozen it gradually deteriorates under low temperature. By repeated freezing and thawing, water partially separates from starch as drainage, evaporating in the daytime, which results in a porous structure. When the freezing rate is fairly high, starch in the dried product remains in galatinized form. Therefore, there are kori-mochi that can be reconstituted to the original fresh mochi by dipping in hot water.

Another type of kori-mochi is made by freezing gelatinized paste of glutinous rice powder. After repeated freezing and thawing it can be dried to a very light, porous and layer-structured cake. It is used in this form as an ingredient for cake, and as paste by mixing with hot water.

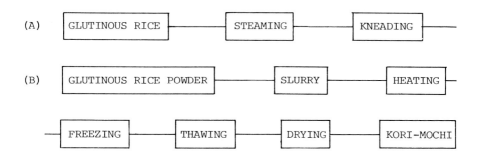

Fig. 4. Flowchart of kori-mochi production. Note: Natural freezing and thawing are repeated.

VI. KORI-KONNYAKU

The last of the frozen-and-dried foods, kori-konnyaku is made of konnyaku tuber, of which the main component is glucomannan, a polymer of glucose and mannose. Konnyaku is a characteristic gel, made by heating a viscous solution of mashed raw konnyaku tuber in calcium hydroxide solution. Powdered konnyaku tuber is also available and, if necessary, it is refined by removing starch and other lighter components by air.

Konnyaku usually contains 96% moisture. It is an elastic and heat-irreversible gel. Because microorganisms cause deterioration and because konnyaku is not preservable it must be prepared and sold daily.

Kori-konnyaku was discovered by chance, like other frozen-and-dried foods, when fresh konnyaku gel was left outdoors in winter. It is frozen at night and thawed in daytime. Water is poured from it every day. Gradually the texture changes to that of a sponge. Finally it is frozen and thawed without pouring off water. It is now dried into a product that has a porous texture. Figure 5 is a flowchart of production. It is cooked with vegetables and

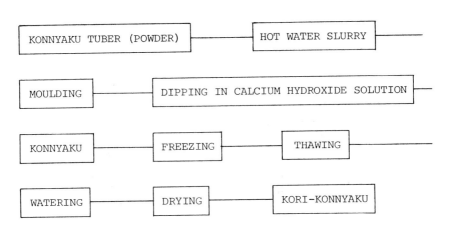

Fig. 5. Flowchart of kori-konnyaku production. Note: Natural freezing, thawing and watering are repeated.

seasonings. Its physical properties greatly differ from the original material. It has a unique chewiness. The separation of water
by freezing and thawing of konnyaku may be brought about by some
conformational changes of glucomannan, if one considers retrogradation of frozen starchy foods, but the details of this process
are not yet clear and should be elucidated in future studies.

VII. SUMMARY

This chapter discusses several traditional frozen-and-dried
foods that have been produced in Japan. In contrast to the use
of refrigeration in modern frozen-food processing, freezing in
our case has been used as a way of drying. It is not necessary
for the product to retain its original character. Easy and
simple drying is more important for these foods. It is understood
that the mild treatment in food processing which makes use of natural energy (freezing and drying outdoors, etc.) should be re-
evaluated from the viewpoint of food safety, resource economy, and
energy saving. A number of possibilities exist for developing new
foods in the future by applying freezing-and-drying.

Acknowledgments

I wish to thank Dr. T. Matsuhashi, Nagano Prefectural Laboratory of Food Technology, Dr. E. Yamamura, Kagoshima Prefectural
Agricultural Experimental Station, and Drs. R. Takahashi and N.
Nagashima, Central Laboratory of Ajinomoto Co., for their useful
advice.

References

Hashizume, K., Kakiuchi, K., Koyama, E., and Watanabe, T. (1971).
    Agri. Biol. Chem. *35,* 449.
Matsuhashi, T. (1974). *Refrigeration (Tokyo) 49*, 559.
Yamamura, E. (1966). *J. Food Sci. Technol. (Japan) 13,* 322.
Yamamura, E. (1969). *Food Industry (Korin, Tokyo) IB,* 30.

STATE OF WATER IN SEA FOOD

Taneko Suzuki

.

I. INTRODUCTION

Fish and shellfish foods are prepared in many different ways.
However, they may be classified into two categories: First fish
may be prepared to retain its original flavor and taste by frying,
roasting, or even being left raw, which is common in Japan.
Second, the original shapes and textures of fish may be completely
changed by processing it into fish cake, fish sausage, or "kamaboko", an elastic meat jelly common to Japan.

Freshness of raw material is extremely important. In the
second category, the rate of protein denaturation is related to
rheological properties of fish cake, sausage, and kamaboko. A
number of studies on fish freshness have been published, but only
a few have related the state of water in fish muscle to freshness
(Suzuki, 1973; Blanshard and Derbyshire, 1974). The preparation
of high-quality kamaboko is closely related to properties of the
myofibrillar protein in fish muscle (Okada, 1963; Kawashima *et al.*,
1973; Shimizu, 1974; Shimizu and Nishioka, 1974; Niwa, 1973; Niwa
and Nakajima, 1975; Iwata *et al.*, 1977).

In kamaboko, water estimated by a primitive method, has been
used for evaluating quality (Okada, 1963). The state of water in
kamaboko or water behavior during processing has been reported

(Niwa, 1973; Suzuki, 1973; Takagi, 1973). In this chapter water content in fish muscle and its relation to freshness and rheological properties are discussed.

II. WATER CONTENT IN FISH AND SHELLFISH

Water in health fish muscle usually ranges from 75 to 80%. Water and lipid content in fish muscle are inversely related so that the sum is approximately constant. Their relationship was noted by Brandes and Dietrich (see Love, 1970), who showed a "lipid-water line" for Atlantic herring (*Clupea harengus*), ocean perch (*Sebastes marinus*), and Catfish (*Anarhichas*). For recent work, Horiguchi and Aminaka (1978) and Hasegawa (1977) showed lipid-water lines for sardine muscle. Their graphs revealed an obvious inverse relationship. Horiguchi and Aminaka (1978) reported that dark muscle contains less water than ordinary muscle for sardine, where water in ordinary muscle averages 71.4% and dark muscle 47.7%.

Water in fish muscle increases in spawning season and starvation. Love (1970) reviewed the increase in water content during starvation, and found that cod (*Gadus morhua*) contained 86-88% water when in that condition. Water in carp (*Crypinus carpio*) reached 78.9-91% after 8 months starvation at 20°C, and water in crucian carp (*Carassinus carassinus*) rose from 78.9 to 93.3% after 105 days at 27°C. Tanaka (1969a) observed that water content in carp muscle increased after 4 months starvation at 2-8°C, 10° and 20°C, respectively. He also noted that the water-holding capacity of the muscle decreased and weep or drip loss greater when it was frozen. Protein denaturation during drying of sea bass muscle (*Lateolabrax japonicus*) was more remarkable in the spawning season, when it contained more water (Kanna *et al.*, 1972). Freeze denaturation of muscle protein has been reported to be considerable in the spawning season for watery muscle of Alaska pollack (*Theragra chalcogramma*) (Tanaka, 1969b).

Creatures with extremely high muscle water content are jellyfish (97.5%) and sea cucumber (91.6%). Deep sea fish muscle is usually high in water: about 88% for *Liparis tanakai,* about 86% for *Podonema longipes.* Muscle of *Zestichtys tanakai* is the most most "jellied" when cooked, and was found to consist of 6.6% protein, 0.4% lipid, and 91.4% water (Kakehata, 1974). Meat of Antarctic krill *(Euphausia superba)* contains 84-90% water when peeled by a roller-type peeler (Suzuki *et al.,* 1978).

III. STATE OF WATER IN FISH MUSCLE OF VARYING FRESHNESS

A. *Observation by NMR*

We have used high-resolution nuclear magnetic resonance (NMR) spectrometry to study the state of water in fish muscle. The NMR method is well suited to this task because the width of the signal produced by water hydrogen is dependent on the motional freedom of water molecules. As the mobility of water molecules increases, the linewidth decreases. A advantage of the NMR method is that the sampling is easy. The foods or tissues are finely cut and packed into a sample tube without changing their original properties, and the determination is completed within a few minutes.

Hazelwood and Nichols (1969) reported for skeletal muscles of rats and mice that the linewidths of NMR spectra are considerably broader than distilled water. A comparison of the integrals of the water spectrum in muscle with pure water showed that about 8% of water in muscle was bound too firmly to macromolecules of the muscle to be detected in the high-resolution scan. They also found that the linewidth of NMR spectrum of rat gastroenemius

muscle was broader than pure water, but the line becomes narrow
following heating. They explained that line broadening was due
to specific interaction of water molecules with only the native
protein, and ordered water was released by protein denaturation.
They concluded that water exists in at least two ordered phases
in the muscle. More recently, Belton *et al*. (1972) and Hazelwood
et al. (1974) reported that three different stages of water existed
in living muscles by spin-spin relaxation time of water proton.

We used a JMN-MH-100 (Japan Electron Optic Lab. Co.) high-
resolution NMR to determine the spectra of water protons in fish
and shellfish muscles. Postrigor stage skeletal muscle of Alaska
pollack was taken, its dark muscle and connective tissues were
removed, and these muscles were packed into sample tubes. The
sample tubes were capped to prevent evaporation. All estimations
were made at 25°-27^OC. Figure 1 shows the NMR spectra of water
in Alaska pollack muscle and distilled water, and also the associated integrals of both spectra. The linewidths of water spectra
in the fish muscle are much broader than those of pure water.
The widths at one-half amplitude of water in the fish muscle are
15-16 Hz, almost the same as muscles of rat and mice in the
results by Hazelwood and Nichols (1969).

The total content of water in Alaska pollack muscle was
measured to be 75% by common method. If all the muscle water
were visible to the high-resolution NMR spectrometer, the
muscle water signal should be 75% of the pure water signal. The
integral of the muscle water spectrum was in fact only about 60%
of the pure water signal. About 15% of the water in the muscle
is bound firmly to macromolecules and the water signal was too
broad to be detected in the high-resolution scan. Belton *et al*.
(1972) have found approximately 20% of water in muscle strongly
bound to proteins, which does not freeze.

Figure 2 shows NMR spectra of the jellyfish umbrella (species
unidentified) and the muscle of sea cucumber *(Stichopus japonicus)*

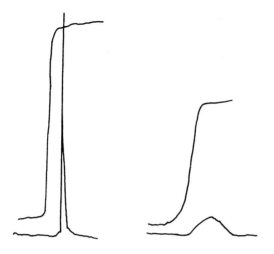

Pure water   Alaska pollack muscle

Fig. 1. High-resolution NMR spectra and associated integrals of pure water and water in Alaska pollack muscle. Both spectra obtained using the same sample tube and machine settings. Sweep width, 1080 H_z; machine, 100 MC (JEOL).

measured by cutting immediately from the living tissues. Line widths of both were 20-28 Hz which was far broader than pure water and slightly broader than fish muscle in prerigor and postrigor stages. A clear conclusion cannot be drawn because of an insufficient number of experimental samples and no information on paramagnetic impurities in these samples.

Figure 3 shows linewidths of NMR spectra of water in flat fish *(Kareius bicoloratus)* in prerigor, rigor, and postrigor stages. The linewidth in rigor stage was broader compared with prerigor and postrigor stages, but in the postrigor stage it became narrower than the prerigor stage. The same has also been observed in sea bass, but only in the case of starved carp,

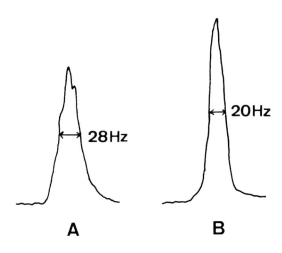

Fig. 2. High-resolution NMR spectra of jelly fish umbrella
(A) and sea cucumber muscle (B). Sweep width, 1080; sweep time,
250 sec; temp, 25°C; machine, 100 MC (JEOL).

where rigor mortis was obscure and the change in the width of
lines was not clear.

Bratton et al. (1965) reported that about 20% of bound water
was released when the muscle contracted, but it contradicted the
fact that the line became broader in the rigor stage, where the
muscle apparently contracted and it seemed some explanation was
necessary. Blanshard and Derbyshire (1974) investigated the
state of water in muscles of various animals as a function of
time after death by spin-lattice relaxation. They obtained the
results that the population of the bound water phases was calculated to be constant, but during rigor there was a net transfer
of water from the free phase of the region, giving rise to a more
rapidly relaxing signal, to the free phase in the region producing
the more slowly relaxing signal.

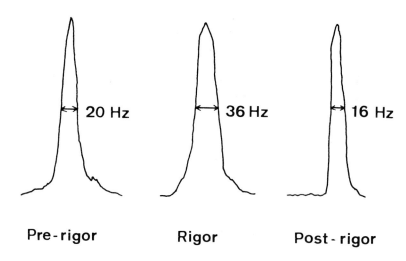

Fig. 3.   High-resolution NMR spectra of flat fish muscle on different freshness.   For conditions of measurement, refer to Fig. 2.

Change in linewidth of NMR spectrum was observed using a muscle slice of sea bass in prerigor stage, where it was gradually dehydrated over silica gel at 2^O-3^OC (Suzuki and Takeuchi, 1971). The result is shown in Fig. 4. As the gains of the spectra differed at each estimation, the integrals of the spectra were not comparable with each other, but the widths of lines were observed to be broader as the sample dehydrated, but it becomes narrower when the water content became less than 20%. The motional restriction of the water molecule became maximum when the water content was 20% in the dehydrated muscle. Furthermore, as seen in Fig. 5, the spectra shifted to a higher magnetic field as the dehydration of the sample proceeded, and the shape of the spectra became asymmetric. An explanation for the asymmetrization and shifting of the spectra as the dehydration proceeds may be that water exists in muscle at various different states, and water in each state is removed as the dehydration

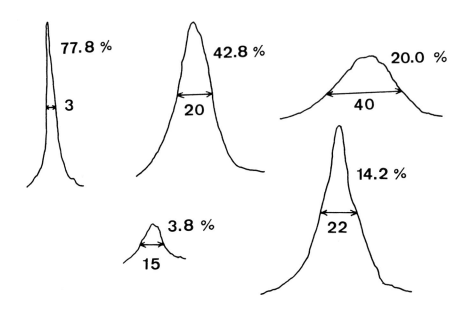

Fig. 4. High-resolution NMR spectra of sea bass minced muscle with various moistures by dehydration at 2-3ºC. F, fresh muscle; machine, 60 MC (JEOL). Widths of spectra indicated in cps.

proceeds, resulting in asymmetrization in the shape of spectra giving a shifted appearance.

In general, the state of water in muscle has so far been discussed in two or three phased models, but more variations in the state of water can be suspected.

B. *Observation with Cryomicroscope*

The state of water may be observed experimentally under a cryomicroscope on a thinly sliced fish muscle, where the water in the tissue has gradually cooled to freezing. If the boundary between frozen and unfrozen parts proceeds slowly at a speed less 1 mm/min the behavior of water can be observed rather precisely

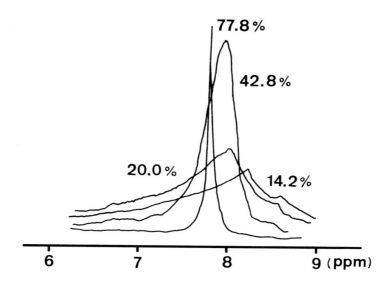

Fig. 5. High-resolution NMR spectra of sea bass minced muscle with various moistures by dehydration at 2-3°C. Right side, high magnetic field; machine, 60 MC (JEOL).

(Tanaka, 1973a). According to Tanaka, at first water in spaces between bundles of muscle cells starts to freeze and immediately runs, spreading along the extracellular spaces branching from the bundles of muscle fibers. Thus fine threads of ice are networked from one place to the next. As this stage, however, the water inside muscle cells does not yet freeze. As the ice network among the cells grows thicker, the cells gradually shrivel and eventually are buried under the ice. Tanaka and Okada (1971) succeeded in photographing the sequence on 16 mm film. From these observations the water existing between the spaces among the bundles of muscle cells or muscle fibers may be classified as free water.

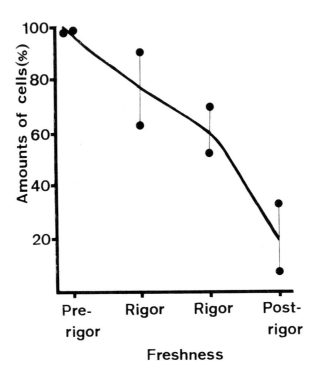

Fig. 6. Relation between amounts of cells with intracellular crystalization and freshness of raw material (Alaska pollack muscle). (from Tanaka, 1969a).

As muscle cells induce nuclei that form ice crystals and contains more solute than in extracellular locations (Asahina, 1971), it does not freeze rapidly and eventually reaches a state of supercooling. Some of the water may then be released through the cell wall and frozen. If proteins in the cell hold water firmly, the water stays inside forming ice crystals within the cells (Tanaka, 1969b). Consequently one can conclude that the degree of ice formation in the cell at a definite freezing rate is an indicator of the state of water in a sample.

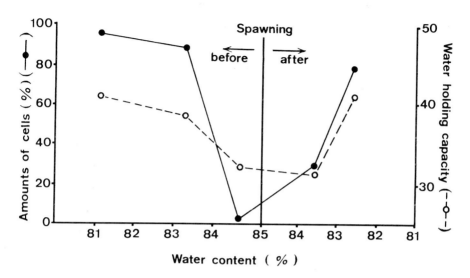

Fig. 7. Amounts of cells with intercellular crystalization and water-holding capacities of muscles before and after spawning of Alaska pollack. (from Tanaka, 1969b).

Figure 6 shows the relation between the number of muscle cells undergoing intercellular crystallization and the freshness of Alaska pollack (Tanaka, 1969b). As the freshness of the muscle decreases, more ice crystals are observed outside the cell, which indicates that free water in the cell increases more than before as the cell wall becomes more permeable (Love and Haraldsson, 1961; Tanaka, 1969b). Figure 7 shows an example of Alaska pollack in spawning season (Tanaka, 1969b). Only a small number of muscle cells are able to form ice inside the cells, and the water-holding capacity decreases more than in any other season. When frozen muscle is stored better quality may be found in those showing higher intercellular crystallizations compared with fresh samples of the same raw material (Tanaka, 1973b).

In the case of muscle of high water and less protein content, e.g., in such species as horse hair crab (*Erimacrus isenbeckii*)

(Tanaka, 1963), king crab (*Paralithodes camtshaticus*) (Tanaka, 1965), tiger prawn (*Peraeus japonicus*) (Tanaka *et al.*, 1968), certain fish species in spawning season (Tanaka, 1969b), starved carp (Tanaka *et al.*, 1969), and crab soon after exuviation (Tanaka, 1965) maintaining quality during long freeze storage is difficult at the temperature held in commercial cold storage. The effect of freezing rate on intracellular ice crystal formation in fish meat has been studied by Tanaka (1965, 1973b) and Suzuki *et al.* (1964). Beneficial effects of high freezing rates on fish quality and hydration properties have been reported by Piskarev and Bomovalova (see Duckworth, 1974) and Crawford *et al.* (1969). With regard to the above-mentioned fish and shellfish species, in most cases fast freezing or ultrarapid freezing is sufficient to guarantee high-quality products.

IV.   STATE OF WATER IN KAMABOKO

A. *Principle of Making Kamaboko*

Kamaboko is a kind of fish meat jelly. Fish meat is ground with salt kneaded, shaped, and then heated to make it a solid, elastic material. Similar types of foods are known as fish cakes in Europe and fish balls in southeast Asia. Kamaboko has greater elasticity, which is preferred in Japan. The elasticity is technically termed as *ashi*.

An outline of kamaboko manufacture is given in Fig. 8. Fish meat is ground in a mortar with salt. During grinding, myofibrillar protein (mainly actomyosin), which is soluble in a salt solution, is dissolved from the meat to form "sol." When adhesive raw meat paste is heated, the sol is converted into a gel, forming a ternary network. The elasticity of kamaboko is due to a network formation by actomysin molecules. The strength

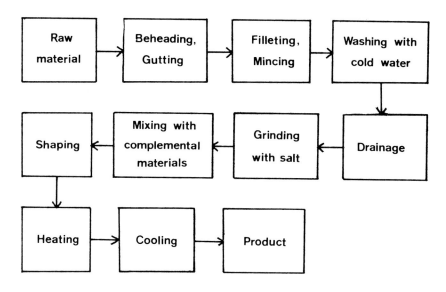

Fig. 8. Flow diagram for making kamaboko.

of *ashi* may be partly expressed by measuring its gel strength.
It is more often organoleptically judged.

B. *Relation between Ashi and State of Water in Kamaboko*

Ashi becomes weaker if the kamaboko contains more water,
though in some cases it may differ in *ashi* strength while they
are equal in water content (Okada, 1963). Strength of *ashi* has
been known to correlate strongly with the amount of water exuding
out of a piece of kamaboko at pressure of 10 kg/cm^2 for 20 sec
(Iwata, 1969). Figure 9 shows the relation between amount of
water exuded and *ashi* strength, which was evaluated organoleptically. A microscopic view indicates that kamaboko possesses a
network and porous structures that retain water (Okada and Migita,

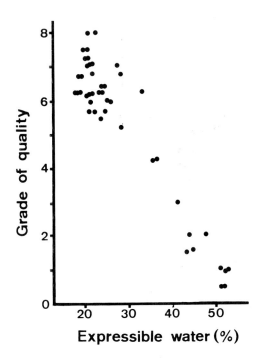

Fig. 9. Relation between qualities of kamaboko measured by sensory evaluation and amounts of expressible water. Material, Alaska pollack, expressible water, 10 kg/cm², 20 sec, 25°C. (from Iwata, 1969).

1956). The immobilized water among the structures exude from kamaboko if it is pressed.

1. *Width of NMR Spectra Determinations.* The author pre-pared samples of kamaboko equal in total water content but differing in *ashi* strength, and studied the relationship between *ashi* strength and water activity following the procedure of Okada (1963), who found that fish meat after being ground with salt will make kamaboko samples equal in total water but differing

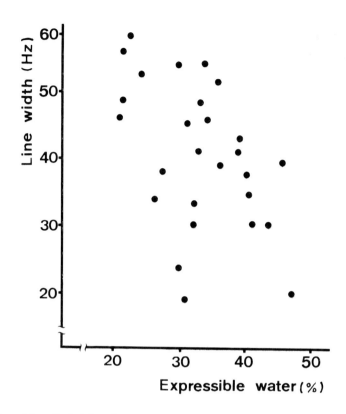

Fig. 10. Relation between amounts of expressible water and widths of high-resolution NMR spectra of water in kamaboko. (r = -0.5566, p < 0.01). Material, minced meat of Alaska pollack with various qualities. All kamaboko samples contain no starch. For parameters of expressible water, refer to Fig. 9.

in elasticity, if such conditions as standing periods, heating temperature, and duration are purposely controlled.

Figure 10 indicates the relationship between linewidths of NMR spectra and expressible water content in kamaboko prepared from different qualities of Alaska pollack. Figure 11 shows a positive relationship between the expressible water or jelly strengths and linewidths of NMR spectra among the kamaboko

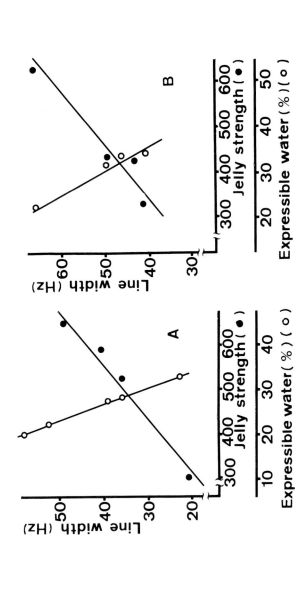

Fig. 11. Relationship of jelly strength, amounts of expressible water, and widths of high-resolution NMR spectra of water in kamaboko. Material, Meat of croaker (A) and minced meat of Alaska pollack (B). All kamaboko samples contain no starch. For parameters of expressible water, refer to Fig. 9. Jelly strength (g cm) was calculated by stress-strain curves obtained by Okada's gelometer (1963).

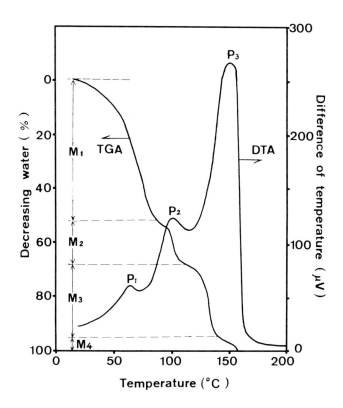

Fig. 12. DTA and TGA thermograms of kamaboko. Material,
croaker meat; moisture, 74.0%; conditions of thermal analysis;
sample, 10 mg; speed of increasing temperature, 5.0°C/min;
sample holder, 6 mm Ø X 6 mm open pan cell. (from Takagi, 1973).

samples prepared from frozen minced meat of Alaska pollack
or meat of croaker (Nibea mitsukurii). Testing with a series of
kamaboko samples prepared from the same fish material, gel
strengths and expressible water highly correlated with linewidths
of NMR spectra.

2. *Thermal Investigations*. Takagi (1973) measured thermograms of DTA (differential thermal analysis) and TGA (thermogravimetric analysis) on prepared samples of kamaboko equal in
water but differ in *ashi*. Figure 12 shows the results. According to TGA results, water in kamaboko seems to be classified into
four fractions, namely, M_1 fraction occupying about 55% of the
total and converted into vapor if the sample is heated to about
80°C; M_2 fraction, about 15%, which vaporize around 100°C; M_3
fraction, about 25%, which vaporize around 130°C, and M_4 fraction, about 5%, which vaporizes if heated over 150°C. Thus, it
seems that water exists in kamaboko at least four states differing in its freedom. On the other hand DTA results showed three
peaks: P_1 at about 60°C, P_2 at about 100°C, and P_3 at about 150°C.
Takagi reported that peaks P_1 and P_2 correspond to free water,
while P_3 corresponds to water firmly bound to protein in the
kamaboko sample.

Figure 13 presents studies by Takagi (1973) on the amount of
water in each fraction of TGA on five kamaboko samples differing
in *ashi* strength. Kamaboko with strong *ashi* contains more
bound water, which corresponds to M_3 plus M_4, and less free
water, which corresponds to M_1 plus M_2, as shown in Fig. 13.
After studying the relation between the expressible water, which
has been regarded as one of the measurements for the quality of
kamaboko, and the fraction M_1 and M_2 on a TGA thermogram,
Takagi (1973) reports that free water will be released when the
sample is expressed as shown in Fig. 14.

The water activity (a_w) of kamaboko must be less than 0.94
in order to prevent growth of botulism bacillus. The desirable
attributes for a kamaboko of better quality may be related to the
strength of the *ashi* and its lower water activity.

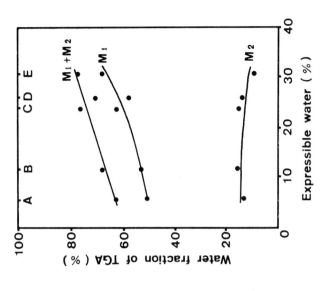

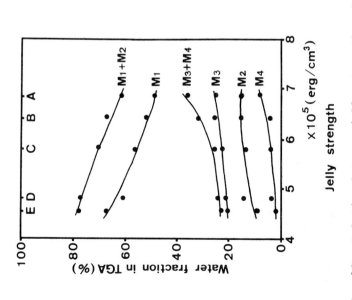

Fig. 14. Relationship of expressible water and amounts of M_1, M_2 fractions of TGA in Kamaboko. Expressible water, 10 kg/cm^2, 10 kg/cm^2, 20 sec, 25°C. For relation among kamaboko samples, refer to Fig. 13. (from Takagi, 1973).

Fig. 13. Relation between jelly strength and amounts of water fractions in TGA. Five kamaboko samples have different jelly strengths (extension toughness, erg/cm^3), respectively, as follows: (A) (6.9 x 10^5) > (B) (6.5 x 10^5) > (C) (5.8 x 10^5) > (D) (4.8 x 10^5) > (E) (4.5 x 10^5) (from Takagi, 1973).

ACKNOWLEDGMENTS

The author wishes to acknowledge the suggestions of
Dr. T. Tanaka and also to thank Dr. H. Hirao for critical reviews
of the manuscript.

REFERENCES

Asahina, E. (1971). *In* "Cryobiology" (K. Nei, ed.), p. 21.
    Tokyo Univ. Press, Tokyo.
Belton, P. S., Jackson, R. R., and Packer, K. J. (1972). *Bio-
    chim. Biophys. Acta 286,* 16.
Blanshard, J. M. V., and Derbyshire, W. (1974). *In* "Water
    Relations of Foods" (R. B. Duckworth, ed.), p. 568. Academic
    Press, London
Bratton, C. B., Hopkins, A. L., and Weinberge, J. R. (1965).
    Science 147, 738.
Crawford, L., Finch, R., and Daly, J. J. (1969).   *Food Technol.
    23,* 549.
Duckworth, R. B., ed. (1974).   "Water Relations of Foods", p. 513.
    Academic Press, London.
Hasegawa, K. (1977).   *Proc. 11th Congr. Processing Utilization
    of Marine Products,* p. 61.   Fishery Agency, Tokyo.
Hazelwood, C. F., and Nichols, B. L. (1969). *Nature 222,* 747.
Hazelwood, C. F., Chung, D. C., Nichols, B. L., and Woessner,
    D. E. (1974). *Biophys. J. 14,* 583.
Horiguchi, T., and Aminaka, J. (1978). *Proc. 12th Congr. Process-
    ing Utilization of Marine Products, p. 69.*   Fishery Agency,
    Tokyo.
Iwata, K., (1969). *New Food Ind. 11,* 12.
Iwata, K., Kanna, K., and Okada, M. (1977). *Bull. Japan Soc.
    Sci. Fish 43,* 237.
Kakehata, K. (1974).   *JAMARC,* no. 6, 4.
Kanna, K., Kakuda, K., and Sakuraba, M. (1972). *Bull. Tokai Reg.
    Fish. Res. Lab. 69,* 125.
Kawashima. T., Ohba, A., and Arai, K. (1973). *Bull. Japan Soc.
    Sci. Fish 39,* 1201.
Love, R. M. (1970).   "The Chemical Biology of Fishes", p. 29.
    Academic Press, London.
Love, R. M., and Haraldsson, S. B. (1961). *J. Sci. Food Agri.
    12,* 442.
Niwa, E. (1973). *In* "Water in Foods" (Japan Soc. Sci. Fish., ed.),
    p. 83.   Kosei-sha Kosei-kaku Ltd., Tokyo.
Niwa, E., and Nakajima, G. (1975). *Bull. Japan Soc. Sci. Fish.
    41,* 579.
Okada, M. (1963). *Bull. Tokai Reg. Fish. Res. Lab. 30,* 21.

Okada, M., and Migita, M. (1956). *Bull. Tokai Reg. Fish. Lab.* *22*, 265.

Shimizu, Y. (1974). *Bull. Japan Soc. Sci. Fish 40*, 175.

Shimizu, Y., and Nishioka, F. (1974). *Bull. Japan Soc. Sci. Fish.* *40*, 231.

Suzuki, T. (1973). "Water in Foods", p. 25. Kosei-sha Kosei-kaku Ind., Tokyo.

Suzuki, T., and Takeuchi, M. (1971). *J. Japan Soc. Sci. Technol.* *17*, 110.

Suzuki, T., Kanna, K., and Tanaka, T. (1964). *Proc. FAO Symp. The Technology of Fish Utilization (R. Kreuzer, ed.).* Fishing News (Books) Ltd., London.

Suzuki, T., Kanna, K., Suzuki, M., Okazaki, E., and Morita, N. (1978). Paper presented at *5th Int. Congr. Food Sci. Technol., Kyoto, Japan.*

Takagi, I. (1973). *In* "Water in Foods" (Japan. Soc. Sci. Fish., ed.), p. 95. Kosei-sha Kosei-kaku Inc., Tokyo.

Tanaka, T. (1963). Paper presented at *Annu. Mtg. Japan Soc. Sci. Fish., Otaru.*

Tanaka, T. (1969a). *Bull, Tokai Reg. Fish. Res. Lab. 59*, 29.

Tanaka, T. (1973a). *In* "Water in Foods" (Japan Soc. Sci. Fish., ed.), p. 63. Kosei-sha Kosei-kaku Inc. Tokyo.

Tanaka, T. (1973b). *Sakana, Tokai Reg. Fish. Res. Lab. 11*, 1.

Tanaka, T., and Okada, J. (1971). *Japan. Soc. Res. Freeze Dry 17*, 106.

Tanaka (1969b). *Bull. Tokai Reg. Fish Res. Lab. 60*, 143.

Tanaka, T., Konagaya, S., and Okada, Y. (1968). Paper presented at *Ann. Metg. Japan Soc. Sci. Fish., Tokyo.*

Tanaka, T., Takeuchi, M., and Takeshita, T. (1969). Paper presented at *Ann. Mtg. Japan Soc. Sci. Fish., Tokyo.*

ROLE OF WATER IN WITHERING
OF LEAFY VEGETABLES

Toshimasa Yano, Ikuo Kojima,
and Yasuo Torikata

Due to the physical separation between food production and
consumption, foods are transported a longer distance and require
prolonged storage. Many vegetables, especially fresh leafy
vegetables as spinach, lettuce, and leek, are transported within
an intermediate distance under natural conditions, or precooled
at best. One of the problems involved in the transporation and
storage of fresh leafy vegetables is their high water content
which decreases the efficiency of transportation and the storage
capacity. However, a decrease in the water content of leafy
vegetable causes withering that is generally taken as a loss of
freshness. Thus, even when the vegetables are precooled
immediately after harvest, to avoid withering, care must be
taken so that loss of water does not exceed about 3% of the
initial weight.

From a modern scientific view point, however, a loss of water
itself may not necessarily be equivalent to a loss of freshness.
The dehydrated vegetables can be rehydrated as shown for freeze-
dried and compressed foods by Rahman *et al.* (1969, 1978). Leafy
vegetables are known to recover from withering more easily than

other vegetables or fruits, and so there may be a way to take advantage of the withering of leafy vegetables for more efficient transportation and storage. Much work has been done on the transportation and storage of vegetables. (Fennema, 1975; Ogata, 1977). However, little attention has been paid to the forced withering of fresh leafy vegetables.

In this chapter, some physical phenomena related to the withering of leafy vegetables are studied. Materials used were spinach (Hoyo) and leek (Green Belt). This is not the culmination, but only the beginning of research in this area.

I. FRESH VS. COMPRESSED DRY VEGETABLES

First of all, the two typical states of vegetables must be defined.

Fresh leafy vegetables, spinach or leek, for example, contain as much as 88 to 95% of water on a wet basis when they are transported. Precooling to 3^0 or 5^oC is recommended, but care must be taken so that the water loss does not exceed about 3% of the initial weight. They are transported fresh and consumed usually within 2 or 3 days.

The other typical state of vegetables, blanched, freeze-dried, and compressed vegetables developed by the U. S. Army Natick Laboratories, has only 12% moisture effecting a 75 to 94% volume reduction, 60 to 90% weight reduction, and has a storage life longer than 3 months. Table I gives a comparison between the two typical states of vegetables. In Table I the bulk volume of compressed vegetables was estimated by the authors.

Between the two states of vegetables, however, there could be other states optimal for the transporation and storage of leafy vegetables, depending on individual situations. Reduction of water content without blanching, freeze-drying, or compressing

TABLE I.   Comparison between Precooled Spinach and Compressed
Freeze-Dried Vegetables

	Precooled spinach	Compressed freeze-dried vegetables
Water content (% wet basis)	88	12
Bulk volume (m^3/k_g dry matter)	0.15	0.005
Distribution life	3 Days	> 3 Months
Physiological state	Alive	Dead

may be worth considering as one of the goals for the transportation and storage of fresh leafy vegetables.

Figure 1 shows the relation between water content on a wet
basis and the total weight for water-rich materials.   The
ordinate is the dimensionless total weight, where W and D are
the weights of water and dry matter, respectively.   The relation
is hyperbolic, and the decrease in weight is most obvious in the
range of high water content.   If a fresh vegetable containing
95% water is reduced to 90% water content, the total weight of
the vegetable becomes exactly one-half the initial weight.   If
it is reduced to 80%, the total weight becomes one-fourth.   Thus,
if the initial 95% or 90% water content could be reduced to 80%
or less without unfavorable effects, it will solve a major problem in the transportation and storage of fresh leafy vegetables.

Figure 2 shows the relations between water content and bulk
volume of spinach and leek.   When transported, the fresh spinach
is very loosely packed to avoid possible mechanical damage, the
bulk volume being more than 0.15 (m^3/kg dry matter).   If it could
be more tightly packed without mechanical damage, the bulk volume

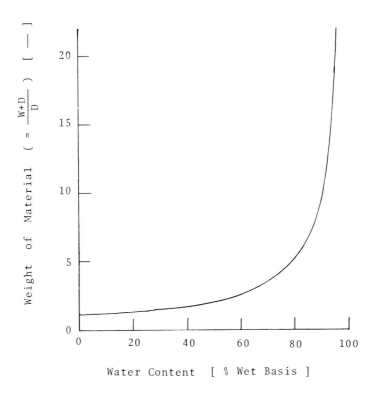

Fig. 1.   *Relation between water content and total weight.*

for transportation could be reduced to one-half, and by withering
to one-third, of the present bulk volume. An increase in the
respiratory heat evolution per unit volume would be suppressed
by precooling the vegetables. Leek contained much more water
than spinach, but its bulk volume was about the same as spinach.

Examination of the critical water content, from which the
leafy vegetable recovered on soaking one day in water, showed
that the spinach and leek could be reversibly dehydrated down to
70 and 80%, respectively, under vacuum, while to only 78% and
86%, respectively, with warm air, as shown in Table II. Kuroda

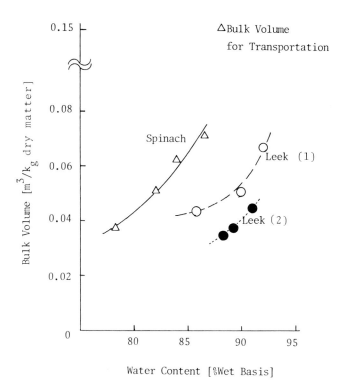

Fig. 2.   Relation between water content and bulk volume.

TABLE II.   Critical Water Content (%) for Reversible Dehydration

	Fresh	Method of dehydration		
		Vacuum	Air(20°C)	Air(40°C)
Spinach	88%	70%	78%	—
Leek	92%	80%	86%	87%

et al., (1978) found the critical water loss for spinach to be about 20% of the initial weight, or to 87.5% from 90% on a wet basis, after four days' storage at 2°C. More water could be reversibly removed from spinach through the quick dehydration under vacuum.

II. VISCOELASTIC PROPERTIES OF WITHERED LEAFY VEGETABLES

Since quite a large amount of water was removed reversibly from the leafy vegetables, changes in their viscoelastic properties were studied in relation to withering.

Viscoelastic properties have been measured numerous on fruits, tomato, potato, etc. (Mohsenin *et al.*, 1963; Morrow and Mohsenin, 1966; Finney *et al.*, 1964, 1967, Finney and Hall, 1967; Finney and Norris, 1968; Finney and Massie, 1975; Ishibashi *et al.*, 1970; Chuma *et al.*, 1971) and a few leafy vegetables (Kuroda *et al.*, 1978). Although mechanical damage tests will be needed from a practical viewpoint (Kaufman *et al.*, 1950; Break-iron, 1959; O'Brien *et al.*, 1963, 1965; Chuma *et al.*, 1967), measurement of fundamental viscoelastic properties will give insight into what happens in vegetables.

Breaking stress was measured with a commercially available rheometer. Static viscoelastic properties were studied through the stress relaxation with the rheometer, and the creep behavior with the authors' hand-made assembly shown in Fig. 3. In Fig. 3, the sample leaf piece was immersed in the liquid paraffin to avoid water evaporation, the strain being measured with a differential transformer. Dynamic viscoelastic properties were measured by the vibrating-reed method as shown in Fig. 4. The oscillator created a sinusoidal voltage change, the frequency was counted, power was amplified, and the leaf piece was vibrated with an exciter. Ther terminal amplitude of oscillation was

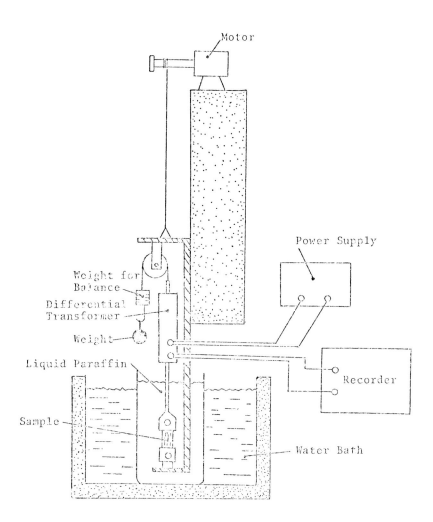

Fig. 3.   Assembly for creep test.

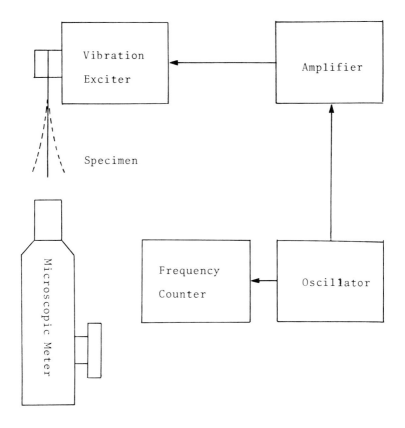

Fig. 4.   Block diagram of the vibrating-reed method.

was measured with a micrometer.   All the measurements were made
on a rectangular specimen, 8 x 15 mm wide and 30 mm long, cut
off from the leaf, avoiding the nervure.

The tensile breaking stress increased with a decrease of
water content as shown in Figs. 5 and 6.

Stress relaxation showed the existence of two relaxation
processes with the relaxation times as shown in Table III.   The
relaxation times decreased when the vegetables were withered.

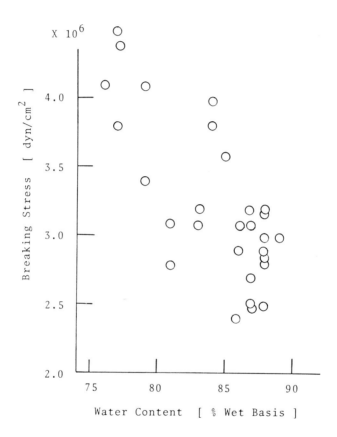

Fig. 5. Effect of water content on tensile breaking stress (spinach).

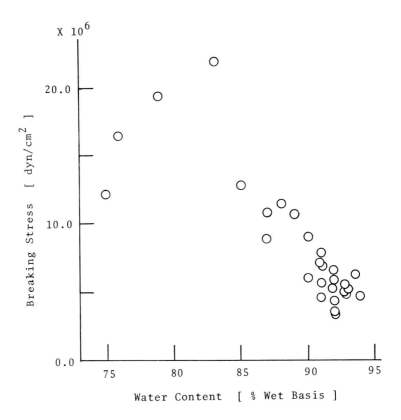

Fig. 6. *Effect of water content on tensile breaking stress (leek).*

TABLE III. Effect of Temperature on Leaf Length

	Water content (% wet basis)	Change in length (-)	
		$3^{o}C \rightarrow 37^{o}C$	$37^{o}C \rightarrow 3^{o}C$
Spinach	83	-0.0020	+0.0050
	84	-0.0043	+0.0083
	87	-0.0043	+0.0060
	88	-0.0050	+0.0087
	90	-0.0023	+0.0067
Leek	89	-0.0057	+0.0070
	90	-0.0043	+0.0100
	93	-0.0047	+0.0070
	93	-0.0060	+0.0097
	93	-0.0007	+0.0020

The creep behaviors of spinach and leek were describable with the four-element model, combining the Maxwell and the Voigt models. To avoid redundancy, only the results for leek are presented here. Figure 7 shows the effect of water content on the elastic modulus of a Maxwell body. The value of elastic modulus for leek sharply decreased at around 91% water content, independent of temperature, and the change was almost reversible under the experimental conditions as shown by the arrows in Fig. 7. Elastic modulus of a Voigt body changed with the pattern similar to the Maxwell body's as shown in Fig. 8.

Although the values of elastic moduli were independent of temperature at an equilibrium state, a change in length was observed at a transient state when the environmental temperature was changed. The change in length was quite small as shown in

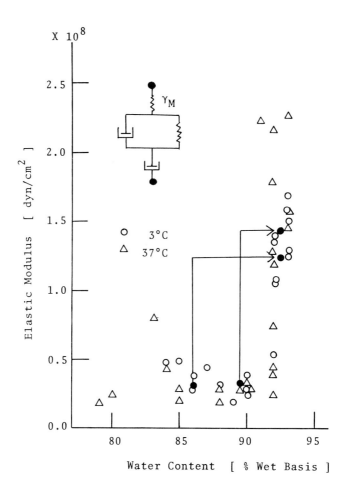

Fig. 7. *Effect of water content on elastic modulus of Maxwell body (leek).*

Table III, but the length definitely decreased at a high temperature and increased at a low temperature. Therefore, the nature
of elasticity seems to be entropic rather than energetic. More
complicated elastic behavior was observed on the cell wall of
Nitella opaca (Haughton *et al.*, 1968).

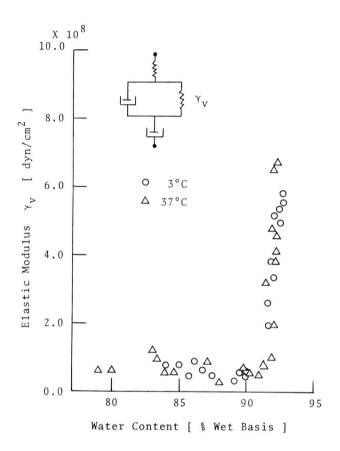

Fig. 8. *Effect of water content on elastic modulus of Voight body (leek).*

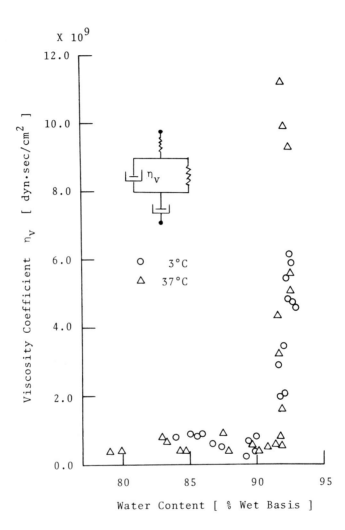

Fig. 9. Effect of water content on viscosity of Voigt body
(leek).

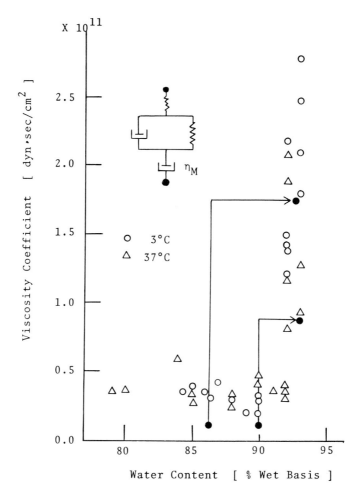

Fig. 10. Effect of water content on viscosity of Maxwell body.

The viscosity of the Voigt body for leek was also decreased at around 91% water content, as shown in Fig. 9. Interestingly, the viscosity was independent of temperature. Therefore, the momentum seems to be transported not through the friction between biological solids. The change in the viscosity of the Maxwell body was almost the same as that of the Voigt body, as shown in Fig. 10. The viscosity change was almost reversible as shown by the arrows in Fig. 10.

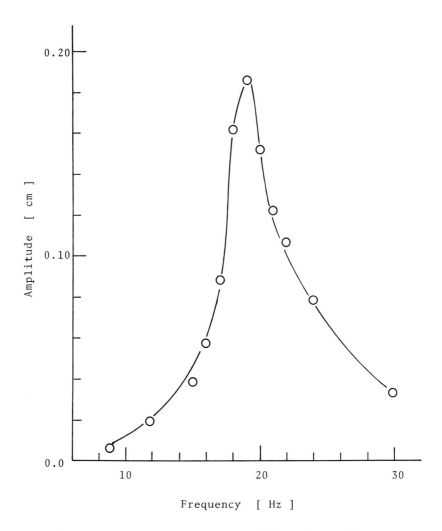

Fig. 11. *Typical resonance curve of leek (vibrating reed method).*

Figure 11 shows a typical resonance curve obtained from the vibrating-reed experiment. From the amplitude of the resonance peak, the dynamic elastic modulus was calculated, and from the width of curve at the 1/ 2 or 1/2 height of the peak, the dynamic viscosity was calculated. The dynamic elasticity of

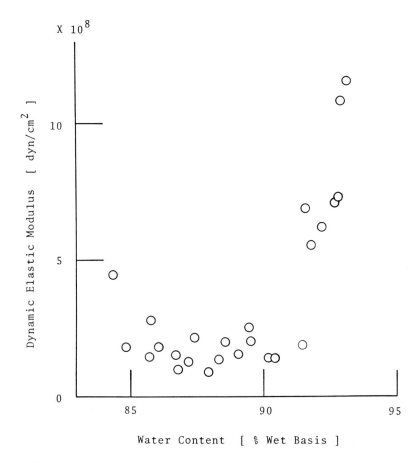

Fig. 12. Effect of water content on dynamic elastic modulus (leek).

leek again sharply decreased at around 91% water content as shown in Fig. 12.

Table IV summarizes the viscoelastic properties of spinach and leek, fresh and critically withered under vacuum to water contents 70 and 80%, respectively. When critically withered, the breaking stress increased, relaxation times decreased, elastic moduli decreased, and viscosities decreased. A big differ-

TABLE IV. Summary of Physical Properties

Property		Spinach Fresh	Spinach Critically withered	Leek Fresh	Leek Critically withered
Breaking stress (dyn/cm^2)		2.5×10^6	4×10^6	5×10^6	15×10^6
Relaxation time (sec)	$\tau_i = 1500$		1000	2000	500
	$\tau_i = 30$		20	100	20
Elastic modulus (dyn/cm^2)	γ_M	3×10^7	1×10^7	20×10^7	2×10^7
	γ_V	10×10^7	1×10^7	60×10^7	6×10^7
	γ_{vib}	–	–	5.2×10^8	2.2×10^8
Viscosity $\left(\dfrac{dyn\ sec/}{cm^2}\right)$	η_M	8×10^{10}	1×10^{10}	40×10^{10}	0.5×10^9
	η_V	10×10^8	1×10^8	10×10^9	0.5×10^9
	η_{vib}	–	–	3.3×10^5	2.7×10^5

ence between the static dynamic viscosities might be a feature of the biological multicellular materials.

These data show how the spinach and leek became flexible viscoelastically be removing water. By forced withering, the fresh leafy vegetables not only reduce their weight and volume but also become more resistant to mechanical damage during handling and transportation.

III. EVAPORATION OF WATER FROM HARVESTED LEAFY VEGETABLES

Much work has been done on the transpiration of living plants in connection with diffusion resistances as well as with stomatal control (Parlange *et al.*, 1971, 1972; Nobel, 1974, 1975; Meidner, 1975, 1976; Kaufmann, 1976; Ishihara *et al.*, 1971;

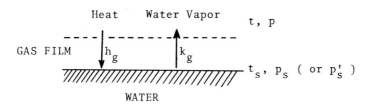

AIR

Fig. 13. Simultaneous transfers of water vapor and heat.

Tsuno, 1975; Takeda *et al.*, 1978; Lewis and Nobel, 1977; Browne and Fang, 1978; Roberts and Knoerr, 1978; Farquhar and Raschke, 1978), but little work has been done on the evaporation of water from the harvested leafy vegetables. Recently Kuroda *et al.* (1977) studied the water loss from spinach during storage as a decreasing-rate drying process. Here, however, the evaporation of water from the leafy vegetable is analyzed in a different way.

Evaporation of water is the simultaneous transportation of heat and mass as schematically depicted in Fig. 13. Water evaporates from the surface through a gas film and, at steady state, the latent heat of evaporation has to be supplied from the atmosphere. Then, for evaporation from pure water, the fluxes of water vapor and heat are given as follows:

$$-\frac{1}{A}\frac{dW}{d\theta} = k_g(p_s - p) \tag{1}$$

$$\beta h_g(t - t_s) = \lambda_w k_g(p_s - p) \tag{2}$$

where

A is the surface area for evaporation (cm^2), h_g the heat transfer coeeficient through the gas film (J/hr cm²), k_g the mass transfer coefficient through the gas film (g/hr cm² °C), p the partial water vapor pressure (mm H), p_S the saturation vapor pressure for pure water (mm H_g), t the temperature in bulk atmosphere (°C), t_S the temperature at evaporation surface (°C), β the correction factor for difference between the surface areas for water vapor and heat transfer, for radiation effect, etc. λ_w the specific latent heat of water evaporation (J/g), and θ the time (hr).

For an air-water vapor system, the following relation is known (Chem. Eng. Assoc., 1968):

$$\frac{p_S - p}{t - t_S} = \frac{\beta\, h_g}{\lambda_w k_g} \doteqdot 0.5 \quad \text{mm } H_g/°C \tag{3}$$

For evaporation from biological material, the saturation vapor pressure will depend on its water activity. Thus, let it be p_S'. Then, similarly to Eqs. (1) to (3),

$$-\frac{1}{A}\frac{dW}{d\theta} = k_g(p_S' - p) \tag{4}$$

$$\beta h_g\,(t - t_S) = \lambda_w k_g\,(p_S' - p) \tag{5}$$

$$\frac{p_S' - p}{t - t_S} = 0.5 \quad \text{mm } H_g/°C) \tag{6}$$

$$a_w = \frac{p_S'}{p_S} \tag{7}$$

where a_w is the water activity and p_s' the saturation vapor
pressure for biological water (mm H_g).

The experimental procedure is as follows: First, the value
of k_g has to be determined, which can be done from an evaporation experiment for pure water. Temperature and vapor pressure
of the atmosphere, t and p, are experimentally given. Then,
from Eq. (3), both the temperature at the evaporation surface,
t_s, and the saturation vapor pressure at that temperature,
p_s, can be determined with the aid of psychrometric tables.
By plotting the measured water vapor flux vs. $p_s - p$ in Eq. (1),
the value of k_g is determined from the slope of the straight
line.

Second, the rate of water evaporation from the biological
material is measured under a relatively low vapor pressure. The
value of p_s' will be determined from Eq. (4) since k_g is known.
Substituting p_s' into Eq. (6), the value of t_s is estimated.

Finally, the water activity is estimated from Eq. (7) after
finding the saturation vapor pressure for pure water, p_s, at
temperature t_s from the psychrometric table.

Figure 14 illustrates a simple method for the evaporation
experiment. Wet filter paper or a leaf sample was hung within
a closed cylinder over sulfuric acid of known concentration,
and the rate of water evaporation was measured using a spring.

An example of the estimated t_s and p_s is shown in Table V.
The temperature of the wet filter surface was estimated to be
17.2^O, $23.3^O C$, and so on, under the corresponding relative
humidity at $30^O C$.

The measured water vapor fluxes were proportional to the
estimated driving forces $p_s - p$ for the wet filter paper as
shown in Fig. 15, from which the mass transfer coefficient
through the gas film k_g was determined to be 1.1×10^{-3}
(g/hr cm^2 mm H_g).

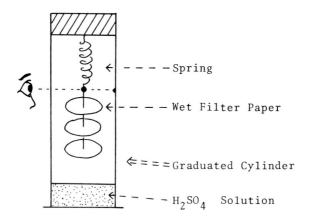

Fig. 14.   Measurement of evaporation rate.

TABLE V.   Example of Estimated t_S and p_S (Wet Filter)

t (°C)	R.H. (%)	p (mm H_g)	t_S (°C)	p_S (mm H_g)
30.0	26.0	8.27	17.2	14.71
	56.6	18.0	23.3	21.45
	66.6	21.2	25.0	23.76
	81.7	26.0	27.4	27.37

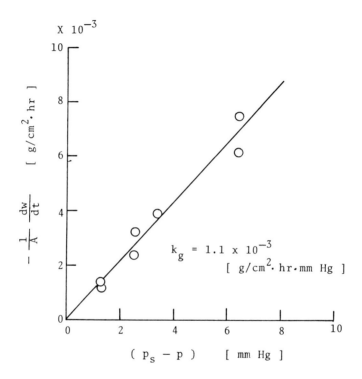

Fig. 15. Determination of k_g (30ºC).

A sample calculation of a_w is given below. The material was leek with more than 89% water content. The temperature of the atmosphere was 30°C.

1. Relative humidity, 75% ($p = 23.87$ mmHg):

   $-(1/A)(dW/dt) = 0.85 \times 10^{-3}/$ g/cm^2 hr

   $p_S' = 24.64$ mmHg

   $t_S = 28.64 \leftarrow$ °C, $p_S = 29.12$ mmHg

   $a_w = 0.85$

2. Relative humidity, 57% ($p = 18.14$ mmHg):

   $-(1/A)(dW/dt) = 2.5 \times 10^{-3}$ g/cm^2 hr

   $p_S' = 20.38$ mmHg

   $t_S = 25.52 \leftarrow$ °C, $p_S = 24.50$ mmHg

   $a_w = 0.83$

At relative humidities 75 and 57%, the water activities of leek were estimated to be 0.85 and 0.83, respectively. These values were much lower than expected.

The rate of water evaporation was assumed to be limited by the boundary air layer. Total resistance for water evaporation, however, will consist of (1) the parallel stomatal and cuticular resistances, and (2) the boundary air layer in series with the stomatal or cuticular resistance (Browne and Fang, 1978), although it is said that partial closure of stomata has little effect on transpiration in quiet air while it greatly reduces the rate in moving air (Kramer, 1974). In moving air the site of water evaporation was shown to be the mesophyll tissue (Meidner, 1975; Farquhar and Raschke, 1978).

If the water activity of leafy vegetable is determined independently, the present method of analysis may be used for analyzing the diffusion resistances within the leaf as well as the stomatal control of water evaporation.

REFERENCES

Breakiron, P. L. (1959). USDA Marketing and Transportation Rep. MTS-134, p. 33.

Browne, C. L., and Fang, S. C. (1978). *Plant Physiol. 61*, 231.

Chemical Engineers' Association (1968). "Chemical Engineering Handbook", 3rd ed., p. 594. Maruzen, Tokyo.

Chuma, Y., *et al.* (1967). *J. Soc. Agr. Mach. Japan 29*, 82, 88, 104.

Chuma, Y., *et al.* (1971). *J. Soc. Agri. Mach. Japan 33*, 304.

Farquhar, G. D., and Raschke, K. (1978). *Plant Physiol. 61*, 1000.

Fennema, O. R. (1975). *In* "Physical Principles of Food Preservation" (Karel, M., Fennema, O. R., and Lund, D. B. eds.), p. 133. Marcel Dekker, New York.

Finney, E. E., Jr., and Hall, W. W. (1967). *Trans. ASAE 10*, 4.

Finney, E. E., Jr., and Massie, D. R. (1975). *Trans. ASAE 18*, 1184.

Finney, E. E., Jr., and Norris, K. H. (1968). *Trans. ASAE 11*, 94.

Finney, E. E., *et al.* (1964). *J. Agri. Eng. Res. 9*, 307.

Finney, E. E., *et al.* (1967). *Proc. Am. Soc. Hort. Sci. 90*, 275.

Haughton, P. M., *et al.* (1968). *J. Exp. Bot. 19*, (58), 1.

Ishibashi, S., and Kojima, T. (1970). *J. Agri. Mach. Japan 32*, 59.

Ishihara, K., *et al.* (1971). *Proc. Crop Sci. Soc. Japan 40*, 491, 497, 505.

Kaufman, J., *et al.* (1950). USDA, Handling, Transportation and Storage Office Rep. 228.

Kaufmann, M. R. (1976). *Plant Physiol. 57*, 898.

Kojima, T., and Ishibashi, S. (1974). *J. Soc. Agri. Mach. Japan 36*, 298.

Kramer, P. J. (1974). *Plant Physiol. 54*, 463.

Kuroda, S., *et al.* (1977). *J. Soc. Agri. Mach. Japan 39*, 199.

Lewis, D. A., and Nobel, P. S. (1977). *Plant Physiol. 60*, 609.

Meidner, H. (1975). *J. Exp. Bot. 26*, 666.

Meidner, H. (1976). *J. Exp. Bot. 27*, 691.

Mohsenin, N. N., *et al.* (1963). *Trans. ASAE 6*, 85.

Morrow, C. T., and Mohsenin, N. N. (1966). *J. Food Sci. 31*, 686.

Nobel, P. S. (1974). *Plant Physiol. 54*, 177.

Nobel, P. S. (1975). *J. Exp. Bot. 26*, 120.

O'Brien, M., *et al.* (1963). *Hilgardia 35*, 113.

O'Brien, M., *et al.* (1965). *Trans. ASAE 8*, 241.

Ogata, K. (1977). "Seika Hozo Hanron", Kenpakusha, Tokyo.

Parlange, J. Y., *et al.* (1972). *Plant Physiol. 50*, 60.

Rahman, A. R., *et al.* (1969). Tech. Rep. 70-36-FL, U. S. Army Natick Labs.

Rahman, A. R., *et al.* (1978). *Abstr. Fifth Int. Congr. Food Sci. Technol., Kyoto, Japan*, p. 112.

Roberts, S. W., and Knoerr, K. R. (1978). *Plant Physiol.* *61*, 311.
Takeda, T., *et al.* (1978). *Japan J. Crop Sci.* 47, 82.
Tsuno, Y. (1975). *Proc. Crop Sci. Soc. Japan 44*, 44.
U. S. Army Natick Laboratories (1977). "Compressed Foods."

INFLUENCE OF WATER ACTIVITY

ON THE MANUFACTURE AND AGING OF CHEESE

M. Rüegg and B. Blanc

I.   INTRODUCTION

The concept of water activity (a_w) (Scott, 1957) has provided
a basis for an increasing understanding of microbial/water relations in food. Thus, for various types of food, control of
water activity has become an important factor for improving its
stability and safety (Karel, 1973; Troller, 1973; Leistner and
Rödel, 1976). Control of growth and survival of microorganisms
that cause spoilage or food poisoning are among the major
reasons for increasing interest in water activity. Comparatively
little attention has been paid to the role of water activity as
a controlling factor for the proper ripening process of fermented
food products such as cheese. In this case, for example, not
much information on the significance of a_w is available in the
literature. It was thus considered useful to determine water
activities of various cheeses and to compare these values with the
a_w requirements of the starter cultures and the enzymes responsible
for the eye formation and ripening process.

This contribution describes the problems encountered in
measuring a_w values in cheese, gives typical a_w values for various
types of cheese, as well as critical a_w limits for the growth of
microorganisms used in starter cultures. A mathematical concept

is presented that enables one to interpret a_w values on the basis
of the chemical composition of cheese and on the other hand allows
one to estimate approximate a_w values on the basis of
a known composition. The results of a study of the relationship
between a_w and defects in gruyere and emmentaler cheese are
presented as an example of the application of this concept.

II. a_w MEASUREMENT IN CHEESE

    Several difficulties are associated with a_w determinations in
cheese. The high a_w range, the possible presence of volatile
compounds (CO_2, NH_3, volatile falvor compounds, etc.), microbial
and enzymatic activity, as well as heterogeneity of cheese limit
the choice of methods suitable for the determination and require
several precautions. The a_w values reported here determined
using an electric hygrometer (modified Sina instrument, Nova-Sina
A. G., Zurich, Switzerland), which allows one to carefully control the temperature of the sample and measuring cell and record
the equilibration process (Rüegg and Blanc, 1977). Constant
humidities in the headspace of the sample cells were achieved
within 2-4 hr. Calibration of the sensors was performed at least
monthly by using saturated salt and sulfuric acid solutions in the
a_w range 0.90 to 0.98 (Greenspan, 1977; Gal, 1967). A second-
order polynomial was fitted to the calibration data. This was
found to be a sufficient approximation to the instrument signal-
a_w curve. a_w values of cheese samples were then determined
mathematically from the calculated calibration curves. The mean
coefficient of variation of a_w values was 0.5%; this was determined using samples of hard cheeses with a_w values in the range
of 0.93 to 0.98.
    Table I shows typical a_w values determined for various types
of cheese, the standard deviation, and range of a_w values. For

TABLE I. Typical a_w Values for Various Cheese

Type	Typical a_w[a]	Standard deviation	a_w range in rind
Appenzeller	0.960	0.011	0.97-0.98
Brie	0.980	0.006	0.98-0.99
Camembert	0.982	0.008	0.98-0.99
Cheddar	0.950	0.010	0.94-0.95
Cottage cheese	0.988	0.006	–
Edam	0.960	0.008	0.92-0.94
Emmentaler[b]	0.972	0.007	0.90-0.95
Fontal	0.962	0.010	0.93-0.96
Gorgonzola	0.970	0.017	0.97-0.99
Gouda	0.950	0.009	0.94-0.95
Gruyere[b]	0.948	0.012	0.92-0.98
Limburger	0.974	0.015	0.96-0.98
Münster	0.977	0.011	0.96-0.98
St. Paulin	0.968	0.007	0.96-0.97
Parmesan	0.917	0.012	0.85-0.88
Quarg	0.990	0.005	–
Sbrinz[b]	0.940	0.011	0.80-0.90
Tilsiter	0.962	0.014	0.92-0.96
Processed cheese	0.975	0.010	–

[a] Measured at 25°C.

[b] Values for emmentaler, gruyere, and sbrinz were measured after ripening periods of 4-5, 6-7, and 10-11 months, respectively. The other values were determined using commerically available samples.

comparison the a_w value of milk estimated from its freezing point depression is about 0.995.

Reporting mean or typical a_w values of cheese can be misleading if one does not consider that a_w changes during the ripening process and may show considerable zonal variation within the cheese body. The influence of sample age or degree of ripening on a_w for two types of hard cheeses are shown in Fig. 1. Fig. 2 shows a_w values in different zones of some cheeses. Similar to the effects of other physicochemical parameters and the concentration of various constitutents (See Steffen, 1975, and Kurmann et al., 1976), a_w values here show significant fluctuation. The following factors are most probably responsible for these zonal variations: (1) temperature gradient in the cheese body during the early stages of the fermentation process, (2) NaCl concentration gradient due to the diffusion of the salt into the body, (3) loss of water, and (4) microbial activity on the rind. Depending on the objectives of a study, a_w values in the rind may be of crucial importance. The sampling method for a_w determination must be adapted to the particular problem.

III. INFLUENCE OF WATER ACTIVITY ON STARTER CULTURES USED IN
      CHEESE MANUFACTURE

During the first stages of cheese manufacture a_w is about 0.99 and thus does not interfere with the growth and activity of the starter culture. However, after salting and during curing the a_w levels encountered are significantly lower than the optimal requirement of starter bacteria and it is likely that a_w contributes to the control of their metabolic activity and multiplication (Brown, 1976).

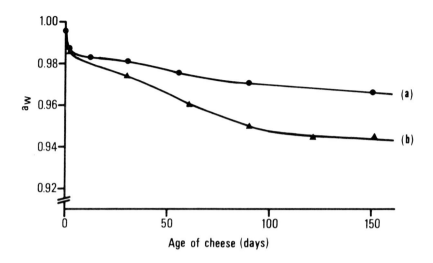

Fig. 1.   Influence of degree of ripening on a_w of emmentaler
(a) and gruyere (b) cheese.   (a_w at time 0, 0.995, corresponds
to the value for milk, which was estimated from its freezing
point.)

Table II lists the minimal a_w required for multiplication of
some strains of bacteria used in cheese manufacture. Water
activity requirements of the strains studied varied with the
type of solute added to adjust a_w. Specific inhibitory effects
and the possibility of serving as additional nutrients may be
responsible for these differences. The lowest a_w limits were
always observed in media containing glycerol and the highest with
glycine. It has been shown that many microorganisms accumulate
free amino acids, especially glutamic acid and proline, inside
the cell in response to increased sodium chloride concentration
in the environment (Christian, 1963; Measures, 1975). The
effects of other solutes on osmoregulatory mechanisms have not
been studied in detail.

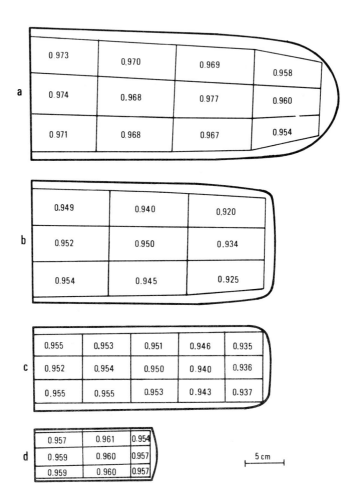

Fig. 2. Typical zonal variation of a_w for slices of emmentaler (a), sbrinz (b), gruyere (c), and appenzeller (d) cheese. Cross sections shown are from the center to the rim. The age of the samples was (a, b) 5, (c) 6, and (d) 5 months, respectively. a_w values for the rind were about (a) 0.90-0.95, (b) 0.80-0.90, and (c, d) 0.92-0.98.

TABLE II. Lower a_w Limits for Multiplication of Bacteria Used in Cheese Manufacture[a]

Strain[b]	Culture medium	(°C)	Solute added to control a_w	Critical solute concentration (mol/liter)	Critical[c] a_w
Streptococcus lactis 1712	EAH	30	Glycine	0.70	0.983
			Glucose	3.00	0.950
			NaCl	1.10	0.965
			Glycerol	3.40	0.924
Streptococcus thermophilus 1706	EAH	37	Glycine	0.25	0.988
			Glucose	1.20	0.985
			NaCl	0.40	0.984
			Glycerol	2.40	0.948
Lactobacillus herveticus 1716	MRS	37	Glycine	1.05	0.976
			Glucose	2.15	0.966
			NaCl	0.95	0.962
			Glycerol	3.30	0.929
Lactobacillus lactis 19	MRS	37	Glycine	0.65	0.983
			Glucose	1.70	0.974
			NaCl	0.90	0.964
			Glycerol	2.10	0.954

TABLE II Con't.

TABLE II. Lower a_w Limits for Multiplication of Bacteria Used in Cheese Manufacture[a]

			NaCl Concentration of culture medium		
Propionibacterium shermanii	Peptone-whey	25		1.15	0.955[d] 0.955[d]

[a] Streit, Rüegg, and Blanc (publication in preparation). The lower limits correspond to a_w values at which no growth was observed after 7 days of incubation.

[b] From the collection of Institute.

[c] Determined from calibration curves, i.e., from regression lines in plots of solute molarity vs. a_w.

[d] Rüegg et al. (1976).

In the case of propionibacteria a more detailed study seemed desirable because of their important role for the manufacture of emmentaler cheese (Rüegg et al., 1978). Propionibacteria are responsible for eye formation in this cheese and must be considered as one of the major factors in the development of the so-called late fermentation, a defect connected with a "blowing" of the cheese after the main eye development (Steffen and Puhan, 1976). Figure 3 shows that multiplication and metabolism of propionibacteria are very sensitive to changes in a_w in the range 0.99 to 0.95. Thus it seems that a_w might contribute to the control of growth and CO_2 production of these bacteria during the ripening process.

Thus, with the exception of the stress imposed by low water activity, the environments chosen for the studies of a_w tolerances should be favorable for bacterial growth. The tolerances for a_w would probably be diminished if other factors in the environment were also rendered unfavorable.

IV. INFLUENCE OF a_w ON HYDROLYSIS OF CHEESE PROTEINS

Proteolytic enzymes in rennet, starter cultures, and other microorganisms as well as the indigenous milk protease contribute to the hydrolysis of proteins during cheese ripening (Visser, 1977). Little and sometimes contradictory information is available concerning the influence of a_w on these enzyme systems. Creamer (1971) concluded from his measurements of α_{s1}- and β-casein hydrolysis in caseinate solutions that a_w is one of the major factors determining the relative rates of their degradation by rennet. He found that with 30% sucrose concentration ($a_w = 0.983$) the rate of β-casein hydrolysis was one-fifth that of α_{s1}-casein. With 60% sucrose ($a_w = 0.963$), β-casein was only slightly hydrolized and α_{s1}-casein underwent initial degradation at a rate of hydrolysis that was about 20 times slower than without

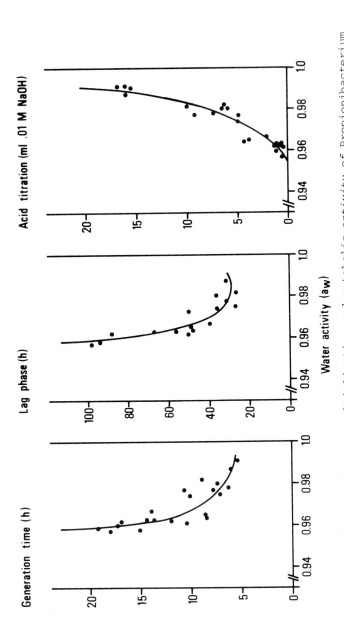

Fig. 3. Influence of a_w on multiplication and metabolic activity of Propionibacterium shermanii. Incubation was carried out at 25°C and a_w was controlled using NaCl. The amount of 0.01 mol/liter NaOH needed to neutralize a steam distillate of 1 ml of the peptone-whey culture medium after 96 hr of incubation was used as a measure of the metabolic activity (Rüegg et al., 1976).

sucrose. Similarly, Fox and Walley (1971) found that rennin and
papsin hydrolysis of β-casein was strongly dependent on NaCl concentration. At 5% NaCl concentration, (a_w 0.97) β-casein
hydrolysis was significantly reduced and at 10% NaCl (a_w 0.94)
was completely inhibited. These authors concluded that the inhibitory effect of NaCl on the proteolysis of β-casein was rather
selective because the rate of $α_{s1}$-casein hydrolysis was found to
be maximal in the presence of 5-10% NaCl and even at 20% NaCl
concentration (a_w 0.87)$α_{s1}$-casein was substantially hydrolyzed.
A stimulating effect of NaCl on paracasein hydrolysis by rennet
had already been observed by Stadhouders (1962). Proteolytic
activity of rennet was found to be higher in the presence of
2-7% NaCl (a_w 0.98-0.96) with an optimum at about 3% NaCl. This
phenomenon was explained by the known higher solubility of paracasein in the presence of salt.

   Phelan *et al*. (1973) reported an increase in β-casein hydrolysis in cheddar cheese when the water content was increased or
when the salt content was reduced from 2.2 to 0.26%. On the
other hand, no change in the extent of hydrolysis of $α_{s1}$-casein
was observed when the salt content was changed within the same
range. Phelan *et al*. (1973) and Fox and Walley (1971) concluded
that changes in the substrate rather than enzyme were responsible
for the reduced proteolysis of β-casein in the presence of NaCl.
Marcos *et al*. (1976) recently reported that a_w in cheese correlates significantly with the amount of unhydrolyzed $α_{s1}$-casein
but not significantly with the amount of remaining β-casein.
Similar correlation coefficients were found between the extent
of casein hydrolysis and the concentrations of various constituents such as water, salt, and ash.

   Noomen (1978), studying protein breakdown in soft cheese,
also found that NaCl stimulated the degradation of $α_s$-casein by
rennet, up to a concentration of about 4% (in the cheese moisture).
β-casein hydrolysis was maximal in the absence and considerably
reduced by the presence of salt. The hydrolysis of both $α_s$- and

β–casein by milk protease in the soft cheeses was stimulated by low concentrations of NaCl but reduced by high concentrations. Maximum breakdown of caseins by milk protease was found at about 2% NaCl content in the cheese moisture.

Bie and Sjöström (1975) pointed out the importance for cheese ripening of the intracellular enzymes of the lactic acid bacteria present, which are set free during autolysis. They observed a strong effect of NaCl on the rate of this autolysis; sodium ions increased, calcium and magnesium ions decreased the rate.

It appears from these reports that a_w controls to some extent proteolytic reactions occurring during cheese ripening but that specific inhibitory effects of solutes (Von Hippel and Schleich, 1969), especially NaCl, on enzymatic reactions may sometimes be more important.

V.   RELATIONSHIPS BETWEEN a_w AND CHEESE COMPOSITION

This relationship has been studied using stepwise multiple regression procedures (Rüegg and Blanc, 1977). Assuming linear relationships between a_w and concentration of constituents, the mathematical procedures revealed some factors that may explain the differences between the observed a_w levels. The results obtained suggested that the a_w in cheese is mainly controlled by low-molecular-weight constituents and water content. Similar calculations have been carried out by Dixon (1975) using more data and a different regression procedure. Furthermore, concentrations were expressed on a water content basis, i.e., in grams per 100 g cheese moisture. Results obtained using data from 82 different cheeses with a_w values in the range 0.87 to about 0.994 are summarized in Fig. 4 and Table III.

Figure 4 shows the dependence of a_w on the concentration of cheese constituents. The direct comparison of a_w with the nitrogen content, ash, and salt-free ash content (ash minus NaCl

TABLE III.   Relationship between a_w and Chemical Composition of Cheese

Variable	Regression coefficient	S. E. of coeff.
NaCl	-0.0064	0.0007
NPN[b]	-0.0077	0.0018
Saltfree ash[c]	-0.0024	0.0008
pH	0.0127	0.0022
Intercept	0.939	
Multiple r^2	0.860[d]	
S.E. of est.	0.010	

[a]*Multiple linear regression analysis using data of 82 different cheeses with a_w in the range 0.87-0.995. Concentrations were calculated relative to the amount of water associated with cheese (grams/100 g water). Calculations were performed using the BMDP2R stepwise regression program (Dixon, 1975). Table includes those variables with significant regression coefficients.*

[b]*"Nonprotein nitrogen", soluble in 12% TCA.*

[c]*Ash minus NaCl content.*

[d]*r^2 is equal to the coefficient of determination and indicates the fraction of the variance, which may be explained by the influence of the variables included in the list.*

content) revealed correlation coefficients in the range -0.8 to -0.9. The corresponding coefficients for the non-protein-nitrogen (NPN) and NaCl content were between -0.7 and -.08. The direct comparison of a_w with the pH values and content of water soluble nitrogen showed no significant correlation.

The result of the stepwise multiple regression analysis, which is more effective in resolving factors that explain the difference between the a_w levels in cheese is shown in Table III. As in earlier calculations (Rüegg and Blanc, 1977) pH gives a positive and the NaCl and NPN contents a negative regression

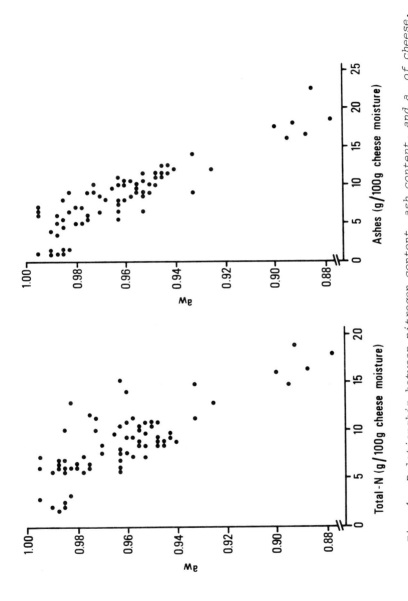

Fig. 4. Relationship between nitrogen content, ash content, and a_w of cheese. The corresponding correlation coefficients are -0.813 and -0.891 (82 samples).

coefficient (Table III). Increasing the data sets resulted in the salt-free ash appearing on the list of variables with significant regression coefficients. This was expected because the inorganic material included in the salt-free ash is likely to have a strong effect on water activity. Moreover, the salt-free ash correlates with protein content (with percentage of total N: $r = 0.932$).

It is interesting to note that the regression coefficient for the NaCl effect may be compared to that of pure sodium chloride solutions in the same a_w range. If a straight line is fitted to the a_wNaCl concentration data reported by Robinson and Stokes (1959), it has a slope of -0.00500 ± 0.00002 (salt concentration expressed in grams NaCl per 100 g of water). The slightly higher value found for water associated with cheese (-0.0064 ± 0.0007) is probably due to the fact that some of the water used to calculate the concentrations is bound and therefore not available as solvent (Geurts et al., 1974).

The intercept value of 0.939 (Table III) has no direct physical meaning. It merely corresponds to the water activity in the hypothetical case with all variables set to zero. However, the intercept value must be considered if one wishes to estimate a_w values using the regression coefficients and the corresponding concentration and pH data.

Regression analysis revealed that about 58% of the variation of the observed a_w values may be explained by the different NaCl water ratios in cheese. The important role of salt in the control of a_w values of camembert cheeses that were left for different times in the brine bath. The a_w values and NaCl concentrations were determined after 15 days of ripening. In the time interval tested, both the NaCl uptake and a_w brining time data pairs, a slope of -6.4×10^{-5} min^{-1} is obtained. The corresponding slope for the NaCl concentration is 0.011 min^{-1}.

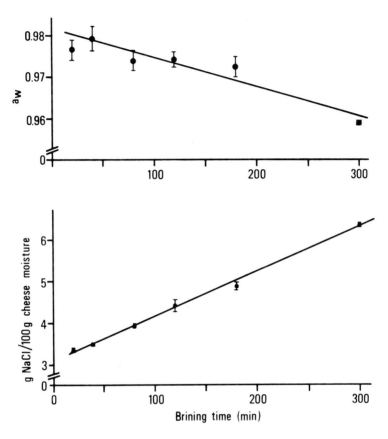

Fig. 5.   Influence of brining on a_w for camembert cheese.
The brine bath had a temperature of $14^{O}C$ and a salt content of
20% (w/w).   NaCl and a_w determinations in cheese samples were
made 15 days after manufacture.   (The length of the bars corresponds to ±1 standard deviation.)

VI.   WATER ACTIVITY AND CHEESE DEFECTS

The influence of a_w on microbial activity and proteolysis and
the relationships between a_w and cheese composition just reviewed
suggests a correlation between this parameter and cheese quality.
The influence of NaCl on casein proteolysis and cheese quality

that has been observed by a number of authors also points to an a_w/quality relationship (Schulz and Kay, 1956; Stadhouders, 1962; Fox and Walley, 1971; Maurer and Stock, 1977).

Two examples of comparisons of a_w values and cheese quality are summarized in Table IV. The data were from studies of two groups of emmentaler and gruyere cheese. For each type of cheese one group consisted of normal, first-quality cheeses and another group of cheeses that showed defects due to late fermentation (Steffen and Puhan, 1976). The data were collected over one year by taking measurements every month five samples. The samples, which had approximately the same age, were purchased from various Swiss cheese factories. The small fluctuations in the sample age, which could not be avoided, were taken into account in the final statistical analysis by introducing the age as a covariate. Table IV shows the results of a statistical analysis of the data. The analysis of covariance considers the variations of the a_w values between the two qualities and the variations between the 12 experiments that were carried out every month. The highly significant F values show that the differences between the a_w values can be explained by the influence of the age, the quality, and the variance between sampling months. In this latter type of variation the possible differences brought about by the monthly calibration of the humidity sensors should also be taken into account, as well as the possible variation in the manufacturing procedures.

Comparison of the sums of squares in Table IV reveals that the contribution of the defect to the variation of a_w was comparatively small. The differences between sample age and between sampling months explained a greater portion of the observed variation than the quality difference. The slope of the covariate indicates that a_w decreases by 0.0001 for gruyere and by about 0.0002 units per day for emmentaler at the mean age of both cheeses. This decrease is consistent with the slope of the curves in Fig. 1.

TABLE IV. Comparison of a_w Values for Gruyere and Emmentaler Cheese with and without Late Fermentation[a]

	Gruyere	Emmentaler
A. Means and standard deviation		
a_w of samples without defects	0.9477 ± 0.0121	0.9719 ± 0.0070
a_w of samples with defects	0.9506 ± 0.0115	0.9681 ± 0.0084

		Gruyere		Emmentaler	
Source of variation	Degrees of freedom	Sum of squares ($\times 10^4$)	F value[b]	Sum of squares ($\times 10^4$)	F value[b]
B. Analysis of covariance					
Covariate					
main effects					
Age	1	10.869	13.66***	1.835	5.59**
Total	12	71.142	7.45***	38.945	9.88***
Quality	1	4.964	6.24***	3.924	11.95***
Sampling month	11	67.274	7.69***	34.976	9.68***
Two-way interaction: quality sample month	11	5.470	0.63 n.s.	2.485	0.69 n.s.
Explained variation	24	87.481	4.58***	43.265	5.49***
Residual variation	95	75.606		31.198	
Total	119	163.087		74.463	
Slope of covariate age (days^{-1})		-0.00017		-0.00010	

[a] The samples from different Swiss factories had approximately the same age (gruyere: 173±20; and emmentaler: 134±14 days). Five samples of each quality were analyzed every month during one year. The a_w values were compared by means of a two-way analysis of covariance with the sample age as the covariate. Calculations were performed using SPSS computer programs (Nie et al., 1975).
[b] Confidence level of F: ***, = >99.9%; **, 99.0-99.9%; n.s., not significant.

It is interesting to note that for gruyere cheeses, the mean a_w values of the two qualities, 0.948 and 0.951, are around the critical level for growth of *Propionibacterium shermanii* and other bacteria (Table II). Also the mean a_w for the samples with defects was slightly higher than 0.95.

In addition to a_w, various other physicochemical and chemical parameters were measured and analyzed in the same way. For gruyere cheeses, it was found that the NaCl and NPN content and pH were also significantly different for the two groups of cheeses. For the samples with defects, the pH and NPN content were higher and the NaCl content lower. Taking into account the sign and significance of the correlations between these parameters and a_w, we may conclude that the lower NaCl content was one of the major factors responsible for the observed a_w difference. The observed a_w difference of 0.004 can be explained by a difference of about 0.6% NaCl in the cheese water. (This corresponds to approximately 0.2% NaCl in the total mass of cheese). The effect on the water activity of the higher NPN content in cheeses with defects was masked by the stronger effects of NaCl and pH. A higher NPN content is usually observed in cheeses with late fermentation defects (Steffen and Puhan, 1976).

For the Emmentaler cheeses the mean a_w of samples with defects was 0.968, which is about 0.004 lower than that of normal cheese. Inspection of the list of variables that were investigated simultaneously with a_w revealed, for example, that the water and NPN content and the pH were significantly different for the two types of samples. Similar to the case discussed before, the NPN content and pH were higher for the cheeses with defects, but also the mean water content was about 0.3% higher than in normal cheese. The difference in NPN in cheese moisture was 0.286%. This accounts for a difference in a_w of about 0.002.

The pH difference was 0.033 and could explain a decrease of a_w of about 0.0004.

The results of the statistical analyses confirm the relationships between a_w and cheese composition that have been discussed and indicate that water activity measurements may be a useful tool for studying the ripening and controlling the quality of cheese.

ACKNOWLEDGMENTS

We would like to thank the Cheese Research Department for the samples studied and for chemical analyses. We are grateful to Miss. U. Moor for her technical assistance and to Dr. P. Rüst and Dr. M. Casey for interesting discussions and liguistics assistance.

REFERENCES

Bie, R., and Sjöström, G. (1975). *Milchwissenschaft 30*, 739.
Brown, A. D. (1976). *Bact. Rev. 40*, 803.
Christian, J. H. B. (1963). *Advan. Food Sci. 3*, 248.
Creamer, L. K. (1971). *N. Zeal. J. Dairy Technol. 6*, 91.
Dixon, W. J. (editor) (1975). Biomedical Computer Programs, Health Sciences Computing Facility, Univ. of California Press, Berkeley.
Fox, P. F., and Walley, B. F. (1971). *J. Dairy Res. 38*, 165.
Gal, S. (1967). "Die Methodik der Wasserdampf-Sorptionsmessungen." Springer, Berlin
Geurts, T. J., Walstra, P., and Mulder, H. (1974). *Neth. Milk Dairy J. 28*, 46.
Greenspan, L. (1977). *J. Res. NBS 81A*, 89.
Karel, M. (1973). *C. Rev. Food Technol. 5*, 329.
Kurmann, J. L., Gehriger, G., Flückiger, E., Steffen, C., and Kaufmann, H. (1976). *Schweiz. Milchztg. 102*, 57.
Leistner, L., and Rödel, W. (1976). *In* "Inhibition and Inactivation of Vegetative Microbes." (Skinner, F. A., ed.). Academic Press, London.
Marcos, A., Asuncion Esteban, M., and Fernandez-Salguero, J. (1976). *Arch. Zootec. 25*, 73,

Maurer, L., and Stock, H. (1977). *Oesterr. Milchw. 32, Wiss. Beilage 1,* 1.

Measures, J. C. (1975). *Nature 257,* 398.

Nie, N. H., Hull, C. H., Jenkins, J. G., Steinbrenner, K., and Bent, D. H. (1975). "Statistical Package for the Social Sciences," 2nd ed. McGraw Hill, New York.

Noomen, A. (1978). *Meth. Milk Dairy J. 32,* 26, 49.

Phelan, J. A., Guiney, J., and Fox, P. F. (1973). *J. Dairy Res. 40,* 105.

Robinson, R. A., and Stokes, R. H. (1959). "Electrolyte Solutions." Butterworths, London.

Rüegg, M., and Blanc, B. (1977). *Milchwissenschaft 32,* 193.

Rüegg. M., Glättli, H., and Blanc, B. (1976). *Schweiz. Milchw. Forsch. 5,* 119.

Rüegg. M., Glättli, H., and Blanc, B. (1978). *Proc. 20th Int. Dairy Congr., Paris,* p. 606.

Schulz, M., and Kay, H. (1956). *Deutsche Molkerei-Z. 40,* 1385.

Scott, W. J. (1957). *Advan. Food Res. 7,* 83.

Stadhouders, J. (1962). *Proc. 16th Int. Dairy Congr., Copenhagen, B,* 353.

Steffen, C. (1975). *Lebensm.-Wiss. Technol. 8,* 1.

Steffen, C., and Puhan, Z. (1976). *Schweiz. Milchw. Forsch. 5,* 1.

Troller, J. A. (1973). *J. Milk Food Technol. 36,* 276.

Visser, F. M. W. (1977). Ph.D. Thesis. Agricultural Univ., Centre for Agricultural Publishing and Documentation, Wageningen, The Netherlands.

Von Hippel, P. H., and Schleich, T. (1969). *In* "Structure and Stability of Biological Macromolecules" (Timasheff, S. N., and Fasman, G. D., eds.). Marcell Dekker, New York.

THERMAL PROPERTIES OF WATER IN RELATION TO MICROBIAL CELLS

S. Koga
Y. Maeda

I. INTRODUCTION

The physical states of water in microbial cells are disclosed
by various measurements, e.g., thermodynamic, mechanical, dielectric, spectroscopic. Data point invariably to the presence of
several states of the cell water, suggesting that part of the cell
water is somewhat "bound" to a variety of hydrophilic sites on
the dry matter components and loses some of its observed responses
when it is in the bulk phase. The mode and the extent of this
response reduction depend upon the kind of binding sites available
to the water molecules in microbial cells.

Experimental data obtained in the authors' laboratory using a
series of partially dehydrated cells of *Saccharomyces cerevisiae*
suggest the existence of three regions of the cell water as indicated by thermodynamic, dielectric, and NMR measurements (Koga
et al., 1966, 1969; Echigo *et al.*, 1966; Maeda *et al.*, 1968).
Since the cell water below 10% on a wet basis did not appreciably
contribute to the dielectric increment, water molecules seem to be
irrotationally held at a number of charged binding sites in this
region of low water content. Meanwhile, water molecules above a
water content of 25% were freezable and gave an approximately
constant heat of melting by weight in the course of thawing.
This part of the cell water seems to be in a liquid phase, making

a continuous solvent medium inside the cells. The intermediate
region of 10-25% seems to correspond to water molecules that are
rotationally trapped at uncharged polar sites, giving rise to a
value of the differential heat of vaporization similar to that in
bulk, its dielectric characteristics remaining basically the same
as in the continuous phase mentioned above. The water molecules
in these low and intermediate regions may be distributed in a discrete way among the separate binding sites in the cells.

This chapter deals with the specific heat of partially dehydrated microbial cells with special regard to the behavior of
the cell water. Specific heat is a basic thermodynamic parameter,
yet few data are available so far with microbial specimens.

II. EXPERIMENTAL

The sample microbes used in the present study are the vegetative cells of *Bacillus cereus* IFO 3131. They were harvested after
6 hr growth in aerobic shaking culture at 30°C using a synthetic
medium supplemented with yeast extract (Foerster and Foster, 1966).

This microbe was used as a spore-forming bacterium in order
to make a comparative physical study of the vegetative and spore
cells of the same species. However, data shown in this article
are related only to its vegetative form.

A microcalorimeter of the differential temperature scanning
type was used in its stepwise holding mode. The temperature was
varied from 30° to 35°C with an air-tight sealed sample vessel
packed with the vegetative cells harvested as above, washed several times with water, and dehydrated to varied extents. (Experimental details are cited elsewhere.)

The heat capacity of the partially dehydrated cells of
B.cereus is plotted in Fig. 1 against the amount of water added
to 1 g of the dried specimen. The dotted line is the theoretical
curve expected for the mixtures of the dried specimen and pure

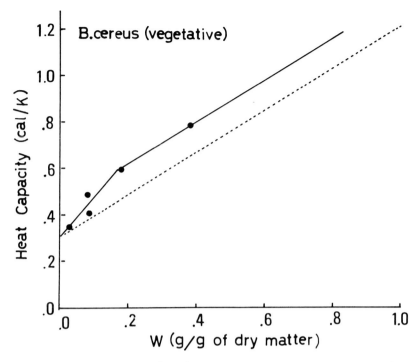

Fig. 1. Heat capacity increment vs. water content. The heat capacity increment of the vegetative cells of B.cereus is plotted against the amount of water added to 1 g of the dry matter (W). The full line is drawn by using data for higher values of the water content (W > 1).

water in which there are no interactions between the two components. The deviation of data from the theoretical curve indicates that there are some interactions between those two components in the region of lower water contents, and possibly the cell water takes somewhat different physical states in this water content region.

To examine the whole range of the water content, the specific heat as obtained from the heat capacity is plotted in Fig. 2 against the water content (wet basis) of the partially dehydrated cells as well as the cells suspended in water. The curve seems to break at x_1 and x_2, corresponding to 15 and 62%, respectively. Since the water content of the native cells is estimated as about

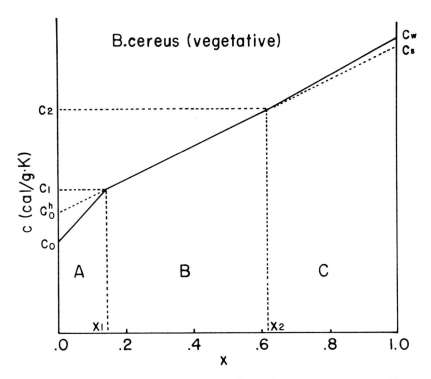

Fig. 2. Specific heat as function of water content. The specific heat (c) of the vegetative cells of B.cereus is plotted against the water content on a wet basis (x). The observed curve breaks at $x = x_1$ and x_2. $c_0 = 0.31$ cal/g K, $c_0^b = 0.88$ cal/g K, $c_0^h = 0.41$ cal/g K, $c_s = 0.97$ cal/g K (see text).

65%, the specific heat at x_2 corresponds to that of the native cells of B.cereus prepared as mentioned above. Region C therefore gives the specific heat of the native cells suspended in water at various concentrations, while region B and A give the specific heat of the partially dehydrated microorganism.

III. THEORETICAL

In this simple theory of the specific heat of microbial cells, we propose a two-component mixture model assuming that:

(1) So long as the specific heat of partially dehydrated cells is concerned, the microbial cells can be considered as consisting of the cell water and the dry matter component that has an average thermal property.

(2) All dry matter is perfectly dehydrated at $x = 0$ and fully hydrated at $x = x_1$. Contributions from unit mass of the dry matter in the cell remain invariable in the water content region x_1 to 1.

(3) The native cells give the value of specific heat found at $x = x_2$ and further addition of water above x_2 causes no change in regard to their intracellular state.

(4) Contributions from water in its vapor phase can be disregarded in analyzing the specific heat of the microorganism under study.

Since the specimens in range C in Fig. 2 are composed of the native cells having water content x_2 and variable amounts of cellular water, the specific heat of this two-component mixture is given by

$$c = c_2 \frac{1-x}{1-x_2} + c_w \frac{x-x_2}{1-x_2} \qquad (1)$$

where c_2 is the specific heat of the native cells as observed at $x = x_2$ and c_w is that of pure water as evaluated by extrapolation.

The specimens in region B are taken as mixtures of the fully hydrated dry matter component and the continuous solvent water. The specific heat is given by

$$c = c_0^h(1-x) + c_s x \qquad (2)$$

where c_0^h and c_s are the specific heat of the fully hydrated
solutes and the continuous solvent water, respectively. These
values can be evaluated graphically as denoted in the figure.

The specimens in region A are regarded as consisting of partially hydrated solutes and discrete hydration water. The contribution from unit mass of each component varies as x increases,
because the conformation of macromolecules changes in accordance
with the water content increase and thereby the hydration water
may have variable degrees of motional freedom as it binds to less
and less hydrophilic sites in the cells. Hence,

$$c = c'(x)(1-x) + c''(x)x \tag{3}$$

where c' and c'' are the specific heat of the partially hydrated
solutes and the discrete hydration water, respectively. The
simplest possible approximation to c' and c'' will be

$$c'(x) = c_0(1+\alpha x), \quad c''(x) = c_0^b + (c_s - c_0^b)x/x_1 \tag{4}$$

where c_0 is the specific heat of the completely dehydrated solute
component, c_0^b the specific heat of water in almost dehydrated
samples, and α stands for the hydration coefficient. From (3)
and (4),

$$c = c_0 + \left[c_0(\alpha-1) + c_0^b\right]x + \left(\frac{c_s - c_0^b}{x_1} - \alpha c_0\right)x^2 \tag{5}$$

By assuming that the observed data fit a straight line in region
A, the approximate values of α and c_0^b will be obtained by use of
(5) as follows:

$$c_s - c_0^b = \alpha x_1 c_0, \quad \left[c_0(\alpha-1) + c_0^b\right]x_1 = c_1 - c_0 \tag{6}$$

IV. DISCUSSION

    In such systems in which water cannot form any continuous
phase as in microbial cells with a small amount of cell water,
water molecules are distributed in a discrete way over a variety
of binding sites. Hydration in partially dehydrated microbial
cells might be classified into several categories: (1) structural,
(2) conformationally buried, (3) irrotationally bound, (4) rotationally trapped. The structural water supports the molecular or
supermolecular structures in a stoichiometric manner, while the
conformationally buried water is kept in less hydrophilic environments where it is localized to variable extents in the course of
molecular-conformational or supermolecular-structural changes during the period of, e.g., drying and/or freezing. Although the
state of the buried water is neither stoichiometric nor thermodynamically stable, it seems to be difficult with some specimens
to dehydrate it completely without causing structural alterations.
Irrotational hydration is observed with charged binding sites and
its interaction energy is much larger than that for water molecules in the bulk phase, while rotational hydration seems to occur
on uncharged polar sites and its interaction is of the same magnitude as that of water molecules in bulk.
    Addition of unit mass of water to the microbial cells in region
A causes, as shown in Fig. 1, a heat capacity increment large as
compared with that of pure water given as the slope of the theoretical curve in the figure. In terms of the present mixture model,
this fact must lead to either the statement that the added cell
water has an abnormally large specific heat for some reason or
the conclusion that the cell dry matter undergoes state changes
that result in an increase of its contribution to the specific
heat of the cells. Since dielectric data with some other microbes
and physical data with some proteins (Rao and Bryan, 1975; Nomura
et al., 1977; Lerreque and Garson, 1977; Celaschi and Mascarenhas,
1977; Sturtevant, 1977) favor the latter choice, discussion will

be made on possible conformational events that may account for
the steep slope in region A on the basis of the latter alternative.

In this connection hydration water in the microbial cells is
distributed among a variety of their constituent molecules. However, proteins are the component most abundant in amount and most
flexible in their conformation as affected by a varying water content. In this chapter we discuss in a qualitative manner one way
of conformational changes in proteins that may result in changes
of specific heat as depicted in Fig. 2. Proteins have polar sites
both on their backbone chains and on their side chains. However,
since the state of the side-chain end seems most effective for
conformational changes in protein molecules, we confine our present
discussion to polar sites at the end of their side chains.

As the hydration energy is larger for ionic sites by several
times than for nonionic polar sites to which water molecules are
bound through the hydrogen bond, the hydration-dehydration events
are likely to take place mainly at the binding sites of the
former type in the left half of region A, where protein molecules
possibly have a very small amount of hydration water. These hydration-dehydration events can be visualized by use of a charge-pairing model for simple ions in solution. Suppose we have an
aqueous solution of a strong electrolyte that dissociates therein
completely into the pair of a simple cation and a simple anion.
Taking a single pair of these ions, the energy of the mutual interaction is plotted as a function of their mutual distance in Fig.
3, where curve A is the hypothetical one for the unhydrated ion
pair and curve B for the fully hydrated pair. Reference line (1)
indicates the energy state for infinite separation of the ion pair
in vacuum, while reference (2) is the energy state for the fully
hydrated pair infinitely separated from each other in water. As
referred to (1), the reference (2) is shifted downward by an
amount of hydration energy averaged for this ion pair. The detailed shape of these curves is not relevant in the present discussion and the descending parts on the right-hand side of the po-

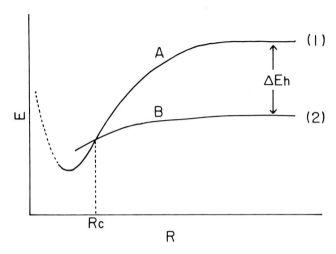

Fig. 3. Potential profile for the pair of oppositely charged ions. Curves were obtained in terms of the electrostatic potential of the small ion pair in vacuo (A) and in water (B). Those curves are separated from each other for R → ∞ by an amount of the hydration energy (ΔE$_h$) averaged for the ion pair.

tential minimum are due to the attractive electrostatic force exerted on each other in vacuum and in water, respectively. The electrostatic evaluation of these potential energies reveals that there is a critical distance R_c at which these two curves cross each other. For the pair of typical small ions, R_c is less than 1×10^{-7} cm as of center-to-center distance, which suggests that this single pair of ions in water may be spontaneously dehydrated and coupled to each other when the number of water molecules between the ion pair becomes less than several.

Likewise, the ionic linkage between oppositely charged side chains will be energetically favored with a protein molecule having a very small amount of water, whereas the hydration of separate charged sites might result when the protein is added with a larger amount of water. It is now possible for each segment on the side chains to assume a number of different conformational states, and thus the increase in the degree of freedom as to the intramolecular conformation may result in an increase of specific heat of the protein molecule as a whole.

Similar hydration-dehydration events are likely to happen in the microbial cells in the left half of region A, which make c' vary as a function of x, while c'' varies as a result of changes in the average motional degree of freedom on the part of the hydration water molecules.

In the right half of region A, hydration-dehydration events seem to take place at nonionic polar sites, where the energy of interaction is similar in magnitude to that of water in bulk. Taking, as above, protein molecules as representative of the dry matter of the cells, rotationally trapped water molecules in this range up to x_1 are distributed among nonionic polar sites that are separated from each other by an average mutual distance of at least 5 Å as estimated from the moisture content usually observed with fully hydrated protein molecules. This means that the hydration molecules are distributed in a discrete manner, the nearest-neighbor distance being several times as large as in pure water liquid. Since the distance of 10^{-7} cm order is far smaller than the length of the mean free path in the vapor phase, water in this discrete hydration state might be said to take an "expanded liquid" state in which molecules are required to overcome a larger energy barrier in hopping around from one site to another as compared with those in normal liquid state. Therefore, the diffusion coefficient is much smaller in this expanded liquid state than in the normal liquid state. This reduced value of the diffusion coefficient presumably makes the cell water in a discrete region kinetically remain "unfrozen" when all the cell specimens are cooled at a relatively rapid rate down to subzero temperature ranges, even if the trapped state is thermodynamically unstable as referred to the frozen phase in bulk at the freezing temperature.

References

Celaschi, S., and Mascarenhas, S. (1977). *Biophys. J. 20*, 273.
Echigo, A., Fujita, T., and Koga, S. (1966). *J. Gen. Appl. Microbiol. 12*, 91.
Foerster, H. F., and Foster, J. W. (1966). *J. Bacteriol. 91*, 1168.
Koga, S., Echigo, A., and Nunomura, K. (1966). *Biophys. J. 6*, 665.
Koga, S., Maeda, Y., and Echigo, A. (1969). "Freezing and Drying of Microorganisms," p. 143. Univ. of Tokyo Press, Tokyo.
Lerreque, J. L., and Garson, J. C. (1977). *Biopolymers 16*, 1725.
Maeda, Y., Fujita, T., Sugiura, Y., and Koga, S. (1968). *J. Gen. Appl. Microbiol. 14*, 217.
Nomura, S., Hiltner, A., Lando, J. B., and Baer, E. (1977). *Biopolymers 16*, 231.
Rao, P. B., and Bryan, W. P. (1975). *J. Mol. Biol. 97*, 119.
Sturtevant, J. M. (1977). *Proc. Natl. Acad. Sci. USA 74*, 2236.

SPECIFIC SOLUTE EFFECTS ON

MICROBIAL WATER RELATIONS

J. H. B. Christian

I.   INTRODUCTION

The application of the concept of water activity (a_w) to the
water relations of microorganisms makes one major assumption:
When the aqueous solution in the organism's environment is concentrated by addition of solutes or by removal of water, the
consequences for microbial growth result solely from the change
in a_w.

When Scott introduced this concept 25 years ago, he supported
it with very convincing data obtained with 14 strains of the
food-poisoning bacterium, *Staphylococcus aureus* (Scott, 1953).
Whether the a_w was varied by addition of certain electrolytes or
sucrose to laboratory media or by adjusting the water content of
dried foods, *S. aureus* grew at levels of a_w down to 0.88-0.86
(Table I). Water content certainly was not the factor limiting
growth, for it ranged from 16 to 375% of the dry weight in the
various foods and media at the limiting levels of a_w. Specific
effects due to the presence of particular solutes were small.
The inhibitory conditions that led to the cessation of growth
within this relatively narrow range of a_w were clearly a direct
consequence of reduced a_w, or some property closely related to it.

TABLE I. Minimum a_w and Water Contents Permitting Growth of
S. aureus in Various Media at 30°C[a]

Nutrient substrate	Solutes added to control	Minimum a_w	Water as % dry wt.
Casamino acids, yeast extract, casitone	Salts mixture[b]	0.88	375
Nutrient broth	Salts mixture	0.86	315
Nutrient broth	Sucrose	0.88	60
Nutrient broth	Sucrose 3.44 m + salts mixture[b]	0.86	75
Milk	None	0.86	16
Meat	None	0.88	23
Soup	None	0.86	63

[a]*From Scott (1962).*

[b]$NaCl:KCl:Na_2SO_4$, *5:3:2:moles.*

Experiments with 16 serotypes of *Salmonella* confirmed the
usefulness of the a_w concept (Christian and Scott, 1953). In
several complex media, growth proceeded to levels of a_w between
0.94 and 0.95, whether a_w was controlled by additions of salts
or sucrose or by adjusting the water content of dehydrated media.
Salmonella oranienburg, which grew at a_w levels down to 0.946 in
brain heart broth, grew to 0.93-0.94 in partially rehydrated dried
milk, meat, or soup. In the author's experience, other Enterobacteriaceae show similar tolerances to salt and sucrose solutions
when their concentrations are expressed on an a_w basis.

These data indicate that the growth response to a_w can be, at
least for two bacteria differing in morphology, gram reaction, and
salt tolerance, largely independent of the types of solutes present in the medium.

There are, of course, many toxic solutes for which such
relationships cannot be expected to hold, but there are, in
addition, cases where microbial response differs widely at a
particular a_w when the latter is obtained with such relatively
bland solutes as sodium chloride, sucrose, or glycerol. These
are considered to be specific solute effects, and it is with them
that this chapter is concerned.

The first problem is to define specific solute effects. If
the response to two solutes differs at the same a_w, how does
one decide which, if either, is the "normal" response to a_w and
which solute is responsible for a "special" effect? Sometimes
this is relatively simple, as with *S.oranienburg* growing in a
simple glucose-salts medium at a range of a_w levels achieved by
additions of salts, sucrose, glucose, or glycerol (Christian,
1955b). The growth responses to the salt mixture and to the two
individual sugars (Fig. 1) are sufficiently similar to indicate
that these solutes affect growth predominantly through their
influence of a_w. The dosage-response curve for these solutes
might therefore be taken as the normal water relations of this
bacterium. In contrast, the influence of glycerol at low levels
of a_w differs greatly from this and it may be assumed that there
is here a specific solute effect that counteracts some of the inhibition caused by reduced a_w. As will be seen below, it is
rarely that specific solute effects can be as easily ascribed as
in this example.

Most of the published data yielding information about solute
effects report simply the minimum a_w in various nutrient solutions at which growth was observed. The solutes most commonly
tested have been sodium chloride, sucrose, and glycerol, but more
recent studies have tended to include other salts, sugars, and
polyols. Unfortunately most reports compare sodium chloride with
sucrose or sodium chloride with glycerol, comparisons of all
three solutes being rare. When only two or three solutes are

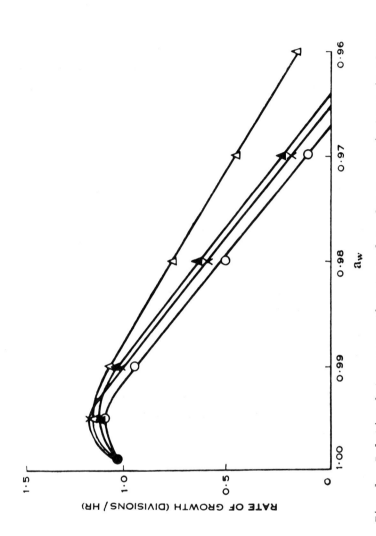

Fig. 1. Relation between growth rate and a_w for S. oranienburg in glucose-salts medium using four methods for controlling a_w: o, NaCl:KCl:Na$_2$SO$_4$ mixture in ratio of 5:3:2 moles; x, sucrose; △, glucose; △, glycerol. (Christian, 1955b).

compared and results differ, it is difficult to conclude which, if any, are acting solely on a_w and which are not. In this report, only data in which more than one solute is tested are discussed. Published data on bacteria, yeasts, and molds are reviewed for evidence of specific effects and suggestions are made as to where explanations for these effects may be sought.

II. BACTERIA

The prime examples of specific solute effects in bacteria are found among the halphiles. These are, by the most common definitions, salt (sodium chloride) requiring as well as salt tolerant. *Halobacterium salinarium* is typical of the red extreme halphiles. It grows in media containing from 3 M to saturated NaCl, equivalent to an a_w range of about 0.89-0.75. One-half of the 3 M NaCl requirement can be replaced by KCl, some 2.5 M KCl being required to give equivalent growth (Brown and Gibbons, 1955). The sodium requirement cannot be substituted by nonelectrolytes. Such organisms clearly have a very low water requirement but the specificity of the solute requirement makes it impracticable to consider it in terms of a_w. The mechanism of this effect has been studied extensively and is mentioned later.

A similar, but less specific, effect of solutes is observed with the so-called moderate halophile *Vibrio costicolus,* which grows over a much lower range of salt concentration. Although it has been claimed that all of the sodium requirement (0.3 M) can be replaced by potassium (Flannery *et al.,* 1952), there is a very small obligate requirement for 0.01-0.02 M NaCl, which can be provided by some basal media (Christian, 1956). Nonelectrolytes cannot substitute for sodium chloride.

Other species of the genus *Vibrio* are also sensitive to high concentrations of nonelectrolytes. *Vibrio metchnikovii* has a

clear need for reduced a_w if growth is to be rapid (Fig. 2). Although sucrose will provide this stimulation in low concentrations,
it is completely inhibitory at concentrations reducing a_w to 0.997
(Marshall and Scott, 1958). The food-poisoning bacterium *V.*
parahaemolyticus, of marine origin, is halophilic, although its
requirement for sodium chloride is very low. In a medium containing 0.5% NaCl to satisfy the halphilic requirement, Beuchat
(1974) recorded lower limits for growth of this organism of
a_w = 0.948 in NaCl-adjusted medium, a_w = 0.957 in sucrose and
a_w = 0.937 in glycerol.

The effects of substituting sucrose for sodium chloride on
the growth of 48 strains of halotolerant and halphilic bacteria
were reported by Ishida (1970). Some substitution was possible
at high a_w levels with halotolerant strains but very little
could be effected at any a_w with halphiles. The genera of the
isolates were not given.

Thus halophilic bacteria and vibrios, all of which are
susceptible in varying degrees to lysis on dilution, show specific solute effects on growth at reduced a_w, nonelectrolytes being
more inhibitory, or less protective, than sodium chloride. Marine
bacteria have a specific requirement for sodium ions and in some
cases for chloride ions also (MacLeod and Onofrey, 1956).
Possibly other bacteria with requirements for low concentrations
of salt, e.g., some *Brucella* spp. (Koser *et al.,* 1941), will also
prove to be intolerant of nonelectrolytes. Note, however, that
the low requirement of some bacteria for sodium ions may be
obscured by the sodium content of the growth medium. The latter
can vary greatly among complex media and is rarely measured in
studies of the salt relations of growth.

The first demonstration of a specific solute effect in nonhalophilic bacteria was probably that shown in Fig. 1 for *S.*
oranienburg (Christian, 1955b). Subsequently, Baird-Parker and
Freame (1967) reported that for germination and growth of

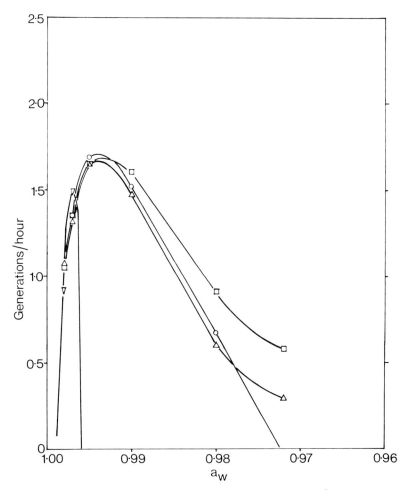

Fig. 2. Relation between rate of growth at 30°C and a_w for
V. metchnikovii *in brain broth with added solutes (Marshall and*
Scott, 1958).

Clostridium botulinum (types A, B, and E) also, glycerol was less
inhibitory than sodium chloride at equivalent levels of a_w. These
findings were not surprising, as the permeability of certain bacteria to glycerol is well known and has been quantitated for

Escherichia coli by Alemohammad and Knowles (1974). Such a permeable solute, having only a transitory effect on cell turgor, would not be expected to have the inhibitory effect of a plasmolyzing solute such as sodium chloride.

However, a comparison, on an a_w basis, of the sodium chloride and glycerol tolerances of 16 species of nonhalophilic bacteria (Marshall *et al.*, 1971) revealed that, while most of the bacteria tested were more tolerant to glycerol, the more salt-tolerant bacteria, predominantly cocci, were more sensitive to glycerol than to salt (Fig. 3). If one deletes *V.metchnikovii,* on the grounds of its almost halphilic behavior, this range of limiting a_w values in salt for the other 15 species was 0.97-0.83 and in glycerol was 0.95-0.89; the average minimum a_w was 0.92 in both solutes. For bacteria where specific effects exist, it cannot be assumed that a particular solute will always be more or always be less inhibitory than another.

A summary of some of the reported data on the water relations of food-poisoning bacteria is given in Table II. Note that the lower limits of a_w permitting growth are generally similar when media are adjusted with sodium chloride or with sucrose, solutes most likely to be encountered in high concentrations in conventional foods of low a_w. The greater tolerance shown toward glycerol by many of the bacteria might therefore be of concern, in view of the interest in this and other polyols as humectants in intermediate-moisture foods (IMF). However, the bacterium posing the greatest potential hazard to IMF is *S.aureus,* which is more sensitive to glycerol than to NaCl (Fig. 3).

The lowest a_w permitting growth of an organism is of paramount importance when a food is to be preserved by low a_w alone. However, in many instances several factors, none by itself totally inhibitory, combine to ensure the safety or stability of a food. In such circumstances rates of growth at intermediate levels of a_w may be relevant. Bacteria with similar a_w limits

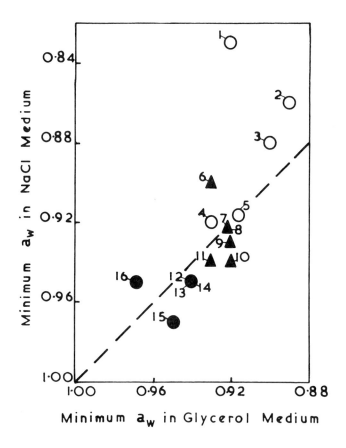

Fig. 3. Relation between the minimum a_w levels for growth of 16 species of bacteria in NaCl- and in glycerol-adjusted medium. o, Gram-positive cocci; Δ, gram-positive rods; o, gram-negative rods. The numbers identify data for individual species: 1, Micrococcus sp. MR1; 2, Staphylococcus aureus; 3, S.albus; 4, M.lysodeikticus; 5, Sarcina sp. 2b; 6, Bacillus subtilis W168; 7, B.spehaericus; 8, B.cereus C3; 9, B.cereus 5B; 10, B.megaterium KM; 11, Clostridium botulinum type A; 12, Salmonella newport; 13, S.oranienburg; 14, Escherichia coli; 15, Pseudomonas fluorescens; 16, Vibrio metchnikovii. Points for species equally sensitive to NaCl and glycerol fall on the dashed line (Marshall et al., 1971).

TABLE II.   Limiting a_w Values for Growth of Food-Poisoning
Bacteria in Complex Media Adjusted with Various Solutes

Species	Solute	Limiting a_w	Species	Solute	Limiting a_w
B. cereus	NaCl	0.94^d	C. perfrin-	NaCl	0.97^e
	Glucose	0.95^d	gens	Sucrose	0.97^e
	Glycerol	0.92^d		Glycerol	0.95^e
C. botu-			S. oranien-	Salts	0.95^c
linum			burg	Sucrose	0.95^e
Type A	NaCl	0.96^a		Glycerol	0.94^f
	Salts	0.95^a			
	Glycerol	0.92^a	S. aureus	Salts	0.86^h
Type B	NaCl	0.96^a		Sucrose	0.88^h
	Salts	0.94^a		Glycerol	0.89^f
	Glycerol	0.92^a	V. parahae-	NaCl	0.95^b
Type E	NaCl	0.97^a	molyticus	Sucrose	0.96^b
	Salts	0.97^a		Glycerol	0.94^b
	Glycerol	0.94^a			

[a] Baird-Parker and Freame (1967).

[b] Beuchat (1974).

[c] Christian and Scott (1953).

[d] Jakobsen and Murrell (1977a).

[e] Kang et al. (1969).

[f] Marshall et al. (1971).

[g] Ohye and Christian (1966).

[h] Scott (1953).

for growth in different solutes may respond differently to the
same solutes at higher a_w levels. *Salmonella oranienburg* growing
in a nutritionally rich medium (in contrast to the minimal medium
used in studies summarized in Fig. 1) is a case in point (Fig. 4)
(Christian and Scott, 1953). Although salts and sucrose both
prevented multiplication when a_w was reduced below 0.95, growth
at 0.97 was twice as rapid in sodium chloride as in sucrose.
Thus a suboptimal pH, for example, would be more likely to prevent growth of *S. oranienburg* at 0.97 if sucrose were the added

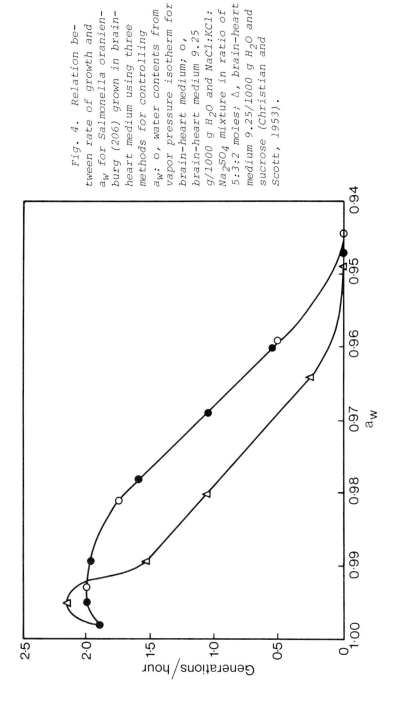

Fig. 4. Relation between rate of growth and a_w for Salmonella oranienburg (206) grown in brain-heart medium using three methods for controlling a_w: o, water contents from vapor pressure isotherm for brain-heart medium; o, brain-heart medium 9.25 g/1000 g H_2O and NaCl:KCl: Na_2SO_4 mixture in ratio of 5:3:2 moles; △, brain-heart medium 9.25/1000 g H_2O and sucrose (Christian and Scott, 1953).

solute than if it were NaCl. In this particular case, the shapes of the curves in Fig. 4 suggest that the specific effect is being provided by sucrose. This example is discussed later.

Somewhat similar specific effects are seen in Fig. 5 (Christian, unpublished). At 30° and 50°C, limiting a_w levels for growth of *Bacillus subtilis* lie in the ranges 0.90-0.92 and 0.92-0.94, respectively, but at higher a_w values rates of growth in NaCl and in glycerol differ manyfold. In contrast, with *B. stearothermophilus* at 50°C, differences in response to glycerol on the one hand and salt and sucrose on the other increase as a_w is lowered. Note that, in keeping with the trend in Fig. 3, the more salt-tolerant species (*B. subtilis*) is relatively glycerol intolerant, while the reverse applied to the less salt tolerant *B. stearothermophilus*. It is not obvious what is the normal response of these bacteria to a_w.

The water relations of germination and outgrowth of bacterial spores have been extensively investigated. For both *Bacillus* spp. and *Clostridium* spp. there is evidence that spores may initiate germination at a_w levels appreciably lower than those at which outgrowth or growth from vegetative inocula can occur (Gould, 1964; Baird-Parker and Freame, 1967).

Specific effects of solutes on spore germination are common. In general, ionic solutes (NaCl, KCl) are the most inhibitory in terms of a_w, followed by disaccharides and monosaccharides. Least inhibitory to germination are solutes of lower molecular weight, such as glycerol, ethylene glycol, and urea (Jakobsen *et al.*, 1972; Jakobsen and Murrell, 1977b). The latter authors discuss these effects in relation to events known to occur during spore germination.

III. YEASTS

Some yeasts are recognized as capable of growth in environments of much lower a_w than are tolerated by most bacteria. For

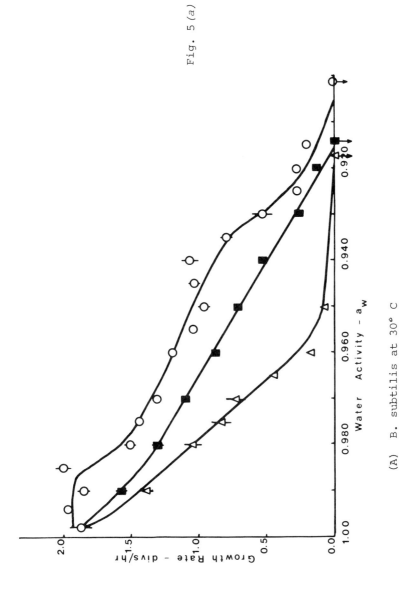

Fig. 5 (a)

(A) B. subtilis at 30° C

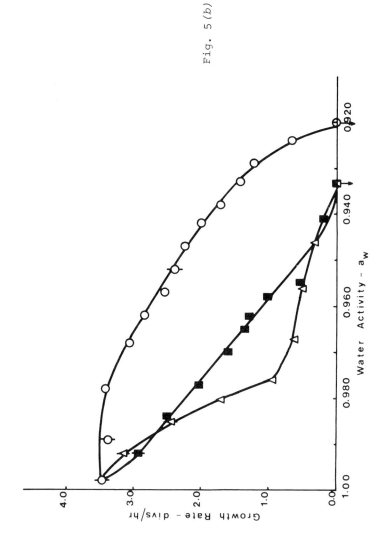

Fig. 5 (b)

(B) B. stearothermophilus at 50°C.

838

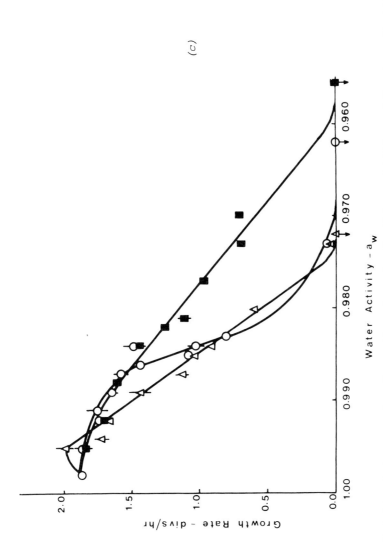

Fig. 5. Relation between rate of growth and a_w for Bacillus subtilis and B. stearothermo-philus grown in quarter-strength brain-heart broth using three methods to control a_w: O, sodium-chloride; Δ, sucrose; , glycerol. (a) B.subtilis at 30^oC, (b) B.subtilis at 50^oC, (c) B. stearothermophilus at 50^oC (Stewart and Christian, unpublished).

example, *Saccharomyces cerevisiae* will multiply down to about $a_w = 0.90$ in sucrose and $a_w = 0.92$ in sodium chloride (Christian, unpublished), levels tolerated by few bacterial species. Onishi (1963) quotes limiting salt and sucrose concentrations for *S. cerevisiae* that are close to an a_w value of 0.94.

The osmophilic (or xerophilic; Pitt, 1975) yeasts have been the most studied in respect of their water relations. Eight of nine osmophilic yeasts tested by Anand and Brown (1968) grew at a_w levels down to 0.765 in media adjusted with sucrose, while only one of six isolates considered to be nonosmophilic grew below $a_w = 0.917$. However, when a_w was controlled by additions of polyethylene glycol, there was no significant difference between the average minimum a_w tolerated by the two groups, which was, particularly for the osmophiles, much higher than in media adjusted with sucrose.

von Schelhorn (1950) reported the growth of isolates of *Saccharomyces rouxii* at a_w down to 0.62 in a fructose syrup and in a pear concentrate. A large number of osmophilic yeasts previously given other names, some coming from such highly saline environments as soy sauce mashes, have been classified as *S. rouxii* by Lodder and Kreger-van Rij (1952) and later authorities. *Saccharomyces rouxii* is thus among the most salt tolerant as well as sugar tolerant of yeasts.

Salt and sugar tolerances of a number of osmophilic yeasts have been determined by Onishi (1963). He concluded that strains of *S. rouxii* isolated from high sugar environments were less salt tolerant than strains from soy sauce and miso paste, although all had similar sugar tolerance. However, the highest limiting concentrations mentioned by Onishi (1963) for osmophilic yeasts were 22% M/V sodium chloride and 90% M/V sucrose, each equivalent to $a_w = 0.85-0.86$. Three of seven strains studied by Krömer and Krumbholz (1932) grew in the presence of 4 *M* potassium chloride or of 5.7 *M* glycerol, about $a_w = 0.84$ and 0.82, respectively (Scott, 1957).

There is evidence that the origin of *S.rouxii* isolates can
influence their relative salt and sugar tolerance, but where
sucrose is the test sugar, the lowest a_w that can be obtained
(i.e., at saturation) is about 0.86, and from von Schelhorn's
data, some strains grow in nonelectrolytes at a_w values much
lower than this. There appear to be no comparative data on the
salt and sugar tolerances of salt-tolerant species of *Pichia,*
Debaryomyces, Torulopsis, and *Hansenula.* Indeed, the limits of
salt tolerances of these species have not been reported, and it
seems unlikely that any grows in saturated sodium chloride
(a_w = 0.75).

IV.   MOLDS

Extensive studies on the growth of molds at reduced relative
humidities have shown that many species are xerophilic by the
definition of Pitt (1975), being capable of growth at a_w below
0.85. These data, reviewed by Scott (1957), were obtained mostly
in experiments where the desired levels of relative humidity were
obtained by equilibration. Thus little can be deduced of the
nature and concentration of the major solutes in the substrates
and hence of the existence of significant specific solute effects.
Many xerophilic molds do not grow on dilute substrates.
Scott (1957) demonstrated with *Aspergillus asmtelodami* that this
reflected a requirement for reduced a_w. Although each of five
solutes used to adjust a_w supported optimal growth at about a_w =
0.96, the rate of radial growth at that a_w varied between
solutes by up to twofold. Specific solute effects appeared
to lessen at lower a_w levels except that inhibition by
sodium chloride became pronouned at a_w = 0.92.

Pitt and Hocking (1977) confirmed for several species of xerophiles that the optimum a_w for growth was largely independent of the nature of the predominant solute, except when sodium chloride was used at pH 4.0. However, species varied greatly in response to solutes (Fig. 6). *Aspergillus flavus* showed least variation in response, while *Xeromyces bisporus* was markedly intolerant of sodium chloride at both levels of pH tested.

The water relations of germination of the same two fungi are contrasted in Fig. 7 (Pitt and Hocking, 1977). Although intolerance to sodium chloride is again evident in the data for *X.bisporus,* there is otherwise remarkably little solute specificity in the germination reactions of either species to reduced a_w.

For eight species of *Penicillium,* the effect of solute on growth at various a_w levels was slight (Pitt and Hocking, 1978). Like *A.flavus,* these penicillia are less xerophilic than *Eurotium chevalieri, X.bisporus,* and *Chrysosporium fastidium,* and these authors note that the more highly adapted molds become, the more likely they are to become intolerant of solutes other than sugars as a_w is decreased.

At various levels of a_w, glycerol was more inhibitory than sucrose to the growth of *Aspergillus niger* and *A.glaucus,* but the reverse effect was found for a species of *Rhizopus* (Horner and Anagnostopolous, 1973). The rates of growth of *Neruospora crassa* at a_w = 0.91 in four solutes decreased in the order glycerol > NaCl > glucose > sucrose (Charlang and Horowitz, 1971).

Following these rather selective examples of specific solute effects in bacteria, yeasts, and molds, it is appropriate to consider the possible mechanisms of these effects on microorganisms.

V. MECHANISMS

It must be stated at the outset that the reasons for solute-specific effects are known in very few cases. The problem is

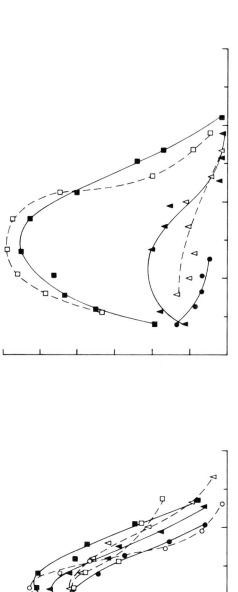

Fig. 6. Effect of a_w and solute on the radial growth rate of two xerophilic fungi: (a) Aspergillus flavus; (b) Xeromyces bisporus. ○, NaCl, pH 4.0; ●, NaCl, pH 6.5; ■, glucose/ fructose, pH 4.0; □, glucose/fructose, pH 6.5; ▲, glycerol, pH 4.0; △, glycerol, pH 6.5 *(Pitt and Hocking, 1977).*

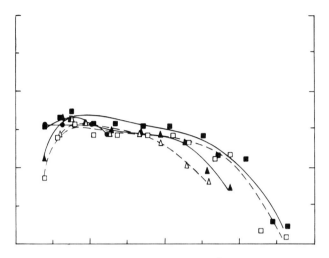

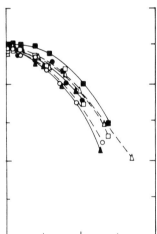

Fig. 7.   Effect of a_w and solute on
the time taken for germination of
spores of two xerophilic fungi: (a)
Aspergillus flavus; (b) Xeromyces
bisporus.   O, NaCl, pH 4.0; O, NaCl,
pH 6.5;   , glucose/fructose, pH 4.0;
     , glucose/fructose, pH 6.5; Δ, gly-
cerol, pH 4.0; Δ, glycerol, pH 6.5
(Pitt and Hocking, 1977).

difficult, partly because the same effect may have a different
basis in different organisms and partly because where a specific
effect is demonstrated on some characteristic in addition to
growth it may be impossible to distinguish cause and effect.   The
object here is to report what is known and to comment upon possible lines of investigation that might throw light on the
causes of specific effects.

If an organism has a requirement for a specific solute in high concentrations, any substitution by another will not support growth and a specific solute effect will result. The classical example of this is the salt requirement of extreme halophilic bacteria. Salt is essential to maintain the structural integrity of halphile envelopes (Brown, 1964), which carry an excess of protein carboxyl groups provided by aspartic and glutamic acids. These groups are neutralized by cations contributed by salt in the aqueous environment, giving stability and rigidity to structures that would otherwise disaggregate with resultant lysis of the membranes. The rapidly developing picture of the halophilic salt requirement has been reviewed in detail by Brown (1976), who commented that "the instability of the cell envelope reflects a very special solute requirement that can be described as 'water relations' only by severely stretching definitions and concepts." No requirements for such high concentrations of specific solutes have been reported for other types of microorganisms. For instance, xerophilic yeasts that grow to a_w levels approaching 0.6 have no such absolute requirement (Brown, 1976).

If the influence of high solute concentrations on microbial cells were osmotic in origin, inhibition at reduced a_w levels might be lower with permeating solutes such as glycerol, than with relatively impermeant solutes that could produce plasmolysis. This might explain the data of Fig. 1, but, as no bacteria have been reported to be impermeable to glycerol, it cannot explain the results obtained with S.aureus and some other cocci shown in Fig. 3. However, the level of polyol (including glycerol) accumulation in osmophilic yeasts indicates that some restrictions on glycerol permeation exist in these organisms (Brown, 1974).

A specific solute effect upon the rate of respiration would be expected to be reflected in the rate of growth of the organism. Indeed, glucose oxidation rates of washed cells of S.oranienburg incubated at low a_w levels in solutions of glycerol, glucose, sucrose, and salts showed the same order of inhibition as was

found in the growth rate measurements of Fig. 1 (Christian, 1955b),
with glycerol causing no inhibition of respiration over the a_w
range tested. Prior (1978) obtained very similar results when
examining the influence of salt, sucrose, and glycerol on growth
and respiration of *Pseudomonas fluorescens*. However, comparable
experiments (Christian, unpublished) with *S.aureus* also showed
less respiratory inhibition by glycerol than by salt, although
the reverse was found with growth (Fig. 3).

When cells are subjected to reduce a_w, they experience an
osmotic stress that, if the bulk solute in the environment per-
meates the cell readily, is short-lived. If the solute is
relatively impermeable, the protoplast contracts and remains
shrunken until the solute diffuses in, some component of the
external environment is actively accumulated, or a solute is
synthesized within the cell, resulting in some level of cell
rehydration. In either case, the cell contents have become
more concentrated and in most cases of different solute composition. The intracellular solute pools thus formed are usually
dominated by one or two solutes. As intracellular solute accumulation appears to be prerequisites for growth, it can occur
in the absence of protein synthesis provided that appropriate
solute or solute precursors and an energy source are available
(Christian and Hall, 1972). Thus similar information on solute
pools can often be obtained in both growth and respiration experiments.

The solutes known so far to be accumulated by particular microorganisms during growth at reduced a_w levels are of restricted
types. In glucose-salts media, nonhalophilic gram-negative bacteria concentrate potassium, presumably neutralized largely by
organic anions derived from glucose (Epstein and Schultz, 1965).
Nonhalophilic bacteria concentrate from complex media proline,
glutamic acid, or γ-aminobutyric acid (Measures, 1975). In both
S.oranienburg and *S.aureus,* potassium is the dominant internal

solute at levels of a_w in the top two-thirds of the growth range. At lower a_w values, amino acids predominate (Christian and Hall, 1972; Christian and Waltho, 1964). Moderately halophilic bacteria also contain substantial amounts of both potassium and amino acids, while in extreme halphiles the dominant solute, present in very high concentrations, is potassium chloride (Christian and Waltho, 1962). In osmophilic yeasts, polyols (including glycerol and arabitol) are accumulated to high levels (Brown, 1974).

Unfortunately few of the data in the references cited above give comparative information derived when different solutes were used to control a_w, let alone when specific solute effects were evident on growth. Intracellular concentrations of sodium and potassium in *S.aureus* obtained after growth in several solutes are shown in Table III. No specific solute effects were observed on growth rate. It was concluded (Christian and Waltho, 1964) that water content was a function of a_w, not of the solute used to control it. Cell sodium remained constant at all three a_w values when sodium chloride had not been added to the medium, and hence it depended primarily on the external salt concentration, not the a_w value. In contrast, cell potassium was influenced both by the a_w value and by high concentrations of potassium chloride in the medium. Two intracellular solute compartments were postulated, one containing the specifically accumulated solute (potassium) and the other containing solutes equilibrating nonspecifically with the environment.

With *S.oranienburg* there are differential effects of salt and sucrose on growth at intermediate levels of a_w in complex (Fig. 4) but not in minimal media (Fig. 1). Nutritional studies in NaCl-adjusted media showed that proline was a prerequisite for stimulation of growth rate by other amino acids and vitamins at reduced a_w (Christian, 1955a). Growth rates in media adjusted to $a_w = 0.97$ by sodium chloride or sucrose with various nutrients are listed in Table IV.

TABLE III. Water, Sodium, and Potassium Contents of *Staphylo-coccus aureus* grown in Media[a] Adjusted to Several a_w with NaCl, KCl, or Sucrose[b]

a_w value	Adjusting solute	Cell water (g/g dry wt)	Sodium (mM)	Potassium (mM)
0.993	nil[a]	1.66	100	671
0.97	NaCl	1.35	382	837
0.97	Sucrose	1.37	84	841
0.92	NaCl	0.86	930	1020
0.92	KCl	0.90	78	1922

[a] *Basal medium was brain + heart infusion broth.*

[b] *From Christian and Waltho (1964).*

TABLE IV. Growth Rate of *Salmonella oranienburg* (Generations/hr) in Various Media Adjusted to a_w = 0.97 by Addition of NaCl or Sucrose[a]

Growth medium	Adjusting solute	
	NaCl	Sucrose
Glucose + salts (G.S.)	0.10	0.20
G.S. + proline (P)	0.27	0.45
G.S. + P + methionine	0.52	0.50
G.S. + 5 amino acids (AA)[b]	0.58	
G.S. + vitamin-free casaminio acids	0.61	
G.S. + 5 AA + 8 vitamins	0.65	
Casamino acids-yeast extract-casitone	0.76	0.54
Nutrient broth	1.00	0.56
Brain-heart broth (quarter strength)	1.08	0.63

[a] *From Christian (1955a; and unpublished).*

[b] *Proline, methionine, glutamic acid, serine, histidine.*

Addition of proline to glucose salts media doubled growth
rate in both solutes. Subsequent addition of methionine doubled
the rate in salt, but provided little stimulation in sucrose.
Further supplements and substitution of richer media, which again
doubled the rate in salt-adjusted media, increased the growth
rate in sucrose by only one-quarter. Thus cells in sucrose can
make little use of supplementary nutrients beyond proline. It
is not a specific effect on respiration, as QO_2 values at a_w =
0.97 with glucose, proline, and glucose + proline as substrates
are similar in salt and sucrose (Christian and Waltho, 1966).
Indeed addition of methionine and other amino acids to glucose
+ proline does not stimulate respiration further. Whether sucrose
interferes with the uptake of these nutrients or with their
subsequent metabolism is not known.

The tolerance of *S.oranienburg* to glycerol compared with that
to salt in a simple medium (Fig. 1) is also seen in a complex
medium (Fig. 3). Experiments similar to those discussed above
have been performed using salt and glycerol as adjusting solutes.
In glycerol, proline did not play the basic role in respiration
and growth that has been demonstrated in salt and sucrose-adjus-
ted media (Christian, unpublished). The effects of added proline
and methionine were additive, not synergistic as before. It
appears therefore that penetration of the cell membrane by glycerol removes the need to accumulate proline for restoration of
cell turgor, so that the mechanism of regulation of intracellular
a_w is apparently different.

These data suggest that the changes immediately following
transfer of growing cells to a lower a_w value may throw light
on the influence of different solutes. Information on such
changes is incomplete, but as a baseline, changes in respiration
and growth rate of exponentially growing *S.oranienburg* that
follow addition of NaCl, sucrose, or glycerol to glucose-salts
medium may be of interest (Fig. 8).

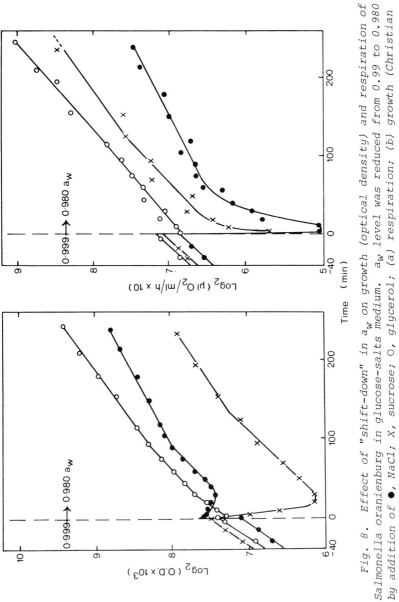

Fig. 8. Effect of "shift-down" in a_w on growth (optical density) and respiration of Salmonella oranienburg in glucose-salts medium. a_w level was reduced from 0.99 to 0.980 by addition of ●, NaCl; X, sucrose; O, glycerol; (a) respiration; (b) growth (Christian and Stewart, unpublished).

In all cases, respiration and optical density decreased initially, most in sucrose and least in glycerol. Optical density decrease resulted from cell shrinkage. Respiration recovered more rapidly than growth, the ultimate rate being achieved immediately in glycerol but not for about 2 hr in NaCl. Three phases of growth could be identified in all three solutes. The rates in the second ("fast") growth phase of 80-120 min duration were identical in all solutes and were presumably a consequence solely of a_w. The first ("adjustment") phase and the third ("final") growth phase were strongly influenced by solute.

Accelerated uptake of solutes such as potassium and proline occur predominantly during the first phase, but also in the second. The second phase is not simply one of increasing optical density resulting from the hydration that accompanies this accumulation, since it is prevented by chloramphenicol while accumulation is not.

"Shift-down" experiments of this nature can provide comparisons of the simultaneous effect of various solutes on respiration, growth, solute accumulation, and synthesis. They show promise of demonstrating the basis of solute specificity in microorganisms amenable to this type of experimentation.

VI. CONCLUSIONS

It is not intended to give the impression that the a_w concept is inadequate for describing the water requirements of microorganisms. Instances where the a_w interpretation fits experimental observations closely have been mostly ignored here. In view of the differences, e.g., in ionic strength, viscosity, and molecular size, that exist in properties of the concentrated solutions that have been discussed, it is surprising that some of

the differences in the response of nonhalophilic bacteria at the
lower a_w levels are not larger. What is more puzzling is that
the differences described, e.g., between salt and glycerol, are
not consistent among nonhalophilic species. In this particular
instance, a pointer to the explanation may lie in the very small
range of lower limits of a_w in glycerol-adjusted media (Fig. 3).
Brown (1974) reported that glycerol equilibrated in a nonosmophilic
yeast at about the same concentration as in the suspending solution with an upper internal level of about 1 M. If a number of
bacterial species have similar internal concentration limits,
their growth might be inhibited over a relatively narrow range
of glycerol concentrations or a_w levels.

Although the salt and sugar tolerances of most food-related
microorganisms are known, the increasing interest in intermediate-
moisture foods is focusing attention on other humectants such as
glycerol and glycols. These are among the solutes most likely to
cause specific solute effects. More knowledge is therefore needed
of the microbiological consequences of their use in controlling
a_w and of the physiological bases for specific solute effects.

Several possible approaches to the investigation of the
specific solute responses are suggested by data reviewed above.
All would involve the use of microorganisms lacking any major
specific requirement for the a_w-controlling solutes to be tested.

While respiration measurements on nongrowing cells suspended
in various solutes have not yielded specific solute effects that
correlate with those observed on growth, respiration studies on
growing cells show promise. These might reveal whether, for
example, a higher level of respiration is required to support a
particular rate of growth in one solute than another. The
existence of differing requirements of maintenance energy might
be demonstrated.

Decrease of the a_w of a growing culture by solute addition
yields information about the adjustment the cell must make to

grow in the new environment. So does transfer from one medium to another of the same low a_w but controlled by a different solute. Respiration, accumulation, synthesis, growth, and division can be measured. The succession of growth rates depicted in Fig. 8 has yet to be explained.

The interference of sucrose with nutrient uptake or utilization by salmonellae at intermediate-a_w levels (Table IV) warrants further investigation. A similar situation appears to exist with certain *Bacillus* spp. (Fig. 5).

Finally, studies are needed to define which of the solutes accumulated, especially by nonhalphilic bacteria, are "compatible solutes" (Aitken *et al.*, 1970). These are solutes in the intracellular pools that at high concentrations cause little inhibition of the cellular enzymes. Potassium chloride in halphilic bacteria and polyols in osmophilic yeasts have this property (Brown, 1976).

REFERENCES

Aitken, D. M., Wicken, A. J., and Brown, A. D. (1970). *Biochem. J. 116*, 125.

Alemohammad, M. M., and Knowles, C. J. (1974). *J. Gen. Microbiol. 82*, 125.

Anand, J. C., and Brown, A. D. (1968). *J. Gen. Microbiol. 52*, 205.

Baird-Parker, A. C., and Freame, B. (1967). *J. Appl. Bacteriol. 30*, 420.

Beauchat, L. R. (1974). *Appl. Microbiol. 27*, 1075.

Brown, A. D. (1964). *Bacteriol. Rev. 28*, 296.

Brown, A. D. (1974). *J. Bacteriol. 118*, 769.

Brown, A. D. (1976). *Bacteriol. Rev. 40*, 803.

Brown, H. J., and Gibbons, N. E. (1955). *Can. J. Microbiol. 1*, 486.

Charlang, G. W., and Horowitz, N. H. (1971). *Proc. Nat. Acad. Sci. 68*, 260.

Christian, J. H. B. (1955a). *Aust. J. Biol. Sci. 8*, 75.

Christian, J. H. B. (1955b). *Aust. J. Biol. Sci. 8*, 490.

Christian, J. H. B. (1956). Ph. D. dissertation. Univ. of Cambridge.

Christian, J. H. B., and Hall, J. M. (1972). *J. Gen. Microbiol. 70*, 497.

Christian, J. H. B., and Scott, W. J. (1953). *Aust. J. Biol. Sci. 6*, 565.

Christian, J. H. B., and Waltho, J. A. (1962). *Biochim. Biophys. Acta 65*, 506.

Christian, J. H. B., and Waltho, J. A. (1964). *J. Gen. Microbiol. 35*, 205.

Christian, J. H. B., and Waltho, J. A. (1966). *J. Gen. Microbiol. 43*, 345.

Epstein, W., and Schultz, S. G. (1965). *J. Gen. Physiol. 49*, 221.

Flannery, W. L., Doetsch, R. N., and Hanson, P. A. (1952). *J. Bacteriol. 64*, 713.

Gould, G. W. (1964). *Proc. 4th Int. Symp. Food Microbiol.*, p. 17.

Horner, K. J., and Anagnostopolous, G. D. (1973). *J. Appl. Bacteriol. 36*, 713.

Ishida, Y. (1970). *Bull. Japan Soc. Sci. Fisheries 36*, 481.

Jakobsen, M., and Murrell, W. G. (1977a). *J. Appl. Bacteriol. 43*, 239.

Jakobsen, J., and Murrell, W. G. (1977b). *In* "Spore Research, 1976" (A. N. Barker *et al.*, eds.), Vol. 2, p. 819. Academic Press, London.

Jakobsen, M., Filtenborg, O., and Bramsnaes, F. (1972). *Lebens.-Will. Technol. 5*, 159.

Kang, C. K., Woodburn, M., Pagenkopf, A., and Cheney, R. (1969). *Appl. Microbiol. 18*, 798.

Koser, S. A., Breslove, B. B., and Dorfman, A. (1941). *J. Infect. Dis. 69*, 114.

Krömer, K., and Krumbholz, G. (1932). *Arch. Mikrobiol. 3*, 384.

Lodder, J., and Kreger-van Rij, N. J. W. (1952). "The Yeasts, a Taxonomic Study." North-Holland Publ. Co., Amsterdam.

MacLeod, R. A., and Onofrey, E. (1956). *J. Bacteriol. 71*, 661.

Marshall, B. J., and Scott, W. J. (1958). *Aust. J. Biol. Sci. 11*, 171.

Marshall, B. J., Ohye, D. F., and Christian, J. H. B. (1971). *Appl. Microbiol. 21*, 363.

Measures, J. (1975). *Nature (London) 257*, 398.

Ohye, D. F., and Christian, J. H. B. (1966). *Proc. 5th Intl. Symp. Food Microbiol., Moscow*, p. 217.

Ōnishi, H. (1963). *Advan. Food Res. 12*, 53.

Pitt, J. I. (1975). *In* "Water Relations of Foods" (R. B. Duckworth, ed.). p. 273. Academic Press, London.

Pitt, J. I., and Hocking, A. D. (1977). *J. Gen. Microbiol. 101*, 35.

Pitt, J. I., and Hocking, A. D. (1978). In press.

Prior, B. A. (1978). *J. Appl. Bacteriol. 44*, 97.

Scott, W. J. (1953). *Aust. J. Biol. Sci. 6*, 549.

Scott, W. J. (1957). *Advan. Food Res. 7*, 83.

Scott, W. J. (1962). *Proc. Low Microbiol. Symp. Campbell Soup Co., New Jersey*, p. 89.

von Schelhorn, M. (1950). *Lebensm.-Unters.-Forsch. 91*, 117.

MICROBIOLOGY OF MEAT AND MEAT PRODUCTS
IN HIGH- AND INTERMEDIATE-MOISTURE RANGES

L. Leistner
W. Rödel
K. Krispien

I. INTRODUCTION

Scott (1953, 1957) gave fresh impetus to the field of water activity (a_w). Some of his results and ideas still govern our concept of the a_w of meat and meat products in relation to micro-organisms. On the other hand, considerable progress has been made in the last decade. The prerequisite was developments in instrumentation. As soon as reliable instruments for measuring the a_w of products and surfaces became available, knowledge of the a_w of meat and meat products increased. First the water activities of "normal" meat and a large variety of meat products were established. The changes in a_w during the ripening of meat and the processing of meat products were investigated. Recently also the a_w of surfaces of meat carcasses and meat products has attracted attention. Furthermore, a_w limits are increasingly used in food and feed regulations. Initial steps were made to influence and optimize the a_w of meat and meat products in order to improve shelf life. Concepts for new foods (IMF, SSP) have been obtained from a better understanding of a_w and other "hurdles" in foods. In addition, urgent problems, such as the necessary reduction of nitrate and nitrate addition to meat products, as well as future challenges, such as energy conservation

855

in food processing, may be better understood in terms of a_w and moisture limitation.

In earlier papers (Leistner and Rodel, 1975b, 1976a,b), we discussed the current knowledge of the impact of a_w of meat and meat products on microorganisms and their growth. In this chapter, this subject is summarized and emphasis is placed on results and concepts that have emerged recently.

II. HURDLE EFFECT AND FOOD PRESERVATION

The primary objective of traditional and newly developed food preservation processes is the inhibition or inactivation of microorganisms. Table I lists these processes and the parameters that govern them. This table indicates that more processes than parameters are applied in food preservation, and that most processes are based on several parameters or hurdles (Leistner, 1978). In these processes one or two main hurdles are involved. Minor or additional hurdles are, however, needed to accomplish the expected microbial stability achieved by a particular process. Of the processes used for the preservation of meat products a_w is a main hurdle in drying, curing, salting, freezing, and freeze-drying. Furthermore, for some types of fermented sausages, for oriental meat products with high sugar content, as well as for traditional and newly developed intermediate-moisture meats, a_w is a main hurdle too. For other preservation processes, such as chilling, smoking, heating, or canning of meat and meat products, a_w is often an additional hurdle.

Since most processes used in food preservation are based on several hurdles, most processed foods have several inherent hurdles that accomplish the desired microbial stability of the product. Figure 1 illustrates the hurdle effect in foods, using seven examples. Example (1) is a food containing six hurdles, which the microorganisms present cannot overcome. Therefore, this

TABLE I. Processes used in food preservation and parameters or hurdles they are based on (Leistner, 1978)

Parameters \ Processes	Heating	Chilling	Freezing	Freeze drying	Drying	Curing	Salting	Sugar addition	Acidification	Fermentation	Smoking	Oxygen removal	IMF (f)	Radiation
F(a)	X(c)	*	*	*	*	*	o	*	o	*	*	*	*	o
t(b)	*(d)	X	X	o	o	*	*	o	*	*	*	*	*	o
a_w	*	*	X	X	X	X	X	X	o	X	*	o	X	o
pH	*	*	o	o	*	*	*	X	X	X	*	*	*	o
Eh	*	*	*	*	o	*	*	*	*	*	*	X	*	o
Preservatives	*	*	o	o	*	X	*	*	*	X	X	*	X	o
Competitive flora	o(e)	o	o	o	o	*	o	o	*	X	o	*	*	o
Radiation	*	o	o	o	o	o	o	o	o	o	o	o	o	X

(a) High temperature;   (b) Low temperature;   (c) Main hurdle;   (d) Additional hurdle;   (e) Generally not important for this process;   (f) Intermediate moisture foods.

(a) High temperature; (b) Low temperature; (c) Main hurdle; (d) Additional hurdle; (e) Generally not important for this process; (f) Intermediate moisture foods.

product has sufficient microbial stability. In this example, all hurdles have the same height, i.e., the same intensity; however, this is only theoretically possible. A more likely situation is presented in example (2). This product could, for example, be a raw ham. The microbial stability of this product is based on five hurdles of different intensity. The main hurdles are the water activity and preservatives (nitrite and smoke in raw ham), and additional hurdles are the storage temperature, the pH, and the redox potential. These five hurdles are sufficient to inhibit the usual types and numbers of organisms associated with such a product. Example (3) represents the same product, but in a superior hygienic condition, i.e., only a few microorganisms are

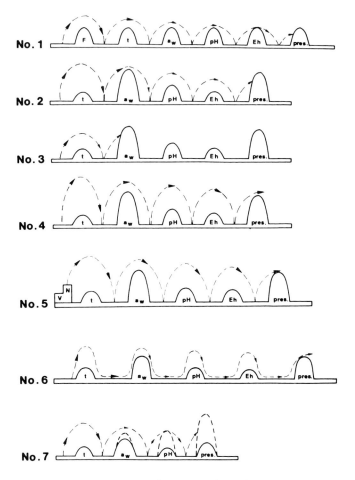

Fig. 1. Seven examples illustrating the hurdle effect on
which the microbial stability of foods is based. F, high temperature; t, low temperature; a_w, water activity; pH, acidity; E_h,
redox potential; pres., preservatives; V, vitamins; N, nutrients.

present at the start. Therefore, in this raw ham only two hurdles
would be enough. On the other hand, in example (4), due to bad
hygienic conditions, too many undesirable organisms are initially
present. Therefore, the same hurdles inherent in this product
cannot prevent spoilage. Example (5) is a food superior in nutrients and vitamins. Therefore, even for the usual types and
numbers of organisms the same hurdles as in examples (2)-(4) are

not sufficient. However, the stability of foods is also related to storage time. For instance, in a canned meat product the inherent hurdles could be overcome with time [example (6)]. Whether this is due to the phenomenon that surviving organisms can surmount the hurdles better if more time is available if it is due to the dormancy of spores, or if there is a decline of some hurdles with time (e.g., caused by a depletion of nitrite), needs more study. There is some indication that the product of the hurdles rather than their number determines the microbial stability of a food. Example (7) illustrates the synergistic effect that hurdles in a food might have on each other. The synergistic effect of hurdles deserves further research.

The hurdle effect (Leistner and Rödel, 1976b; Leistner, 1977a, 1978) is of fundamental importance for food preservation, since the hurdle concept may govern microbial food poisoning as well as spoilage and fermentation of foods. Table I and Fig. 1 illustrate the hurdle concept and the long-known fact (Stille, 1948a,b; Scott, 1957; Wodzinski and Frazier, 1960, 1971a,b; Riemann, 1963; Iandolo *et al.*, 1964; Peterson *et al.*, 1964; Segner *et al.*, 1966; Genigeorgis and Sadler, 1966; Ohye and Christian, 1967; Ohye *et al.*, 1967; Baird-Parker and Freame, 1967; Marland, 1967; McLean *et al.*, 1968; Alford and Palumbo, 1969; Emodi and Lechowich, 1969a, Genigeorgis *et al.*, 1971a,b; Segner *et al.*, 1974; Matches and Liston, 1972; Roberts and Ingram, 1973; Tomčov *et al.*, 1974; Roberts *et al.*, 1976; Leistner and Rödel, 1976b) that complex interactions of a_w, pH, temperature, preservatives, etc. are significant in the bacterial stability of foods. Further research is needed, which should lead to a quantitative understanding of the hurdle concept. As a first step, research should be carried out on the inhibition or inactivation of important genera and species of bacteria, yeasts, and molds, by a particular hurdle (e.g., a_w, pH, or temperature). The effect of a hurdle should be studied under otherwise optimal conditions for growth, metabolic activity, and survival of the organisms under investigation. A body of relevant

data is already available. The next step is to study the interaction of hurdles in the inhbiition or inactivation of microbes. Information of this kind has been published (Roberts and Ingram,
1973; Roberts *et al.*, 1976). A further research step could be to
computerize quantitative data on the hurdles in foods. Once this
has been accomplished, some of the current food microbiology
could become obsolete. The computerized data could be used to
predict what genera and species of microorganisms might occur in
a food of which only the physical and chemical hurdles are
measured. The initial microbial population would be a variable
more difficult to predict; however, it could be compensated by a
predetermined margin of safety.

Using the hurdle concept, based on quantitative data, not only
could the present processes used in food preservation become better understood, but they could be optimized and new processes
could be developed.

III. a_w AND MICROORGANISMS

The a_w of foods influences the multiplication and metabolic
activity (including toxin production) of microorganisms, as well
as their resistance and survival. This is true not only for organisms that cause spoilage and food poisoning, but also for those
which are desirable for the fermentation of certain foods, such
as salami.

A. *Multiplication of Microorganisms*

Microbial spoilage, food poisoning and fermentation take place
if the a_w of the substrate is favorable for the multiplication and
metabolic activity of the microorganisms involved. The a_w influences the lag growth, and stationary phases, as well as the death
rate of a culture (Troller and Christian, 1978). Most organisms

associated with foods grow best at a relatively high a_w; only a few require a low a_w for growth. Hence, if a_w decreases, then fewer genera of microorganisms are able to multiply on or in a food.

In Table II the minimal a_w required for the growth of a number of genera of bacteria, yeasts, and molds recovered from foods is listed. This table has been compiled from the data reported by numerous authors (Stille, 1948a; Snow, 1949; Burcik, 1950; Bullock and Tallentire, 1952; Christian and Scott, 1953; Scott, 1953, 1957; Williams and Purnell, 1953; Christian, 1955a, Beers, 1957; Clayson and Blood, 1957; Ohnishi, 1957; Wodzinski and Frazier, 1960, 1961b; Christian and Waltho, 1962, 1964; Lanigan, 1963; Riemann, 1963; Blanche Koelensmid and von Rhee, 1964; Gough and Alford, 1965; Hobbs, 1965; Kim, 1965; Matz, 1965; Brownlie, 1966; Limsong and Frazier, 1966; Segner et al., 1966, 1971; Baird-Parker and Freame, 1967; Ohye and Christian, 1967; Ohye et al., 1967; Kushner, 1968; McLean et al., 1968; Pitt and Christian, 1968; Pivnick and Thatcher, 1968; Ayerst, 1969; Emodi and Lechowich, 1969; Kang et al., 1969; Mossel, 1969; Bem and Leistner, 1970; Strong et al., 1970; Troller, 1971, 1972; Jakobsen et al., 1972; Rödel et al., 1973; Tomčov et al., 1974; Beuchat, 1974; Pitt, 1975; Leistner and Rödel, 1975b, 1976a,b; Jakobsen and Murrell, 1977; Troller and Christian, 1978; Christian, 1979; Ruegg and Blanc, 1979).

Compared with earlier versions (Leistner and Rödel, 1975b, 1976a,b), this table has become more complex. As additional data emerge, more isolates of a particular genus will be found that somewhat differ in their minimal a_w from other representatives of the same genus. Therefore, the data in Table II should be considered as a guideline and not as a rule. However, some principles apparently are well established, which are significant from a practical point of view. In general, molds are more tolerant of a decreased a_w than yeasts, and yeasts are more tolerant than bacteria. Of the bacteria Pseudomonadaceae and Enterobacteriaceae

TABLE II. Minimal a_w for multiplication of microorganisms associated with foods

a_w	Bacteria	Yeasts	Molds
0.98	Clostridium (1), Pseudomonas*	-	-
0.97	Clostridium (2), Pseudomonas*	-	-
0.96	Flavobacterium, Klebsiella, Lactobacillus*, Proteus*, Pseudomonas*, Shigella	-	-
0.95	Alcaligenes, Bacillus, Citrobacter, Clostridium (3), Enterobacter, Escherichia, Propionibacterium, Proteus, Pseudomonas, Salmonella, Serratia, Vibrio	-	
0.94	Bacillus*, Clostridium (4), Lactobacillus*, Microbacterium, Pediococcus, Streptococcus*, Vibrio	-	Stachybotrys
0.93	Bacillus (5), Micrococcus*, Lactobacillus*, Streptococcus	-	Botrytis, Mucor, Rhizopus
0.92	-	Pichia, Rhodotorula, Saccharomyces*	
0.91	Corynebacterium, Streptococcus		-
0.90	Bacillus (6), Lactobacillus*, Micrococcus, Pediococcus, Staphylococcus (7), Vibrio*	Hansenula, Saccharomyces	-
0.88	-	Candida, Debaryomyces, Hanseniaspora, Torupolis	Cladosporium

a_w			
0.87	—	Debaryomyces*	—
0.86	Micrococcus*, Staphylococcus (8), Vibrio (9)	—	—
0.84	—	—	Alternaria, Aspergillus*, Paecilomyces
0.83	Staphylococcus*	Debaryomyces*	Penicillium*
0.81	—	Saccharomyces*	Penicillium
0.79	—	—	Penicillium*
0.78	—	—	Aspergillus Emericella
0.75	Halobacterium, Halococcus	—	Aspergillus*, Wallemia
0.70	—	—	Aspergillus*, Chrysosporium,
0.62	—	Saccharomyces*	Eurotium*
0.61	—	—	Monascus (Xeromyces)

*Some isolates; (1) Clostridium botulinum type C; (2) C. botulinum type E and some isolates of C. perfringens; (3) C. botulinum type A and B and C. perfringens; (4) Some isolates of C. botulinum type B; (5) Some isolates of Bacillus stearothermophilus; (6) B. subtilis under certain conditions; (7) Staphylococcus aureus anaerobic; (8) S. aureus aerobic; (9) Some isolates of Vibrio costicolus.

require $a_w \geq 0.95$. For *Clostridium* spp., the minimum a_w is
generally 0.95-0.94. Some isolates of *Bacillus* might grow at
$a_w = 0.93 - 0.90$. Several Lactobacillaceae still multiply at
0.90 and the a_w limits of Micrococcaceae range from 0.93 to 0.83.

Most data in Table II were obtained by investigating the a_w
tolerance of the studied organisms under otherwise optimal growth
conditions using an artificial substrate with the a_w adjusted with
additives or by desorption. The a_w tolerance of the organisms
will decrease if other hurdles (such as suboptimal temperature,
pH, E_h, preservatives, competitive organisms) are present, as is
often the case in foods. Therefore, from a practical point of
view, the data in Table II contain a certain margin of safety.

It also has to be remembered that the minimum a_w requirement
is apparently influenced by the method of preparation of the food
system. Foods prepared to a given a_w by desorption or adsorption
processes exhibit different characteristics for growth of microorganisms (Labuza *et al.*, 1972a,b; Plitman *et al.*, 1973. A food
prepared by desorption has a higher moixture content at a given
a_w than a food prepared by adsorption and organisms have a higher
a_w requirement for growth if the food is prepared by adsorption
(Labuza *et al.*, 1972a, Acott and Labuza, 1975). The adsorption
process, i.e., the food is first freeze-dried and then adjusted
to the desired a_w by adsorption of water, could improve the
stability of foods, but it is more expensive than the desorption
process. Meat products are generally prepared by desorption.

The minimum a_w requirements of microorganisms may vary somewhat depending upon the solute used to adjust the a_w of the substrate (Baird-Parker and Freame, 1967; Kang *et al.*, 1969,
Kushner, 1971; Marshall *et al.*, 1971; Jakobsen *et al.*, 1972;
Beuchat, 1974; Jakobsen and Murrell, 1977; Christian, 1979).
Most of the data listed in Table II were obtained by adjusting
the a_w with NaCl, and therefore they apply to meat products. If
instead of sodium chloride as a humectant sucrose is used, then
the differences in the limiting a_w values for growth are generally

small.   However, in the presence of glycerol some organisms,

such as *Clostridium botulinum* types A, B, and E. (Baird-Parker

and Freame, 1967), *C. perfringens* (Kim, 1965; Kang *et al.*, 1969),

Bacillus cereus (Jakobsen *et al.*, 1972; Jakobsen and Murrell,

1977), *Salmonella oranienburg* (Christian, 1955b; Marshall *et al.*,

1971), and *Vibrio parahaemolyticus* (Beuchat, 1974) grow, germi-

nate, and sporulate at a lower a_w than if NaCl and sucrose are

the humectants.   This has to be remembered if glycerol is used as

a humectant in the preparation of intermediate-moisture meats.

   If only salts and sugars are taken into consideration as hu-

mectants, then of the food-poisoning bacteria, representatives of

the genus *Shigella* are inhibited by $a_w < 0.96$ (Tomčov *et al.*,

1974) and most other gram-negative rods by $a_w < 0.95$.   The latter

was demonstrated for *Salmonella* (Christian and Scott, 1953;

Christian, 1955a; Hansen and Riemann, 1962; Blanche Koelensmid

and von Rhee, 1964; Tomčov *et al.*, 1974), and enteropathogenic

Escherichia coli (Tomčov *et al.*, 1974).   However sometimes growth

of *Salmonella* spp. occurs at levels of a_w between 0.95 and 0.94

(Christian and Scott, 1953).   Clayson and Blood (1957), using sur-

face equilibration techniques, even reported $a_w = 0.92$ as minimal

for *Salmonella* growth.   *Vibrio parahaemolyticus* is a food-poisoning

organism that requires at least some NaCl for its growth.   Most

isolates of *V. parahaemolyticus* have, in the presence of NaCl, a

limiting a_w of 0.95 (Beuchat, 1974), but some require $a_w < 0.94$

for inhibition (Rödel *et al.*, 1973).   Another food-poisoning

organism, *B. cereus*, is also inhibited by $a_w < 0.95$ (Jakobsen *et*

al., 1972).   Thus this a_w limit is not exceeded by the majority

of all known food-borne pathogens.   The growth of *Clostridium*

botulinum type C is inhibited for $a_w < 0.98$ (Segner *et al.*, 1971),

of type E for $a_w < 0.97$ (Segner *et al.*, 1966; Ohye and Christian,

1967; Pivnick and Thatcher, 1968; Emodi and Lechowich, 1969a,b),

and of types A and B (Scott, 1955b; Greenberg *et al.*, 1959; Ander-

ton, 1963; Ohye and Christian, 1967) and *C. perfringens* (Kim,

1965) for $a_w < 0.95$.   However, under extreme conditions a somewhat

lower limiting a_w is observed, e.g., growth of *C. botulinum*
type B occurred at 0.94 after 17 days at 40°C (Ohye and Christian,
1967). On the other hand, Baird-Parker and Freame (1967) found
minimal a_w levels for growth of *C. botulinum* types A, B, and E,
using NaCl as humectant, to be a_w 0.96, 0.96, and 0.98, respectively.

Undoubtedly, of the food-poisoning bacteria, the lowest a_w
is tolerated by *Staphylococcus aureus*. Under anaerobic conditions this organism is reported to be inhibited for $a_w < 0.91$
(Scott, 1953) or 0.90 (Troller and Christian, 1978), but aerobically $a_w < 0.86$ (Scott, 1953; Christian and Waltho, 1962, 1964;
Matz, 1965) is minimal. Labuza *et al.* (1972a) observed growth of
S. aureus at $a_w = 0.84$ in a glycerol-containing intermediate-
moisture meat prepared by desorption. Troller (1976), confirming
an observation of Hill (1973), reported that one out of seven
strains of *S. aureus* tested did grow at $a_w = 0.83$ in a protein
hydrolyzate medium supplemented with beef extract and adjusted
with NaCl.

B. *Metabolic Activity of Microorganisms*

Toxin production is the metabolic activity of microorganisms
that cause public health-related concerns, and toxins produced in
foods by *Clostridium botulinum, Staphylococcus aureus*, and toxinogenic molds are of particular interest. Table III contains some
data on the minimal a_w for toxin production by the organisms
mentioned. However, Table III, even more than Table II, has to
be regarded as incomplete and is subject to change when additional
data become available.

Ohye and Christian (1967) observed that where growth of
C. botulinum occurs, toxin formation is also demonstrable, and
conversely, in the absence of growth, cultures of this organism
are nontoxic. Minimal a_w levels supporting growth and toxin formation by types A, B, and E were 0.95, 0.94, and 0.97, respectively.

TABLE III. Minimal a_w for toxin production by microorganisms

Clostridium botulinum	
Type C	0.98
Type E	0.97
Type A	0.95
Type B	0.94
Staphylococcus aureus	
Enterotoxin C	0.94
Enterotoxin B	0.90
Enterotoxin A	0.87
Mycotoxins	
Penitrem A	0.94
Citrinin	0.90
Patulin	0.90
PR-Toxin	0.90
Cyclopiazonic acid	0.87
Roquefortine	0.87
Citreoviridin	0.86
Ochratoxin A	0.85
Griseofulvin	0.85
'S-toxin'	0.84
Aflatoxin	0.83
Penicillic acid	0.80

On the other hand, Riemann (1967) observed that in brain-heart infusion broth, toxin formation by *C. botulinum* type E ceased at a
higher a_w than growth. Growth and toxin production by *C. botulinum*
type C were reported to be inhibited for a_w < 0.98 (Segner *et al.*,
1971).

Several authors (Genigeorgis and Sadler, 1966; McLean *et al.*,
1968; Genigeorgis *et al.*, 1969, 1971b; Hojvat and Jackson, 1969;
Markus and Silverman, 1970; Troller, 1971, 1972, 1973a; Troller
and Stinson, 1978) related both staphylococcal growth and enterotoxin production to NaCl concentration or a_w. It was found that
the production of enterotoxin B (Genigeorgis and Sadler, 1966;

McLean *et al.*, 1968; Genigeorgis *et al.*, 1969; Troller, 1971) and
enterotoxin C (Genigeorgis *et al.*, 1971b) by *S. aureus* ceases for
a_w < 0.94. Hojvat and Jackson (1969) noted that with increasing
NaCl concentrations growth and enterotoxin B production were both
suppressed. However, the effect was greater on toxin formation
than on growth. Troller (1971, 1972) observed that the production
of enterotoxin B was more affected by a decreasing a_w than enterotoxin A. Enterotoxin A was detected in a NaCl-containing culture
medium at a_w = 0.90 (Troller, 1972), and very low levels of enterotoxin B were recovered at a_w = 0.90 from a culture medium adjusted with a salt mixture (Troller, 1973a). Troller and Stinson
(1975) could detect no enterotoxin A in glycerol-treated shrimp
slurries at a_w = 0.93, the same level where growth of *S. aureus*
had decreased considerably. In contrast, in glycerol-treated
potato doughs, enterotoxin A production occurred at an a_w level
of 0.93 but not at 0.88 even with 10^7 viable cells/ml recovered.
These authors detected enterotoxin B at a_w = 0.93 in shrimp slurries and at 0.97 in potato doughs. It has to be remembered that
S. aureus grows at lower a_w if media are adjusted with salts or
sucrose than with glycerol (Marshall *et al.*, 1971). Lotter and
Leistner (1978) studied the minimal a_w for growth and enterotoxin
A formation by *S. aureus* using a salt mixture broth, containing
NaCl, KCl, and Na_2SO_4 in a ratio of 5:3:2 moles. Within 7 days
at 30°C both of the investigated *Staphylococcus* strains grew and
formed enterotoxin A minimally between a_w = 0.864 and 0.867. In
contrast, at 25°C, the minimal a_w for enterotoxin A production
was increased to between 0.870 and 0.887 after a 2 week incubation.
Hill (1973) noted enterotoxin A formation at a_w = 0.86 in pork
containing 3% dextrose and 100 ppm sodium nitrite, and held at
35°C. Apparently no data are so far available on the limiting a_w
for the formation of enterotoxin D. It should be mentioned that
with *S. aureus* not only could growth occur without detectable enterotoxin formation, but also enterotoxin production without
growth. This could be of significance in intermediate-moisture

foods where contaminating staphylococci may be present but do
not multiply (Pawsey and Davies, 1976).

At least some mycotoxins are produced by molds at low a_w.
Xerophilic molds ($a_w < 0.85$) so far reported to be mycotoxic belong primarily to *Aspergillus* (including *Eurotium* and *Emericella)*
and *Penicillium genera*, but *Paecilomyces* and *Alternaria* are of
interest too.   Information on the a_w requirements for mycotoxin
formation and stability is still scarce (Leistner and Pitt, 1977).
Mycotoxins are frequently produced at low growth temperatures
(Jarvis, 1971), and sometimes are also favored by low a_w (Bacon
et al., 1973).   In peanuts aflatoxins are produced within 14 days
at 25°C at a_w = 0.86 (Sanders *et al.*, 1968).   Diener and Davis
(1970) observed 84% as limiting RH (relative humidity) for aflatoxin production in peanuts after long storage (84 days) at 20°C.
a_w levels were not reported, but were probably in the 0.83-0.84
range.   Growth of *Aspergillus flavus* on laboratory media is optimal at a_w = 0.98 and minimal at a_w = 0.78 (Ayerst, 1969).   Northolt
et al. (1976) noted that the growth of *A. parasiticus* was more
rapid at a given a_w if glycerol, rather than a mixture of salts,
was used to adjust the a_w of the medium.   In the glycerol medium,
growth was observed at a_w = 0.82, but not at 0.79, and toxin production occurred at a_w = 0.87, but not at 0.82.   In a study by
Bacon *et al.* (1973) with poultry feed inoculation with *A. ochraceus*
it was observed that penicillic acid began to accumulate at
a_w = 0.80 at 22° and 30°C, whereas ochratoxin A began to accumulate at a_w = 0.85 at 30°C; maximum production of penicillic acid
occurred at 22°C and a_w = 0.90 and of ochratoxin A at 30°C and
a_w = 0.95.   Lötzsch and Trapper (1979) studied 16 *Penicillium*
isolates that were known to produce 10 mycotoxins.   The results
of this work are shown in Table III.   Northolt and van Egmond
(1978) detected the production of penicillic acid, patulin,
ochratoxin A, and aflatoxin B$_1$ at 0.99, 0.95, 0.87 and 0.83,
respectively.   From the data available, it is possible to conclude
that the formation of several mycotoxins is possible in the a_w

range of intermediate-moisture foods. However, more work is
needed to establish the minimal a_w levels for the production of
the many mycotoxins known. The same mycotoxin is often synthesized by different species, sometimes belonging even to different
genera (e.g., *Aspergillus* or *Penicillium*), which could have quite
variable a_w requirements. Therefore, the limiting a_w for a particular mycotoxin should be studied using the most xerophilic
isolates that synthesize this toxin. Generally toxinogenic molds
grow at lower a_w than these at which toxin formation is detectable.
However, the difference in measurable growth and toxin formation
depends on the sensitivity of the methods used. If, for example,
the detection limits for patulin and aflatoxin B_1 are 20 and 0.3
ppb, respectively, then the lower limiting a_w of aflatoxin, compared with patulin, is influenced by this fact (Lötzsch and Trapper, 1978).

C. *Resistance of Microorganisms*

The influence of a_w on the thermal resistance of microorganisms
is an important concern. The heat resistance of bacteria is known
to depend on a number of factors, which include the strain of the
organism, its growth conditions and age, the number of cells
heated, and the physical and chemical composition of the heating
menstruum. The influence of a_w on the heat resistance of different
groups of microorganisms has been reviewed recently (Troller,
1973b; Corry, 1975, 1976; Pawsey and Davies, 1976; Troller and
Christian, 1978). Generally, the heat resistance of bacterial
spores increases with decreasing a_w and is maximal in the a_w range
0.2-0.4. An increase in thermal resistance as the a_w of the suspending medium decreases has also been demonstrated with vegetative
microorganisms, and sucrose proved more protective than glycerol.
However, sometimes the heat resistance of bacteria is lower in the
presence of high solute concentrations than in dilute systems, because the solute itself or substances deriving from it during

heating contribute to the heat damage of the cells. Further it has been shown that for given solutes a certain concentration of each gives maximal heat protection whereas other concentrations result in greater sensitivity (Horner and Anagnostopoulos, 1975). In general, the survival after heating cannot be directly related to the a_w of the heating medium, but its magnitude depends to a great extent on the kinds of solutes present (Baird-Parker et al., 1970). Nevertheless, it is always safe to assume that the heat resistance of microorganisms could increase in foods with decreasing a_w, and that somewhat higher heat treatments are needed for intermediate-moisture foods (IMF) than for high-moisture foods in order to achieve the expected microbial stability. Hsieh et al. (1975) recommended prepasteurizing of high a_w, intermediate-moisture food components before combination with humectants to assure maximal destruction of microorganisms. An increase in the heat resistance with decreasing a_w has been demonstrated for food-poisoning organisms, such as C. botulinum (Murrel and Scott, 1966), S. aureus (Kadan et al., 1963; Calhoun and Frazier, 1966; Hsieh et al., 1975; Bean and Roberts, 1975), and Salmonella spp. (Gibson, 1973; Corry, 1974), as well as for meat spoilage organisms such as Streptococcus faecium and S. faecalis (Vrchlabský and Leistner, 1970b).

D. Survival of Microorganisms

In general, osmotic stress may only temporarily incapacitate and prevent the growth of microbes for a relatively short period of time (Rose, 1976). On the other hand it is also a fact that microorganisms that are unable to grow gradually die. Therefore, if the water activity of a food is below the minimal or above the maximal a_w for growth of the organisms present, their number slowly decreases. Generally, gram-positive bacteria are more resistant than gram-negative bacteria to inactivation by low a_w (Fry, 1954; Hale, 1957; Davis and Bateman, 1970; McDade and Hall, 1963, 1964;

Uzelac, 1976). The survival of microorganisms increases with decreasing a_w (Scott, 1958, 1960; Davis and Bateman, 1960; Sinskey
et al., 1967; Riemann, 1968; Liu *et al.*, 1969; Carlson and
Snoeyenbos, 1970; Christian and Stewart, 1973; Uzelac, 1976).
Thus their survival during long storage is most likely in freeze-
dried foods (a_w < 0.10), followed by low-moisture foods (a_w < 0.60).
However, it is less likely in intermediate-moisture foods
(a_w < 0.90) and should be least likely in high-moisture foods
(a_w > 0.90), if the microorganisms present cannot grow in the
available substrate. Of course, mere storage could not produce a
consistent elimination in any of these product groups (Mossel,
1963; Brockmann, 1970; Pawsey and Davies, 1976; Corry, 1976). It
should be remembered that survival rates in glycerol may be somewhat different from those occurring in salts and sugars (Rose,
1976; Pawsey and Davies, 1976).

Sometimes it has been observed that microorganisms survive
best not at a_w = 0.00 but in the a_w range 0.05-0.25 (Watts, 1945;
Higginbottom, 1953; Scott, 1958; Christian and Stewart, 1973),
which is to say that a small amount of moisture (0.5-1.5%) favors
survival (Fry, 1966; Davies, 1968). Protective substances, such
as certain proteins or sugars, increase survival during storage
(Greaves, 1960; Sinskey *et al.*, 1967; Strange and Cox, 1976),
while others, such as Maillard reaction compounds (Scott, 1960;
Marshall *et al.*, 1974), free fatty acids, or preservatives, may
enhance destruction. Also low pH, oxygen, and possibly light increase destruction of microorganisms in foods and other dried substrates during storage (Scott, 1958, 1960; Fry, 1966; Cox, 1966;
Sinskey *et al.*, 1967; Christian and Stewart, 1973; Strange and
Cox, 1976). Furthermore, storage of dried foods at elevated temperatures results in a higher death rate of organisms, although
this may result in unacceptable changes in organoleptic or functional properties in some products (Banwart and Ayres, 1956;
Fry, 1966; Licari and Potter, 1970; Strange and Cox, 1976).

Not only drying but also freezing results in a decrease of a_w

in a food (Leistner and Rödel, 1975a, 1976a; Rödel and Krispien, 1977). Gram-positive bacteria seem more resistant than gram-negative bacteria to inactivation during frozen storage (Major *et al.*, 1955; Kereluk and Gunderson, 1959). Generally, the lower the storage temperature (and the less it fluctuates), the lower the rates of inactivation during frozen storage (Ingram and MacKey, 1976). With a number of species of bacteria it was observed that at storage temperatures of -70°C or below, death rates are low or zero. At temperatures between -60° and 0°C the survival of most species decreased with time, the rate of the decrease depending on the species, the storage temperature, the freezing menstruum, and sometimes the cell concentration (Mazur, 1966). The superior survival of microorganisms at low storage temperatures might be caused by the reduced a_w of substrates, including foods, held at very low freezing temperatures. In both frozen and dried foods low a_w apparently favors the survival of microorganisms. If this is true, it could be advantageous, from a microbiological point of view, to store frozen foods first at about -10°C (i.e., at a relatively high a_w), to reduce the number of undesirable organisms (e.g., Salmonellae), and subsequently at very low freezing temperatures (i.e., -30°C) for optimal quality preservation (Leistner, 1977a). Similarly, if dried foods are stored for some time with a relatively high a_w and subsequently with a low a_w, an improved decline of undesirable microorganisms is likely.

For food-poisoning organisms, the survival in relation to a_w is of particular interest; however, this is also true for agents of food-borne parasitic infections. For instance, the viability of trichinae could be a function of a_w, since for an inactivation of these parasites drying, salting, or freezing of meat and meat products is recommended. Therefore, in our laboratories the survival of *Trichinella spiralis* in fermented sausages and raw hams in relation to a_w was studied. In model experiments muscle tissue of rats containing larvae of trichinae was incorporated into fermented sausages (Lötzsch and Rödel, 1974), and in major experi-

TABLE IV. Inactivation of Trichinella spiralis in meat products by water activity (Lötzsch and Leistner, 1977)

Investigated products	Inactivating a_w	
	experimentally observed	considered safe
Fermented sausages (four varieties)	0.949 to 0.931	≤ 0.90
Raw hams (five varieties)	0.948 to 0.904	≤ 0.87

ments hogs were infected with trichinae and slaughtered after 9-15 months (Lötzsch and Leistner, 1977). Meat products originating from these hogs were formulated, processed, and stored under commercial conditions. Initially, the fermented sausages contained 200-800 and the raw hams 400-700 trichina larvae per gram. The viability of the larvae in the sausages was followed for 32 days and in the hams for 73 days using mice and a chemical digestion method. The investigation indicated that trichinae lost their invasive ability, depending on the type of product and the applied technology, in fermented sausages (four varieties tested) between 6 and 14 days (a_w = 0.949-0.931, and in raw hams (five varieties tested) between 10 and 57 days (a_w = 0.948-0.904. Therefore, trichinae were inactivated in fermented sausages and raw hams within a relatively small a_w range and it appeared that the viability of trichinae can be judged by the a_w of these products. Taking into account quite a large margin of safety, it was concluded that meat products that are consumed raw can be considered harmless with respect to trichinae if the a_w of fermented sausages is $\leq$ 0.90 and of raw hams $\leq$ 0.87 (Table IV). Further experiments will determine whether these large margins of safety are indeed necessary.

IV. a_w OF MEAT SURFACES

Due to contamination during slaughter many more microorganisms are present on the surface than in the interior of carcass meat. During storage of meat at approximately $0°C$ nearly all microbial growth occurs on the surface and gram-negative bacteria are predominant (Ingram and Dainty, 1971). Bacteria are confined to the surface of meat during the logarithmic phase of growth (Dainty *et al.*, 1975). When proteolytic organisms approach their maximum cell density, extracellular proteases are secreted, which break down connective tissues between muscle fibers, allowing proteolytic bacteria to penetrate the meat. At $<5°C$ this penetration rate is very slow (Gill and Penny, 1977).

The multiplication of organisms on the surface of meat is governed by temperature, pH, oxygen, and a_w. It is common knowledge that a dry (low a_w) surface improves the shelf life of carcass meat, including poultry. For this reason the surface of carcasses is sometimes dried after the slaughter process or during storage; however, weight losses and discoloration of the meat surface are limiting factors for this procedure. Structural changes in slaughter houses as well as in the handling of meat have contributed to the fact that carcass meat now often enters processing plants with a high bacterial load (Leistner, 1977b). If this happens then the large investments made in the last decade with respect to improved refrigeration, replacement of other materials by stainless steel, and emphasis on general sanitary measures are jeopardized. Therefore, the number of microorganisms on the surface of carcass meat is still one of the major problems in meat hygiene. To tackle this problem "hygiene grades" for carcass meat have been suggested and a simple sampling instrument was introduced for a quick estimation of bacterial counts on the surface (Leistner *et al.*, 1977, 1979). With this instrument a certain surface area is washed thoroughly and the turbidity of the rinsing fluid is measured.

As an additional step for improving the hygienic quality of carcass meat, an investigation of the a_w of meat surfaces was initiated. Little is known of the actual a_w of carcass surfaces even though the significance of this factor was stressed by several authors (Heiss, 1933, 1938; Schwartz and Loeser, 1934,35; Kaess and Schwartz, 1934; Scott and Vickery, 1939; Scott, 1955; Hicks *et al.*, 1957; Csiszár and Biró, 1967; Karmas, 1975; Wirth *et al.*, 1976). Hicks *et al.* (1955) removed thin surface layers from lamb carcasses and measured a_w by using a gravimetric method. These authors observed 84-85% relative humidity and surface a_w's in the range 0.933-0.862, after 10 days storage at $3°C$, depending on the type of tissue. However, this method is cumbersome and could lead to erroneous results, since the investigated surface layers must be detached from the material they cover. Reliable instruments for measuring surface a_w *in situ* are required for an exact investigation of the a_w of meat surfaces.

Most methods used for measuring the a_w of food products are based on the determination of the equilibrium relative humidity of the sample, i.e., the sample is put into a closed container and, after an equilibrium between the a_w of the sample and the relative humidity of the enclosed air is established, the relative humidity is measured with different devices. The equilibrium relative humidity divided by 100 is assumed to be the a_w of the sample. If a closed capsule is put on a meat surface the relative humidity of the enclosed air does not solely depend on the a_w of the surface but soon reflects the a_w of tissue below this surface. The surface a_w of many foods is in a dynamic equilibrium between the relative humidity of the surrounding air and the a_w of the foodstuff covered by this surface. The moisture flows from the food to the air or vice versa, depending on the relative humidity of the air. If the surface a_w is in a dynamic state it should not be disturbed during measurement. Sensors used to determine surface a_w therefore are open systems, i.e., a tube (e.g., height 25 mm) that is put on the meat surface. Inside this open tube are two sensors that measure the relative humidity in different distances from the surface (e.g., 4 and 8 mm above the surface). About 30 min after the tube is placed on the surface, the relative humidities in the different distances from the surface are recorded. Since the distance of the sensors from the surface is known, by mathematical extrapolation the instrument attached to the sensors indicates the relative humidity in the distance of zero, and this corresponds to the a_w of the surface.

In our laboratories, Krispien (1978) tested such an instrument and developed a new model. Furthermore, he determined the surface a_w of different tissues of beef and hog carcasses during storage in a slaughter house as well as under controlled climatic conditions in a precision incubator. The a_w of meat and meat products is little influenced by the temperature in the range from 25°C to the freezing point of meat or of a particular meat product (Krispien and Rödel, 1976; Rödel and Krispien, 1977). However, the a_w of several concentrated salt solutions or slurries used for calibration of the sensors is temperature dependent, and also the characteristic of the sensors might change with temperature. Therefore, if the surface a_w, e.g., of meat stored at 5°C, is measured, the calibration of the sensors should be done at the same temperature, and for calibration preferably solutions of NaCl should be used, because they are little influenced by the temperature in the range from 25° to 0°C (Krispien and Rödel, 1976).

Krispien (1978) also attempted to adjust the a_w and pH of meat surfaces simultaneously by using a spray of an aqueous solution of 11.7% citric acid and 21.6% NaCl. If this spray is applied a few hours after slaughter it maintains the a_w and pH of the surface of the carcasses for several days in a range that is unfavorable for the growth of gram-negative bacteria (Table V).

It is reasonable to expect that such a treatment would improve the shelf life of carcasses since the minimal temperature of the spoilage organisms could increase and their lag phase would be prolonged (Leistner, 1977b). By decreasing the a_w and pH on the

TABLE 5. *Surface a_w and pH of carcass meat stored 3 days at 1.5°C ± 0.5°C and 89.0% ± 2.0% relative humidity and 0.1 - 0.3 m/ sec air velocity (Krispien, 1978)*

Tissue/species	Without treatment a_w	pH	With treatment (a) a_w	pH
Muscle (b)/beef	0.986	5.8	0.959	4.7
Muscle (c)/beef	0.961	6.5	0.936	4.8
Fat (d) /beef	0.912	7.9	0.925	4.3
Skin /pork	0.926	7.9	0.921	4.4

(a) Sprayed 5 hours after slaughter with an aqueous solution of 11.7% citric acid and 21.6% sodium chloride; (b) cut surface; (c) uncut muscle; (d) with facia.

surface of meat during storage and so increasing these hurdles for the microorganisms present, the storage temperature is no longer solely decisive for the shelf life of meat carcasses. This is of interest in industrialized countries with respect to their current meat handling practices as well as for developing countries with their lack of refrigeration. It might be of more concern in the future with respect to energy conservation (Leistner, 1978).

In the last decade artificial casings are increasingly used for meat products such as Bologna-type sausage, liver, and blood sausage which are not penetrated by water vapor. Therefore, these products do not lose weight during storage and have a low surface a_w, which considerably improves their shelf life. While sausages stuffed in natural casings even under refrigeration quickly develop a sticky surface due to microbial growth, the same products in artificial casings impermeable to water can be stored for some weeks under the same conditions without this defect. For other foods, like air-chilled poultry, cheese, or bakery goods, the surface a_w is probably of similar importance as for red meats, and investigations could be initiated with the instruments now available.

V. HIGH-MOISTURE MEATS

Table VI lists the minimal and maximal as well as the modal a_w of fresh meat and representative meat products. Detailed data on the a_w of German meat products have been published by Rödel (1975). The table indicates that meat and most meat products are in the a_w range 0.99–0.90. Hence they are high-moisture foods. However, some meat products, especially air-dried fermented sausages, unboned hams, and dried beef, are intermediate-moisture foods, since their a_w < 0.90.

The lean portion of fresh meat has a_w > 0.99. Several authors (Heiss, 1933; Schwartz and Loeser, 1934/35; Scott, 1936, 1937, 1938; Scott and Vickery, 1939) are assuming that 0.993 is the a_w of fresh meat. Leistner *et al.* (1971a) found little dif-

Product	Minimal	Maximal	Modal
Fresh meat	0.98	0.99	0.99
Bologna sausage	0.87 (a)	0.98	0.97
Liver sausage	0.95	0.97	0.96
Blood sausage	0.86 (b)	0.97	0.96
Raw ham	0.80 (c)	0.96	0.92
Dried beef (d)	0.80	0.94	0.90
Fermented sausage	0.65 (e)	0.96 (f)	0.91

(a) Tiroler; (b) Speckwurst; (c) Country cured ham; (d) Bündner Fleisch; (e) Hard sausage; (f) Frische Mettwurst.

ference between the a_w of porcine and bovine muscles; the average a_w's were 0.9904 and 0.9912, respectively. Also there is no statistically significant correlation between the water-holding capacity and the a_w either in pork or in beef muscle (Vrchlabský and Leistner, 1970a). The a_w of minced meat is <0.99, if it contains some fat or is dried during mincing and storage (Leistner *et al.*, 1971a).

Meat products have a lower a_w than fresh meat and therefore a better shelf life. The factors most influential in changing a_w during processing of meat products are (1) the withdrawal or addition of water, (2) the addition of NaCl or other salts, and (3) the content or addition of fat. Fat only influences the a_w of meats indirectly because it contains little water and therefore favors the concentration of water-soluble substances such as added salts, in the water of the lean portion of the product. As Table VI indicates, meat products with a generally high a_w, e.g., 0.97, are Bologna-type sausages. Of these products more than 100 varieties are distinguishable in Europe, ranging from Bierschinken, Lyoner, Gelbwurst, Göttinger, and Tiroler, to frankfurters and luncheon meat. Most of these items are highly perishable and have to be kept under refrigeration. However, there are a few exceptions, such as genuine Italian Mortadella, which is produced with relatively high amounts of salt, fat, and milk powder, but little addition of water. This results in an a_w in the range 0.96-0.93. Some slowly dried Bologna-type sausages (Tiroler, Polnische) have even lower a_w (0.95-0.87). Some of these products are in the a_w range of intermediate-moisture foods. The a_w in liver and blood sausages, generally is lower than in Bologna-type sausages, however, these products also are easily perishable. The a_w of liver sausage, due to a generally high fat content, is in the range 0.97-0.95. This also is true for the a_w of blood sausages. These are cooked products that contain blood and often pieces of cured meat and tongue as well as fat. In some blood sausages, i.e., German Speckwurst, not only many fat pieces are

included but they are also dried to a low a_w, e.g., 0.93-0.86,
and thus have an exceptional shelf life.

In order to secure the required stability and safety, the a_w
of meat products that are not heated during processing and are
consumed raw has to be much lower than in the heated products
mentioned previously. However, here again there is much variation according to the type of product. The a_w of raw ham is in
the range 0.96-0.80; a low a_w is found, e.g., in country cured
hams, especially on the surface, while products with a short
ripening time have a much higher a_w. In the production of raw
hams a dry cure and/or cover brine is used for the withdrawal of
water and the penetration of curing salts. The curing process,
at a temperature of $6°-8°C$, takes, depending on the product, two
weeks to several months and is often succeeded by drying and
smoking processes, which lead to a further a_w decrease of the
hams. It should be emphasized that in raw hams, as in some other
meat products, the a_w of the lean and the fat portion do not come
to equilibrium either in short- or in long-ripened products. In
Table VII data pertaining to commercial samples of Südtiroler
Bauernspeck, a speciality in the north of Italy, are listed. In
every product the a_w of the lean meat was considerably higher than
that of the fat (Leistner and Rödel, unpublished). Dried beef,
e.g., Bündner Fleisch, typical for Switzerland, also has a low a_w,
but again the actual a_w of these products depends on the drying
process. To avoid rancidity, dried beef is void of fat.

In addition to pH, a_w is of utmost importance for the processing of fermented sausages. This product is preserved by a withdrawal of water, i.e., a decrease of the a_w, and a lowering of the
pH, caused by lactic acid bacteria and sometimes furthered by the
addition of glucono-δ-lactone. In products processed and consumed within a week, the a_w is just below 0.95, but products with
a ripening time of several months have an a_w that is much lower,
e.g., the a_w of Hungarian salami is about 0.83. The microbial
stability of short-ripened products depends primarily on a low pH

TABLE VII. *Water activity of the lean and fat portions of six samples raw ham with mild or strong flavor, i.e., a shorter or longer ripening time*

Flavour	Lean			Fat			A_w difference lean to fat
	A_w	H_2O	NaCl	A_w	H_2O	NaCl	
Mild	0.931	51.1	4.1	0.883	3.0	0.4	0.048
	0.928	46.9	3.8	0.885	1.8	0.3	0.043
	0.909	53.6	6.5	0.875	3.0	0.6	0.034
Strong	0.863	45.7	7.5	0.840	4.4	0.8	0.023
	0.857	40.9	7.2	0.815	1.6	0.4	0.042
	0.824	38.4	7.8	0.734	1.5	0.5	0.090

(4.8-5.2), while in fermented sausages with a long ripening time and an agreeable flavor, the a_w is low (0.90-0.80) and the pH is relatively high (5.3-5.8). Process control is particularly important in the manufacturing of fermented sausages, since a controlled withdrawal of water is essential for the desired microbial fermentation. Hence the proper flavor, appearance, and stability of the product can be attained. An investigation of the a_w of fermented sausages (Leistner *et al.*, 1971b), and the a_w changes during the ripening process (Rödel, 1973; Klettner and Rödel, 1978) improved the process control for these products. Rodel (1973) has recommended keeping the relative humidity in the ripening chamber low until the temperature of the sausages (initially about $0°C$) is equal to that of the environment, i.e., $20-25°C$. This takes about 6 hr and is followed by the drying period in which the relative humidity of the ripening room is adjusted to four units below the a_w of the sausage meat (e.g., if the a_w of the sausage reaches 0.96 the relative humidity in the ripening chamber should be 92%). Manufacturers of fermented sausage in Germany are now aware of the importance of the water activity and use the a_w-Wert-Messer (Rödel *et al.*, 1975) to control the a_w of their products and follow the recommended guidelines (Wirth *et al.*, 1976) for the ripening process.

The a_w is important not only for meat but also for fat. Krispien *et al.* (1979) investigated animal fat and observed a_w in the range 0.996-0.242. Fresh fat removed from warm pork or beef carcasses has, depending on the location in the animal, an a_w of 0.996-0.991; the water and NaCl content amount to 27-10% and 0.2-0.1%, respectively. During storage the a_w of fat quickly decreases due to evaporation of water and a corresponding increase in salt content. If fat, e.g., back fat of hogs, is salted and smoked or even melted to lard, then the a_w further decreases (Table VIII). The fat portion of raw ham consistently has a lower a_w than the lean (Table VII). Fat with a high a_w undergoes quick microbial spoilage. Because water is easy removable from fat by drying and rehydration is difficult, the fat obtained from pork or beef carcasses often has a low a_w and therefore is rarely spoiled by microorganisms. Also, the a_w of fat

TABLE VIII. A_w of porcine and bovine fat (Krispien, Rödel and Leistner, 1979)

Species & location	Treatment	A_w	H_2O %	NaCl %
Pork, back fat	Fresh	0.991	9.6	0.1
	2 days chilled	0.982	6.5	0.1
	5 days dryed	0.884	4.8	0.8
	Salted & smoked	0.724	2.4	1.2
	Lard	0.280	0.6	1.7
Beef, tallow	Fresh	0.993	13.6	0.2
	2 days chilled	0.984	4.2	0.2

remaining on the surface of carcasses decreases during storage
more than the a_w of meat surfaces (Table V), and therefore is
less suitable as a growth medium.

Meat manufacturers are well aware of the fact that some meat
products are more perishable than others and therefore need better
refrigeration during storage. The meats that are less perishable
are in general characterized by a low a_w. However, the multiplication of microorganisms on or in meats does not solely depend on
the a_w, but on other factors too, such as temperature, pH, E_h,
nitrite, and the competitive flora (Leistner, 1974). Even in
most instances the shelf life of a meat product is related to the
combined effect of several hurdles, which inhibit microbial
growth. Nevertheless, it is feasible to predict the shelf life
of meat products if only the a_w, pH, and temperature, which are
easy to measure in routine work, are taken into consideration.

According to a regulation of the Norwegian Health Department,
issued on 15 November 1969, meat and fish products not canned or
frozen, are considered easily perishable if the pH is above 4.5
and at the same time the a_w is higher than 0.90; these easily
perishable products have to be kept under refrigeration and must
be labeled with the date of packaging and the date to which they
are usable without risk. This Norwegian concept is too strict
to be applied to meats common in Germany.

In our laboratories the a_w of representative German meat
products as well as their usual shelf life was studied, and related to the a_w, pH, and temperature requirements reported in the
literature of food-poisoning and spoilage organisms. From this
study a concept derived for grouping meat products into three
categories based on the a_w and pH of the product (Rödel, 1975;
Rödel *et al.*, 1976). Each category demands an appropriate storage
temperature. This concept covers bacteria that cause spoilage as
well as those that cause food-poisoning, but not yeasts and molds,
which usually grow slower than bacteria, and may be inhibited by
chemical fungistats (Leistner *et al.*, 1975). Table IX indicates

TABLE IX. *Storage categories of meat products based on the a_W and the pH of the product, with recommended storage temperatures (Rödel, 1975; Leistner and Rödel, 1976a)*

Category	Criteria	Temperature of storage
Storable	$a_W \leqslant 0.95$ and pH $\leqslant 5.2$ or $a_W \leqslant 0.91$ or pH $\leqslant 5.0$	no refrigeration required
Perishable	$a_W \leqslant 0.95$ or pH $\leqslant 5.2$	$\leqslant +10°$
Easily perishable	$a_W > 0.95$ and pH > 5.2	$\leqslant +5°$

that according to the concept of Rödel (1975), easily perishable meat products have $a_W > 0.96$ and pH > 5.2 and must be stored at or below $+5°C$. The perishable meat products have either $a_W \leq 0.95$ or pH ≤ 5.2 and must be stored at or below $+10°C$. The storable meat products have $a_W \leq 0.95$ and pH ≤ 5.2 or $a_W \leq 0.91$ and pH ≤ 5.0; these products need no refrigeration, and their shelf life is often not limited by bacterial factors, but is subject to chemical or physical spoilage, especially rancidity and discoloration.

Concepts of this type, which are based on measurable hurdles important for food-poisoning and spoilage bacteria, and not on empirical experience alone, could probably be helpful for meat manufacturers as well as food inspection services, and would improve the quality of meats. The EEC regulation "Meat Products" of December 1976 has adopted Rodel's criteria of storable meat products for a definition of preserved meat products that are not canned. However, if the stability rests primarily on pH this regulation demands pH < 4.5, while according to Rödel's concept pH ≤ 5.0 is sufficient. It has to be remembered that there are practically no meat products with pH < 4.5 (Wirth et al., 1976).

Since a_W determinations and limits are increasingly introduced into quality control and regulations of foods, reliable in-

struments for measuring the a_w of meat products, as well as of
other foods, are necessary. In the last decade much progress
has been made in this respect, e.g., instruments have been developed for a determination of the a_w of products and surfaces.
The latter have been mentioned above. For measuring product a_w
under quality control conditions, the a_w-Wert-Messer (Lufft,
Stuttgart, West Germany) has proved suitable (Rödel *et al.*, 1975)
and is widely accepted. In this instrument (Leistner and Rödel,
1975b), a polyamide thread with a specific precision in the range
85-100% relative humidity is incorporated. This a_w-cup is rather
inexpensive, but comparable in accuracy with other instruments.
Rödel *et al.* (1978) recently described electronic a_w instruments
that are in use in our laboratories. We adapted a dewpoint hygrometer (EG&G, Cambridge Systems, Waltham, Massachusetts), to which
four sensors were attached. Another dewpoint hygrometer (Walz,
Effeltrich, West Germany) is also now available. Instruments
using immobilized salt solution sensors, based on lithium salts
or similar electrolytes, have been likewise improved. Examples
are the Hygroskop D, with six sensors, and the Hygroskop DT
(Rotronic, Zurich), the hygrometers type JAN-A and type DAL 02
(Novasina, Zurich), as well as the ISO instrument (ISO, Ottobrunn,
West Germany) with 22 Sina-Sensors (Fig. 2). The accuracy of the
immobilized salt solution and dewpoint instruments is $a_w = \pm 0.005$.
However, a strict temperature control ($25° \pm 0.1°C$) during measurement and a careful calibration of the sensors are essential (Rödel
et al., 1978). Stoloff (1978) concluded from a collaborative
study that measurements with instruments using immobilized salt
solution can be made with an accuracy and reproducibility of
$a_w = \pm 0.01$. A particular problem is posed by the a_w determination
of intermediate-moisture foods (IMF) and feeds (IMPF), since immobilized salt solution, dewpoint, and thread sensors are disturbed by humectants such as glycols and glycerol. This is not
true for Pope sensors and capacitive sensors. Therefore, the hygrometer type 400 D (General Eastern, Watertown, Connecticut) or

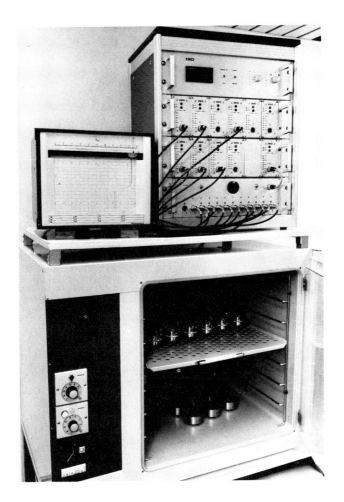

Fig. 2. ISO a_w instruments (ISO, Ottobrunn, West Germany), composed of amplifiers for SINA sensors. Below, in the precision temperature incubator set at 25° 0.2 C, are located 22 electrolytic SINA sensors (NOVASINA, Zürich, Switzerland). The ISO instrument is automatically called on by the recorder (Hartmann and Braun, Frankfurt). This design is suitable for measuring product a_w of 22 samples simultaneously.

the Evaporimeter Ep 1 (Servomed, Stockholm) are suitable for measuring IMF and IMPF (Rödel et al., 1978). Since the Evaporimeter was designed for determination of surface a_w, the sensor had to be adapted for product measurement. For Sina-sensors special filters are now available that prevent the interference of humectants, and so these sensors now have applicability for IMF and IMPF. Since the formulation of IMF is attracting increasing attention and because a_w limits for IMPF have been introduced into German feed regulation, suitable instruments for these products are in demand.

VI. FROZEN MEATS

The a_w of meat and meat products decrease with temperature from 25°C to the chilling and freezing range. Above the freezing point the decrease is only slight (∿0.00015/°C), however, below the freezing point the decrease is considerable (∿0.008/°C). This means that the a_w drop per 1°C below the freezing point, is 50 times what it is at above freezing point (Rödel and Krispien, 1977). As the temperature is reduced below the freezing point of a food and progressively more of the water is converted into ice, the concentration of dissolved solids rises in that portion of the water remaining unfrozen; hence the a_w falls correspondingly (Ingram and Mackey, 1976). The a_w of frozen foods can be calculated from the vapor pressure of ice divided by the saturation vapor pressure of water, and is known from several investigations (Moran, 1936; Storey and Stainsby, 1970; Wolf *et al.*, 1973; Fennema and Berny, 1974). Table X lists the a_w of frozen foods with an initial a_w > 0.990. All foods, including water, have at the same freezing temperature the same a_w, provided the temperature is below the freezing point of the food item (Table XI). At what temperature a particular food freezes depends on its a_w. Therefore, if the a_w of meat or meat products is measured at 25°C, it is possible to predict the freezing point of these foods by using Table X. On the other hand, the a_w at 25°C can be deduced by determining the freezing point of meat and meat products (Rödel and Krispien, 1977).

TABLE X. A_W *of foods, including meats, with an initial* a_W *above 0.990 at various freezing temperatures. Calculated data from Moran (1936), Storey and Stainsby (1970), Fennema and Berny (1974)*

Temp. ($^{\circ}$)	a_W	Temp. ($^{\circ}$)	a_W	Temp. ($^{\circ}$)	a_W
- 1	0.990	- 11	0.899	- 21	0.815
- 2	0.981	- 12	0.889	- 22	0.807
- 3	0.971	- 13	0.881	- 23	0.799
- 4	0.962	- 14	0.873	- 24	0.792
- 5	0.953	- 15	0.864	- 25	0.784
- 6	0.943	- 16	0.856	- 26	0.776
- 7	0.934	- 17	0.847	- 27	0.769
- 8	0.925	- 18	0.839	- 28	0.761
- 9	0.916	- 19	0.831	- 29	0.754
-10	0.907	- 20	0.823	- 30	0.746

TABLE XI. A_W *of meat and meat products at various freezing temperatures (Rödel and Krispien, 1977)*

Temperature $^{\circ}$C	a_W of product examples			
	Fresh meat	Bologna sausage	Liver sausage	Fermented sausage
+ 25	0.993	0.980	0.970	0.870
+ 5	essentially unchanged ↓	essentially unchanged ↓	essentially unchanged ↓	
± 0				
- 1				essentially unchanged
- 2	0.981			
- 3	0.971	0.971		
- 4	0.962	0.962	0.962	
- 5	0.953	0.953	0.953	
- 10	0.907	0.907	0.907	
- 15	0.864	0.864	0.864	0.864
- 20	0.823	0.823	0.823	0.823

Many microorganisms are inhibited in chilled or frozen foods solely by temperature. For instance, growth of *Staphylococcus aureus* is inhibited <7°C (Angelotti *et al.*, 1961; Riemann *et al.*, 1972) and inhibition of the growth of salmonellae occurs at <5°C (Hess, 1970). However, the growth of some microorganisms in frozen foods is probably not limited by temperature but by a_w (Scott, 1957; Leistner and Rödel, 1975a, 1976a). Representatives of the latter are psychrophilic gram-negative rods and numerous yeasts and molds. In general, of the microorganisms associated with foods, molds are more tolerant of decreased a_w than yeasts, and yeasts are more tolerant than bacteria (Table II). Therefore, the spoilage of frozen foods held at inadequate temperatures is most frequently caused by molds. If foods are stored at less than -18°C, all growth of microorganisms is inhibited. In Table XII the minimal a_w for growth of psychrophilic bacteria, as well as yeasts and molds, is correlated with the known minimal growth temperatures (Ingram, 1951; Schmidt-Lorenz, 1970) for these organisms. Even though the agreement between the minimal a_w and the a_w corresponding to the minimal growth temperature of these organisms is not perfect, a relationship is evident, particularly in the case of bacteria and yeasts (Leistner and Rödel, 1975a, 1976a). Indeed, for some microorganisms, the a_w of frozen foods might inhibit their multiplication rather than the temperature. If minimal a_w and minimal growth temperature for molds do not relate too well as for bacteria and yeasts, then this might be due to the fact that some xerophilic molds are mesophilic.

VII. INTERMEDIATE MOISTURE MEATS

Intermediate-moisture foods (IMF) should be sufficiently plastic to eat without further hydration or cooking and should be storable without refrigeration. The a_w range of IMF has been defined as 0.85-0.70 (Brockmann, 1970), 0.85-0.60 (Plitman *et al.*,

TABLE XII. *Correlation between minimal a_w and minimal temperature for growth of micro-organisms (Leistner and Rödel, 1976a)*

Organisms	Minimal a_w	Minimal temperature
Bacteria	0.90	$-10^{\circ}C$ (i.e. a_w 0.91)
Yeasts	0.88	$-12^{\circ}C$ (i.e. a_w 0.89)
Moulds	0.70	$-18^{\circ}C$ (i.e. a_w 0.84)

1973), 0.90-0.70 (Karel, 1973), and 0.90-0.60 (Collins *et al.*, 1972). From a microbiological point of view it is reasonable to divide foods into three categories: high-moisture foods (HMF), intermediate-moisture (IMF), and low-moiusture (LMF), with an a_w range of 1.0-0.90, 0.90-0.60, and 0.60-0.00, respectively (Leistner and Rödel, 1976b). For the spoilage of most HMF, microorganisms are decisive, while for IMF only certain types of organisms are relevant and in LMF microbial growth is impossible.

It is useful to differentiate between traditional or fabricated and new or newly developed IMF. Both product groups are in the a_w range 0.90-0.60. However, the "traditional IMF," such as certain meat and bakery products, as well as dried fruits and jams, are processed by the withdrawal of water and/or the addition of conventional legally approved additives (Salts, sugar, fat, etc.). On the other hand, "new IMF" are processed by using humectants such as glycerol, glycols, sorbitol, and antimycotics, such as sorbate, parabens, pimaricin, and sometimes antioxidants. These substances are not, at the present time, approved by the food laws of several countries (Lück, 1977). New IMF are processed either by desorption or adsorption of water or a combination of both processes (Brockmann, 1970; Kaplow, 1970; Bone, 1973; Lück, 1973; Labuza, 1974; Sloan *et al.*, 1976; Troller and Christian, 1978). The desorption process is less costly; however, microbial growth may be more rapid in desorbed IMF, than in those

prepared by adsorption, even when parity of a_w levels occurs
(Labuza, 1972b; Acott and Labuza, 1975). In Tables XIII and XIV

TABLE XIII. A_W *range and inherent hurdles of some traditional intermediate moisture foods (Leistner and Rödel, 1976b)*

Products	A_W range	Inherent hurdles
Jams and jellies	0.90 - 0.80	a_w, pH, Eh, F, pres.*
Meat products	0.90 - 0.65	a . pH, (Eh), t, pres.* c.m.**
Cake and pastry	0.90 - 0.60	a_w, (t), F, pres.*
Dried fruits	0.75 - 0.60	a_w, pH,(F), pres.*
Frozen foods	0.90 - 0.60	a_w, t

Preservatives (such as sorbic acid, nitrite);
**Competitive microflora (Lactobacillaceae, Streptococcaceae,
molds.*

TABLE XIV. A_W *and inherent hurdles of some newly-developed
intermediate moisture foods (Leistner and Rödel, 1976b)*

Products	A_W	Inherent hurdles
Sweet and sour pork (1)	0.85	a_w, pH, (Eh), F, pres.*
Chicken dish, ready-to-eat (2)	0.85	a_w, F, pres.*
Sliced and dried bologna (3)	0.85	a_w, Eh, F, pres.*
Hennican (4)	0.85	a_w, (pH), Eh, (F), pres.*
Diced carrots (5)	0.77	a_w, (F), pres.*

*(b) Brockmann (1970); (c) Luck (1973b); (d) Pavey (1972);
(e) Labuza (1974); (f) Kaplow (1970); * Preservatives (such as
sorbic acid, propylene glycol, glycerol, nitrite).*

for some traditional and new IMF the a_w range and the inherent hurdles are listed. To the traditional IMF certain fish products, which contain a high amount of sugar and salt (Suzuki, 1979), as well as Parmesan cheese (Rüegg and Blanc, 1979) should be added. Meat is frequently used for traditional and new IMF.

As Tables XIII and XIV indicate, the inhibition of microorganisms also in IMF does not solely depend on the a_w, but also on pH, E_h, F value (heating), t value (chilling), preservatives, and competitive microorganisms. Traditional and new IMF alike should be protected against spoilage and food-poisoning organisms by a combination of hurdles that ensure the necessary microbial stability of the products (Leistner and Rödel, 1976b). The number of these hurdles and their intensity depend not only on the types but also on the number of organisms present, as Fig. 1 illustrates. It is possible to inhibit a low number of organisms present in an IMF with fewer and much lower hurdles than are required for the inhibition of a large number of organisms of the same type. Therefore, the raw material used for the production of traditional and new IMF should contain low counts of organisms, especially of those types of bacteria, yeasts, and molds that tolerate a low a_w. The preparation, especially of new IMF, should be done by handling the raw materials under hygienic conditions and refrigeration to ensure a low initial count of a_w-tolerant organisms. Where feasible, the raw material used in the preparation of IMF should be heat-processed to inactivate undesirable organisms as well as enzymes that deteriorate the products. a_w-tolerant organisms are not heat resistant, and therefore a low F value, i.e., an internal temperature of about 85°C, is sufficient for elimination of these organisms in the raw materials. However, this temperature is required, because the heat resistance of spoilage and food-poisoning organisms, as has been mentioned above, may increase somewhat with decreasing a_w. For example, sucrose is more protective than glycerol. Since some organisms, including *Salmonella* spp. and *Staphylococcus aureus* may have their maximum heat resistance in the a_w range of IMF, it was suggested (Labuza, 1974; Hsieh *et al.*, 1975) that components such as meat and eggs, which are at high a_w should be heat-pasteurized prior to combination with the a_w-lowering agents and other dry components of IMF. This should ensure maximum heat inactivation of organisms by a relatively mild heat process.

The aim of IMF technology is to reduce the a_w of a food to a range in which most bacteria will no longer grow (Plitman *et al.*, 1973). In considering the data available and by taking into account that some species or strains of microorganisms are more tolerant of a lower a_w than other members of the genus, it may be concluded that in the a_w range of IMF (0.90-0.60) the bacteria, yeasts, and molds listed in Table XV are of potential significance. These bacteria are able to multiply in the a_w range 0.90-0.75 and yeasts and molds in the range 0.90-0.60. Most organisms listed in Table XV cause spoilage, some produce toxins (*Staphylococcus, Alternaria, Paecilomyces, Penicillium, Aspergillus, Emericella, Eurotium*), and a few species of the genus *Candida* may be pathogenic. On the other hand, well-adapted, nontoxinogenic strains of the genera *Pediococcus, Streptococcus, Lactobacillus, Staphylococcus, Debaryomyces, Saccharomyces, and Penicillium* might be desirable in IMF as competitive microflora. In some additional IMF these competitive organisms multiply and are metabolically active at remarkably low a_w levels. The microbial stability of traditional IMF allows one to store these products without refrigeration for a sufficient time. Perhaps much could be learned from the empirical hurdles built into traditional products for the protection of new IMF against spoilage and food-poisoning. Competitive microorganisms that serve as starter cultures are an effective hurdle in some traditional IMF, such as fermented sausages. The usefulness of starter cultures for new IMF should be explored, since the competitive microflora could support the microbial stability and in addition could render the taste and appearance of new IMF more similar to traditional products (Leistner and Rödel, 1976b).

TABLE XV. *Microorganisms of potential significance for inter-mediate moisture foods (Leistner and Rödel, 1976b)*

Bacteria	Yeasts	Moulds
Lactobacillus	Hansenula	Cladosporium
Streptococcus	Hanseniaspora	Alternaria
Pediococcus	Candida	Paecilomyces
Vibrio	Torulopsis	Penicillium
Micrococcus	Debaryomyces	Aspergillus
Staphylococcus	Saccharomyces	Emericella
Halobacterium		Wallemia
Halococcus		Eremascus
		Eurotium
		Chrysosporium
		Monascus

Preliminary results obtained in our laboratories indicate that selected strains of *Pediococcus halophilus* might be suitable as a starter culture for intermediate moisture meats (Holzapfel *et al.*, 1976).

Humectants, i.e., safe and effective agents for reducing a_w levels, are of the utmost importance for toxicological approval and consumer acceptance of IMF. NaCl and sucrose are commonly used as humectants. NaCl is more effective on a weight-for-weight basis but sucrose is more easily tolerated in high concentrations. If both are used in a well-balanced combination, the organoleptic properties of IMF are improved. However, new IMF with a desired low a_w need additional humectants, such as glycerol, propylene glycerol, or 1,3-butanediol. Glycerol, which often is used for

IMF, has a sweet taste, not familiar with meat products. Furthermore, it was recently reported that feeding a high level of dietary
glycerol to rats resulted in an accumulation of lipids in blood
serum and the liver (Narayan *et al.*, 1975, 1977). Apparently, a
humectant suitable in every respect for new IMF is still unknown.
From the microbiological point it has to be remembered that glycerol and perhaps other glycols cause specific solute effects. As
pointed out in previous chapters, glycerol influences the multiplication, metabolic activity, resistance, and also the survival
of microorganisms somewhat differently than salts and sugars.
Since the a_w of IMF is <0.90 the effect of glycerol on the growth
of food-poisoning bacteria such as *Clostridium botulinum, C.per-
fringens, Bacillus cereus, Salmonella spp.* and *Vibrio parahaemolyticus* is of little concern, because these organisms, even in the
presence of glycerol, will not multiply in the a_w range of IMF.
However, *Staphylococcus aureus* and toxinogenic molds may grow and
synthesize toxins in IMF (Table III). Labuza *et al.* (1972a) observed growth of *S. aureus* in a glycerol-containing intermediate-
moisture meat, prepared by desorption, at a_w = 0.84, while Marshall *et al.* (1971) reported a higher limiting a_w (0.89). Northolt
et al. (1976) noted more rapid growth of *Aspergillus parasiticus*
at a given a_w if glycerol, rather than a mixture of salts, was used
to adjust the a_w. More knowledge is needed of the microbiological
consequences of the use of glycerol or glycols in controlling the
a_w of IMF or IMPF. In appropriate studies, reliable instruments,
as described in a previous chapter, for measuring the a_w of substrates containing glycerol or glycols should be used, since contradictory results on the microbiological effect of these humectants may be caused by faulty methodology.

   In addition to a_w, pH and E_h also are of major importance for
the stability of IMF. If the pH is close to 5.0, microbiological
stability may be achieved with an a_w of about 0.90. Therefore,
the pH of IMF and IMPF should be as low as palatability permits.
A pH < 5.0 inhibits growth and toxin production by *S. aureus*

(Reiser and Weiss, 1969), but some representatives of the genus *Micrococcus* are inhibited only at pH < 4.5 (Lerche, 1957). For the inhibition of other a_w-tolerant bacteria (*Pediococcus, Streptococcus, Lactobacillus*) as well as for yeasts and molds, a much lower pH would be necessary. It should be mentioned that the growth of lactic, acetic, and butyric acid bacteria as well as the growth of yeasts and molds is favored in the pH range 6.0-3.0 (Böhringer, 1962). Only representatives of the genes *Micrococcus* are inhibited by a low redox potential since other a_w-tolerant bacteria are facultatively anaerobic. However, it should be remembered that *S. aureus* is under anaerobic conditions already inhibited by a_w < 0.91 (Scott, 1953) or 0.90 (Troller and Christian, 1978). Preliminary results of our laboratories indicate that *S. aureus* is inhibited in brain-heart infusion broth by 1.5% sodium citrate or 2.0% sodium tartrate (Lotter and Leistner, 1976); this finding may be applicable for IMF. Like *S. aureus* yeasts also tolerate a much lower a_w level aerobically rather than anaerobically. Most mold growth is prevented by a low E_h. Therefore, IMF and IMPF should be packaged in evacuated containers or pouches that are impermeable to oxygen. The enclosure of IMF in packages that are also moisture-proof would further prevent interchange of water between the food and the atmosphere. If mold growth occurs, the shelf life of IMF is not only jeopardized, but also the possibility arises, as has been proved in traditional IMF such as country-cured hams and fermented sausages, that mycotoxins are penetrating the products (Leistner and Pitt, 1977). In general, yeasts and especially molds are the major spoilage factors for traditional IMF. Yeasts cause spoilage in syrups, confectionery products, jams, and dried fruits (Walker and Ayres, 1970), while molds spoil meats, jams, cake, and pastry as well as dried fruits. Molds are also the main microbial problem for new IMF (Labuza, 1974). Since vacuum packaging often is not practical, antimycotic substances, such as sorbic, propionic, or benzoic acid, must be incorporated into IMF. For meat products potassium

sorbate is often employed (Lück, 1970, 1972; Leistner *et al.*,
1975), but pimaricin also proved effective (Moerman, 1966, 1967;
Hechelmann and Leistner, 1969). In new IMF humectants possessing
some degree of antimycotic activity, such as 1,3-butanediol,
propylene glycol, or glycerol, contribute to the inhibition of
molds (Labuza, 1974).

It should be possible to store IMF without refrigeration.
However, it is obvious that not only the chemical and physical
but also the microbial stability of these products would be improved and their shelf life prolonged if IMF were stored below
room temperature. On the other hand, organisms unable to multiply
in IMF will be inactivated sooner if these products are stored for
some time at elevated temperatures. Finally it should be mentioned that frozen foods are in the range of IMF (Table XIII),
since freezing is equal to drying. As pointed out in a previous
chapter, the a_w of frozen foods is in the range 0.90-0.60 at temperatures between -11° and -53°C; thus these items could be regarded as IMF.

Traditional IMF for human consumption are well-accepted
products, which often have a high nutritional and organoleptic
quality as well as an excellent microbial stability. On the other
hand, despite impressive developments of the market for intermediate-moisture pet foods (IMPF), the prospectives of new IMF for
human consumption are viewed pessimistically (Troller and Christian, 1978). The major reason for this opinion is the lack of
humectants that are safe, cheap, effective, and without flavor,
color, or nutritional drawbacks. Since such humectants are not
yet in sight the concept for new IMF should be modified. An
overloading of new IMF with humectants, which leads to organoleptic and toxicological objections, could be avoided if the a_w is
less depressed. An important concern for new IMF is growth and
toxin production by *S. aureus*, and therefore $a_w \leq 0.85$ was suggested (Table XIV) for these products. However, if in addition
to a_w other hurdles, such as pH, E_h, citrate, and/or competitive

microorganisms, are introduced for the inhibition of *S. aureus*, probably a depression of the a_w to 0.90, which requires lower levels of humectants, could be sufficient. Also other potentially deleterious effects, such as color changes, lipid oxidation, and nutritional losses, would probably be less at a_w = 0.90 than at 0.85. An a_w of 0.90 could be achieved in new IMF with humectants that are legally approved and are tolerable with respect to the organoleptic properties of the product. Of course, for the inhibition of molds and yeasts in new IMF with a_w = 0.90, as well as for those products with a_w = 0.85, a nontoxic antimycotic agent, such as potassium sorbate, is necessary.

VIII. SHELF-STABLE PRODUCTS

Intermediate-moisture products, e.g., pet foods, are enjoying increasing success, but newly developed IMF encounter difficulties when they are introduced for general use. The reasons for this are an unfamiliar taste (sweet or bitter), toxicological considerations ("chemical overloading of the food"), and legal problems (necessity for regulatory approval of additives). To overcome these hindrances a new generation of foods could be envisaged, which should be named "shelf stable products" (SSP) (Leistner, 1977c).

SSP are heated in hermetically sealed containers sufficiently to inactivate all vegetative microorganisms; surviving spores cannot spoil the food since the a_w is adjusted to <0.95. The a_w, preferably in combination with other hurdles, should be low enough to inhibit the multiplication of Bacillaceae, i.e., representatives of the genera *Bacillus* and *Clostridium*. Other organisms are of no concern since they are inactivated by heat and so recontamination after heat processing is prevented by the sealed container (can, pouche, casing, etc.). The mild heat treatment (retort temperature <100°C, core food temperature 65°-95°C) provides favorable retention of organoleptic and nutritional properties of the product. Little energy is required for processing and refrigeration is not required for storage. Neither preservatives (e.g., nitrite) nor antimycotics are needed to inhibit microorganisms and antioxidants are not necessary to prevent rancidity.

Leistner and Karan-Djurdjić (1970) introduced the SSP concept. In their experiments batches of liver sausage with a_w in the range 0.97-0.93 were prepared by adding increasing amounts of NaCl and fat. These products were inoculated with spores of *C. sporogenes* PA 3679, then canned and heated to a core temperature of 95 C, and subsequently stored at 37°C for one month. The products (pH $\approx$ 6.2) with a low spore level proved stable at a_w < 0.960 and those batches with a high spore level were stable at a_w < 0.945 (Table XVI). The data in Table XVI indicate that not only spore outgrowth but also the survival of clostridia was influenced by the a_w. Apparently the spores germinated at low a_w, but if the vegetative cells were unable to multiply they died. An a_w < 0.950 was obtained in the liver sausage by adding 2.5% NaCl and 44% fat. Thus the stable products were organoleptically acceptable. Reichert and Stiebing (1977) confirmed that liver sausage processed according to the SSP concept is shelf stable if the a_w < 0.955. These authors achieved a_w = 0.955 in liver sausage by the addition of 2.0% NaCl and 50% fat.

It has to be remembered that commercial meat products empirically based on the SSP concept are already available. An example is genuine Italian Mortadella. This product has an a_w = 0.96-0.93, is sufficiently heated in a closed casing to inactivate vegetative organisms, and so it can often be stored without refrigeration. In Germany some of the Bologna-type sausages (Gottinger, Tiroler, Polnische) as well as some types of liver sausage (fette Leberwurst) and blood sausage (Speckwurst) are traditional SSP, if their a_w < 0.95.

Since SSP are more promising than IMF, their further developments should be pursued. In this context several questions have

TABLE XVI. Stability of Canned Liver Sausage, Heated to 95°C Core Temperature, in Relation to Water Activity

Liver sausage a_w	Clostridium sporogenes PA 3679 organisms/g		
	After processing	14 days storage*	30 days storage*
0.970	35	>1.000.000	Spoiled
0.967	35	>1.000.000	Spoiled
0.962	110	>1.000.000	5.400.000
0.961	17	32	0
0.959	24	10	0
0.957	92	2	0
0.954	54	1	0
0.947	170	2	0
0.977	6.400	Spoiled	Spoiled
0.974	3.500	Spoiled	Spoiled
0.965	9.200	35.000.000	Spoiled
0.957	11.000	16.000.000	1.700.000
0.956	27.000	7.900.000	Spoiled
0.947	17.000	18.000.000	>24.000
0.942	22.000	3.100	0
0.933	19.000	11	0

[a]From Leistner and Karan-Djurdjić, 1970.
[b]Storage temperature 37°C.

to be studied: (1) What are the limiting a_w in SSP prepared with
salt, sugar, or glycerol for the growth of Bacillaceae? (2) Is
there an a_w equilibrium in SSP? (3) By what means (e.g., water
removal, addition of humectants or fat) could the a_w in SSP be
depressed below the critical level? (4) What heat treatment is
necessary to inactivate all vegetative microorganisms in SSP?

   According to accepted technology, canned acid foods (pH < 4.5),
especially fruits, are only pasteurized, i.e. the non-spore-forming
bacteria and fungi are normally inactivated by heat and the growth
of Bacillaceae is inhibited by the low pH.   To improve the microbiological stability of such products with respect to butyric
anaerobes (*C. pasteurianum, C. butyricum,* etc.), the a_w could be
adjusted in addition to pH.   For example, canned pears proved
stable at pH = 4.5 and $a_w \leq 0.97$ or with pH < 4.0 and $a_w < 0.98$
(Jakobsen and Jensen, 1975).   The pH of most heated meat products
is in the range 6.0-6.3 (Wirth *et al.*, 1976).   Genuine Italian
Mortadella has a pH range of 5.3-6.0; Göttinger, Tiroler, and
Polnische 5.7-6.1; fette Leberwurst 6.1-6.5, and Speckwurst
6.7-7.1.   Therefore, in some of these traditional SSP the pH is
a hurdle.   However, the stability of meat-based SSP should be
achieved primarily by a depression of the a_w.

   Generally, the sporulation, germination, and vegetative cell
growth of Bacillaceae are inhibited at $a_w < 0.95$.   The limiting
a_w for the formation of mature spores of *B. cereus* T. proved to
be about 0.95 for NaCl, glucose, and sorbitol, whereas it was
about 0.91 for glycerol (Jakobsen and Murrell, 1977a).   These
authors suggested that the a_w of the sporulation medium determines
the quantity of spores rather than the spore properties.   Kang *et*
al. (1969) observed that the lower limit for sporulation of *C.*
perfringens was about 0.98 when the a_w was controlled by NaCl,
sucrose, or glycerol.   Apparently somewhat higher a_w are required
for sporulation than for growth of Bacillaceae (Murrell, 1961;
Hagen *et al.*, 1967).   On the other hand, there is general agreement that spores of Bacillaceae may initiate germination at a_w

levels appreciably lower than those which will permit vegetative
cell growth (Bullock and Tallentire, 1952; Gould, 1964; Hagen *et
al.*, 1967; Baird-Parker and Freame, 1967; Jakobsen *et al.*, 1972;
Jakobsen and Murrell, 1976b; Troller and Christian, 1978). Germination of *C. botulinum* types A, B, and E. (Baird-Parker and
Freame, 1967) and *B. cereus* (Jakobsen *et al.*, 1972) does occur at
lower a_w levels with glycerol than NaCl. In a study by Jakobsen
and Murrell (1976b) the germination of clostridia was completely
inhibited at a_w = 0.95 when NaCl was the solute, whereas no significant inhibition was observed when glycerol or glucose was
used at the same a_w. It is significant that spores germinate in
SSP below the critical a_w for growth, because the number of surviving spores in such products could decrease during storage
(Table XVI.

Of most concern for SSP is the vegetative cell growth of
Bacillaceae, i.e., that surviving spores develop into a culture
that might cause food poisoning or spoilage. As Table XVII indicates, in substrates adjusted with salts or sugars, the growth
of organisms of the genera *Clostridium* and *Bacillus* generally is
inhibited at a_w < 0.95. Only few exceptions to this rule are
known, and occurred when high incubation temperatures and artificial substrates were used. In foods the a_w tolerance of Bacillaceae will decrease if other hurdles are present, such as pH,
E_h, and storage temperatures. For instance a reduction of temperature from 40° to 20°C increased (Ohye and Christian, 1967)
markedly the water requirement of *C. botulinum* type A (from
a_w = 0.950-0.970) and B (from a_w = 0.940-0.970. However, what
margin of safety is required for SSP, i.e., which a_w level perhaps
in combination with other hurdles (especially pH) is indispensable,
still needs to be defined. Certainly foods adjusted with glycerol
(or similar substances) need special precautions since, e.g., the
limiting a_w values for growth of *C. botulinum* types A and B
(Baird-Parker and Freame, 1967) as well as for *C. perfringens*
(Kang *et al.*, 1969) and *B. cereus* (Jakobsen *et al.*, 1972; Stewart

TABLE XVII. Limiting a_W for vegetative cell growth of the genera Clostridium and Bacillus in substrates adjusted with salts or sugars

TABLE 17. Limiting a_w for vegetative cell growth of the genera <u>Clostridium</u> and <u>Bacillus</u> in substrates adjusted with salts or sugars

Organism	Limiting a_w	References
<u>C. botulinum</u> type A	0.96 0.95	Baird-Parker and Freame (1967) Scott (1955b); Ohye and Christian (1967)
<u>C. botulinum</u> type B	0.96 0.95 0.94(a)	Baird-Parker and Freame (1967) Scott (1955b); Kim (1965) Ohye and Christian (1967)
<u>C. botulinum</u> type C	0.98	Segner et al. (1971)
<u>C. botulinum</u> type E	0.98 0.975 0.97 0.965	Baird-Parker and Freame (1967) Emodi and Lechowich (1969b) Segner et al.(1966);Ohye and Christian (1967) Ohye et al. (1967)
<u>C. perfringens</u>	0.95	Kim (1965)
<u>B. cereus</u>	0.95 0.95	Jakobsen et al. (1972) Stewart and Jakobsen (1977)
<u>B. subtilis</u>	0.90(b)	Christian (1979)
<u>B. stearothermophilus</u>	0.97	Christian (1979)

(a) In artificial substrate at 40 °C; (b) In brain heat infusion broth at 30 °C

(a) In artificial substrate at 40°C; (b) In brain heat infusion broth at 30°C.

and Jakobsen, 1977) are in the range a_w = 0.93-0.91. The significant problem with most of the studies on the limiting a_w for growth of Bacillaceae is the relatively small number of strains that have been examined, and the question of whether the properties of these "laboratory" strains are representative of those strains occurring naturally (Roberts and Smart, 1976). This could be remembered in the future development of SSP.

The rate and extent of equilibration in stored foods has not been investigated thoroughly in spite of the practical importance of this question. In the experiments of Jakobsen and Jensen (1975) with canned pears, equilibrium between the brine and the fruit was fully obtained when the cans had been stored at room temperature for four days after heating. In certain meat products, an a_w equilibrium is not likely to occur between the lean and the fat portion. In raw ham the a_w of the fat portion is lower than in the lean, even after a long ripening time (Table VII). In pasteurized canned ham, Leistner and Karan-Djurdjić (1970) reported a_w in the fat, lean, and jelly portions of 0.976, 0.982, and 0.986, respectively. Again the a_w of the fat was lower. Also in a sample of genuine Italian Mortadella the fat pieces had an a_w = 0.924 and the surrounding sausage emulsion of 0.933. The a_w of homogeneous sausage products, such as finely ground Bologna-type sausage, as well as liver, blood, and fermented sausage, could be regarded as uniform. After a long ripening period the surface layers of large fermented sausages and raw hams often have a much lower moisture content and a_w than the center of the products. The a_w differences within a meat product require additional study. However, it may be safe to conclude that the portion of a product of a product with the highest a_w is generally decisive for the microbial stability of the food.

Preferably the a_w of SSP should be adjusted to <0.95 by controlling the water, fat, and salt or sugar content of the food, since glycerol and glycols are not legally permitted in several countries. In our laboratories the depression of the a_w of ground

meat caused by 1% of the additives listed in Table XVIII was
studied (Leistner and Karan-Djurdjić, 1970; Leistner and Rödel,
1975b, 1976a). This 1% value can be used to estimate the resulting
depression of a_w when these additives are incorporated in the usual
amounts into meat products. The data in Table XVIII indicate that
NaCl and fat are the most effective legal substances for a depression of the a_w of meat. The addition of 3% NaCl causes about the
same depression of the a_w of meat as 30% fat. Reichert and
Stiebing (1977) reported in their studies on liver sausage an a_w
depression of only 0.00044 caused by a 1% addition of fat.

In traditional SSP meat products an a_w < 0.95 is achieved by
different technologies. Genuine Italian Mortadella is produced
with relatively high amounts of salt, fat, and milk powder, but
little addition of water. In liver sausage (fette Leberwurst) the
a_w is adjusted by the addition of fat and salt, in blood sausage
(Speckwurst) and certain Bologna-type sausages (Göttingen, Polnische, Tiroler) pieces of fat are added and water is removed by
slow drying.

In SSP all vegetative microorganisms must be inactivated by
heat. The heat process should be mild but effective. Bologna-
type sausages are heated to a core temperature of $65°$-$70°C$, liver
and blood sausages to $75°$-$80°C$. This also is true for products
that are subsequently dryed to become SSP. Since Italian Mortadella already has a relatively low a_w during the heating process,
this product is empirically heated to a somewhat higher core temperature ($70°$-$75°C$) than other Bologna-type products. As discussed in a previous chapter, the heat resistance of microorganisms
increases with decreasing a_w. Thus food products with an
a_w < 0.95 should be heated more than those with a higher a_w.
Gram-negative bacteria, including salmonellae, have a relatively
low heat resistance, which is influenced by the a_w of the substrate (Corry, 1976). For *S. aureus* heated to $60°C$ the addition
of NaCl (Calhoun and Frazier, 1966) and sugar (Kadan *et al.*,
1963) to buffer the system and skimmed milk to give an a_w = 0.95

TABLE XVIII. Depression of the a_w of meat caused by additives (Leistner and Rödel, 1976a)

TABLE 18. Depression of the a_w of meat caused by additives (Leistner and Rödel, 1976a)

1%-value(a)	Additive	Depression of a_w caused by using additives at (%)							
		0.1	0.3	2.0	3.0	5.0	10	30	50
0.0100	Lithium chloride(b)	0.0010	0.0030	–	–	–	–	–	–
0.0062	Sodium chloride	0.0006	0.0019	0.0124	0.0186	–	–	–	–
0.0061	Polyphosphate	0.0006	0.0018	–	–	–	–	–	–
0.0050	1,2-Propylene glycol(b)	0.0005	0.0015	0.0100	0.0150	–	–	–	–
0.0047	Sodium citrate x 5.5 H$_2$O	0.0005	0.0014	–	–	–	–	–	–
0.0041	Ascorbic acid	0.0004	–	–	–	–	–	–	–
0.0040	Glucono-δ-lactone	0.0004	0.0012	–	–	–	–	–	–
0.0037	Sodium acetate, cryst.	0.0004	–	–	–	–	–	–	–
0.0033	Sodium hydrogen tartrate	0.0003	–	–	–	–	–	–	–
0.0030	Glycerol(b)	0.0003	0.0009	0.0060	0.0090	0.0150	0.0300	0.0900	–
0.0026	Potassium sorbate(b)	0.0003	0.0008	–	–	–	–	–	–
0.0024	Glucose	0.0002	0.0006	–	–	–	–	–	–
0.0022	Lactose	0.0002	0.0006	0.0044	0.0066	–	–	–	–
0.0019	Sucrose	0.0002	0.0006	–	–	–	–	–	–
0.0013	Milk protein	0.0001	0.0004	0.0026	0.0039	–	–	–	–
0.00062	Fat	0.0001	0.0002	0.0012	0.0019	0.0031	0.0062	0.0186	0.0310

(a) Depression of a_w caused by 1 % additive;
(b) Not legally permitted in West Germany.

(a) Depression of a_w caused by 1% additive; (b) Not legally permitted in West Germany.

was very protective. Apparently, of the non-spore-forming bacteria occurring in meat, the enterococci are the most heat resistant. It was demonstrated also that the heat resistance of enterococci increases with decreasing a_w. If the a_w was adjusted with NaCl the highest heat resistance of these organisms was observed at 0.95 (Vrchlabský and Leistner, 1970b). Therefore, in further studies into the necessary heat treatment for SPP, enterococci could be used as indicator organisms. Possible a_w changes in foods during the heating process and their effect on the heat resistance of microorganisms should also be explored.

References

Acott, K. M., and Labuza, T. P. (1975). *J. Food Technol. 10,* 603.

Alford, J. A., and Palumbo, S. A. (1969). *Appl. Microbiol. 17,* 528.

Anderton, J. I. (1963). *Sci. Tech. Surveys 40,* 1. The British Food Manufacturing Industries Research Assoc., Leatherhead, Surrey, England.

Angelotti, R., Foter, M. J., and Lewis, K. H. (1961). *Am. J. Publ. Health 51,* 76.

Ayerst, G. (1969). *J. Stored Prod. Res. 5,* 127.

Bacon, C. W., Sweeney, J. G., Robbins, J. D., and Burdick, D. (1973). *Appl. Microbiol. 26,* 155.

Baird-Parker, A. C., and Freame, B. (1967). *J. Appl. Bacteriol. 30,* 420.

Baird-Parker, A. C., Boothroyd, M., and Jones, E. (1970). *J. Appl. Bacteriol. 33,* 515.

Banwart, G. J., and Ayres, J. C. (1956). *Food Technol. 10,* 68.

Bean, P. G., and Roberts, T. A. (1975). *J. Food Technol. 10,* 327.

Beers, R. J. (1957). *In* "Spores" (H. O. Halvorson, ed.). Am. Inst. Biol. Sci., Washington, D.C.

Bem, Z., and Leistner, L. (1970). *Fleischwirtschaft 50,* 492.

Beuchat, L. R. (1974). *Appl. Microbiol. 27,* 1075.

Blanche Koelensmid, W. A. A., and van Rhee, R. (1964). *Ann. Inst. Pasteur 15,* 85.

Bohringer, P. (1962). *In* "Die Hefen," Band II, Technologie der Hefen (F. Reiff, R. Kautzmann, H. Lüers, and M. Lindemann, eds.). Verlag Hans Carl, Nürnberg.

Bone, D. (1973). *Food Technol. 27* (4), 71.

Brockmann, M. C. (1970). *Food Technol. 24,* 896.

Brownlie, L. E. (1966). *J. Appl. Bacteriol. 29,* 447.

Bullock, K., and Tallentire, A. (1952). *J. Pharm. Pharmacol. London 4*, 917.

Burcik, E. (1950). *Arch. Microbiol. 15*, 203.

Calhoun, C. L., and Frazier, W. C. (1966). *Appl. Microbiol. 14*, 416.

Carlson, V. L., and Snoeyenbos, G. H. (1970). *Poultry Sci. 49*, 717.

Christian, J. H. B. (1955a). *Aust. J. Biol. Sci. 8*, 75.

Christian, J. H. B. (1955b). *Aust. J. Biol. Sci. 8*, 490.

Christian, J. H. B., and Scott, W. J. (1953). *Aust. J. Biol. Sci. 6*, 565.

Christian, J. H. B., and Stewart, B. J. (1973). *In* "The Microbiological Safety of Food" (B. C. Hobbs and J. H. B. Christtian, eds.), p. 107. Academic Press, London.

Christian, J. H. B., and Waltho, J. A. (1962). *J. Appl. Bacteriol. 25*, 369.

Christian, J. H. B., and Waltho, J. A. (1964). *J. Gen. Microbiol. 35*, 205.

Clayson, D. H. F., and Blood, R. M. (1957). *J. Sci. Food Agri. 8*, 404.

Collins, J. L., Chen, C. C., Park, J. R., Mundt, J. O., McCarty, I. E., and Johnston, M. R. (1972). *J. Food Sci. 37*, 189.

Corry, J. E. L. (1974). *J. Appl. Bacteriol. 37*, 31.

Corry, J. E. L. (1975). *In* "Water Relations of Foods" (R. B. Duckworth, ed.), p. 325. Academic Press, London.

Corry, J. E. L. (1976). *In* "Intermediate Moisture Foods" (R. Davies, G. G. Birch, and K. J. Parker, eds.), p. 215. Applied Science Publishers, London.

Cox, C. S. (1966). *J. Gen. Microbiol. 43*, 383.

Csiszár, V., and Biró, G. (1967). *Fleischwirtschaft 47*, 125.

Dainty, R. H., Shaw, B. G., De Boer, K. A., and Scheps, E. S. J. (1975). *J. Appl. Bacteriol. 39*, 73.

Davies, J. D. (1968). *In* "Low Temperature Biology of Foodstufs" (J. Hawthorn and E. J. Rolfe, eds.), p. 177. Pergamon Press, Oxford and London.

Davis, M. S., and Bateman, J. B. (1960). *J. Bacteriol. 80*, 577.

Diener, U. L., and Davis, N. D. (1970). *J. Am. Oil Chem. Soc. 47*, 347.

Emodi, A. S., and Lechowich, R. V. (1969a). *J. Food Sci. 34*, 78.

Emodi, A. S., and Lechowich, R. V. (1969b). *J. Food Sci. 34*, 82.

Fennema, O., and Berny, L. A. (1974). *Proc. 4th Int. Congr. Food Sci. Technol., Madrid,* Work Doc. Topic 2, p. 12.

Fry, R. M. (1954). *In* "Applications of Freezing and Drying" (R. J. C. Harris, ed.), p. 215. Academic Press, New York.

Fry, R. M. (1966). *In* "Cryobiology" (H. T. Meryman, ed.), p. 665. Academic Press, London.

Genigeorgis, C., and Sadler, W. W. (1966). *J. Bacteriol. 92*, 1383.

Genigeorgis, C., Riemann, H., and Sadler, W. W. (1969). *J. Food Sci. 34*, 62.

Genigeorgis, C., Savoukidis, M., and Martin, S. (1971a). *Appl. Microbiol. 21*, 940.

Genigeorgis, C., Foda, M. S., Mantis, A., and Sadler, W. W. (1971b). *Appl. Microbiol. 21*, 862.

Gibson, B. (1973). *J. Appl. Bacteriol. 36*, 365.

Gill, C. O., and Penney, N. (1977). *Appl. Environ. Microbiol. 33*, 1284.

Gough, B. J., and Alford, J. A. (1965). *J. Food Sci. 30*, 1025.

Gould, G. W. (1964). *Proc. 4th Int. Symp. Food Microbiol., 1963*, p17.

Greaves, R. I. N. (1960). *In* "Recent Research in Freezing and Drying" (A. S. Parkes and A. U. Smith, eds.), p. 203. Blackwell Scientific Publ., Oxford.

Greenberg, R. A., Silliker, J. H., and Fatta, L. D. (1959). *Food Technol. 13*, 509.

Hagen, C. A., Hawrylewicz, E. J., and Erlich, R. (1967). *Appl. Microbiol. 15*, 285.

Hale, C. M. F. (1957). *Exp. Cell Res. 12*, 657.

Hansen, N.-H., and Riemann, H. (1962). *Fleischwirtschaft 14*, 861.

Hechelmann, H., and Leistner, L. (1969). *Fleischwirtschaft 49*, 1639.

Heiss, R. (1933). *Die Wärme, Z. Dampfkessel Maschinenbetrieb 56*, 72.

Heiss, R. (1938). *Z. Fleisch. Milchhyg. 48*, 333.

Hess, E. (1970). *Alimenta, Sonderausgabe Mikrobiol. 9*, 35.

Hicks, E. W., Scott, W. J., and Vickery, J. R. (1955). *Commun. Papers, Commissions 4 and 5, Madrid, 1954, Annexe 1955-1, Suppl. Bull. Inst. Int. Froid*, p. 51.

Hicks, E. W., Scott, W. J., and Vickery, J. R. (1957). *Bull. Inst. Int. Froid 37*, 582.

Higginbottom, C. (1953). *J. Dairy Res. 20*, 65.

Hobbs, B. C. (1965). *J. Appl. Microbiol. 28*, 74.

Hofvat, S. A., and Jackson, H. (1969). *Can. Inst. Food Technol. J. 2*, 56.

Holzapfel, W., Hechelmann, H., and Leistner, L. (1976). *Jahresber. Bundesanstalt Fleischforsch., Kulmbach, 1976*, p. C.29.

Horner, K. J., and Anagnostopoulos, G. D. (1975). *J. Appl. Bacteriol. 38*, 9.

Hsieh, F., Acott, K., Elizondo, H., and Labuza, T. P. (1975). *Lebensm.-Wiss. Technol. 8*, 78.

Iandolo, J. J., Ordal, Z. J., and Witter, L. D. (1964). *Can. J. Microbiol. 10*, 808.

Ingram, M. (1951). *J. Appl. Bacteriol. 14*, 243.

Ingram, M., and Dainty, R. H. (1971). *J. Appl. Bacteriol. 34*, 21.

Ingram, M., and Mackey, B. M. (1976). *In* "Inhibition and Inactivation of Vegetative Microbes" (F. A. Skinner and W. B. Hugo, eds.), p. 111. Academic Press, London.

Jakobsen, M., and Jensen, H. C. (1975). *Lebensm.-Wiss. Technol. 8*, 158.

Jakobsen, M., and Murrell, W. G. (1977a). *J. Appl. Bacteriol. 43*, 239.

Jakobsen, M., and Murrell, W. G. (1977b). *In* "Spore Research 1976" (A. N. Barker *et al.*, eds.), Vol. 2, p. 819. Academic Press, London.

Jakobsen, M., Filtenborg, O., and Bramsnaes, F. (1972). *Lebesm.-Wiss. Technol. 5,* 159.

Jarvis, B. (1971). *J. Appl. Bacteriol. 34,* 199.

Kadan, R. S., Martin, W. H., and Michelsen, R. (1963). *Appl. Microbiol. 11,* 45.

Kaess, G., and Schwartz, W. (1934). *Arch. Microbiol. 5,* 443.

Kang, C. K., Woodburn, M., Pagenkopf, A., and Cheney, R. (1969). *Appl. Microbiol. 18,* 798.

Kaplow, M. (1970). *Food Technol. 24,* 889.

Karel, M. (1973). *CRC Crit. Rev. Food Technol. 3,* 329.

Karmas, E. (1975). "Fresh Meat Technology," p. 174. Noyes Data Corp., Park Ridge, New Jersey and London.

Kereluk, K., and Gunderson, M. F. (1959). *Appl. Microbiol. 7,* 327.

Kim, C. H. (1965). Ph.D. Thesis. Purdue Univ., Lafayette, Indiana (University Microfilms No. 65-8624).

Klettner, P.-G., and Rödel, W. (1978). *Fleischwirtschaft 58,* 57.

Krispien, K. (1978). Thesis. Free University, Berlin.

Krispien, K., and Rödel, W. (1976). *Fleischwirtschaft 56,* 709.

Krispien, K., Rödel, W., and Leistner, L. (1979). *Fleischwirtschaft,* in press.

Kushner, D. J. (1968). *Advan. Appl. Microbiol. 10,* 73.

Kushner, D. J. (1971). *In* "Inhibition and Destruction of the Microbial Cell" (W. B. Hugo, ed.). Academic Press, London.

Labuza, T. P. (1974). "Storage Stability and Improvement of Intermediate Moisture Foods" Phase II. Contract NAS 9-12560. NASA, Food and Nutrition Office, Houston, Texas.

Labuza, T. P., Cassil, S., and Sinskey, A. J. (1972a). *J. Food Sci. 37,* 160.

Labuza, T. P., McNally, L., Gallagher, D., Hawkes, J., and Hurtado, F. (1972b). *J. Food Sci. 37,* 154.

Lanigan, G. W. (1963). *Aust. J. Biol. Sci. 16,* 606.

Leistner, L. (1974). *Fleischwirtschaft 54,* 1036.

Leistner, L. (1977a). *In* "How Ready Are Ready-to-serve Foods?" (K. Paulus, ed.), p. 260. Karger, Basel.

Leistner, L. (1977b). *Proc. 7th Int. Symp. World Assoc. Vet. Food Hygienists, Garmisch-Partenkirchen,* Part II, p. 17.

Leistner, L. (1977c). *Jahresbericht der Bundesanstalt für Fleischforschung, Kulmbach, 1977,* C.39.

Leistner, L. (1978). *In* "Food Quality and Nutrition" (W. K. Downey, ed.), p. 553. Applied Science Publishers, London.

Leistner, L., and Karan-Djurdjić, S. (1970). *Fleischwirtschaft 50,* 1547.

Leistner, L., and Pitt, John, I. (1977). *In* "Mycotoxins" (J. V. Rodricks, C. W. Hesseltine, and M. A. Mehlman, eds.), p. 639. Pathotox Publ., Park Forest South, Illinois.

Leistner, L., and Rödel, W. (1975a). *Deutsche Z. Lebensmittel-technol. 26,* 169.

Leistner, L., and Rödel, W. (1975b). *In* "Water Relations of Foods" (R. B. Duckworth, ed.), p. 309. Academic Press, London.

Leistner, L., and Rödel, W. (1976a). *In* "Inhibition and Inactivation of Vegetative Microbes" (F. A. Skinner and W. B. Hugo, eds.), p. 219. Academic Press, London.

Leistner, L., and Rödel, W. (1976b). *In* "Intermediate Moisture Foods" (R. Davies, G. G. Birch, and K. J. Parker, eds.), p. 120. Applied Science Publ., London.

Leistner, L., Herzog, H., and Linke, H. (1971a). *Fleischwirtschaft 51,* 578.

Leistner, L., Herzog, H., and Wirth, F. (1971b). *Fleischwirtschaft 51,* 213.

Leistner, L., Maing, I. Y., and Bergmann, E. (1975). *Fleischwirtschaft 55,* 559.

Leistner, L., Bem, Z., and Dresel, J. (1977). *Mitteilungsblatt Bundesanstalt Fleischforsch., Kulmbach, 1977,* 3070.

Leistner, L., Bem, Z., and Dresel, J. (1979). *Fleischwirtschaft 59,* in press.

Lerche, M. (1957). *Berliner Münchener Tierarztl. Wschr. 70,* 13.

Licardi, J. J., and Potter, N. N. (1970). *J. Dairy Sci. 53,* 871.

Limsong, S., and Frazier, W. C. (1966). *Appl. Microbiol. 14,* 899.

Liu, T. S., Snoeyenbos, G. H., and Carlson, V. L. (1969). *Poultry Sci. 48(2),* 1628.

Lötzsch, R., and Leistner, L. (1977). *Proc. 24th Eur. Congr. Meat Res. Work., Moscow,* Session I.

Lötzsch, R., and Rüdel, W. (1974). *Fleischwirtschaft 54,* 1203.

Lotzsch, R., and Trapper, Doris (1978). *Fleischwirtschaft 58,* in press.

Lötzsch, R., and Trapper, Doris (1979). *Fleischwirtschaft 59,* in press.

Lotter, L. P., and Leistner, L. (1976). *Jahresbericht Bundesanstalt Fleischforsch., Kulmbach, 1976,* 31.

Lotter, L. P., and Leistner, L. (1978). *Appl. Environ. Microbiol. 36,* 377.

Lück, E. (1970). *In* "Sorbinsäure," Band 3, Technologie. B. Behr's Verlag, Hamburg.

Lück, E. (1972). *In* "Sorbinsäure,: Band 2, Biochemie-Mikrobiologie. B. Behr's Verlag, Hamburg.

Lück, E. (1973). *Ernährungswirtschaft/lebensmitteltechnik 20,* 346.

Luck, E. (1977). "Chemische Lebensmittelkonservierung." Springer-Verlag, Berlin, Heidelberg, New York

McDade, J. J., and Hall, L. B. (1963). *Am. J. Hyg. 78,* 330.

McDade, J. J., and Hall, L. B. (1964). *Am. J. Hyg. 80,* 192.

McLean, R. A., Lilly, H. D., and Alford, J. A. (1968). *J. Bacteriol. 95,* 1207.

Major, C. P., McDougal, J. D., and Harrison, A. P., Jr. (1955). *J. Bacteriol. 69,* 244.

Markus, Z. H., and Silverman, G. J. (1970). *Appl. Microbiol. 20,* 492.

Marland, R. E. (1967). *Health Sci.,* 3165-B.

Marshall, B. J., Ohye, D. F., and Christian, J. H. B. (1971). *Appl. Microbiol. 21,* 363.

Marshall, B. J., Coote, G. G., and Scott, W. J. (1974). CSIRO Aust. Div. Food Res., Tech. Pap. No. 39, p. 1.

Matches, J. R., and Liston, J. (1972). *J. Milk Food Technol. 35,* 39.

Malz, S. A. (1965). *In* "Water in Foods," p. 249. Avi Publ. Co., Westport, Connecticut.

Mazur, P. (1966). *In* "Cryobiology" (H. T. Meryman, ed.), p. 214. Academic Press, London.

Moerman, P. C. (1966). *Vlees 8,* 73.

Moerman, P. C. (1967). *Vleesdistr. Vleestechnol. 2,* 243.

Moran, T. (1936). "Report of the Food Investigation Board for the Year 1935," p. 20. H. M. Stationery Office, London.

Mossel, D. A. A. (1963). *Ann. Inst. Pasteur, Paris 104,* 551.

Mossel, D. A. A. (1969). *Alimenta 8,* 8.

Murrell, W. G. (1961). *In* "Microbial Reaction to the Environment" (G. G. Meynell and H. Gooder, eds.), p. 100. University Press, Cambridge.

Murrell, W. G., and Scott, W. J. (1966). *J. Gen. Microbiol. 43,* 411.

Narayan, K. A., McMullen, J. J., Butler, D. P., Wakefield, T., and Calhoun, W. K. (1975). *Nutr. Rep. Int. 12,* 211.

Narayan, K. A., McMullen, J. J., Wakefield, T., and Calhoun, W. K. (1977). *J. Nutrition 107,* 2153.

Northolt, M. D., and van Egmond, H. P. (1978). *Int. Conf. Mycotoxins, August 14-15, 1978, Munich,* Abstr., p. 19.

Northolt, M. D., Verhulsdonk, C. A. H., Soentoro, P. S. S., and Paulsch, W. E. (1976). *J. Milk Food Technol. 39,* 170.

Ohnishi, H. (1957). *Bull. Agri. Chem. Soc. Japan 21,* 137.

Ohye, D. F., and Christian, J. H. B. (1967). *Proc. 5th Int. Symp. Food Microbiol., Moscow, 1966,* p. 217.

Ohye, D. F., Christian, J. H. B., and Scott, W. J. (1967). *Proc. 5th Int. Symp. Food Microbiol., Moscow, 1966,* p. 136.

Pavey, R. L. (1972). Tech. Rep. 73-17-Fl. US Army Natick Laboratories, Natick, Massachusetts.

Pawsey, R., and Davies, R. (1976). *In* "Intermediate Moisture Foods" (R. Davies, G. G. Birch, and K. J. Parker, eds.), p. 182. Applied Science Publ., London.

Peterson, A. C., Black, J. J., and Gunderson, M. F. (1964). *Appl. Microbiol. 12,* 70.

Pitt, J. I. (1975). *In* Water Relations of Foods" (R. B. Duckworth, ed.), p. 273. Academic Press, London.

Pitt, J. I., and Christian, J. H. B. (1968). *Appl. Microbiol. 16,* 1853.

Pivnick, H., and Thatcher, F. S. (1968). *In* "Safety of Foods" (J. C. Ayres, F. R. Blood, C. O. Chichester, H. D. Graham, R. S. Mccutcheon, J. J. Powers, B. S. Schweigert, A. D.

Stevens, and G. Zweig, eds.), p. 212. The Avi Publ. Co., Westport, Connecticut.

Plitman, M., Park, Y., Gomez, R., and Sinskey, A. J. (1973). *J. Food Sci. 38,* 1004.

Reichert, J. E., and Stiebing, A. (1977). *Fleischwirtschaft 57,* 910.

Reiser, R. F., and Weiss, K. F. (1969). *Appl. Microbiol. 18,* 1041.

Riemann, H. (1963). *Food Technol. 17,* 39.

Riemann, H. (1967). *Proc. 5th Int. Symp. Food Microbiol., Moscow, 1966,* p. 150.

Riemann, H. (1968). *Appl. Microbiol. 16,* 1621.

Riemann, H., Lee, W. H., and Genigeorgis, C. (1972). *J. Milk Food Technol. 35,* 514.

Roberts, T. A., and Ingram, M. (1973). *J. Food Technol. 8,* 467.

Roberts, T. A., and Smart, J. L. (1976). *In* "Intermediate Moisture Foods: (R. Davies, G. G. Birch, and K. J. Parker, eds.), p. 203. Applied Science Publ., London.

Roberts, T. A., Jarvis, B., and Rhodes, A. (1976). *J. Food Technol. 11,* 25.

Rödel, W. (1973). *Fleischwirtschaft 53,* 27.

Rödel, W. (1975). Thesis, Free University, Berlin.

Rödel, W., and Krispien, K. (1977). *Fleischwirtschaft 57,* 1863.

Rödel, W., Herzog, H., and Leistner, L. (1973). *Fleischwirtschaft 53,* 1301.

Rödel, W., Ponert, H., and Leistner, L. (1975). *Fleischwirtschaft 55,* 557.

Rödel, W., Ponert, H., and Leistner, L. (1976). *Fleischwirtschaft 56,* 417.

Rödel, W., Krispien, K., and Leistner, L. (1978). *Proc. 24th Eur. Meeting Meat Res. Workers, Kulmbach, Volume II, p. G21:1.*

Rose, A. H. (1976). *In* "The Survival of Vegetative Microbes" (T. R. G. Gray and J. R. Postgate, eds.), p. 155. Cambridge Univ. Press, Cambridge.

Sanders, T. H., Davis, N. D., and Diener, U. L. (1968). *J. Am. Oil Chem. Soc. 45,* 683.

Schmidt-Lorenz, W. (1970). *Alimenta 9,* 32.

Schwartz, W., and Loeser, E. (1934/35). *Z. Bakt. Abt. II, 91,* 395.

Scott, W. J. (1936). *J. Council Sci. Ind. Res. 9,* 177.

Scott, W. J. (1937). *J. Council Sci. Ind. Res. 10,* 338.

Scott, W. J. (1938). *J. Council Sci. Ind. Res. 11,* 266.

Scott, W. J. (1953). *Aust. J. Biol. Sci. 6,* 549.

Scott, W. J. (1955a). *Proc. Conf. Beef Export Ind., Brisbane,* p. 30.

Scott, W. J. (1955b). *Ann. Inst. Pasteur Lille 7,* 68.

Scott, W. J. (1957). *In* "Advances in Food Research," Vol. 7 (E. M. Mrak and G. F. Stewart, eds.), p. 83. Academic Press Inc., New York.

Scott, W. J. (1958). *J. Gen. Microbiol. 19,* 624.

Scott, W. J. (1960). *In* "Recent Research in Freezing and Drying" (A. S. Parkes and A. U. Smith, eds.), p. 188. Blackwell Scientific Publ., Oxford.

Scott, W. J., and Vickery, J. R. (1939). *Council Sci. Ind. Res. Bull. 129,* p. 1.

Segner, W. P., Schmidt, C. F., and Boltz, J. K. (1966). *Appl. Microbiol. 14,* 49.

Segner, W. P., Schmidt, C. F., and Boltz, J. K. (1971). *Appl. Microbiol. 22,* 1025.

Sinskey, T. J., Silverman, G. J., and Goldblith, S. A. (1967). *Appl. Microbiol. 15,* 22.

Sloan, A. E., Waletzko, P. T., and Labuza, T. P. (1976). *J. Food Sci. 41,* 536.

Snow, D. (1949). *Ann. Appl. Biol. 36,* 1.

Stille, B. (1948a). *Z. Lebensm. Unters. Forsch. 88,* 9.

Stille, B. (1948b). *Arch. Mikrobiol. 14,* 108.

Stoloff, L. (1978). *JAOAC 61,* 1166.

Storey, R. M., and Stainsby, G. (1970). *J. Food Technol. 5,* 157.

Strange, R. E., and Cox, C. S. (1976). *In* "The Survival of Vegetative Microbes" (T. R. G. Gray and J. R. Postgate, eds.), p. 111. Cambridge Univ. Press, Cambridge.

Strong, D. H., Foster, E. F., and Duncan, C. L. (1970). *Appl. Microbiol. 19,* 980.

Tatmi, S. R. (1973). *J. Milk Food Technol. 36,* 559.

Tomčov, D., Bem, Z., and Leistner, L. (1974). *RIM 6(4),* 3.

Troller, J. A. (1971). *Appl. Microbiol. 21,* 435.

Troller, J. A. (1972). *Appl. Microbiol. 24,* 440.

Troller, J. A. (1973a). *Acta Aliment. Acad. Sci. Hung. 2,* 351.

Troller, J. A. (1973b). *J. Milk Food Technol. 36,* 276.

Troller, J. A. (1976). *J. Milk Food Technol. 39,* 499.

Troller, J. A., and Christian, J. H. B. (1978). "Water Activity and Food," p. 215. Academic Press, New York.

Troller, J. A., and Stinson, J. V. (1975). *J. Food Sci. 40,* 802.

Troller, J. A., and Stinson, J. V. (1978). *Appl. Environ. Microbiol. 35,* 521.

Uzelac, Gordana (1976). Thesis University of Bonn, Bonn.

Vrchlabský, J., and Leistner, L. (1970a). *Fleischwirtschaft 50,* 967.

Vrchlabský, J., and Leistner, L. (1970b). *Fleischwirtschaft 50,* 1237.

Walker, H. W., and Ayres, J. C. (1970). *In* "The Yeasts," Vol. 3, Yeast Technology (A. H. Rose and J. S. Harrison, eds.). Academic Press, London.

Watts, P. S. (1945). *J. Pathol. Bacteriol. 57,* 191.

Williams, O. B., and Purnell, H. G. (1953). *Food Res. 18,* 35.

Wirth, F., Leistner, L., and Rödel, W. (1976). "Richtwerte der Fleischtechnologie." Verlagshaus Sponholz, Frankfurt.

Wodzinski, R. J., and Frazier, W. C. (1960). *J. Bacteriol. 79,* 572.

Wodzinski, R. J., and Frazier, W. C. (1961a). *J. Bacteriol. 81,* 353.

Wodzinski, R. J., and Frazier, W. C. (1961b). *J. Bacteriol. 81,* 359.

Wolf, W., Spiess, W. E. L., and Jung, G. (1973). *Lebensm.-Wiss. Technol. 6,* 94.

Index